EXPLORING THE INVISIBLE

UNIVERSE

From Black Holes to Superstrings

EXPLORING THE INVISIBLE UNIVERSE

From Black Holes to Superstrings

Belal E Baaquie
Frederick H Willeboordse
National University of Singapore

World Scientific
NEW JERSEY • LONDON • SINGAPORE • BEIJING • SHANGHAI • HONG KONG • TAIPEI • CHENNAI

Published by

World Scientific Publishing Co. Pte. Ltd.
5 Toh Tuck Link, Singapore 596224
USA office: 27 Warren Street, Suite 401-402, Hackensack, NJ 07601
UK office: 57 Shelton Street, Covent Garden, London WC2H 9HE

Library of Congress Cataloging-in-Publication Data
Baaquie, B. E., author.
Exploring the invisible universe : from black holes to superstrings / Belal E. Baaquie (National University of Singapore, Singapore), Frederick H. Willeboordse (National University of Singapore).
pages cm
Includes bibliographical references and index.
ISBN 978-9814618670 (hardcover : alk. paper)
1. Physics--Popular works. I. Willeboordse, Frederick H., author. II. Title.
QC24.5.B33 2015
523.1--dc23

2014032724

British Library Cataloguing-in-Publication Data
A catalogue record for this book is available from the British Library.

The cover image is an adaptation of two images.
Credit of the original images goes to NASA and ESO/MPE/Marc Schartmann.

In-house Editor: Ng Kah Fee

Typeset by Stallion Press
Email: enquiries@stallionpress.com

Printed in Singapore

Preface

The Universe surrounds and permeates us in all possible ways. We go about our daily lives concentrating mainly on matters that directly affect us, and for good reasons since the relentless challenges for survival have to be constantly surmounted. Those of us who look upon the Universe only as it directly appears to us — and from the perspective of our survival needs — are only apprehending the visible manifestation and appearance of the Universe. From this point of view, we will not be able to grasp the underlying universal laws of Nature and neither will we encounter the mysterious and invisible realms of the Universe.

The invisible and unseen domains and phenomena of the Universe are beyond our direct perception and do not appear in the manifest Universe. However, if we turn our attention and thinking towards the invisible Universe, towards the unseen, imperceptible and hidden realms of the Universe, then these realms will become real and comprehensible to us.

As the title of the book indicates, we explore the invisible Universe, and especially those hidden and imperceptible phenomena that can never, in principle, be directly perceived by our five senses. The guiding lights for making a scientific journey to these invisible domains are our conceptual and mathematical notions of the underlying universal laws of Nature — which in turn are founded on experimental and observational results.

The laws of Nature, and in particular, the laws of Physics, are our main theoretical instruments in exploring and grasping the invisible realms and phenomena of the Universe. It is for this reason that we expend considerable effort in discussing and elaborating the theoretical edifice of Physics. Except for the first chapter on fields, the analysis is carried out in the context of a specific problem so that the reader's attention remains grounded on phenomena.

Black holes are astrophysical objects that, in principle, cannot be observed directly. The flow of time ceases on the boundary of the black hole: this is just one of the mysterious properties of black holes that are difficult to comprehend or accept. We explore the properties of black holes from a theoretical point of view and then show how our concepts lead to the interpretations of observations that indirectly point to the remarkable properties of these otherwise invisible objects.

In journeying from black holes to superstrings, we encounter many unexpected properties of our Universe, such as the fact that gravity is geometry, and that the Universe at the cosmic scale comprises almost entirely of dark energy and dark matter — entities that are unseen and cannot be directly perceived. The coming into existence of the Universe throws light on the eternal question of "how it all came to be?" and unlocks many of the mysteries relating to the formation and evolution of galaxies, stars and planets.

On a journey inwards, to the microscopic essence of materiality, we peel away layers upon layers of phenomena to finally encounter the point-like constituents of matter, namely the quarks and leptons as well as the four fundamental forces. We are then faced with the enigma and paradox that the constituents of the nuclei of all atoms, namely the quarks, are permanently confined inside the nucleus.

A further journey inwards — to ultra-microscopic realms — leads us to the unveiling of another hidden entity, namely the superstring. Superstrings exist in higher dimensions of spacetime that are hidden and invisible but have observable and tell-tale signals in our four dimensional Universe. At the frontiers of Physics, superstrings provide a consistent mathematical theory that purports to show that all of physical reality — from quarks to quasars, and including spacetime itself — is the manifestation of the hidden and unseen realm of the superstring. The Universe that we observe and live in is a vanishingly small reflection of the truly immense and colossal higher dimensional Universe that superstrings inhabit.

With the complete superstring unification and synthesis of all aspects of our invisible, mysterious and enigmatic Universe, we bring our book to a closure.

Witnessing, observing and marveling at the invisible, subtle and intriguing Universe that we live in — full of mysteries, surprises and enigmas — by using our conceptual and mathematical constructs is one of the great abilities of mankind. To develop the possibility of this theoretical consciousness into reality is, of course, up to every individual. It is our hope that this book will demonstrate how fascinating it is to unlock some of the mysteries of the invisible Universe, and that it will lead the reader, especially the younger generation, to continue on the path of further exploring the invisible Universe.

Belal E. Baaquie and Frederick H. Willeboordse

Acknowledgments

The writing of this book has been a long project, going through many phases. We would like to record our deep appreciation to Shih Choong Fong for encouraging us to embark on this project and we were inspired by his reading of science — supported by mathematics and experiments — that is driven by intuition and physical insights.

We would like to thank our colleagues for numerous comments and extensive help in the preparation of this book. We thank Phua Kok Khoo for his warm support and encouragement. Without pretense of completeness and with apologies for the many omitted, we express our gratitude to Thomas Osipowicz, Phil Chan, Kenneth Hong, Ngee Pong Chang, Munawar Karim, Kerson Huang, Tan Meng Chwan, Bal Menon, Xin Du and a special thanks to Edward Teo for many fruitful discussions on gravity. They have been indefatigable interlocutors, always willing to generously share their valuable knowledge of Physics that has greatly helped to clarify many concepts. We are indebted to Rafi Rashid for a careful reading of the draft; his prescient observations and incisive comments brought many ideas into a sharp focus and are greatly appreciated. We would like to thank Ng Kah Fee for his valuable contributions in the publication of this book.

One of us (BEB) would like to acknowledge discussions with late Jamal N. Islam, ranging over many years, on Einstein's theory of gravity; he clarified the idea of how local coordinate patches define spacetime curvature, which is achieved in one single stroke by directly working with the metric tensor. He explained in detail the rotating gravitational fields that constitute the Kerr solution and its remarkably simple form when written in the Schild coordinates. My thanks to Jayant V. Narlikar for many discussions on cosmology, inflation and the early Universe. Thanks are also due to Zahur Ahmed and Kamal Z. Islam for their encouragement and support. And lastly, my late father Muhammad Abdul Baaquie, who was a Gold Medalist in Physics, has been a life long inspiration to do science and no words can express my indebtedness to him.

We would like to thank our families for their unflagging and constant support for an undertaking that has taken many years to complete.

Contents

Chapter 1

Synopsis

The book is written for a broad scientific audience with an interest in the leading theories about the Universe. The focus is on the physical Universe, and the laws of Physics are taken to be the guiding light in all our analysis. Starting from first principles and using self-evident reasoning, all the fundamental ideas that are employed in exploring the hidden and invisible realms of the Universe are shown to arise quite naturally, once one adopts the outlook that has come to light with the advances in Physics.

Symbols and elementary mathematics go a long way in being able to precisely state an idea and we sometimes work out, using elementary arithmetic, results that are required to clarify a chain of reasoning. For this reason we have sparingly used formulas and equations — as often this gets to the heart of the matter. The reader should not be put off by the use of simple mathematics and will, hopefully, find the derivations to be straight forward and intelligible.

Only *minimal* use is made of calculus anywhere in the main text. Knowledge of calculus is not required to read this book and formulas can, in principle, be skipped if so desired. Nevertheless, we do encourage the reader to be open minded about basic mathematics since, ultimately, Nature is inherently mathematical and mathematics is enormously helpful for its understanding. All the more involved derivations have been placed in boxes that are called Noteworthy. All the Noteworthy parts of the book are meant to give greater insight to the reader and are *optional*; none of the discussions in the main text require a reading of the Noteworthy portions.

The logical organization of the topics is given in the chapter dependency flowchart. The key-link in the study of the invisible Universe is the concept of the *field*. There is a vast variety of fields in Physics, but the only fields that concern us are the fundamental fields of Nature: these are physical entities that are spread out *throughout* space and time.

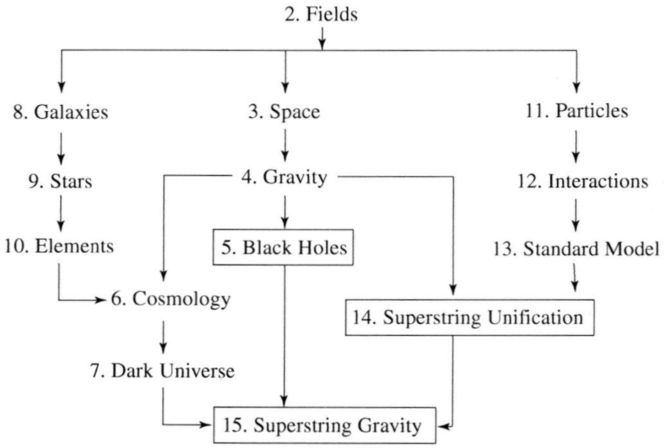

Fig. 1.1 Chapter dependency flowchart.

Chapter 2 on fields analyzes the general features of a field; classical and quantum fields are both briefly discussed as these provide the theoretical basis for much of our later discussions. The concept of the field opens many avenues for exploring the invisible Universe; the main branches that one can take in the study of the various chapters are shown in the chapter dependency flowchart (Fig. 1.1).

- Fields exist in spacetime and this leads, in Chapter 3, to developing the conceptual tools for describing the geometry of spacetime. A remarkable result, discussed in Chapter 4, is that gravity is the manifestation of the geometry of spacetime, which is described by the gravitational field. The geometrical concepts that emerge from gravity are applied, in Chapter 5, to study one of the more exotic objects of gravity, namely the black hole — which has an interior that is invisible. From gravity, one is led, in Chapter 6, to the study of cosmology, which describes the large scale structure of the Universe. Chapter 7 discusses how the Universe is largely composed of dark matter and dark energy: physical entities that cannot be directly observed.
- The formation of galaxies, stars and planets, studied in Chapter 8, results from the study of cosmology but can also be investigated directly after studying the chapter on fields. Chapter 9 discusses the various stages in the life of stars and Chapter 10 analyzes the synthesis of many of the elements of the Universe that takes place in the cores of stars via stellar nucleosynthesis and related processes. From this branch, one can go back to Chapter 6 on cosmology and continue.
- The study of the various elements leads one to analyze, in Chapter 11, the fundamental constituents of matter. Chapter 12 analyzes the interactions between the constituents of matter and Chapter 13 *unifies* all the building blocks of matter in what is called the Standard Model. The concept of the field underpins

the discussion since all the ingredients of the Standard Model are themselves described by quantum field theory.

- A comprehensive synthesis of the leading ideas on what constitutes the physical Universe is the objective of superstring theory, and is discussed in Chapters 14 and 15. Although superstring theory is still being formulated and its final form is not clear — including whether it is testable by present day experiments — the ideas are wide ranging and radical, and this gives the reader a flavor of one of the leading currents in theoretical physics. The synthesis of *all* forms of matter-energy, which includes the unification of matter with spacetime, into one entity is discussed in Chapter 14 on superstring unification. All of physical reality is postulated to be the manifestation of the invisible and fundamental entity, namely the superstring.
- The last Chapter 15 on superstring gravity discusses the quantum theory of gravity. Gravity, from being a manifestation of geometry as described by the classical gravitational field, is shown to have a deeper formulation: classical gravity is replaced by quantum gravity that emerges from superstring theory. A superstring explanation is given on the structure of black holes and the chapter then goes on to propound a cyclic and endless Universe, which is a possible cosmology that is allowed by superstring theory.

All the chapters and most of the analysis is focused on revealing and discovering realms of the Universe that are invisible and hidden. Chapter 5 on black holes and Chapters 14 and 15 on superstrings are particularly important, since the paradoxes and enigmas of the Universe appear most clearly and compellingly for these objects. In fact, the conceptual machinery and framework of theoretical physics has to be developed and exploited in new and novel ways to grapple with and grasp the ideas that explain these objects. This is the reason that these chapters have been placed in boxes and appear in the title of the book as well.

In the process of analyzing and studying natural phenomena, we find that the Universe is an interconnected, interlocked, interrelated and unified entity, with all the disparate and myriad phenomena being the appearance and manifestation of the same invisible and hidden underlying fabric of physical reality.

Chapter 2

Fields

What is the Universe made out of?

2.1 The Question

What is the substance, the 'stuff', that Nature is made out of? The present view of Physics is that every physical entity is a manifestation of energy, which is an *a priori* reality that has no explanation in science. As famously observed by leading theorist Richard P. Feynman, no one knows what is energy *as such*; all that we know are the various *forms* of energy. Science studies the different forms of energy.

The substance of the Universe is energy. By inquiring about the 'stuff' composing the Universe, one is searching for the *specific form(s)* of energy that constitute the Universe. There is a common view — not entirely incorrect — that the Universe is made out of atoms. This view is only partially true since light, which has an ubiquitous presence in the Universe, is not composed of atoms; and neither is gravity — the force that holds us to the Earth.

Atoms, light and gravity are all forms of energy. But what is it that makes light and gravity so different from atoms? The answer starts with Maxwell's equations, discovered in the 1860's, that explain light to be just one of the many manifestations of the *electromagnetic field*; and the concept of the *field* reached its classical pinnacle in 1916 with Einstein's theory of gravity, which postulates that the force of gravity is also a manifestation of a field, namely the *gravitational field*.

It was discovered by the early 20th century that atoms are made out of a nucleus, composed of protons and neutrons, and electrons bound to the nucleus due to the electromagnetic field. For a while, it seemed matter is made out of atoms and that (inter- and intra-atomic) forces were fields. With the development of high energy physics, it was then discovered that protons and neutrons are themselves made out of quarks, and finally, by the 1970's, all the constituents and forces of Nature, except for gravity, were successfully brought together in the Standard Model of particle physics. One of the rather remarkable conclusions of the Standard Model is that for each constituent of Nature — be it a quark or the electromagnetic field — there is an underlying *quantum field*, which is described by what is known as quantum field theory.

The two pillars of contemporary physics are the theory of gravitation and the Standard Model of particle physics. Both these pillars are based on the concept of the field, with gravity being a nonlinear classical field and the Standard Model being described by quantum fields. As of now, we can conclude that fields are the material, the 'stuff', that Nature is made out of.

So we need to address the question: what is a field?

2.2 What is a Field?

A classical particle, say a piece of stone, can be directly apprehended by our five senses. We can pick up a stone and feel whether it is light or heavy. We can throw the stone, and the stone travels a distance that depends on how much effort we put into throwing it. From this daily experience of objects, Newton's laws were formulated to provide a quantitative description of the dynamics of a particle. In the classical Newtonian point of view, a particle's defining property is that it occupies a particular point of space x and has a mass m — namely how light or heavy it is. Furthermore, throwing the particle means giving it energy of motion, called kinetic energy, which is proportional to its mass as well as the square of its velocity v. It was soon realized that a stone on a mountain top has potential energy $U(x)$ due to its very position; this potential energy is manifested as kinetic energy when, for example, the particle rolls down the mountain and acquires a non-zero velocity.

Summing up all these properties of a particle, one finds that the energy E and momentum p of a particle are given by the following equations

$$E = \frac{1}{2}mv^2 + U(x); \quad p = mv. \tag{2.1}$$

The particle's energy and momentum are entirely localized at the point x where the particle is located. As the particle moves, its energy and momentum are conserved (are constant) and move with it.

In contrast to a classical particle, all the fundamental fields in Nature *exist* throughout space and time and are as 'real' as a particle. The field existing at all points of space means that the energy and momentum of the field is, in principle, *spread* all over space. The strength (numerical value) of the field at different points of spacetime is a measure of the field's energy and momentum at that point.

If the field energy is zero at a point, then it means that at that point the field is not energized and is in its lowest energy state. As the field evolves in time, the energy and momentum of the field can be redistributed from one point of space to another by the variation in the field's strength. Just as is the case for particles, the time evolution of a field exactly conserves its total energy and momentum.

The field can be coupled to particles depending on the nature of the field. The electromagnetic field couples to electric charge and charged particles can exchange energy and momentum with the electromagnetic field. In particular, an accelerating charge can generate propagating electromagnetic fields. The term 'propagating field' is a shorthand for describing the redistribution of the field's energy and momentum at different points of space; this redistribution can take place in many ways, with the most commonly studied case being the wave-like oscillations of the field's strength at different points of space. And conversely a propagating electromagnetic field impinging on a charged particle can transfer energy to it and cause the particle to accelerate. The total energy and momentum of the field coupled to charged particles, taken together, are exactly conserved.

2.3 Pressure Field

To understand the key properties of fields, we need to study a physical system that is spread over space. We start with the study of the **pressure field** and examine the propagation of energy and momentum that appears in the form of sound waves. Although it is not a fundamental field of Nature, nevertheless the pressure field has all the general features of a field and hence is a good place to start.

Pressure is defined as force per unit area; the dimension of pressure P is $[P] = \mathrm{N/m^2}$, where the unit of newton is $\mathrm{N} = \mathrm{kgms^{-2}}$. The pressure field is a *scalar field*, which is fully specified by only one number at every point x for each instant t, namely $P(t, x)$.

Consider for example air that is trapped inside an inflated balloon; if one squeezes the balloon and reduces its volume, there is a force opposing this change; we intuitively know that the force originates in the fact that the air in the balloon 'resists' a reduction in its volume. More precisely, the air inside the balloon exerts a force per unit area, namely *pressure*, on the (inner) surface of the balloon. The pressure field is constant in the balloon, being equal to the pressure P exerted on the balloon. If one squeezes the balloon, then the pressure field changes from point to point inside the balloon.

An **ideal gas** is defined to be a dilute gas such that its molecules are non-interacting. At temperature T, pressure P and occupying volume V, the ideal gas law states that

$$PV = Nk_BT \qquad \text{: Ideal Gas Law} \tag{2.2}$$

N is the number of molecules in volume V and k_B is Boltzmann's constant. The Earth's atmosphere behaves like an ideal gas.

Sound is the perception by the human ear of the propagation of energy in a pressure field. To study the propagation of sound, consider a piston that produces sound waves in a cylinder by oscillating back and forth about its equilibrium position as depicted in Fig. 2.2.

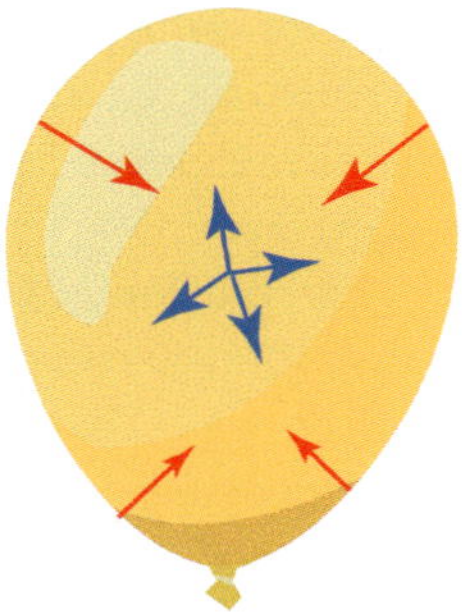

Fig. 2.1 Pressure inside a balloon.

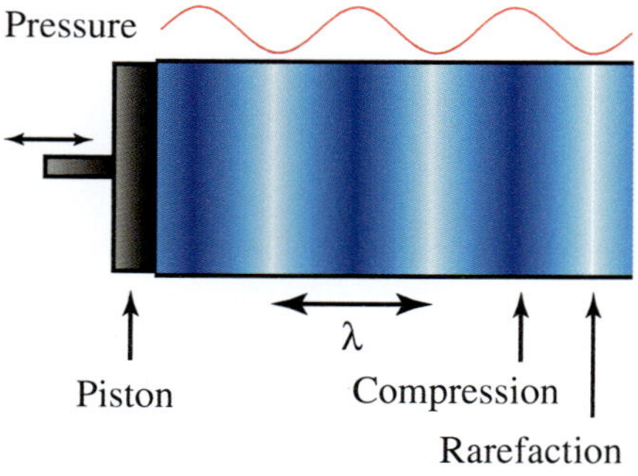

Fig. 2.2 A piston generating sound by creating a pressure wave.

Oscillating the piston with **frequency** f — which is the number of times per second that the piston oscillates back and forth — generates a wave in the cylinder traveling at speed v, which is fixed by the ideal gas law and given later. The wave has a wavelength, with dimension of m, given by

$$\lambda = \frac{v}{f}. \tag{2.3}$$

The *pitch* of the sound is determined by the *frequency* f, with unit of hertz $= s^{-1}$. A frequency of say 100 hertz means that the piston goes back and forth about its equilibrium position 100 times in one second.

Noteworthy (optional content) 2.1: Velocity of Sound

The velocity of propagation of the wave depends on the properties of the medium. In particular, the speed is related to how strongly a gas resists a change in volume, something that is measured with the so-called **bulk modulus** B. We also expect the velocity of sound to be inversely proportional to the density of the medium, denoted by ρ, since the heavier the medium the greater the inertial resistance to motion. It can be shown that

$$v = \sqrt{\frac{B}{\rho}}. \tag{2.4}$$

For an ideal gas, when there is no inflow or outflow of energy (called an adiabatic change), it can be shown that

$$B = \gamma P, \tag{2.5}$$

where γ is the adiabatic index. Combining Eqs. (2.4) and (2.5) with Eq. (2.2) we then obtain the useful equation

$$v = \sqrt{\frac{\gamma RT}{M}}, \tag{2.6}$$

where M is the mass of one mole of gas molecules and R is the gas constant ($R = 8.31\ \mathrm{JK^{-1}mol^{-1}}$).

2.3.1 *Propagating pressure field*

The simplest and best-known periodic functions are the sine and cosine functions. Hence it is natural to attempt to describe the motion of the piston and sound with these functions. Indeed, the position of the piston about its equilibrium position can be written as

$$\Delta s(t) = s_0 \cos(2\pi f t) \tag{2.7}$$

where s_0 is the amplitude (maximum displacement) of the oscillating piston.

Suppose the pressure of the air in the piston's cylinder is P_0. When the piston oscillates, the air in front of the piston also oscillates. This causes a change in the pressure field near the piston, which we denote by ΔP. As the piston oscillates, the compression and rarefaction of air create a sequence of compressed and rarefied air that propagates down along the cylinder of the piston as shown in Fig. 2.2 — causing a variation in the pressure field that is perceived as a sound wave.

Note the important fact that there is *no net transport* of air: as the piston moves back and forth, the air also moves back and forth about its equilibrium position. In other words, although the pressure field occupies all of space, it is the *variation* in the value of the pressure field that results in the transport of energy from point to point in the field.

Sound is a longitudinal wave.

The air in the cylinder oscillates *along* the direction of motion of the produced wave. Such waves are called **longitudinal waves**. The change in pressure field, at some *fixed point* in the cylinder, at time t follows from Eq. (2.7) and can be shown to yield

$$\Delta P(t) = P_m \sin(2\pi f t). \tag{2.8}$$

This is illustrated in Fig. 2.3. It should be noted that, in general, the maximum change in pressure P_m is much less than the ambient air pressure P_0.

The sensation of sound is caused by the pressure changes at the ear drum. The periodic nature of the pressure changes given in Eq. (2.8) results in the perceived sound. The higher the frequency, the more shrill is the sensation of sound and hence the higher the pitch; similarly the larger the P_m, the louder is the volume of sound.

When the air is compressed by the forward stroke of the piston, as illustrated Fig. 2.4, the region of compression travels forward with a speed v.

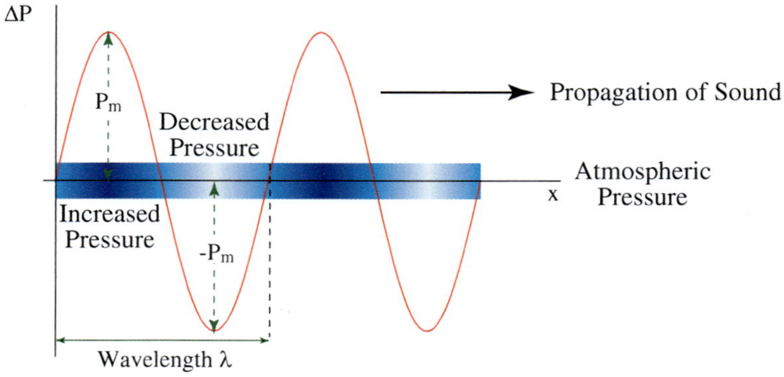

Fig. 2.3 Amplitude and wavelength of a pressure wave.

Fig. 2.4 Propagating pressure field.

Consider moving with the pressure wave at velocity v; if we were to move with the compressed portion of air, then clearly the air pressure for us would remain constant (ignoring friction). Mathematically speaking, this means that in Eq. (2.8) ΔP should not change if t were replaced by $t - x/v$.

Hence, at points along the cylinder surrounding the piston, we expect — as shown in Fig. 2.3 — the generalization of Eq. (2.8) to be given by

$$\Delta P(t,x) = P_m \sin\left(2\pi f\left(t - \frac{x}{v}\right)\right) \equiv P_m \sin(\omega t - kx)$$

where the angular frequency ω and the wave number k are given by

$$\omega = 2\pi f; \quad k = \frac{2\pi f}{v}.$$

Recall both $\Delta P(t,x)$ and P_m have the dimension of pressure, which is Nm^{-2}. Since the argument of a sine function is dimensionless, the parameter k must have a dimension equal to inverse of length, that is $[k] = \mathrm{m}^{-1}$; this then makes kx dimensionless.

2.4 Electric and Magnetic Fields

We now study electric and magnetic fields as these are fundamental **vector fields**, specified by a *magnitude and direction* at every spacetime point, denoted by $\mathbf{E}$ and $\mathbf{B}$ respectively. A study of atoms and molecules shows that it is primarily due to electrical forces that atoms and molecules are bound together into stable objects. Hence, in a real sense, all the varied materials around us exist due to the workings of the electric field.

Consider two point charges q_0 and q_1 separated by distance r. Coulomb found that the force between the charges is given by

$$F_e = k_e \frac{q_0 q_1}{r^2} \equiv q_0 \mathbf{E} \qquad : \textbf{Coulomb's Law} \tag{2.9}$$

where $k_e = 9 \times 10^9\ \mathrm{Nm}^2/\mathrm{C}^2$ is a universal constant, and the *direction* of the force is along the line joining the two charges. The C in k_e is the unit of charge called the *coulomb* (C). The unit of the electric field is force per unit charge, namely $[\mathbf{E}] = \mathrm{NC}^{-1} = \mathrm{Vm}^{-1}$, where V is the unit of volt. The electric field due to charge

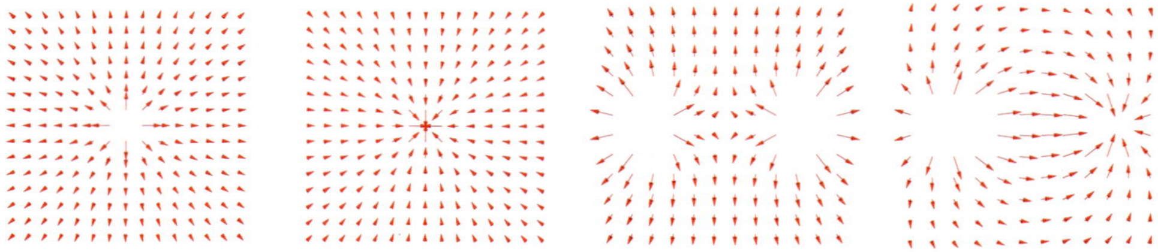

(a) (b) (c) (d)

Fig. 2.5 Fields for one and two point charges. (a) Electric field of a positive point charge. (b) Electric field of a negative point charge. (c) Electric field for like charges. (d) Electric field for opposite charges.

q_1 is given by

$$\mathbf{E} = \frac{F_e}{q_0} = k_e \frac{q_1}{r^2}, \tag{2.10}$$

and this shows that the electric field $\mathbf{E} = \mathbf{E}(r)$ is spread over space, having a different value for each r.

The electric field in general is a vector $\mathbf{E}$. Figure 2.5(a) shows the electric field for a positive charge q by lines *emanating* from it, with the arrows on the lines indicating that the electric field $\mathbf{E}$ points radially outwards. Conversely, the electric field due to a negative charge $-q$ points *inwards* as shown in Fig. 2.5(b), indicating an inward attractive force on a positive test charge brought close to $-q$. Figures 2.5(c) and 2.5(d) show, respectively, the electric field of two like and opposite charges.

From the graphical representation of the electric field, we can see the following:

(1) The electric field varies from point to point, so the energy density at different points of space also varies; the total energy contained in the electric field is obtained by integrating (summing) the energy density over all of space. The energy contained in the electric field is proportional to $\mathbf{E}^2$. The higher the density of electric field lines, the higher is the density of energy in that region of space.
(2) The intensity of the electric field is proportional to the amount of charge that is generating the field.
(3) If two positive charges are brought close to each other, they will repel each other. This because the field's lines repel each other, as shown in Fig. 2.5(c). Consequently, the field has a lower energy if the charges are moved further apart, and hence like charges repel.
(4) Consider a positive charge and negative charge being brought close to each other. The electric field lines emerging from the positive charge then converge on the negative charge, as shown in Fig. 2.5(d). Since some of the electric field lines start and end on the positive and negative charges, the field's energy is reduced by bringing the charges close to each other, and hence they will attract.

Electric fields obey the principle of superposition since the underlying (Maxwell's) equations are linear. If there are a number of charges, namely, q_1, q_2, q_3 and so on, that generate electric fields $\mathbf{E}_1$, $\mathbf{E}_2$, $\mathbf{E}_3$, ... and so on respectively, then the net electric field due to all the charges is given by

$$\mathbf{E} = \mathbf{E}_1 + \mathbf{E}_2 + \mathbf{E}_3 + \cdots . \tag{2.11}$$

2.4.1 *Magnetic field*

The magnetic field **B** is ubiquitous in Nature. The Earth's mild magnetic field has been known since antiquity, and magnetic fields manifest themselves in more subtle forms such as permanent magnets, in electric motors and electric power generators, in permanent magnetic information storage devices, and down to the quantized magnetic fields that are trapped in superconducting materials. Magnetic fields appear everywhere in astrophysics. Spinning neutron stars generate very strong magnetic fields that cause intense X-ray radiation from charged objects that fall into the stars. Magnetic fields determine the properties of quasars and neutron stars; powerful magnetic storms erupt during solar flares.

The magnetic field of a bar magnet, as shown in Fig. 2.6(a), can easily be illustrated by sprinkling some fine iron powder on a thin glass plate below which a magnet is placed; the corresponding field lines are shown in Fig. 2.6(b). Lines of magnetic fields emanate from the north pole and converge on the south pole.

The SI unit of the magnetic field is the **tesla** (T) and has dimensions of $\text{T} = \text{Vsm}^{-2}$. The strengths of some typical magnetic fields found in nature are given in Table 2.1.

Magnetic fields are more difficult to study than electric fields. Unlike the electric field that, for the stationary case, can be reduced to the much simpler electrostatic potential which is a scalar function of spacetime, magnetic fields, even for the simplest stationary cases, can only be represented by vectors.

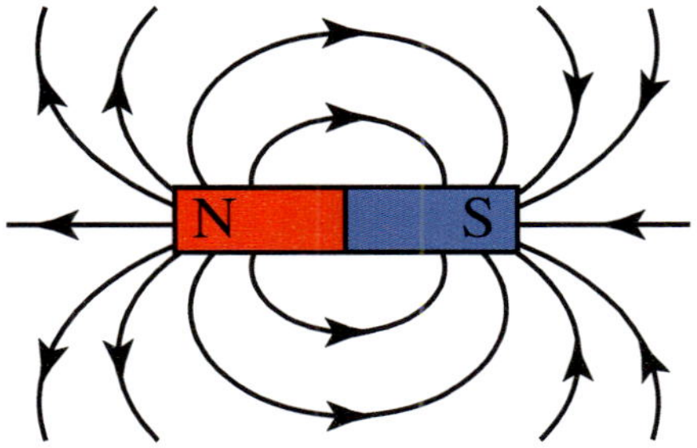

(a)

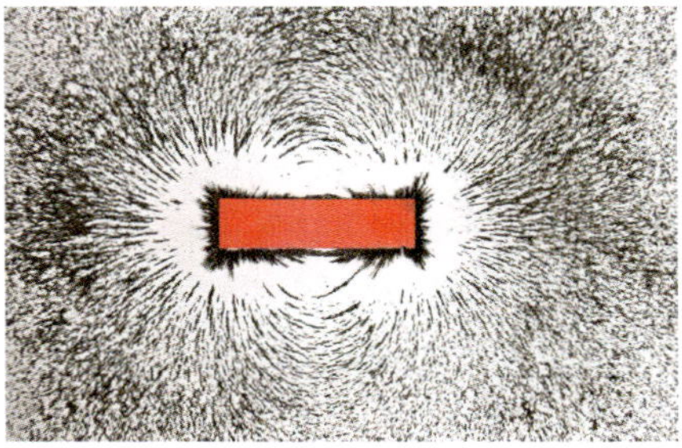

(b)

Fig. 2.6 (a) Magnetic field of a bar magnet. (b) Visualization of the magnetic field with the help of iron filings.

Table 2.1 Strengths of magnetic fields.

Phenomena	Magnetic field (T)
At neutron star's surface	10^6
Large electromagnet	1.5
Small bar magnet	10^{-2}
At Earth's surface	10^{-4}
Detectable magnetic field	10^{-14}

The underlying reason for the asymmetry between electric and magnetic fields is that, unlike electric charges, Nature chooses not to have any "magnetic charges" — otherwise known as magnetic monopoles. Instead of being generated by magnetic charges, all magnetic fields in Nature, in fact, turn out to be generated by moving electric charges.

2.5 The Electromagnetic Field

From our discussion in the previous section, both $\mathbf{E}(t,x)$ and $\mathbf{B}(t,x)$ are vector fields that depend on space and time, and are spread all over space. For every instant t and at every point x of space there are three real numbers that specify the electric field, and similarly three real numbers that specify the magnetic field. In other words, an infinite collection of vectors, for each instant t and at each point x, specify the electric and magnetic fields. In contrast, recall that the pressure field is a scalar field requiring only one number, denoted by $P(t,x)$, for each point of spacetime.

The EM (electromagnetic) field is generated by a source, which consists of charged particles; a static charge creates a static electric field; a moving charge creates a moving electric field and a static magnetic field; an accelerating charge, in contrast, is the source of *propagating* electromagnetic waves. The electromagnetic field itself is electrically neutral and hence does not couple to itself. What is worth noting is that even when the sources are switched off, the disturbances created — as in the case of EM waves — can continue to exist as physical entities and propagate according to Maxwell's field equations.

The electromagnetic field carries energy and momentum for each volume element of space dV at position x. The **field energy density** $\Delta U(x)$ and **field momentum density** $\Delta\vec{S}(x)$, called the **Poynting vector**, are given by

$$\Delta U(t,x) = \frac{1}{2}\left[\epsilon_0\mathbf{E}^2(t,x) + \frac{1}{\mu_0}\mathbf{B}^2(t,x)\right]dV$$

$$\Delta\mathbf{S}(t,x) = \left[\frac{1}{\mu_0}\mathbf{E}(t,x)\times\mathbf{B}(t,x)\right]dV.$$

It can be shown that the velocity of light is given by $c^2 = 1/(\epsilon_0\mu_0)$.

The total energy of the electromagnetic field is given by integrating the energy per unit volume, namely $\Delta U(x)$, over *all* of space, and yields

$$U = \frac{1}{2}\int dV \left[\epsilon_0 \mathbf{E}^2 + \frac{1}{\mu_0}\mathbf{B}^2\right]$$

and a similar expression can be obtained for the total momentum $\mathbf{S}$ of the electromagnetic field.

Energy propagates along the direction of motion of the EM wave. The *intensity* of light is a measure of how much radiant energy is received at a point, and as can be seen by the formula above, is proportional to $\mathbf{E}^2$ and $\mathbf{B}^2$.

The electromagnetic field can exchange energy and momentum with charge carrying particles, showing that the field's energy and momentum are of the same nature as those carried by mass carrying charged particles. The coupling of the field to charged particles results in the conservation of total energy and momentum of the *combined system* — of the field and charged particles taken together.

2.5.1 *Electromagnetic waves*

The great discovery of Maxwell was the understanding that electric and magnetic fields are really the same entity, and appear to be different only in certain special circumstances. A charge moving with a constant velocity, for example, generates a combination of both electric and magnetic fields. Furthermore, in Einstein's theory of special relativity, by transforming from one inertial frame to another, the values of the electric and magnetic fields change, showing that these are frame dependent, and not frame independent physical quantities.

Electromagnetic waves are transverse.

What Maxwell deduced was that oscillations of the electromagnetic field — obtained by for example oscillating an electric charge — give rise to electromagnetic (EM) waves which have observable consequences. This is similar to our study of sound waves earlier in that the oscillations of the electromagnetic field propagate energy and momentum through space. The *electromagnetic waves are transverse waves* — with the electric and magnetic fields oscillating in a direction that is *orthogonal* to the direction of propagation of the wave. This is in contrast to pressure waves which are longitudinal waves.

Suppose an electromagnetic wave is propagating in vacuum along the x-direction. A propagating electromagnetic wave is often called radiation. Since electromagnetic waves are transverse, the oscillations of the electric and magnetic fields are in the directions perpendicular to the direction of propagation, and hence lie in the yz-plane. Let us denote the unit vectors in the xyz-directions by $\mathbf{e}_x, \mathbf{e}_y$ and $\mathbf{e}_z$ respectively. The simplest example of radiation propagating in the x-direction is given by the electric field **E** always lying along the y-axis, and the magnetic field **B** always lying along the z-axis. The **E** and **B** fields for radiation are transverse waves with the added property that all electromagnetic waves propagate in vacuum at the *speed of light.* From Eq. (2.3), we have for electromagnetic radiation

$$f = \frac{c}{\lambda}. \tag{2.12}$$

A derivation based on Maxwell's equations for electromagnetic waves yields that the electric and magnetic fields that constitute an electromagnetic wave — with wavelength λ and propagating in the x-direction — are given by

$$\mathbf{E}(t, x) = E_0 \sin\left(\frac{2\pi}{\lambda}(x - ct)\right)\mathbf{e}_y \tag{2.13}$$

$$\mathbf{B}(t, x) = \frac{E_0}{c}\sin\left(\frac{2\pi}{\lambda}(x - ct)\right)\mathbf{e}_z. \tag{2.14}$$

The propagation of the radiation given by Eqs. (2.13) and (2.14) is shown in Fig. 2.7, with the direction of propagation being the x-direction, and the sinusoidal oscillations of the **E** and **B** fields drawn in the yz-planes.

2.5.2 *Maxwell's field equations*

Before James Clerk Maxwell, the study of electric and magnetic phenomena was dominated by a number of distinct laws, such as the laws of Coulomb, Faraday, Gauss and Ampere. Maxwell noticed that all these laws for electric and magnetic fields could be summarized by four equations.

On combining the equations, Maxwell also noticed that the four equations, taken together, were incompatible with the principle of charge conservation. Charge conservation was considered a physically reasonable requirement and therefore difficult to ignore. Hence Maxwell decided that one of those four laws had to be modified in order to have charge conservation.

As a consequence of these changes, Maxwell found an astounding new result: the modified equations *predicted* the existence of a new phenomenon, namely propagating *electromagnetic waves.* These electromagnetic waves traveled at the same speed as the then known speed of light! Maxwell then hypothesized that

Fig. 2.7 An electromagnetic wave.

light itself was just one form of electromagnetic wave. Maxwell's predictions were experimentally verified and eventually gave birth to vast new domains of applications and research.

The fundamental laws of electricity and magnetism as summarized by Maxwell are called *Maxwell's field equations*, and describe the unified phenomenon of *electromagnetism*. For example, a charge moving with a constant velocity generates a combination of both electric and magnetic fields (recall that an electric current produces a magnetic field and that a current is simply electric charge in motion).

Maxwell's equations are so elegant that they helped inspire Einstein in formulating his special theory of relativity.

Noteworthy (optional content) 2.2: Maxwell's Equations

So what are Maxwell's field equations? In empty space and in the absence of charges and currents, they can be enumerated in the following manner.

(1) Electric fields are generated by positively and negatively charged particles. Mathematically, this is expressed by **Gauss's law**

$$\nabla \cdot \mathbf{E} = 0. \tag{2.15}$$

(2) There is no magnetic monopole. This is an experimental fact that continues to hold. This is expressed by Gauss's law for magnetism

$$\nabla \cdot \mathbf{B} = 0. \tag{2.16}$$

(3) A changing magnetic field generates an induced electric field. This equation of Maxwell is **Faraday's law** of induction, and yields an electric current from a time varying magnetic field.

$$\nabla \times \mathbf{E} = -\frac{\partial \mathbf{B}}{\partial t}. \tag{2.17}$$

(4) An electric current or a changing electric field gives rise to an induced magnetic field. This is **Ampere's law** as extended by Maxwell

$$\nabla \times \mathbf{B} = \frac{1}{c^2}\frac{\partial \mathbf{E}}{\partial t}. \tag{2.18}$$

Ampere's law did not have the $\partial \mathbf{E}/\partial t$ term, which was added by Maxwell to obtain his complete set of equations; it is this term that results in the propagation of electromagnetic waves.

Note that the third and fourth of Maxwell's equations explain the phenomenon of electromagnetic radiation in the following manner: In empty space, Eq. (2.17) states that an electric field that changes in space gives rise to a time varying magnetic field. Equation (2.18) in turn shows that a time varying electric field induces a magnetic field that changes over space. And this feeds back into Eq. (2.17) and so on and provides the mechanism for the propagation of radiation over infinitely large distances!

The electromagnetic field can be represented by the **electromagnetic potential** $A_\mu(t, x, y, z)$, $\mu = 0, 1, 2, 3$ that *unifies* the electric and magnetic field in the following manner

$$\begin{aligned}\mathbf{E}(t, x, y, z) &= \nabla A_0(t, x, y, z) - \frac{\partial \mathbf{A}}{\partial t}(t, x, y, z) \\ \mathbf{B}(t, x, y, z) &= \nabla \times \mathbf{A}(t, x, y, z).\end{aligned} \tag{2.19}$$

We will see later, in Sec. 11.2, that $A_\mu(t, x, y, z)$ is a gauge field.

2.6 Do Fields Need a Medium?

One may consider the difference between particles and fields as being illusory; one can consider the **medium**, for example a solid in which sound propagates, as being made out of small particles connected by small springs, and the continuum 'field' being only an approximation of this collection of particles. One would then explain the flow of energy and momentum in a field as finally being explained by the transfer of energy and momentum between small Newtonian particles. This is precisely what we did for the case of a pressure wave, namely sound; the discrete atomic nature of a gas is approximated by a continuous medium, which is taken to obey Newton's laws. The equations of fluid mechanics are derived for the motion of this continuous medium; sound waves are one example of fluid mechanics, with other examples being the flow of oceans and rivers and so on.

Until the discovery of Maxwell's equations, the view held by most physicists was that a *material medium*, obeying Newton's laws, was the final explanation of all fields and that the EM waves that one observed in Nature are the oscillations of the underlying medium composing the field.

The electromagnetic field was initially analyzed in the framework of Newtonian mechanics. The electric and magnetic fields were considered to be an elegant way of encoding Newtonian *forces* between charged particles, and not objects having any existence independent of the forces they carry. The discovery that light is a propagating electromagnetic wave created the first crack in the mighty edifice of Newtonian mechanics.

The propagating light wave was held to be the oscillations of a material medium, called **ether** and obeying the laws of Newtonian mechanics. In spite of many attempts, including by Maxwell himself, one could not find this medium. The time evolution of the electromagnetic field, as determined by Maxwell's equation, could not be derived from the properties of any material medium obeying Newton's laws of motion. It was the Michelson–Morley experiment, conducted in 1887, that conclusively showed that ether does not exist.

The understanding of the electromagnetic field required the discovery of the special theory of relativity that showed that there is no ether in Nature. For electromagnetic waves there is *no underlying material medium* whose oscillations result in the wave phenomenon. This is the *crucial difference* between electromagnetic waves and material waves, such as a pressure wave, which are created by the oscillations of an underlying medium.

It is important to realize that the electromagnetic field is not merely a mathematical convenience but a physical entity carrying its own energy and momentum. The electromagnetic field is not made out of any material medium (all of which ultimately consist of atoms and molecules).

Noteworthy (optional content) 2.3: Forces versus Fields

Recall that the fundamental and unique feature of a field, in contrast with a particle, is that the field's energy and momentum are spread *over space*, whereas force directly acts on material particles that occupy a *definite position in space*. This holds for the case of fluid mechanics as well, for which the mass is taken to be a continuous distribution. There is no concept of energy being distributed in space for a force.

Newtonian forces can act instantaneously over arbitrarily large distances. Consider for example, Newton's gravitational force F_G between two masses m_1 and m_2, separated by a distance R; there is an instantaneous force, as shown in Fig. 2.8, given by

$$F_G = G\frac{m_1 m_2}{R^2} \tag{2.20}$$

where G is Newton's gravitational constant.

No medium is required to transmit this force; if one changes the position of one of the masses, then the force of gravity changes instantaneously; the instant action of the force over large distances is called *action at a distance.*

2.6.1 *Action at a distance*

Both Coulomb's law and Newton's law of gravitation describe **action at a distance**. The forces between objects do not seem to require an intervening mechanism and act instantaneously. Although this was already considered bizarre

Fig. 2.8 In classical physics, a force is instantaneous.

by Newton, it was not until much later that a satisfactory formulation appeared in terms of fields.

To deepen our understanding of the idea of a field, we need to re-examine the notion of a material entity. From the time of Newton until the mid-nineteenth century, all material entities were thought to be similar to a solid body. A solid body has a definite position in space and 'carries' its energy and momentum. All solid bodies can be thought of as being composed out of small point-like constituents, and hence all physical entities can be thought of as being composed out of a collection of small massive particles. A continuous medium, in Newton's framework, is a collection of these small classical particles. This view has had great success, including the formulation of fluid mechanics and aerodynamics.

In contrast, a field is not a continuous medium but, nonetheless, is a physical entity continuously spread throughout space. For example, if we introduce an electric charge into space, the electric charge generates an *electric field* throughout space, which then acts on other charges. Figures 2.5(a) and (b) show the electric fields due to positive and negative charges. If *another* charge is introduced, the effect of the initial charge is carried by the field that acts on the other charge and which, in turn, creates an attractive or repulsive force, depending on the signs of the two charges. The field mediates all the interactions of the charged particles: there is no direct interaction of the charges. *Thus there is no action at a distance in the field picture: all forces are local.*

The concept of the field goes further than just making the forces local. The special theory of relativity tells us that no physical entity can travel faster than the speed of light. Hence, there can be no action at a distance since this would require propagation at infinite speed. Suppose we change the position of the source; the change cannot *instantaneously* affect other charges. The field is a necessary *mediator* of interactions, with (relativistic) fields propagating a change in the field's configuration at most at the speed of light — and hence affecting other charges with a time lag required by special relativity.

2.7 Lorentz Transformations

Chapters 2, 3 and 4 are devoted to studying geometry and gravity. In this section, we contextualize these later discussions on gravity by showing that its theoretical foundation lies in the concept of the gravitational field. In the presence of matter and energy, the geometry of the Universe is a *curved spacetime*, and which is described by the gravitational field. Before discussing gravity, we instead start by discussing the structure of *empty* spacetime, as embodied in the special theory of relativity, which is discussed in Sec. 3.7; we foreground that discussion by showing how the properties of fields are closely related to the property of spacetime.

The fundamental postulate of the special theory of relativity is that the speed of light c is constant in vacuum for all observers that are moving at constant velocity, called **inertial observers**. Furthermore, c is the *maximum* possible speed in space. If two observers are moving at different velocities, then special relativity predicts that they will both measure the same speed of light. Note that for empty spacetime light propagates in a *straight line* ($x = ct$ for the wave front) for both observers. This fact will be important in generalizing the path of light in the presence of gravity.

Recall from Eqs. (2.13) and (2.14) that a propagating electromagnetic wave with wavelength λ — for an observer with coordinates t, x — is given by

$$\mathbf{E}(t,x) = E_0 \sin\big(k(x-ct)\big)\mathbf{e}_y; \quad k = \frac{2\pi}{\lambda}$$

$$\mathbf{B}(t,x) = \frac{E_0}{c} \sin\big(k(x-ct)\big)\mathbf{e}_z.$$

Suppose that there is a *second* observer, with coordinates given by t', x', who is moving at velocity v with respect to the first observer. Each observer is also called a **frame of reference**. What are the electric and magnetic fields that the second observer will measure?

In general, if a moving observer — traveling at velocity v — measures the velocity of say a car approaching it at velocity u, Newton's law of motion predicts that the moving observer will measure the car's velocity to be $u + v$. The special theory of relativity, as postulated by Einstein, requires that c, the speed of light in vacuum, is a *universal constant* that is the same for all inertial observers. Clearly, the example of the car's velocity observed in Newtonian mechanics cannot hold for the case of the velocity of light, since this would lead to velocity of light being $c + v$ for the moving observer.

Demanding c be a universal constant requires that, in going from the first to second observer, both the space and time coordinates have to be changed to keep their ratio for the speed of light to be constant for both observers. This is in sharp contrast to the Newtonian view of time, which holds that there is one universal time that flows at the same rate for all entities in the Universe and that all observers measure the same flow of time.

Consider the relative velocity between two observers to be solely in the x-direction. One can show that the velocity of propagation c of the electromagnetic fields is the same for both frames only if a transformation, called a **Lorentz transformation** after the physicist Hendrik Lorentz, connects the two frames as given below

$$x' = \gamma(x - vt) \tag{2.21}$$

$$t' = \gamma(t - vx/c^2) \tag{2.22}$$

$$\gamma = \frac{1}{\sqrt{1-(v/c)^2}}. \tag{2.23}$$

Equations (2.21) and (2.22) are the Lorentz transformation that relates the observations of two inertial observers. In general, Lorentz transformations are defined to be all transformations between frame $(\mathbf{x}, t)$ and another frame $(\mathbf{x}', t')$ such that

$$\mathbf{x}^2 - c^2t^2 = (\mathbf{x}')^2 - c^2(t')^2.$$

Note the above equation factorizes; consider velocity v to be along the x-axis; the orthogonal directions are the same for the two observers and the Lorentz transformation is given by

$$\begin{aligned} x^2 - c^2t^2 &= (x')^2 - c^2(t')^2 \\ \Rightarrow (x - ct)(x + ct) &= (x' - ct')(x' + ct'). \end{aligned} \tag{2.24}$$

The electric and magnetic fields transform in a complicated manner under Lorentz transformations; ignoring the vector aspect of the transformation, the piece relevant to our discussion is the velocity of propagation that is given by the following

$$\mathbf{E}'(t', x') = E_0 \sin\big(\tilde{k}(x' - ct')\big)\mathbf{e}'_y + \cdots \tag{2.25}$$

$$\mathbf{B}'(t', x') = \frac{E_0}{c} \sin\big(\tilde{k}(x' - ct')\big)\mathbf{e}'_z + \cdots \tag{2.26}$$

where $\tilde{k}$ is the new wave vector and $\mathbf{E}'(t', x'), \mathbf{B}'(t', x')$ are the electric and magnetic fields in the moving frame. As required by the constant value of the speed of light, the velocity of propagation in the new frame, given in Eqs. (2.25) and (2.26), is *also* the speed of light, namely c. The result above illustrates the special case that we obtained in Eq. (2.25) that, up to a scaling factor, under a Lorentz transformation, we have that $(x - ct)$ is transformed to $(x' - ct')$.

The lesson from the derivation of Eqs. (2.25) and (2.26) is that the propagation of electromagnetic fields reflects the *structure of spacetime*; the reason being that Maxwell's equations fully respect the postulates of special relativity, although ironically this was not known to Maxwell.

To illustrate the effect of special relativity, consider an electric charge at rest as shown in Fig. 2.9(a). Suppose we accelerate the charge and then, after some time, bring it to rest. As per special relativity, the electric field can only change at points that are within a distance that is covered by a signal traveling at the speed of light. The electric field of the moving charge is schematically shown in Fig. 2.9(b) and we see that the electric field far away from the moving charge is not affected instantaneously; only the electric field close enough to the charge is changed; furthermore, the shaded area shows that the field has acquired energy from the acceleration of charge, and this energy is in the form of an electromagnetic wave that is propagating at the speed of light and changing the electric field's direction. The last Fig. 2.9(c) shows that on coming to rest, the electric field is again pointing out radially, but centered at the new position of the charge.

Energy conservation is realized by the electromagnetic radiation carrying exactly the energy expended in accelerating the charge and bringing it to its new position.

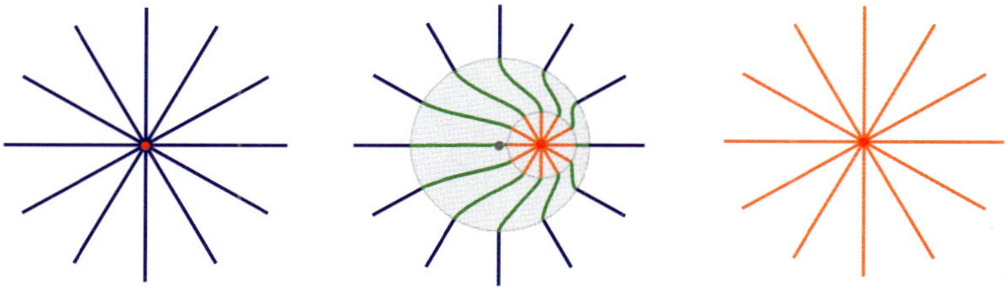

Fig. 2.9 The electric field of a charge moved from initial to final position.

2.8 Gravitational Field

How should we describe spacetime in the presence of gravitating bodies? Einstein's study of Maxwell's equations led him to discover the geometrical theory of gravity. From a modern point of view — and this was not the one taken historically by Einstein — if fields are the fundamental structures that describe the 'stuff' of the Universe, then there should be a field describing gravity as well. The big discovery by Einstein was that in the presence of matter and energy flat spacetime has to be replaced by curved spacetime, and which is described by the **gravitational field**.

How does the gravitational field describe curved spacetime? The principle of special relativity, which states that the speed of light c is constant, must continue to hold for curved spacetime; since the geometry of curved spacetime changes from point to point, one generalizes special relativity by demanding that c is constant *locally*, in the neighborhood of every point. This is discussed in Chapter 3 and we summarize the results in the language of the gravitational field.

The geometry of any curved space, such as the surface of a billiard ball, is completely specified by its **metric**, which determines the distance between any two points of that space. To describe the geometry of curved spacetime, we need a collection of observers at every point of space, with each observer having his own measuring *rod* and *clock*. In the presence of gravitating matter, the measure of distance and the flow of time will vary from point to point; the gravitational field determines how to 'glue' the different measuring rods and clocks together into one seamless whole. The *gravitational field* is determined by gravitating entities and, more mathematically, it is the **metric tensor** that fixes all the distances in spacetime.

Massive bodies cause spacetime geometry to become curved, with the curvature being greater the larger the gravitating mass. To maintain c as the constant speed of light, one needs to postulate that light follows the *shortest* distance between two points, called a **geodesic**. Specifying all the geodesics of spacetime is equivalent to

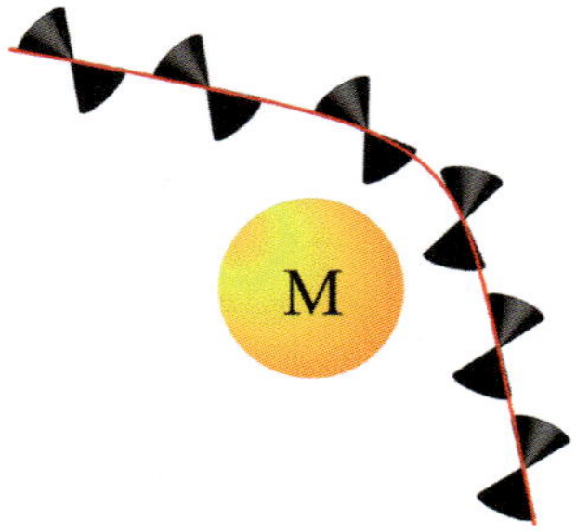

Fig. 2.10 Light follows the curvature of spacetime.

specifying its geometry: the constancy of the speed of light continues to hold locally, even in the curved geometry of spacetime containing gravitating masses.

The gravitational field, at every point of spacetime, carries energy and momentum; since the gravitational field couples to all forms of energy and momentum, it consequently couples to *itself* and hence is a gravitating entity; what this means is the gravitational field interacts with *itself*, and this is the reason that the field equations of gravity are nonlinear. Unlike the electromagnetic field that propagates in spacetime, the gravitational field determines spacetime itself. The spacetime that we live in is not an *a priori* 'stage' in which events take place but rather, spacetime itself is a dynamic entity changing and evolving due to its interaction with all entities that carry energy, including itself.

The curvature of spacetime takes place 'gradually', with space at very small scales always appearing to be flat. The constancy of light is valid at every point of curved spacetime. To have an intuitive understanding of curvature, consider the propagation of light in the presence of a star. Einstein showed that time flows *slower* for points closer to the star; hence the path of light has to change in order to maintain the constancy of the speed of light. As shown in Fig. 2.10, the change of the geometry around the star causes light to be *deflected* from its otherwise straight path; the deflection is precisely in a manner so that, locally, c remains constant.

2.9 Quantum Fields

All the ingredients of the Standard Model of particle physics — to which we devote Chapters 11, 12 and 13 — are summarized in Table 12.1. The fundamental particles and their forces, which are called interactions, are all described by (relativistic) **quantum fields** that are mathematically expressed by quantum field theory. Furthermore, superstring theory, discussed in Chapters 14 and 15, is a continuation of the ideas developed from quantum field theory. We briefly discuss a few of the important properties of quantum fields as this provides the theoretical framework for

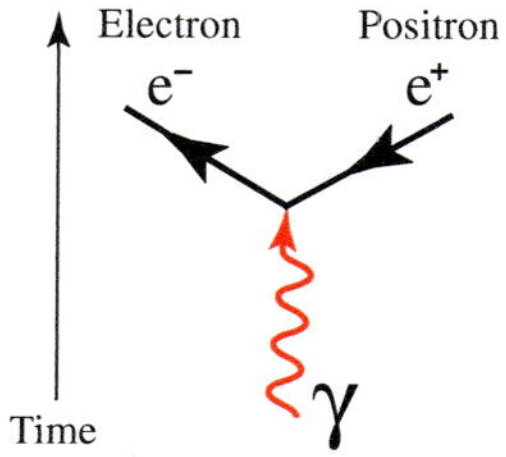

Fig. 2.11 A Feynman diagram for the vertex (coupling) of the electron field with the photon field.

understanding the fundamental particles, their interactions as well as their possible unification in superstring theory.

A classical field, like the electric and magnetic fields, is a physical entity that takes definite values at every point of space and time; for example, electromagnetic potential $A_\mu(t,x,y,z)$, given in Eq. (2.20), is a vector field that is fully specified by four real numbers at every point of space and time. In mathematical terms, a classical field is a **determinate** function of spacetime, and which exists objectively, regardless of whether the field is observed or not.

As a consequence of relativity the number of particles in a system is no longer fixed. The combination of relativity with quantum mechanics is discussed in Sec. 11.8, and is shown to lead to the conclusion that for preserving causality, every particle must have its own corresponding antiparticle. Quantum fields are indeterminate and relativistic quantum fields are the mathematical means of realizing a system for which the number of particles is indeterminate; for example, there are quantum fluctuations, as in Fig. 2.11, in which an electron and a positron are created out of say a photon, and hence increasing by two the number of particles in the system.

Noteworthy (optional content) 2.4: Indeterminate Quantum Field

In contrast to a classical field, a quantum field is intrinsically **indeterminate** and has no fixed value: at each spacetime point (t,x,y,z) the electromagnetic quantum field potential $A_\mu(t,x,y,z)$ has *all possible values*; the mathematical realization of the indeterminacy of the quantum field is that — for *each* spacetime point (t,x,y,z) — $A_\mu(t,x,y,z)$ is an *integration variable*, and $\int dA_\mu(t,x,y,z)$ is how one should conceptualize a quantum field. Physically observable properties of the quantum field are determined by taking averages over all possible values of $A_\mu(t,x,y,z)$. In particular the fundamental mathematical structure of the quantum field $A_\mu(t,x,y,z)$ is given by

$$\prod_{t,x,y,z,}\prod_{\mu}\int_{-\infty}^{+\infty} dA_\mu(t,x,y,z)e^{\text{Action[A]}}.$$

The notation above states that one performs an integration for *every* point of spacetime, with the integration variable being the quantum field $A_\mu(t,x,y,z)$. For the electromagnetic (quantum) field, **Action** [A] is proportional to the time integral of the Maxwell Lagrangian, obtained from Maxwell's equations.

The physically observable quantities of a quantum field are its parameters, such as the electric charge and mass of the electron, as well as *correlation functions* of $A_\mu(t, x, y, z)$, which are functions of spacetime.

2.9.1 *Feynman diagrams*

The diagram drawn in Fig. 2.11 is called a **Feynman diagram** — after the legendary physicist Richard Feynman who invented these diagrams — and represents the **probability amplitude** for a quantum process to take place. The probability amplitude is a notion that is central to quantum physics, the absolute square of which yields the probability of the quantum events.

In a Feynman diagram, each quantum field is represented by a line, with a straight line representing the electron field and the wavy line representing the photon field. Note that the straight line does not end whereas the wavy line appears at the vertex; this is a representation of the fact that the charge of electrons is conserved whereas photons can appear from the vacuum and their number is not conserved. Feynman diagrams for photons and electrons are discussed in Sec. 12.4.1.

Figure 2.11 represents the interaction of the photon field with the electron field. The fundamental vertex given in Fig. 2.11 can be thought of as a virtual process since, taken by itself, the diagram does not conserve energy and momentum for the electron, positron and photon. The creation of a physical electron–antielectron pair from a photon is shown in Fig. 2.12. A useful interpretation of the diagram given in Fig. 2.11 is that it represents the fundamental interaction of the electron field with the photon field and is analogous to the elementary building blocks of a Lego set. There are rules for combining the fundamental vertices into more complicated Feynman diagrams that yield a representation of more complicated processes, examples of which are given in Figs. 2.12 and 2.14.

Feynman diagrams are ubiquitous in the description of quantum fields and their interactions and will be used extensively in the later chapters. More complicated Feynman diagrams are used for describing quantum fields that appear in particle physics. The concepts of quantum fields, antiparticles and Feynman diagrams are discussed in some detail in Chapters 11 and 12.

Feynman diagrams provide a diagrammatic and graphical way of describing the mathematical properties of quantum fields. In addition to providing a direct physical intuition for understanding and describing the complex behavior of quantum fields, Feynman diagrams are also a powerful computational tool, since each diagram corresponds to a precise and well defined mathematical expression. In fact, it would be no exaggeration to state that the study of quantum fields would not have made the progress that it has made if not for the conceptual

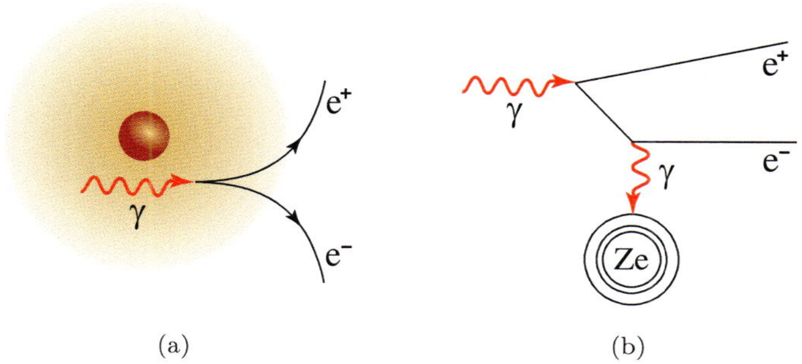

Fig. 2.12 (a) Pair creation in the proximity of a heavy nucleus. (b) A Feynman diagram representation of the pair creation process, with a virtual photon disintegrating into an electron–antielectron, and a secondary photon scattering off a nucleus with charge Ze and in doing so, extracting the energy and momentum required to make the pair have finite energy and momentum.

clarity and mathematical facility that Feynman diagrams have brought to the subject.

2.10 Quantum Vacuum

How does one describe the **state** of an indeterminate quantum field? A fundamental assumption of quantum mechanics, and this carries over to quantum field theory, is that quantum indeterminacy can be described by assigning a **state vector** to the quantum field. The state vector determines the outcomes of measurements carried out on the indeterminate quantum field.

A **fluctuation** is defined to be a specific spacetime configuration of the quantum field. Each specific configuration (of the quantum field) can be thought of as having a certain likelihood of occurrence, which is determined by the quantum field's action. The state vector provides a description of the **excitations** of the quantum field. The excitations are characterized by many quantities, the most important being the energy of the state. In particular, the quantum field can be in a highly energetic state or can be in one of its lower energy states.

The most important quantum state vector of the quantum field is the **vacuum state**, denoted by $|\Omega\rangle$, which is the *lowest* energy state of the quantum field, usually taken to have zero energy; the vacuum state *does not* depend on time.[1] For the electron field, the first state having an energy higher than the vacuum is the **single particle state** of the quantum field and which, in the classical limit, is the electron.

[1] A classical vacuum is the lowest energy configuration for a classical field; the classical vacuum is inert, consisting of one particular and fixed configuration of the classical field.

Noteworthy (optional content) 2.5: Quantum State Vector

How does one describe the statistical regularities of the uncertain outcomes of an experiment carried out on an indeterminate quantum field? The subject of quantum probability arose from the need to describe quantum indeterminacy. A complex valued quantum state vector, also called the state function and denoted by $|\Psi\rangle$, is introduced to describe the observable properties of the quantum field.

The state vector is an element of an infinite dimensional linear vector state space, called a Hilbert space. For a consistent probabilistic interpretation of quantum mechanics, it is necessary that the norm of $|\Psi\rangle$ be unity, namely

$$||\Psi||^2 = 1.$$

More precisely, quantum fields have creation operators, schematically represented by $a_p^\dagger$ and $a_p^\dagger|\Omega\rangle$ is a single particle state that carries momentum p, which for the electron field is equal to a state with one electron having a minimum energy of $m_e c^2$. In symbols,

$$a_p^\dagger|\Omega\rangle = |1 \text{ electron}\rangle; \quad E = \sqrt{m_e^2 c^4 + p^2 c^2}.$$

One says that $a_p^\dagger$ creates a single particle by acting on the vacuum state $|\Omega\rangle$. The excitations of the quantum field, in general, are represented by multiple particles with different energies.

To get a physical feel for the **quantum vacuum**, consider the vacuum state $|\Omega\rangle$ of the photon and electron fields. We can represent the vacuum state as a linear sum of the basis states that are physically meaningful — so as to have an intuitive understanding of the vacuum. The sum of the basis states in the expansion of the vacuum must have the same quantum numbers as the vacuum state; hence, for example, since the vacuum has zero charge the basis states appearing in the expansion must always have an electron together with its antiparticle, namely the positron.

Let γ stand for a photon, e^- for an electron and e^+ for a positron. 'Quantum numbers' are qualities like charge, mass, spin and so on that specify a quantum state and, as expected, are all null for the vacuum state. The photon γ has the same quantum numbers as the vacuum state; on the other hand, an equal number of electrons and positrons need to appear in the expansion of the vacuum state to ensure net zero charge and the sums of all the other quantum numbers are also null. We have the following expansion for the vacuum state of photons and electrons

$$\begin{aligned}|\Omega\rangle &= c_0|\text{No photons; No electrons, No positrons}\rangle \\ &\quad + c_1|\text{One photon; No electrons, No positrons}\rangle \\ &\quad + c_2|\text{No photon; One electron, One positron}\rangle \\ &\quad + c_3|\text{One photon; One electron, One positron}\rangle + \cdots\end{aligned}$$

$$= c_0|0\rangle + c_1|\gamma\rangle + c_2|e^+, e^-\rangle + c_3|\gamma; e^+, e^-\rangle + \cdots$$
$$= \sum_{n_\gamma, n_+, n_-} c_{n_\gamma, n_+, n_-}|n_\gamma, n_+, n_-\rangle$$

where n_γ, n_+, n_- are the number of photons, electrons and positrons in the vacuum with a probability amplitude of c_{n_γ, n_+, n_-}.[2]

One should think of these states as projections of the vacuum onto different possible states, with likelihood of the state occurring in the vacuum being given by $|c_{n_\gamma, n_+, n_-}|^2$. The basis states are not simply book-keeping devices but rather have a physical significance that is measurable.

A dramatic effect of the vacuum of quantum chromodynamics is the permanent confinement of quarks inside the nucleus and is discussed in Sec. 12.14. The Higgs field gives a mass to the weak vector bosons $Z^0, W^\pm$, to all the fermions (quarks and leptons) as well as to the Higgs particle — all of this due to the properties of the Higgs field's quantum vacuum, one of the cornerstones of the Standard Model of particle physics, as discussed in Chapter 13.

One may ask: why does the vacuum seem 'empty'? Why aren't photons and electron–positron pairs popping out of the vacuum all over space and all the time? The answer lies in the nature of the virtual states that constitute the basis of the vacuum. The vacuum has zero energy and zero momentum; the virtual particles have *net zero energy and momentum* and hence cannot be observed since all physical photons, electrons and positrons have finite energy; in particular, to exist physically an electron–positron pair needs a minimum energy of $2m_ec^2$. It *is* possible to create a pair of particles, but to do so one has to supply the required amount of energy and momentum. For instance, if one has an intense electric field emanating from a source, which could be a heavy nucleus, then a photon scattering off the nucleus can produce a physical electron and antielectron, as shown in Fig. 2.12. The heavy atom will need to recoil for conserving momentum and the electric field will lose energy to supply the energy required for creating the pair of an electron and its antielectron.

2.10.1 *Casimir effect and Lamb shift*

There are two famous, experimentally observable effects of the vacuum of quantum electrodynamics (the quantum field theory of photons and electrons), namely the **Casimir effect** and the **Lamb shift**.

[2] A precise expansion of $|\Omega\rangle$ needs to specify the momenta of all particles as well as the polarization of the photon and spin of the electrons and positrons.

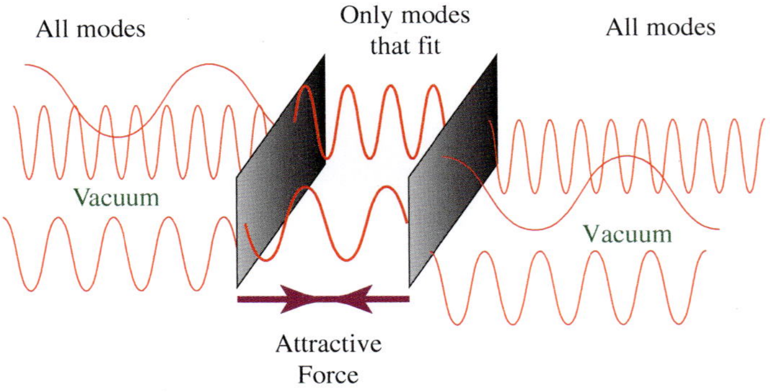

Fig. 2.13 Casimir effect: Two electrically neutral plates, which are electrical conductors, placed in a vacuum can experience a (non-gravitational) attractive force.

To study the Casimir effect, one introduces two parallel plates that are electrical conductors, separated by a distance, say z, as shown in Fig. 2.13. From classical electromagnetism, one expects no effect of the conductors on each other since they are both neutral and gravity is negligible. However, introducing the parallel plates changes the boundary conditions of the vacuum as there always has to be a net zero electric field inside the conductors — since any finite electric field is canceled by the movement of charges in the conductors. The change in the boundary conditions eliminates some of the allowed states for the vacuum state and leads to the observable result that there is an *attractive force* between the two plates. One can also obtain a repulsive **Casimir force** by using a different geometry for the conducting plates.

The fluctuations of the quantum vacuum yield another measurable effect called the *Lamb shift*. In the Lamb shift one studies the energy levels of a single isolated hydrogen atom. The hydrogen atom is the bound state of the proton and electron; the energy levels of the hydrogen atom are obtained using non-relativistic quantum mechanics, which entails the interaction of a quantized non-relativistic electron with the proton via the classical electrostatic potential.

For the case of both the Maxwell field and the electron being relativistic and quantized, the effect of the quantum vacuum modifies the energy levels of the hydrogen atom — which is due to the effect of the vacuum state on the electron and proton, as shown in Fig. 2.14(a); the change in the energy level of the ground state of the hydrogen atom is called the Lamb shift, and its measured result agrees with the predictions of quantum electrodynamics.

The scattering of photons off the hydrogen atom, as shown in Fig. 2.14(b), is also modified due to the quantum fluctuations. This yields an effective (renormalized) interaction between the electrons and alters the charge of the electron from that which is measured in classical physics, and is discussed further in Sec. 12.5.

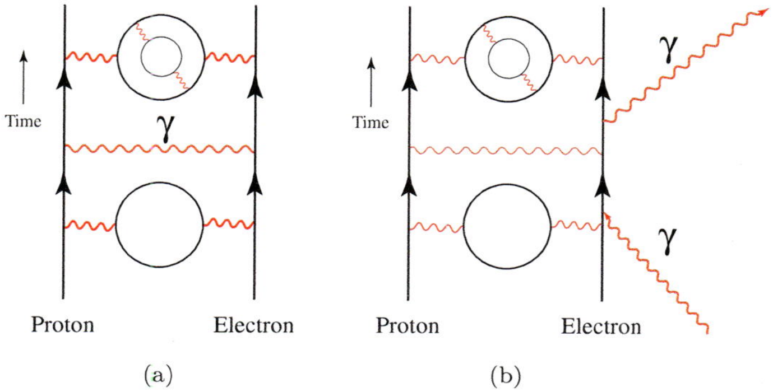

Fig. 2.14 Feynman diagrams of (a) vacuum fluctuations of the hydrogen atom; (b) scattering of photons off the hydrogen atom.

2.11 Unification of Particles and Interactions

Each entry for the particles and forces of the Standard Model, given in Table 13.1, is physically realized by a quantum field. The only difference between the various particles and forces is in the properties of their specific quantum fields. For example, all quarks and leptons are realized by fermionic quantum fields that carry spin half; all forces are realized by spin one bosonic quantum fields; the Higgs field is a scalar boson field.[3]

All observed particles, be they electrons or quarks, have quantum fields that yield a finite mass for the field's excitations. In contrast, some of the quantum fields of forces, such as the photon or gluons, have massless excitations of the respective underlying quantum fields.

Although particles and forces (interactions) are both represented by quantum fields, one needs to explain why they appear to be so different. The difference between particles and forces is due to two distinct reasons. All particles are fermions that have a finite mass and consequently can be brought to rest with respect to our frame of reference — and hence can appear as tangible objects. Secondly, the exclusion principle, which entails that no two fermions can occupy the same quantum state, is obeyed by particles since they are fermions; the exclusion principle is the reason that atoms and bulk matter are stable; in fact, it is the exclusion principle that is the final explanation of why we do not fall through the floor, which is rather remarkable given that 99.99% of the space enclosed by atoms is empty space.

In contrast, forces such as the electromagnetic force or the subnuclear gluonic force between quarks have quantum particles that are massless. This means that these particles always travel at the speed of light and can never be brought to

[3]Bosons and fermions are discussed in Sec. 11.5.

rest in any frame; hence, the intangible nature of light as perceived by our senses. Secondly, due to their bosonic nature, many quanta of these forces are allowed to be present at the same point of space and hence leading to the concept of a force with varying intensity, such as an electric or magnetic field having a low or high strength.

The fact that all particles and forces are simply different kinds of quantum fields opens the way to the effort to unify all particles and forces into a single physical entity. In the Standard Model, the electromagnetic and weak interactions are both manifestations of the underlying electroweak interactions; one can similarly imagine that all particles and forces are the manifestations of a more fundamental object. The first step in this direction is to combine bosons and fermions into a single quantum field, called the supersymmetric quantum field; the next and uptil now incomplete giant step is to combine all particles and interactions, including gravity, into a single quantum theory, with the leading candidate being superstring theory, discussed in Chapters 14 and 15.

2.12 The Answer

The 'stuff' that Nature is made out of is a variety of fields. The key feature of a field — and its most important and defining property — is that the field carries both energy and momentum at *every point* of space. This is the reason that a field is considered to be a *physical entity* — as physical as a classical particle — with the difference that the field's energy and momentum can flow from one part to another, unlike a particle for which its energy and momentum are at the position that it occupies (of course, this position can move). A pressure wave — caused by the oscillations of the pressure field that extends in space — consists of the propagation of the disturbance in the field.

An electric charge has an associated electric field that spreads into nearby space; an electric current creates a magnetic field that emanates into surrounding space. A propagating electromagnetic field can, in principle, spread out over infinite distances.

In general, a classical field is specified by assigning a numerical value (with appropriate dimensions) to every point in space. The values specifying the field at every point can change as it evolves in time. As shown schematically in Fig. 2.15, a particle of classical physics occupies a single point whereas a scalar and vector fields are spread over space. The numerical value of a field can be a single number in the case of a scalar field or it can consist of several numbers, indicating a direction in which the field points, in which case the field is called a vector field. For example, the pressure field is a scalar field specified by a single number whereas the electric field is a vector field requiring three numbers at every point of space. The gravitational field is a tensor field requiring 10 real numbers at every spacetime point.

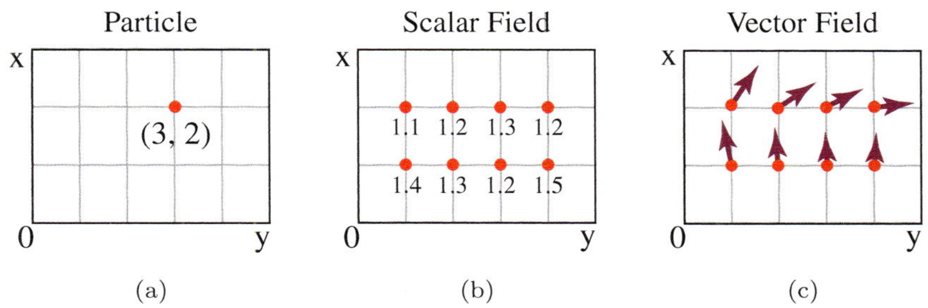

Fig. 2.15 From particle to fields.

A quantum field is a more complex object than a classical field. In contrast to classical fields, a quantum field is *indeterminate* and simultaneously takes all possible allowed values of the field. A quantum particle is seen to be an excitation of its underlying quantum field. In the Standard Model all the elementary particles and forces of Nature are seen to consist of quantum fields.

Chapter 3

The Geometry of Space

How straight is straight?

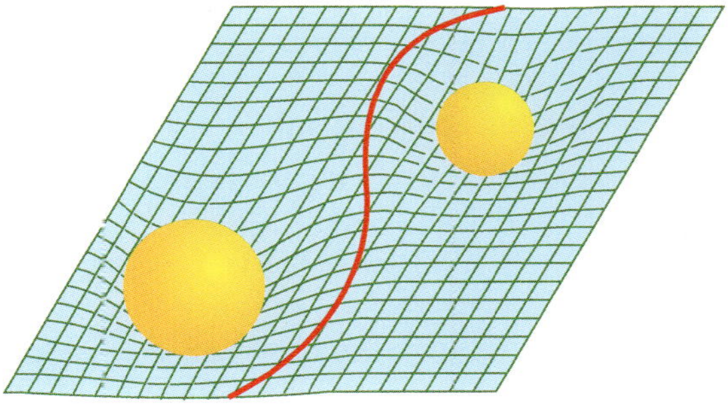

3.1 The Question

Geometry is usually understood to be the study of the shape, size and form of objects. A sphere is the geometrical shape of a billiard ball. One may proceed further and ask about the geometry of space itself. The space we move around in has a geometry, although we may not reflect upon its properties. Many of us remember the rudiments of the geometry of Euclid, in which there is only one straight line between two points, two straight lines intersect at a point and parallel lines do not meet. The rudimentary notions of Euclidean geometry have to be greatly enhanced in order to discuss the more complex geometrical concepts that arise in the study of our Universe's spacetime.

The concept of **dimensionality** is fundamental in the study of geometry. The space that we live in is three dimensional, since one needs three real numbers to specify a point in space; for example, to describe a cube of sugar, we have to provide its length, width and height. A major advance made in our understanding of Nature is that time itself can be described by a real number and clocks measure time in the number of seconds, which is a real number. Time is another dimension of Nature, lying in a direction *perpendicular* to all three space dimensions.

When space and time are taken together, the Universe that we live in turns out to be a four dimensional spacetime, with three space dimensions and one time dimension. In this chapter, we examine the concepts of space and time in some detail with the intention of acquiring the understanding required for studying the structure and properties of our Universe's spacetime.

We ask the questions: what is the geometry of spacetime? How does one describe the geometry of space? What is a curved geometry? How does one define distances for a curved space? What is the geometry of empty spacetime? What determines the curvature of spacetime?

3.2 Is Space Curved?

Figure 3.1 shows light from a distant quasar as it may appear to an observer on Earth. Particularly notable is that there are two images of a quasar that look uncannily similar. The brightness and redshift of the two bright images show that both quasars are at the same distance, usually a few billion light years; redshift analysis of the bright orange image in the middle shows that it is a 'nearby' galaxy a few hundred million light years from Earth.

Why does the nearby galaxy have two identical quasars around it? Or is this actually a single quasar that appears as two quasars? What is the reason for this?

Clearly the two images are not an atmospheric effect since the picture has been taken in outer space. The image is not an optical illusion. Given that the quasars are located billions of light years from Earth, it is highly unlikely that two distinct quasars would appear clustered around a single nearby galaxy.

The explanation is found in the concept of **gravitational lensing**. As light from a single distant quasar passes the lensing galaxy, its path bends around the

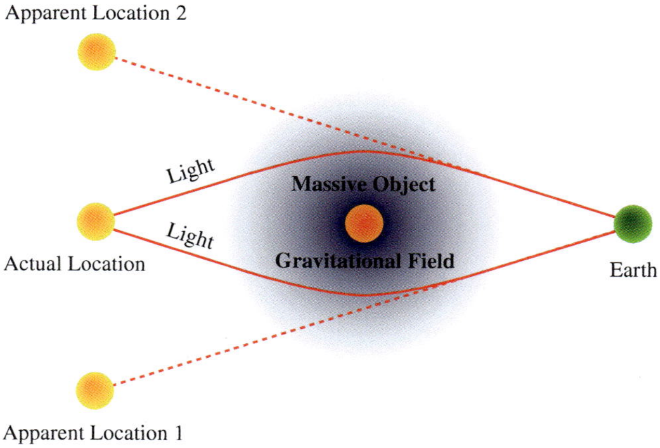

Fig. 3.1 Bending of quasar's light by an intermediate galaxy.

galaxy. The galaxy acts as a gravitational lens, bending light towards the galaxy. Depending on which side of the lensing galaxy the quasar light passes, it is bent in different ways, as shown in Fig. 3.1. An observer on earth extrapolates the light received along one's line of sight and hence, two distinct images of the same quasar are formed on Earth.

Gravitational lensing shows that *space is not flat*, but rather, is curved — and it is the mass of the lensing galaxy that causes the curvature. We are hence led to the idea of curved spacetime, and need to understand how to mathematically represent and analyze the geometry of spacetime.

Sections 3.3 to 3.6 introduce the fundamental ideas required for describing curved geometry — using the specific example of the two dimensional surface of a sphere. Although the surface of a sphere is one of the simplest examples of a curved space, it is sufficient for illustrating all the central concepts of curved space. A key concept is parallel transport, which generalizes the concept of parallel lines in flat space. The concept of the shortest distance between two points is generalized to the concept of geodesics that is valid for curved geometries. And this prepares the ground for discussing the concept of distances in curved space.

It should be remembered that our objective in studying geometry is to understand the workings of gravity in four dimensional spacetime. This chapter on geometry is a preparation for explaining how geometry is the underlying basis of the gravitational force.

3.3 Description of Curved Space

To understand the mathematics of curved geometries, four dimensional spacetime is a complicated place to begin; hence, for the sake of clarity, we consider the two dimensional sphere, denoted by S^2 and shown in Fig. 3.2.

A **manifold** is a collection of points that forms a continuous space which is locally isomorphic ('identical') to a Euclidean space. Mathematically, S^2 is a

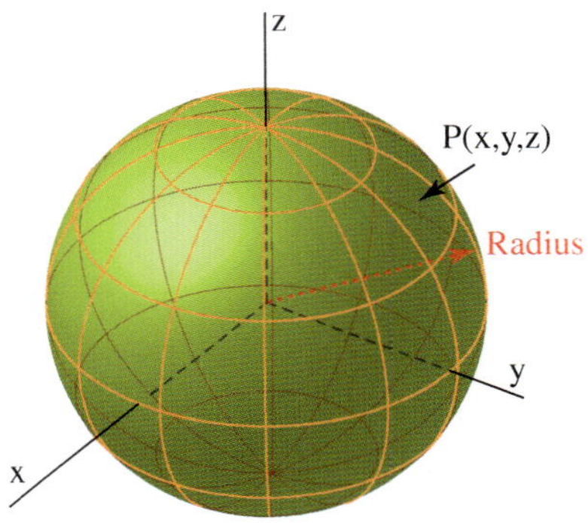

Fig. 3.2 A two dimensional sphere S^2 embedded in three dimensional flat Euclidean space.

collection of points. For a sphere, the points are arranged in such a manner so as to yield a two dimensional space having a *constant* curvature. How does one describe S^2? What do we mean by curvature of space? What does it mean to say that S^2 has constant curvature? We now address these questions.

A **Euclidean space** is a *flat space*; what this means is that the shortest distance between any two points is a straight line and yields that the sum of the angles of a triangle is always 180°.

Einstein's theory of gravity needs no reference to any larger embedding space. And, more importantly, all calculations for gravity can be carried out entirely in four dimensional spacetime. Hence, in general, one need not think of the four dimensional spacetime as being embedded in higher dimensional flat space. However, to understand the properties of a curved manifold, it is often useful to embed it in a flat space.[1] There is a theorem stating that it is always possible to embed any curved spacetime in a sufficiently high dimensional flat Euclidean space.

In order to visualize its geometry, embed S^2, having radius r, in flat three dimensional Euclidean space, denoted by $\Re^3$. As shown in Fig. 3.2, the points $P = (x, y, z)$ on the sphere S^2 form a two dimensional surface in $\Re^3$ which is given by

$$x^2 + y^2 + z^2 = r^2. \tag{3.1}$$

Pythagorus theorem holds for any right angled triangle in flat space; the Pythagoras theorem, as shown in Fig. 3.3, needs to be modified for S^2. For a triangle drawn on S^2, as shown in Fig. 3.4, its three angles add up, unlike for a flat space, to an angle *larger* than 180°. So it is intuitively obvious that S^2 is not a flat but, instead, is a curved surface.

We use the notation that a symbol with an arrow or boldface denotes a vector, namely

$$\vec{x} = \mathbf{x} = (x, y, z).$$

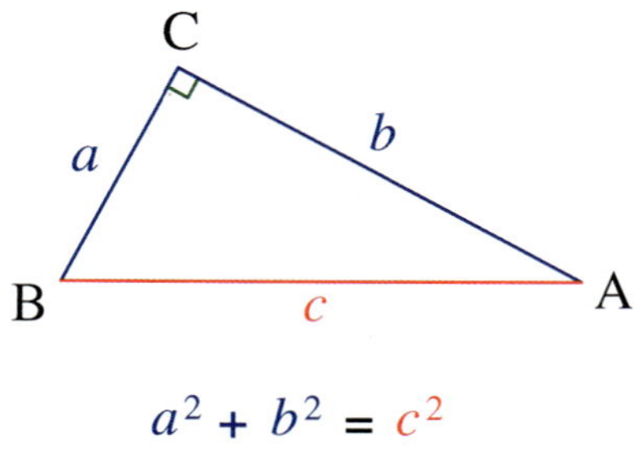

Fig. 3.3 Pythagoras theorem for two dimensional Euclidean space $\Re^2$.

[1]We will later use the idea of embedding to visualize the properties of black holes.

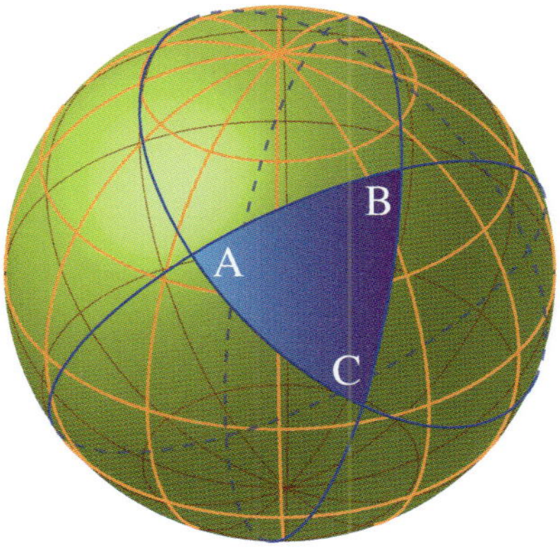

Fig. 3.4 A triangle on a sphere.

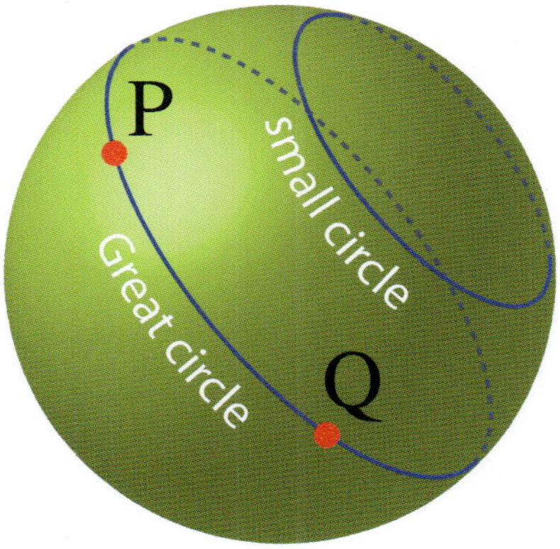

Fig. 3.5 Great and small circles on a sphere.

3.4 Parallel Transport

The shortest distance between any two points P and Q on S^2 is given by the great circle that goes through the two points, as shown in Fig. 3.5: great circles lie in a plane passing through the center of the sphere. How can we characterize the great circles on S^2 in terms of the trajectory of particles?

Consider for starters the motion of a particle in three dimensional flat space $\Re^3$. In the *absence* of any force the particle travels in a straight line, which is the shortest distance between two points. The straight path is also one that has a *constant velocity.* The trajectory can be completely fixed by specifying the starting point and starting velocity. The straight line trajectory is given by[2]

$$\mathbf{x}(t) = \mathbf{x}_0 + \mathbf{v}t; \quad \mathbf{x}(0) = \mathbf{x}_0 \tag{3.2}$$
$$\mathbf{v} \; : \; \text{Arbitrary initial constant velocity.}$$

Figure 3.6 shows the path and the initial position and velocity vector, as well as the velocity vectors along the path. The particle's velocity vector $\mathbf{v}$ is constant along the entire path and this, in effect, means that the velocity vector, at every point of its trajectory, is *parallel* to the starting velocity vector.[3]

---

[2]The straight line trajectory can also be specified by fixing the particle's initial and final positions $\mathbf{x}_1$ and $\mathbf{x}_2$ at times t_1 and t_2 respectively; the straight path $\mathbf{x}(t)$ going between these points is the following straight line

$$\mathbf{x}(t) = \mathbf{x}_1 + \mathbf{v}(t - t_1); \quad \mathbf{v} = (\mathbf{x}_2 - \mathbf{x}_1)/(t_2 - t_1)$$

where $d\mathbf{x}(t)/dt = \mathbf{v}$.

[3]The velocity vector $\mathbf{v}$ also happens to be parallel to the direction of the path itself but this fact does not generalize to a curved space.

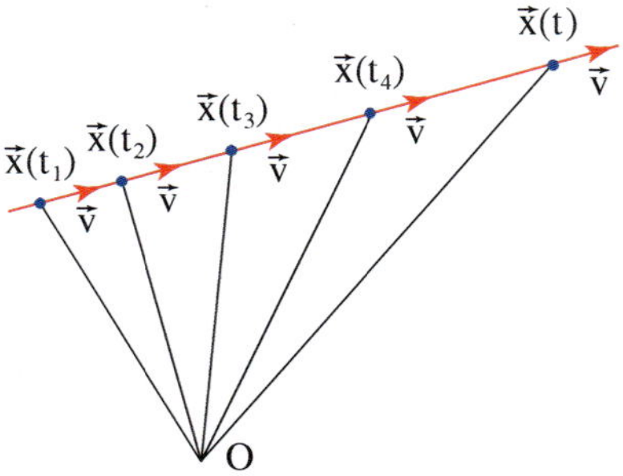

Fig. 3.6 Straight path.

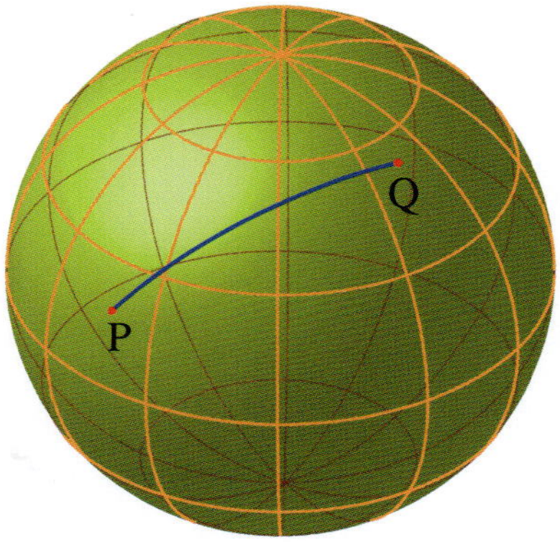

Fig. 3.7 Shortest distance on a sphere.

One would like to generalize the concept of straight line and constant velocity vector to describe 'force-free' motion of particles on a curved space. One can follow, for a curved space, the steps implicit in Eq. (3.2) and define a point $\mathrm{P} = (x, y, z)$ on the sphere as in Fig. 3.7 and try to find the shortest path, lying entirely on S^2, to another point Q. What should be the starting velocity vector? How does one define the shortest path to the point Q?

Noteworthy (optional content) 3.1: Tangent Space

In flat space the velocity vector lies in the same space as the one in which the particle is traveling, as can be seen from Eq. (3.2) where both the position $\mathbf{x}(t)$ and velocity $\mathbf{v}$ are vectors in the same flat space, namely $\Re^3$. For a curved space this is usually not the case. The direction of the velocity vector is tangential to the path and, in mathematics, the velocity vector is called a tangent vector.

For the case of S^2 consider all possible paths through the point P; the allowed velocities are tangents to the curves passing through the point P and all of the

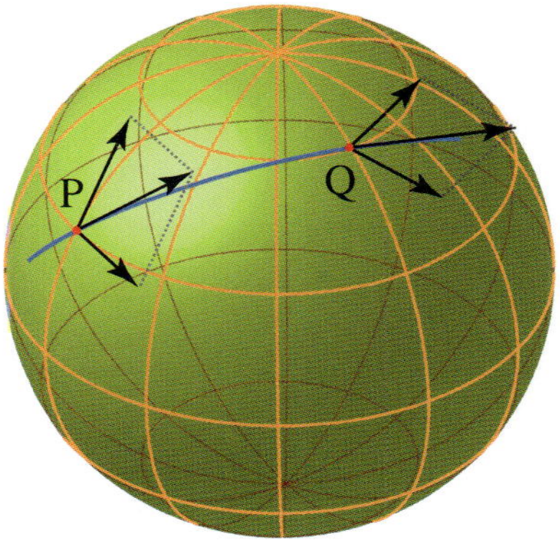

Fig. 3.8 Tangent spaces at P and Q.

tangents lie in a (two dimensional) plane that is tangential to the point P, as shown in Fig. 3.8. Similarly the tangent space at another point Q of S^2 is also a two dimensional flat space that is oriented differently than the tangent space at point P, as seen from Fig. 3.8. In general, the tangent space for a curved space varies from point to point.

The velocity vector at point P, given by $\mathbf{v}_P$, is a **tangent vector** to the particles trajectory at the point P and is shown in Fig. 3.9. One needs to find a path in curved space for which tangent vector $\mathbf{v}_P$ is 'constant' along the trajectory and hence 'parallel' to itself. Hence there are two separate issues that need to be addressed for a curved space: firstly, how does one define the concept of 'parallelism' for tangent vectors at two different points and secondly, how does one keep a tangent vector 'constant' as one moves from one point to another.

The only intuitive understanding we have of parallelism is in flat space, where two vectors are parallel if *all* their components are equal.[4] In flat space, in particular for $\Re^3$, any two (velocity) vectors at two different points can be directly compared since all vectors are in the same space $\Re^3$ and hence any vector can be moved to any point of $\Re^3$ by holding all of its components fixed. For curved manifolds, since the tangent space at P is not equal to the tangent space at Q, a tangent vector at point P cannot be moved — for example, by holding all of its components fixed — to another point, say Q, and still be tangent, as one can clearly see from Fig. 3.9 by inspecting $\mathbf{v}_P$ at P and Q. Hence, holding the tangent vector $\mathbf{v}_P$ to be equal to a constant in going from P to Q makes no sense as $\mathbf{v}_P$ does not belong to the tangent space at Q.

[4]Two vectors are parallel if all their components are equal up to a rescaling by a constant; this subtlety does not concern us and will henceforth be ignored.

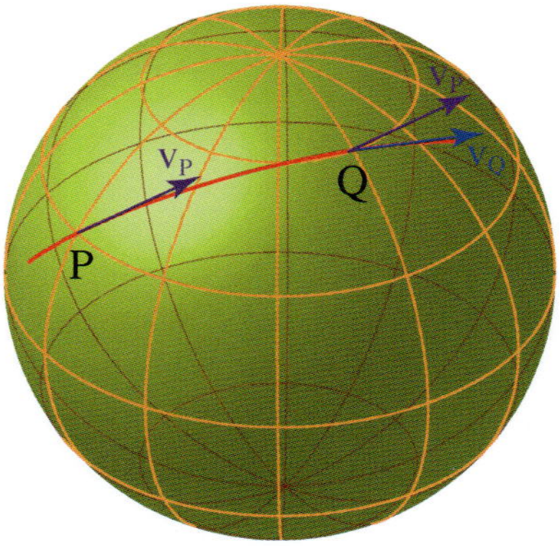

Fig. 3.9 Comparing tangent vectors.

One needs to compare the tangent vectors on the sphere S^2 at P and Q — denoted by $\mathbf{v}_P$ and $\mathbf{v}_Q$ respectively, and shown in Fig. 3.9 — to decide if the two vectors are, in some sense, 'parallel'. The concept of **parallel transport** is introduced for transporting the tangent vector of a curved space from one point to another. Unlike flat space, one needs to define a *path* along which the tangent vector $\mathbf{v}_P$ is transported from P to Q.

Suppose P and Q are infinitesimally close; in this case the path, to lowest order, does not matter and one can directly compare the vectors at P and Q. Consider for now the vector $\mathbf{v}_P$ as a three dimensional vector belonging to the flat three dimensional space $\Re^3$ in which the sphere S^2 is embedded. Let the three dimensional vector $\mathbf{v}_{PQ}$ at point Q be parallel to $\mathbf{v}_P$ in $\Re^3$; since $\Re^3$ is flat we have that $\mathbf{v}_{PQ} = \mathbf{v}_P$. However, in general, the vector $\mathbf{v}_{PQ}$ does not lie in the tangent space of S^2 at point Q; rather, it has a normal component that is orthogonal to ('sticking out of') the tangent space, denoted by $\mathbf{v}_Q^n$, and a tangential component $\mathbf{v}_Q^t$ that lies in the tangent space at point Q, as in Fig. 3.10. Hence

$$\mathbf{v}_P = \mathbf{v}_{PQ} = \mathbf{v}_Q^n + \mathbf{v}_Q^t$$

$$\mathbf{v}_Q^t \ : \ \text{Parallel transport of } \mathbf{v}_P.$$

In essence, the vector $\mathbf{v}_Q^t$ is the parallel transport of the vector $\mathbf{v}_P$. For P and Q infinitesimally close, the length of $\mathbf{v}_Q^n$ is, to first order, negligible and the length of $\mathbf{v}_Q^t$ is equal to $\mathbf{v}_P$.

To define the components of $\mathbf{v}_Q^t$ one needs to set up a complete basis of vectors for the tangent space at Q and then define the orthogonal component $\mathbf{v}_Q^n$ with respect to this basis.

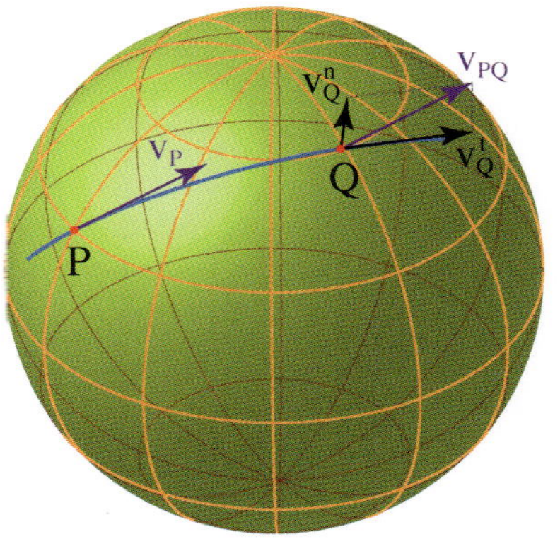

Fig. 3.10 Tangent vectors on a sphere.

3.5 Geodesics

A **geodesic**, in general, is the *shortest path* between two points of a curved space, and is the generalization of the flat space concept of a straight path. The great circle is a special case of a geodesic and yields the shortest path between two points on a sphere S^2.

Recall that for a straight (shortest) path in flat space the velocity vector is a constant. Analogous to keeping the velocity vector a constant in flat space, a geodesic is a path in the curved space that is defined by demanding that the *parallel transport* of the initial tangent vector along the geodesic be a constant. Or equivalently, given a geodesic we can compute the tangent vector at every point of the geodesic; it will turn out that the tangent vector at every point on the geodesic is equal to the initial tangent vector, which in effect has been parallel transported along the geodesic.

3.5.1 *Constructing a geodesic*

Consider an arbitrary curved space. Let us start from a point P and with initial tangent vector $\mathbf{v}_P$, as shown in Fig. 3.11(a). There are infinitely many curves that pass through P; consider points $\mathrm{Q}', \mathrm{Q}'', \mathrm{Q}''', \ldots$ with tangent vectors $\mathbf{v}_{Q'}, \mathbf{v}_{Q''}, \mathbf{v}_{Q'''}, \ldots$ as shown in Fig. 3.11(a). Let us parallel transport the initial tangent vector $\mathbf{v}_P$ to points $\mathrm{Q}', \mathrm{Q}'', \mathrm{Q}''', \ldots$ such that we obtain the parallel transported vectors $\mathbf{v}^t_{Q'}, \mathbf{v}^t_{Q''}, \mathbf{v}^t_{Q'''} \ldots$.

In general, on parallel transporting $\mathbf{v}_P$ along different curves, the vectors we obtain, namely $\mathbf{v}^t_{Q'}, \mathbf{v}^t_{Q''}, \mathbf{v}^t_{Q'''} \ldots$, will *not* be equal to the tangent vector to the curves, denoted by vectors $\mathbf{v}_{Q'}, \mathbf{v}_{Q''}, \mathbf{v}_{Q'''} \ldots$. However, there is a special

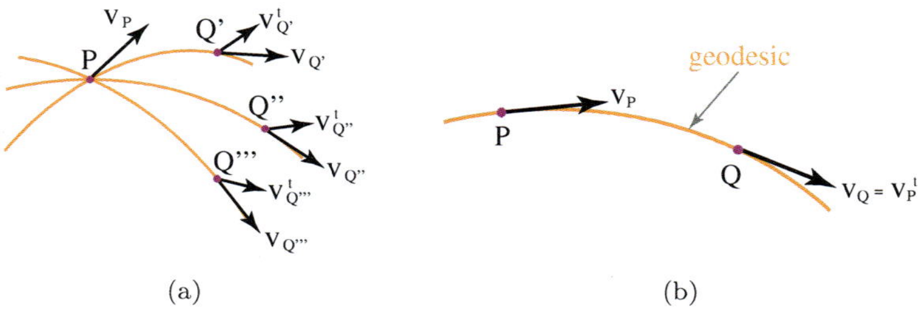

Fig. 3.11 (a) Parallel transport of vector $\mathbf{v}_P$ along paths to points Q$'$, Q$''$, ... (b) A geodesic and a tangent vector parallel transported from P to Q.

Fig. 3.12 A great circle cutting the sphere into two equal halves.

curve such that the parallel transported vector along this special curve is equal to the tangent vector of the curve.

Choose the curve through P and point Q for which the tangent vector at Q, namely $\mathbf{v}_Q$, is equal to the parallel transport of $\mathbf{v}_P$ to the point Q, that is, *equal* to $\mathbf{v}_P^t$ as shown in Fig. 3.11(b). The finite line segment PQ lies on the *geodesic* between P to a point Q not necessarily close to P.

In summary, parallel transport and geodesics are fundamental properties of all curved spaces. A tangent vector can be parallel transported along any path. A geodesic is a very *special path* such that the tangent vector at every point of the path is *equal* to the parallel transport of the initial tangent vector; a geodesic between any two points is also the shortest distance between two points.

3.5.2 *Geodesics on a sphere S^2*

To illustrate our discussion on parallel transport and geodesics we re-examine the great circles on S^2. For concreteness, let S^2 to be the surface of the Earth, and consider a ship that is taking the path of a great circle in sailing from point P to Q. Since a great circle lies on a plane that passes through the center of the sphere it cuts the sphere into two equal halves, as shown in Fig. 3.12.

The ship at P, as shown in Fig. 3.10, has a velocity vector $\mathbf{v}_P$, which is the *tangent vector* to the great circle at point P. Since the ship is sailing on the surface of a sphere, clearly $\mathbf{v}_Q$ cannot be equal to $\mathbf{v}_P$. For example, if we consider $\mathbf{v}_P$ as a three dimensional vector and, keeping all its components fixed, shift it to the point Q, then it can be seen that, at point Q the vector $\mathbf{v}_P$ has a component that is *orthogonal* to the surface of the earth. Hence if $\mathbf{v}_Q$ were equal to $\mathbf{v}_P$ the ship would have a component pointing perpendicular to the Earth's surface and the ship would lift *off* the surface of the Earth!

What is the ship's velocity vector $\mathbf{v}_Q$ at point Q? Since the ship is moving on a geodesic the tangent vector at Q is equal to the parallel transport of $\mathbf{v}_P$ to Q. The ship always sails on the Earth's surface, the parallel transport of the tangent vector $\mathbf{v}_P$ must *rotate* it downwards, as the Earth's surface curves away, in order for it to stay in the tangent space to the sphere; hence $\mathbf{v}_Q$ is equal to the rotated vector $\mathbf{v}_P$. The amount of rotation of $\mathbf{v}_P$ is proportional to the arc length PQ; on going through one complete cycle one starts at point P and ends up again at point P; hence, in one cycle, the tangent vector $\mathbf{v}_P$ rotates through 2π and returns to itself. Furthermore, note that the rotation of the tangent vector is entirely in the plane of the great circle; if the tangent vector had any sideways components the ship would leave the great circle.

The tangent vectors along a geodsic for S^2 are shown in Fig. 3.13. If one looks at the plane of the great circle, parallel transport amounts to rotating the initial tangent vector $\mathbf{v}_P$ in the plane of the great circle, as shown in Fig. 3.14.

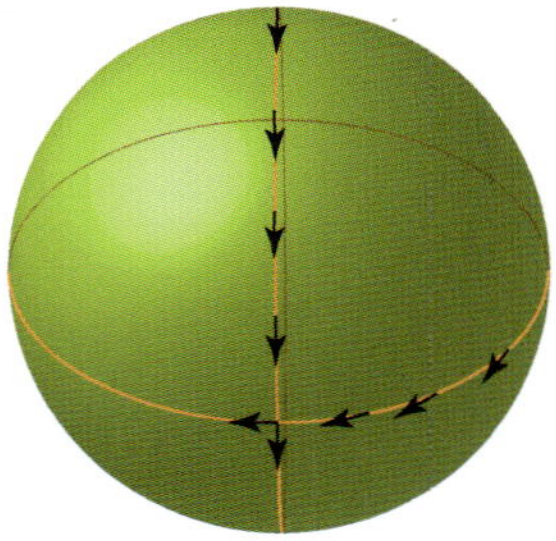

Fig. 3.13 Tangent vectors for the sphere S^2.

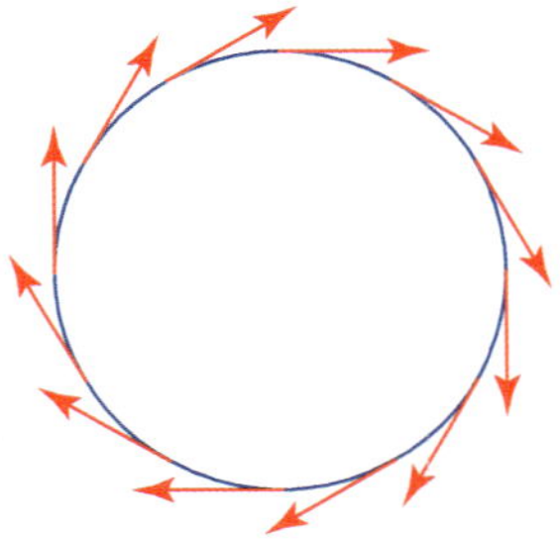

Fig. 3.14 Tangent vectors along the plane of a great circle of S^2.

What is happening from the ship's point of view? The components of the ship's velocity are being continuously changed in a complicated manner so that the ship continues to travel on the great circle and at a constant speed. This way of 'transporting' the ship's tangent vector, while holding its magnitude fixed, so that it stays on the shortest path between two points, is one way of specifying the great circle.

In summary, great circles for the case of S^2 are curves for which the tangent vectors are equal under parallel transport. Anti-podal pair of points, like the north and south pole, are connected by infinitely many geodesics, namely the great circles (longitudes) passing through the poles; all other pairs of points on a sphere are also connected by a unique geodesic.

In flat space, a geodesic reduces to a straight line and parallel transport of the tangent vector reduces to keeping constant all the components of the velocity vector.

Noteworthy (optional content) 3.2: Examples of Geodesics

Recall a geodesic is a trajectory along which an initial velocity undergoes a parallel transport; at every point of the geodesic, the tangent vector at that point is equal to the initial vector. Figure 3.15(a) shows the geodesic for different initial vectors. The geodesic for light, namely the null geodesic, is special since its velocity is c, which is the maximum velocity possible. Figure 3.15(b) shows that the null geodesic of light is the fastest way to go from one point to the next. In Fig. 3.15(c), it is shown that when a particle is thrown into the air, both the Earth and the particle travel on geodesics, with neither the Earth 'attracting' the particle to the Earth or *vise versa.* When the particle hits the Earth's surface, non-gravitational forces come into play and bring the particle's trajectory to a halt.

Figure 3.15(d) shows how the geodesic of a particle can be changed. Consider a football thrown towards a player, which travels on a geodesic; when the football is kicked by the player, a force is exerted on the ball and its velocity is changed, setting if off on another geodesic.

3.6 Distances in Curved Spaces

So far we have discussed parallel transport, geodesics and curvature, but have not used the concept of **distance** that is so fundamental to our understanding of space. Manifolds exist for which parallel transport and curvature are defined without recourse to the concept of distance. However, there is a smaller class of manifolds, called metric spaces, for which the additional structure of the distance between any two points is also defined. It turns out that in Einstein's theory of gravity, spacetime

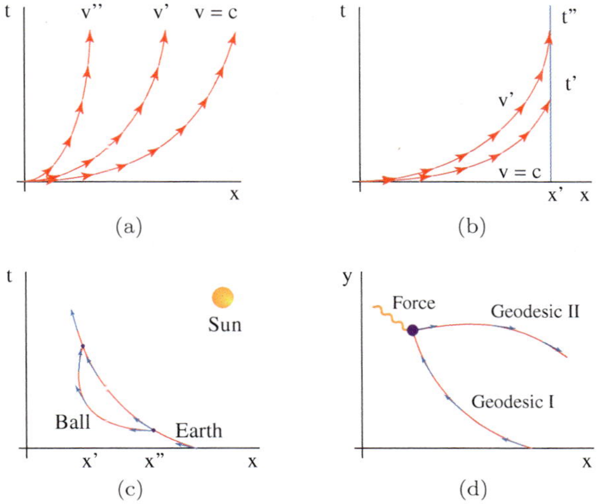

Fig. 3.15 (a) Geodesics for initial speeds v, v', v''. (b) A null geodesic for light with speed $v = c$ compared with another geodesic for a vector with $v' < c$. (c) The geodesics of the Earth and an airborne particle. (d) Geodesic I is changed into Geodesic II by a force.

is always a metric space. Hence we examine the concept of distance for a curved space.

For a flat space in any dimension, the distance Δs between any two points $\mathbf{x}$ and $\tilde{\mathbf{x}}$ is given by

$$(\Delta s)^2 = (\mathbf{x} - \tilde{\mathbf{x}})^2 \tag{3.3}$$

and is shown in Fig. 3.16. Consider the distance between neighbouring (x, y) and $(\tilde{x}, \tilde{y})$ for the two dimensional case; for $\Delta x = x - \tilde{x}$, $\Delta y = y - \tilde{y}$, the Pythagoras theorem yields

$$(\Delta s)^2 = (\Delta x)^2 + (\Delta y)^2.$$

For curved space, the formula for the distance between two points (in terms of the coordinates of their positions) depends on where the points are located and, in fact, changes from point to point.

One can construct a two dimensional flat space by gluing together many small and identical isosceles triangles, as shown in Fig. 3.17.[5] One sees that for a flat two dimensional space the size of the triangle can be kept fixed and we can still cover the entire space. Similarly, for constructing any two dimensional *curved space* one needs to consistently glue together triangles, but now with angles that generally do not add up to 180° as is the case in flat space.

[5]The reason one considers gluing isosceles triangles together is because they are the simplest bounded figures that enclose an area; one can as easily construct a two dimensional space using squares instead of triangles.

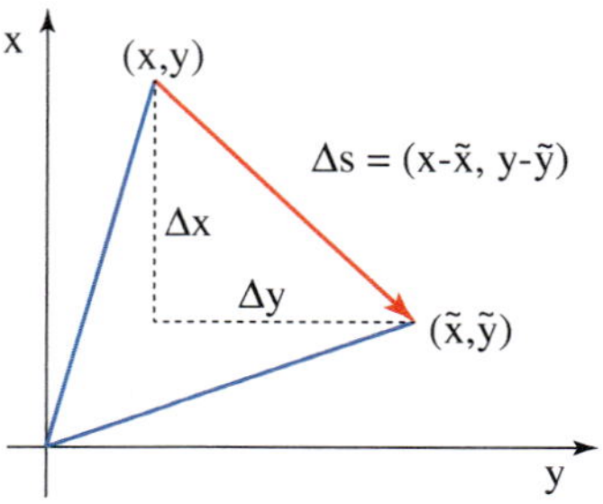

Fig. 3.16 Two dimensional distance.

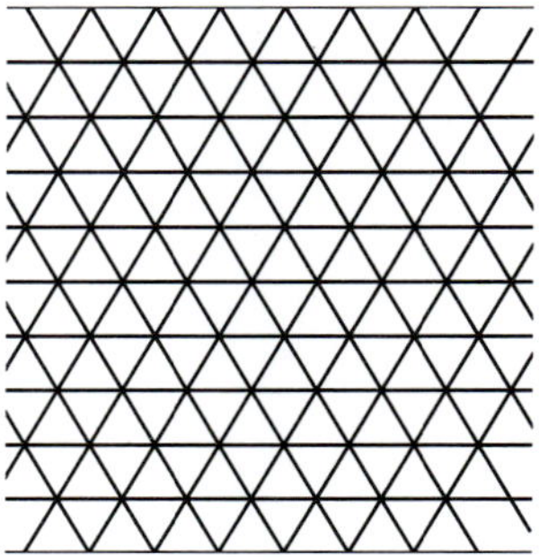

Fig. 3.17 Triangulated plane.

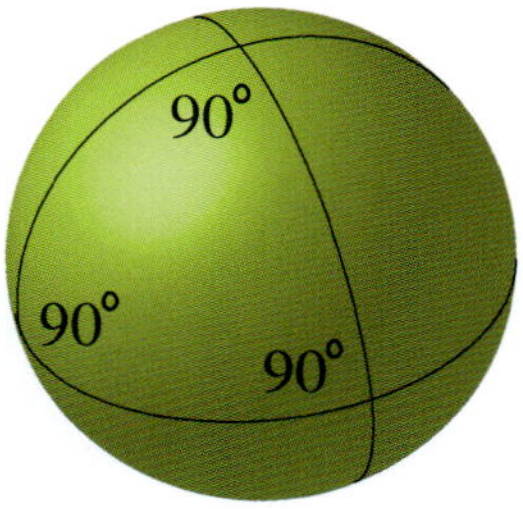

Fig. 3.18 Triangulated sphere. The sum of the angles in the triangle is 270°.

Figure 3.18 illustrates how a two dimensional sphere S^2 can be made by gluing together large spherical triangles. A spherical triangle is drawn on the surface of the sphere; the sum of its three angles is greater than 180°, and is a hallmark of the fact that the sphere is a curved space; for a spherical triangle bounded by geodesics, its angles add up to 270°.

To understand how to describe various two dimensional spaces, such as a sphere and a torus, we need to quantify how to glue the different triangles together. Consider again the two dimensional sphere S^2 with radius r; each point on S^2 is assigned a fixed value of r and angles θ, ϕ using spherical polar coordinates, as shown in Fig. 3.19.[6]

Noteworthy (optional content) 3.3: Spherical Polar Coordinates

A point on the sphere can be specified by giving its distance from the origin, given by the radial distance $r \in [0, \infty)$, and two angles, namely the azimuthal angle

[6]In these coordinates, the great circles between two points on the sphere — which as we discussed earlier are geodesics on the sphere — are given in http://mathworld.wolfram.com/GreatCircle.html.

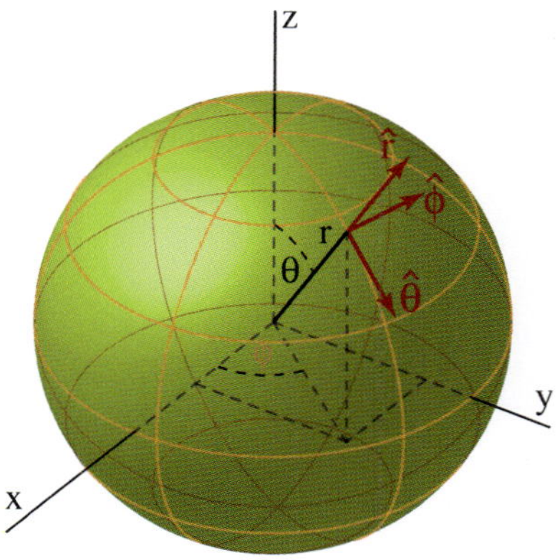

Fig. 3.19 Spherical coordinates for S^2.

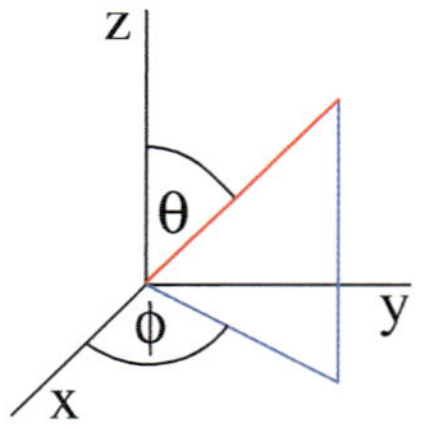

Fig. 3.20 Spherical coordinates for S^2.

$\phi \in [0, 2\pi]$ and the polar angle $\theta \in [0, \pi]$, shown in Figs. 3.19 and 3.20. In Cartesian coordinates, the same point is given by its coordinates (x, y, z). The relation of the two coordinates is given by

$$x = r\sin\theta\cos\phi; \quad y = r\sin\theta\sin\phi; \quad z = r\cos\theta.$$

The inverse mapping is given by

$$r = \sqrt{x^2 + y^2 + z^2}; \quad \theta = \cos^{-1}\left(\frac{z}{r}\right); \quad \phi = \tan^{-1}\left(\frac{y}{x}\right).$$

In vector notation, the unit vectors are given by $\hat{e}_x$, $\hat{e}_y$, $\hat{e}_z$ for Cartesian coordinates; the orthogonal basis vectors for spherical polar coordinates $\hat{r}$, $\hat{\theta}$, $\hat{\phi}$ are shown in Fig. 3.19.

Latitudes are circles drawn on the surface of S^2 with different ϕ values and fixed θ; longitudes are great circles drawn for different θ values and fixed ϕ as shown in Fig. 3.21. Note the radius of the latitude at angle θ is given by $r\sin\theta$ and decreases as one approaches either of the poles; in contrast, all longitudes are great circles

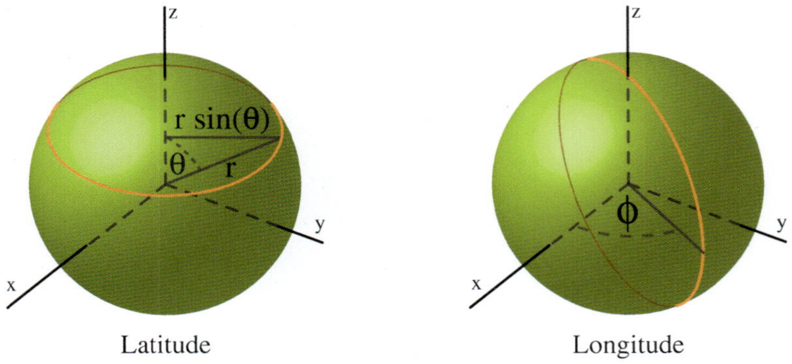

Fig. 3.21 Longitude and latitude.

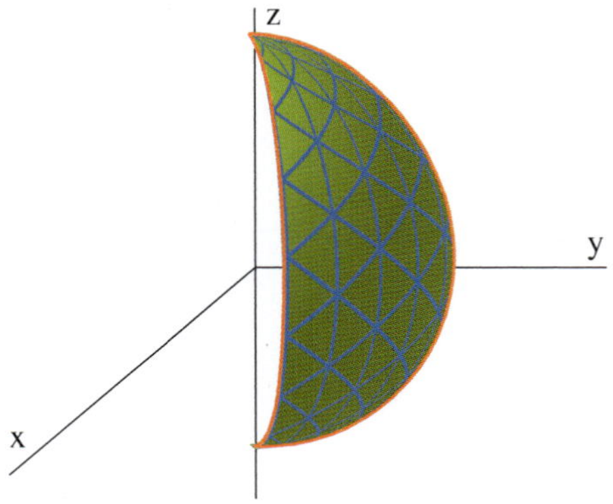

Fig. 3.22 A thin wedge of S^2.

given by different values of ϕ and have radii that are fixed and equal to r: the radius of the sphere.

We divide the sphere S^2 into thin wedges — similar to how we cut a water melon — and consider one slice bounded by two longitudes at ϕ and $\phi + \Delta\phi$, as shown in Figs. 3.22 and 3.23. As indicated in Fig. 3.22 we can completely cover the surface of the wedge of S^2 by covering it using triangles of smaller and smaller size. To quantify the changing size of the triangles consider a triangle near the north pole, as shown in Fig. 3.24. The hypotenuse is the line segment going from point (θ, ϕ) to point $(\theta + \Delta\theta, \phi + \Delta\phi)$; we consider the triangle to be small enough so that the triangle can be considered to be a right triangle.

As seen from Fig. 3.21, the latitude at angle θ has a radius of $r \sin\theta$ as in Fig. 3.24. Hence, for a sphere of radius r the sides of the triangle are $r\Delta\theta$ and $r \sin\theta\Delta\phi$. Then, from Pythagoras theorem, the length of the hypotenuse, namely

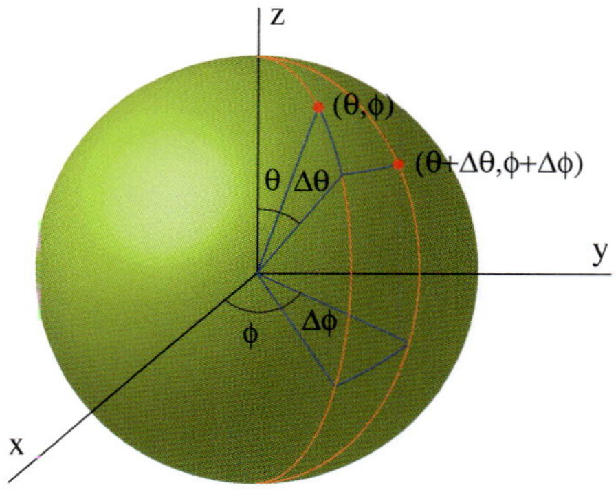

Fig. 3.23 Spherical coordinates for a wedge in S^2.

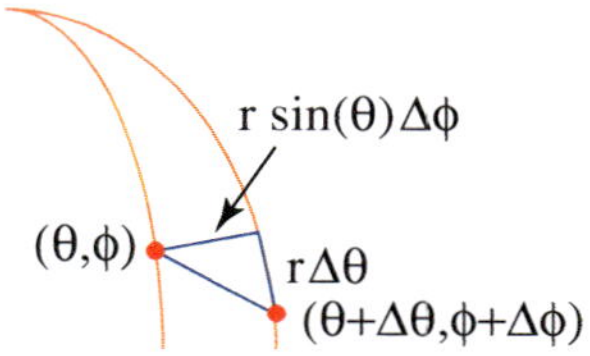

Fig. 3.24 Line segment for S^2.

the line segment connecting point (θ, ϕ) with point $(\theta + \Delta\theta, \phi + \Delta\phi)$, is given by

$$(\Delta s)^2 = [r\Delta\theta]^2 + [r \sin\theta\Delta\phi]^2. \tag{3.4}$$

Note that in Eq. (3.4), we have used Pythagoras theorem to find the length of the line segment; one might reasonably object that Pythagoras theorem holds only for flat space, and so how is one justified in using it for a curved space? The answer is based on one of the fundamental properties of curved spaces, namely that in the *neighborhood* of a point, even a curved space is always *flat*. This is the reason that we are considering a small (infinitesimal) triangle, since the size of the triangle can be made arbitrarily small and hence justifying the use of Pythagoras theorem.

One can see that the size of the triangle becomes smaller and smaller as we approach either the north or the south pole, indicating that the sphere is curved since, unlike a flat space, the size of the triangles filling up the slice of S^2 cannot be kept fixed. In the limit that the triangles are made of infinitesimal size, the distance is denoted by $(\Delta s)^2 \to ds^2$ and hence, from Eq. (3.4) above, one has

$$ds^2 = r^2[d\theta^2 + \sin^2\theta d\phi^2] \tag{3.5}$$

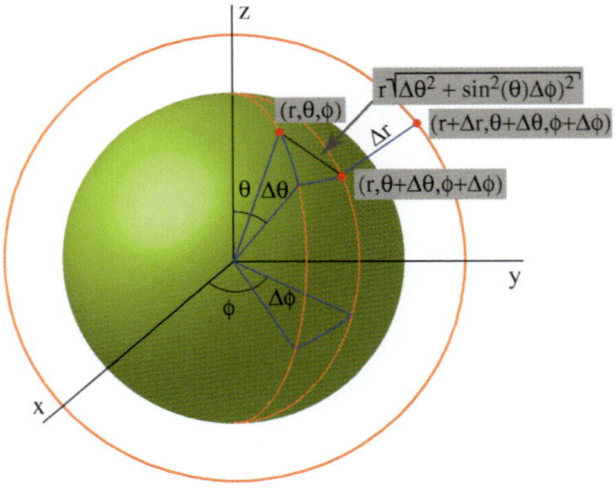

Fig. 3.25 Distance in three dimensions.

Eq. (3.5) yields the distance between two neighboring points separated by **infinitesimal distance** ds. The infinitesimal distance ds is given in terms of the coordinates by what is called the *metric*; in the case of the sphere the components of the metric are r^2 and $r^2 \sin^2 \theta$.

Similar to the case of a two dimensional sphere, one can construct three dimensional curved spaces by filling them up with tetrahedra of varying sizes.[7] One can construct higher dimensional curved spaces by filling them up with higher dimension analogs of the tetrahedra.

One can generalize the distance function obtained for a two dimensional sphere to flat three dimensional space $\Re^3$. We can think of $\Re^3$ as being made out of a smoothly connected collection of concentric (two dimensional) spheres with varying radii. Hence, we choose spherical polar coordinates for three dimensional space exactly in the same way as we did for S^2, and is shown in Fig. 3.25. Let the coordinates of two neighboring points be given by (r, θ, ϕ) and $(r + dr, \theta + d\theta, \phi + d\phi)$. As shown in Fig. 3.25 and using the result obtained earlier for the distance on a sphere, Pythagoras theorem yields that the infinitesimal distance is given by

$$ds^2 = dr^2 + r^2[d\theta^2 + \sin^2 \theta d\phi^2]. \tag{3.6}$$

One can also describe flat three dimensional space using Cartesian coordinates (x, y, z), with neighboring point given by $(x + dx, y + dy, z + dz)$; it can then be shown, doing a change of variables on polar coordinates or directly from Eq. (3.3), that

$$\begin{aligned} ds^2 &= dx^2 + dy^2 + dz^2 \qquad (3.7) \\ &= dr^2 + r^2[d\theta^2 + \sin^2 \theta d\phi^2]. \end{aligned}$$

[7]A tetrahedron has four triangles as boundaries and is the simplest solid that encloses a three dimensional volume; a tetrahedron is the three dimensional analog of a triangle, which encloses a two dimensional surface.

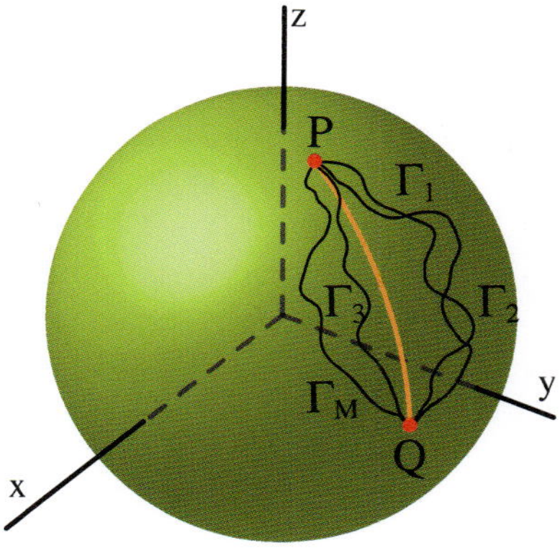

Fig. 3.26 Paths from P to Q, with the geodesic shown in orange.

The distance $\mathcal{D}$ between two points P and Q in an arbitrarily curved space, connected by a path Γ, is given by summing the infinitesimal distances Δs along the path Γ and yields the expression

$$\mathcal{D} = \int_{\Gamma} ds. \tag{3.8}$$

We can now state the precise condition for determining a **geodesic**: for all the paths Γ that go between two points P and Q, the geodesic is the path, denoted by Γ_M, that yields the minimum distance $\mathcal{D}_M$, as shown in Fig. 3.26.

3.7 Special Theory of Relativity

So far we have discussed distances that are all similar to what we experience in ordinary space. Does the concept of 'distance' have any generalization that is consistent with special relativity? In special relativity the position of a point $\mathbf{x}$ is replaced by the concept of an *event*, which is defined by the time and position of an occurrence; hence an *event* is given by $(t, \mathbf{x})$.

An **inertial frame** is defined by an observer moving with constant velocity; any measurement of acceleration would show that the observer has zero acceleration. Every inertial frame consists of observers positioned throughout spacetime, with each observer having an identical length scale and a clock.

The two fundamental postulates of special theory of relativity are that

- All laws of Physics are the same for all inertial observers
- The speed of light is the same for all inertial observers.

To analyze the consistency of the concept of distance in special relativity, we need to know how to go (transform) from one inertial frame to another; only if a quantity

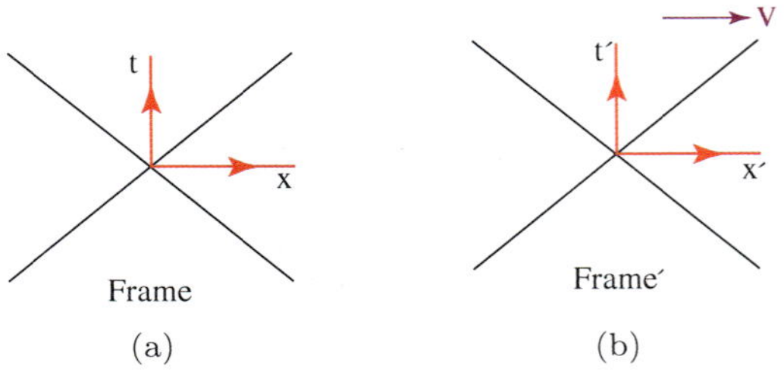

Fig. 3.27 Two inertial frames.

is unchanged in going from one frame to another can we conclude that a particular expression is the same for all observers. The quantity is said to be, for obvious reasons, **relativistically invariant**. We need to define a relativistically invariant generalization of ordinary distance.

Consider an inertial frame moving with velocity v in relation to the another inertial frame as shown in Fig. 3.27; the origin of both inertial frames is taken to be the same; the frame with coordinates (t', x') is moving with a relative velocity v with respect to the frame with coordinates (t, x). For simplicity, we analyze the case of one time and one space dimensions since the results readily generalize to one time and three space dimensions. A **Lorentz transformation**, relating an event observed in the first frame at (t, x) to the same event observed at (t', x') in another frame moving at velocity v, is given by Eqs. (2.21) and (2.22) as follows

$$x' = \gamma(x - vt); \quad t' = \gamma\left(t - \frac{v}{c^2}x\right); \quad \gamma = \frac{1}{\sqrt{1 - v^2/c^2}}. \tag{3.9}$$

We need some combination of the event so that its value as seen by say the first frame is the same as that seen by the second frame: this combination would then be unchanged in going from frame to frame, and would be invariant under Lorentz transformations. The analogy of such a combination is distance in ordinary space: the length of a distance is unchanged if one coordinate system is related to another by a rotation.

What is the analog of ordinary distance for special relativity? Clearly it cannot be the ordinary distance between two points since under a Lorentz transformation the length (distance) between two points in space contracts, as can be seen from Eq. (3.9). Similarly, the clock in a moving frame runs slower compared to the clock in a stationary frame. Hence, taken by themselves, neither space nor time can provide **Lorentz invariance**; the only hope we have is to use some combination of *space* and *time* coordinates of an event. The fact that one has a universal speed, namely speed of light c, allows one to convert time t into space by defining the distance ct, which has the dimension of space. Hence, one can now combine space and time into a single distance function and form an invariant distance given by the following combination

$$-c^2t^2 + \mathbf{x}^2 = -c^2t'^2 + \mathbf{x}'^2 \tag{3.10}$$

where $(t, \mathbf{x})$ and $(t', \mathbf{x}')$ are related by the Lorentz transformation given in Eq. (3.9). Note the far reaching fact that the time coordinate t of a spacetime event comes with a *minus* sign in relation to the space coordinate $\mathbf{x}$; this negative sign is all that separates space from time in Einstein's unification of space and time.

Let $(t_1, \mathbf{x}_1)$ and $(t_2, \mathbf{x}_2)$ be two spacetime events, and $(t'_1, \mathbf{x}'_1)$ and $(t'_2, \mathbf{x}'_2)$ be the same two events as seen in another inertial frame. Then it can be shown that Eq. (3.10) yields the following Lorentz invariant interval

$$\begin{aligned} &\Delta s^2 = -c^2(\Delta t)^2 + (\Delta \mathbf{x})^2 = -c^2(\Delta t')^2 + (\Delta \mathbf{x}')^2 \\ &\text{where} \quad \Delta t = t_1 - t_2; \quad \Delta \mathbf{x} = \mathbf{x}_1 - \mathbf{x}_2 \\ &\Delta t' = t'_1 - t'_2; \quad \Delta \mathbf{x}' = \mathbf{x}'_1 - \mathbf{x}'_2. \end{aligned} \tag{3.11}$$

Note Δs^2, the *spacetime interval* between two events, is relativistically invariant.

For two spacetime events that are infinitesimally separated and given in Cartesian coordinates, as in Eq. (3.7), by (t, x, y, z) and $(t+dt, x+dx, y+dy, z+dz)$, the spacetime interval ds^2 is given by the so-called **Minkowski metric**

$$\begin{aligned} ds^2 &= -c^2dt^2 + (d\mathbf{x})^2 \\ &= -c^2dt^2 + dx^2 + dy^2 + dz^2. \end{aligned} \tag{3.12}$$

The spacetime interval ds^2 defines flat four dimensional **Minkowski spacetime**.

We can also describe three dimensional flat space by spherical polar coordinates as in Eq. (3.6). Two neighboring spacetime points are given in polar coordinates by (t, r, θ, ϕ) and $(t + dt, r + dr, \theta + d\theta, \phi + d\phi)$. The spacetime distance, denoted by ds, between these two points — using Eqs. (3.6) and (3.7) respectively for the three dimensional spatial distance in polar coordinates — is given by

$$ds^2 = -c^2dt^2 + dr^2 + r^2[d\theta^2 + \sin^2\theta d\phi^2]. \tag{3.13}$$

It turns out that both representations given above of flat four dimensional Minkowski spacetime are going to be useful in our later analysis.

Equation (3.13) will be modified in the presence of matter and will be shown to yield the geometry of a black hole. Various cosmological models of our Universe entail the expansion of the distance between any two points, leaving the rate at which time flows unchanged; the **Friedmann–Robertson–Walker metric**, discussed in Noteworthy 6.1, is one such model, and which provides the mathematical basis of Big Bang cosmology. In the Friedmann–Robertson–Walker metric, the distance function given by Eq. (3.12) is modified to yield a spacetime interval that describes the geometry of a four dimensional spacetime which consists of a *flat* three dimensional space, but with the additional feature that the distance between any two points increases as the Universe evolves in time. The Friedmann–Robertson–Walker metric is given by

$$ds^2 = -c^2dt^2 + R^2(t)[dx^2 + dy^2 + dz^2] \tag{3.14}$$

where $R(t)$ is an increasing function of time.

Noteworthy (optional content) 3.4: The Metric Tensor $g_{\mu\nu}$

A complete description of a curved space is given by the distance function, that is by the *metric*, which yields the distance between any two points of the space, as given in Eq. (3.8). Note that the result obtained in Eq. (3.7) is for the metric of a *flat* three dimensional space; the result obtained can be generalized to arbitrary curved N dimensional space by specifying the distance between any two neighboring points in N dimensions. Instead of discussing N dimensions, we analyze only four dimensional spacetime, consisting of one time dimension and three space dimensions, as this is of central importance in the study of black holes and of cosmology.

Consider a curved spacetime and we study how to describe it in the neighborhood of a point (t, x, y, z). For two spacetime events that are infinitesimally separated, even for a curved spacetime, we can always choose the coordinates, as given in Eq. (3.7), by (t, x, y, z) and $(t + dt, x + dx, y + dy, z + dz)$; however, the spacetime interval ds^2 is no longer given by Eq. (3.12), which is only valid for flat Minkowski spacetime. Instead, the spacetime interval ds^2 for a curved spacetime is written in terms of the **metric tensor** $g_{\mu\nu}$ and is given by

$$ds^2 = \sum_{\mu,\nu} g_{\mu\nu} dx^\mu dx^\nu . \tag{3.15}$$

Note that the metric tensor $g_{\mu\nu}$ is dimensionless due to its definition in Eq. (3.15).

To write Eq. (3.15) above, the following notation has been used: $(x^0, x^1, x^2, x^3) = (ct, x, y, z)$. The values of μ, ν range through 0, 1, 2, 3. The metric tensor $g_{\mu\nu}$ is a real and symmetric 4×4 matrix. The infinitesimal distance ds given in Eq. (3.15) is used in Eq. (3.8) for finding the distance between any two points for an arbitrary curved space.

In the notation of the metric tensor, the Minkowski spacetime interval given in Eq. (3.12) is written as

$$ds^2 = \sum_{\mu,\nu} \eta_{\mu\nu} dx^\mu dx^\nu .$$

The metric tensor $g_{\mu\nu} = \eta_{\mu\nu}$ for flat Minkowski spacetime; furthermore, Eq. (3.12) yields that $\eta_{\mu\nu}$ is a diagonal and constant metric tensor, namely

$$\eta_{\mu\nu} = \text{diagonal}(-1, 1, 1, 1). \tag{3.16}$$

The Friedmann–Robertson–Walker metric, from Eq. (3.14), is given by

$$g_{\mu\nu} = \text{diagonal}\big(-1, R^2(t), R^2(t), R^2(t)\big).$$

3.8 Spacetime Intervals

Consider an observer located at the origin, with coordinates given by $(0, 0)$; let an event occur in the observer's frame at spacetime point (t, x). In special relativity the spacetime events occurring at (t, x) can either be a null, timelike or spacelike event; the domains of **null**, **timelike** and **spacelike events** are indicated in Fig. 3.28. The null domain spans out a **light cone** at every point in spacetime, and consists of the points of spacetime on which light travels.

To define timelike, spacelike and null interval more precisely, consider the spacetime interval

$$(\Delta s)^2 = -c^2t^2 + x^2$$

given in Eq. (3.11).

- Two events are *null* if light connects them; hence, two events $(0, 0)$ and (t, x) are null if $(\Delta s)^2 = 0 \Rightarrow x = \pm ct$. The trajectory lies on the light cone.
- Two events $(0, 0)$ and (t, x) are *timelike* if $(\Delta s)^2 < 0 \Rightarrow |x/t| < c$.
- Two events $(0, 0)$ and (t, x) are *spacelike* if $(\Delta s)^2 > 0 \Rightarrow |x/t| > c$.

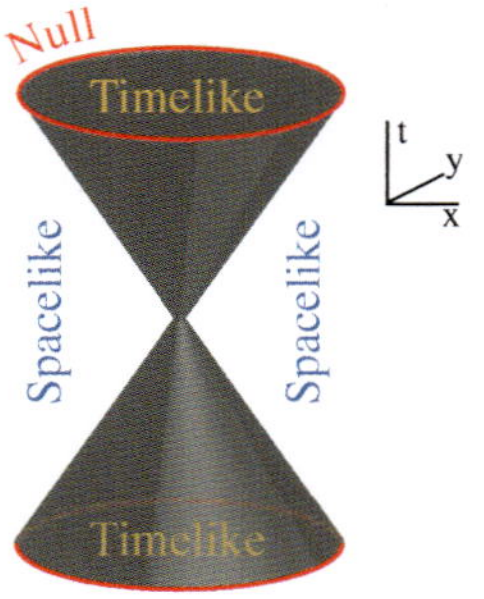

Fig. 3.28 Timelike and spacelike points that are separated by the null cone, also called the light cone.

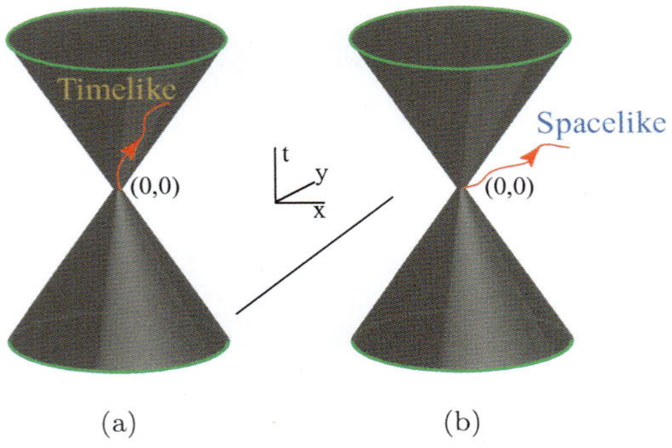

Fig. 3.29 The lightcone is the boundary between the (a) timelike and (b) spacelike intervals. The forward and backward cones are linked by causality.

In other words, for an observer at the origin $(0,0)$, events (t,x) that are timelike have $|x/t| < c$ and are *inside* the light cone; events for which $|x/t| > c$ are spacelike and are *outside* the light cone. The three types of spacetime events break up spacetime into three domains.

Consider an inertial frame with an observer at the origin $(0,0)$ and another observer, in the same inertial frame, located at (t,x). The physical significance of the three spacetime intervals is the following:

- **Timelike interval**
  A timelike spacetime distance (t,x) from the origin consists of all distances that can be reached with velocities less than the velocity of light. This consists of all the normal events we experience in daily life, such as going to work from home and so on, since all velocities we experience in life are less than the velocity of light. Figure 3.29(a) shows a typical timelike trajectory starting at $(0,0)$.

  For a timelike event (t,x) with respect to $(0,0)$, the time ordering of the two events — namely what comes earlier and what comes later — is always the same in *all* inertial frames: in all frames the event (t,x) is either always in the future to $(0,0)$ or always in the past to $(0,0)$. For this reason, only spacetime points that are in the backward timelike domain of the light cone can affect the observer at $(0,0)$; similarly, the observer at $(0,0)$ can affect only spacetime points in the future timelike domain of the light cone. Hence, the light cone defines the points related by causality to the observer at $(0,0)$.
- **Spacelike interval**
  Spacelike spacetime distances from $(0,0)$ consists of all distances that, at time instant $t=0$, are at a space point $(0,z)$, with $z \neq 0$. For simplicity, at instant $t=0$, consider a point at distance of 1 cm from the origin. For a signal from this point to reach the observer will take a finite time; so at instant $t=0$, the point is outside the causal domain of the observer at $(0,0)$. Figure 3.29(b) shows a typical spacelike trajectory starting at $(0,0)$. The trajectory is forbidden by

special relativity since it would require that the entity travel at a speed greater than the speed of light.

In contrast to timelike events, if the event (t, x) is spacelike with respect to the observer located at the origin of the coordinate frame $(0, 0)$, then the time ordering of the event is frame dependent — in some frames being in the future to the observer at $(0, 0)$ and in some other frames being in the past. This should not come as a surprise since an object at a spacelike spacetime distance from $(0, 0)$ is not causally connected to the observer at $(0, 0)$.

- **Null interval: light cone**
  A light signal from $(0, 0)$, traveling at speed of light, reaches (t, x) if it lies on the light cone. The spacetime interval between the two events is called a null event, for which $x = \pm ct$. The null spacetime interval constitutes the light cone and separates the timelike and spacelike events, as shown in Fig. 3.29.

Note in the Newtonian limit of $c \to \infty$ both the surface of the light cone as well as the spacelike region of spacetime collapses and one is left with only the timelike region. In the Newtonian limit there is absolute past and absolute future. In contrast, in Minkowski spacetime of special relativity, two brand new domains of spacetime, namely the null and the spacelike domains are revealed; for the spacelike domain there is no absolute past or absolute future, rather that depends on the observer. For the null cone, light travels on it and nothing with velocity less than that of light can travel on it.

Noteworthy (optional content) 3.5: Lorentz Transformations

We give a more technical description of the various spacetime intervals. The essential difference between timelike and spacelike events can be seen by performing a Lorentz transformation. Recall from Eqs. (2.21) and (2.22) that the event (t, x) in one frame is transformed to another event (t', x') by

$$x' = \gamma(x - vt); \quad t' = \gamma\left(t - \frac{v^2}{c^2}x\right); \quad \gamma = \frac{1}{\sqrt{1 - v^2/c^2}}.$$

The spacetime interval between timelike and spacelike events has the following structure.

(a) Timelike $|x/t| < c$

$$t' = \gamma t \underbrace{\left[1 - \frac{v}{c^2}\frac{x}{t}\right]}_{(positive)}. \tag{3.17}$$

If $t > 0$, then $t' > 0$, that is, the ordering in time is *preserved* in all inertial frames. This fixed time ordering is a necessity for causality since cause must always precede effect in all frames; for example the son cannot be born before his father in any inertial frame.

(b) Spacelike $|x/t| > c$

$$t' = \gamma t \left[1 - \frac{v}{c^2}\frac{x}{t}\right]. \quad (3.18)$$

To change the ordering in time we need to go to a frame where $t' < 0$. We need to find a frame of reference with velocity v such that

$$1 - \frac{v}{c^2}\frac{x}{t} < 0.$$

One can choose a frame moving at $v > c/2$ since this is allowed by relativity and which yields, for $x/t = 2c$, that $t' < 0$. Hence the ordering in time is not necessarily preserved for spacelike events; namely, even if $t > 0$, in one frame, t' can be less that 0 in another frame.

From the above discussion, we see that timelike events can be connected by a signal traveling at a velocity less than c, whereas to connect spacelike events one needs signals traveling faster than light — something forbidden by the special theory of relativity.

For the observer located at the origin $(0, 0)$, since no physical signal can travel faster than the velocity of light, classical causality states that only timelike events are causally linked to the observer, with spacelike events being causally disconnected from it. In other words, all events occurring in the forward and backward light cone, as shown in Fig. 3.29, are causually connected to the observer at $(0, 0)$. All events in the spacelike domain are causally disconnected from the observer at $(0, 0)$; hence the lack of a fixed ordering of events in the spacelike domain does not affect causality as these points are not connected by any physical signal with the observer at $(0, 0)$.

The ordering of events in time is how causality is realized, and continues to be valid, in the special theory of relativity. On combining quantum mechanics with the special theory of relativity, causality turns out to be more subtle — requiring the existence of antiparticles — and is discussed in Sec. 11.8.

3.8.1 *Null, timelike and spacelike*

The Minkowski metric gives a characterization of the trajectory in flat spacetime: light travels on the null cone and massive particles travel in the forward timelike domain of spacetime. One can write the Minkowski metric as

$$ds^2 = -c^2 dt^2 + (d\mathbf{x})^2.$$

Recall that the velocity of light is always equal to c; hence $(d\mathbf{x})^2 = c^2 dt^2$ and which implies $ds^2_{\text{null}} = 0$; similarly for a massive particle its velocity is always less than c and hence $\mathbf{x}^2 < c^2 t^2$, which leads to $ds^2_{\text{timelike}} < 0$. The classification of ds that holds for Minkowski spacetime can be generalized to curved spacetime where the condition on ds is valid *locally*, in the neighborhood of a point. Hence, for curved spacetime, at every spacetime point we have the following classification

$$ds^2_{\text{null}} = 0; \quad ds^2_{\text{timelike}} < 0; \quad ds^2_{\text{spacelike}} > 0.$$

In curved spacetime, light follows a null geodesic along which $ds = 0$; similarly, a massive particle follows a timelike geodesic with $ds < 0$. It can be shown that the geodesic has another important property, namely that if the initial tangent vector is null or timelike, then the parallel transport of it along the geodesic is also null or timelike; furthermore, the geodesic with a null tangent vector is a null geodesic and similarly for a timelike tangent vector.

3.9 Curvature

We have no direct intuitive experience of the curvature of spacetime. Instead what we do have is direct experience of surfaces and volumes that are warped and bent; a surface such as a sphere does not look like a flat sheet of paper and we intuitively attribute the departure from flatness to curvature. Similarly in three (space) dimensions, we see warped shapes that may be twisted and coiled, and we associate these shapes with curvature. What we need is to define, precisely and quantitatively, what is meant by curvature.

The idea of curved space has a long history. As known by many, the postulates of Euclid are adequate to define flat space; the question that was raised by Euclid himself is whether the postulate that parallel lines never meet is required, or whether it can be derived from the other postulates of Euclid. It was only in the 19th century that the answer to this question was given by Riemann, who formulated the general theory of curved spaces. In particular he showed that 'parallel lines can meet' in curved manifolds.

An intuitive way for understanding curvature is to consider a (closed) loop in space and examine the effects of parallel transporting a tangent vector around the loop. Consider for simplicity a sphere S^2 with radius r and parallel transport a vector $\mathbf{v}$ around a triangle on S^2, as shown in Fig. 3.30; the vector $\mathbf{v}$ is transported along three sides of the triangle PQR. Suppose one starts from point P, then parallel transports $\mathbf{v}$ to point Q, and then to point R and then back to P. The vector $\mathbf{v}$ can be shown to have rotated by an angle $\delta\theta$, and taking the limit of the area enclosed ΔA by the triangle as being infinitesimal, we obtain

$$\begin{aligned}\delta\theta &= \text{Area enclosed} \times \text{Curvature at point P} \\ &= \Delta A \times \frac{1}{r^2} \qquad (3.19)\end{aligned}$$

where r is the radius of the sphere. The amount by which the vector $\mathbf{v}$ changes is a precise measure of the curvature at the point where the loop is centered; in the

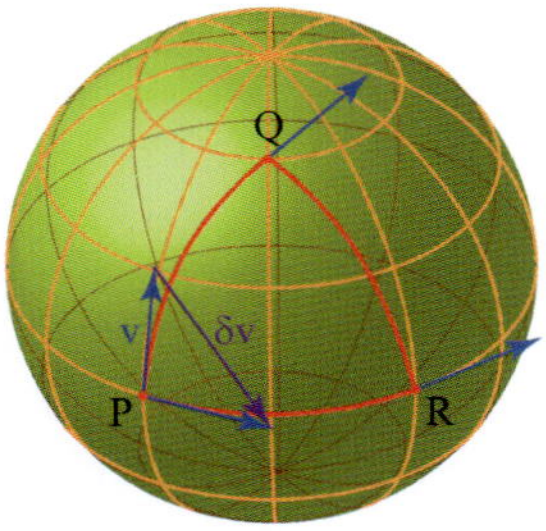

Fig. 3.30 Curvature for sphere. The vector **v** is parallel transported along the spherical triangle PQR.

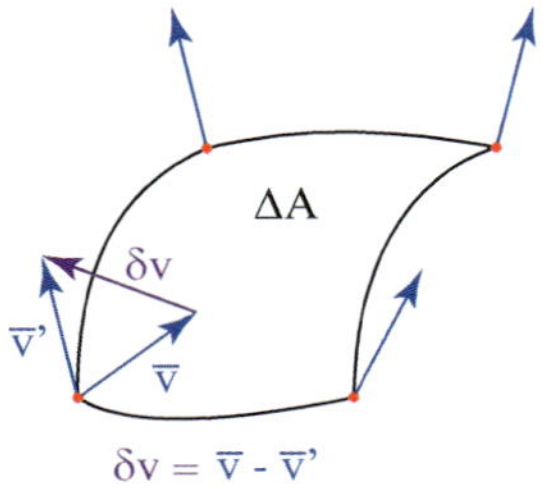

Fig. 3.31 A vector $\bar{v}$ is parallel transported around a closed loop, with final value of $\bar{v}'$. The curvature is proportional to $\bar{v} - \bar{v}'$.

limit that the area goes to zero, the curvature is independent of the size and shape of the area ΔA.

The curvature is the same at all points of the sphere. Hence, the sphere is a space with constant curvature. This is however not the case in general for spaces with arbitrary geometry: the curvature of the space need not be a constant and, in fact, can change from point to point on the manifold.

The curvature of a general higher dimensional manifold is defined in a similar manner. Consider an arbitrary element of area ΔA; two directions are needed to specify the orientation of ΔA, and which is enclosed by a one dimensional boundary; see Fig. 3.31. An arbitrary vector **v** is parallel transported around the closed loop enclosing ΔA. On completing one circuit and returning to its original starting point, the vector is changed by an amount $\delta\mathbf{v}$ that is completely determined by the curvature $\mathcal{R}$ at the location of ΔA. In the limit that ΔA is taken to zero we have

$$\delta\mathbf{v} = \mathcal{R}\mathbf{v}\Delta A. \tag{3.20}$$

The object $\mathcal{R}$ is the famous **Riemann curvature tensor**; it carries all the information about the intrinsic curvature of a space (manifold). From its definition in Eq. (3.20), the Riemann tensor $\mathcal{R}$ has the dimension of $1/L^2$, where L has the unit of length.

From the Riemann curvature tensor, the **Ricci curvature scalar** can be derived. The Ricci scalar does not depend on the coordinates being used and is a quantity that carries a simplified but useful description of space. Notably, the Ricci scalar is zero for flat space and is non-zero for curved space. The geometry of spacetime is completely and uniquely fixed by the value of the metric tensor $g_{\mu\nu}$, discussed in Noteworthy 3.4, and which is given at every point of spacetime for a given geometry. There is a celebrated equation that determines the Riemann curvature tensor from the metric tensor.

For a flat space, parallel transport does not change the vector and hence $\delta \mathbf{v}_{\text{flat}} = 0$, which yields $\mathcal{R} = 0$. For a sphere S^2 with radius r one can show, from Eq. (3.19), that (all the non-zero components of the Riemann tensor are given by)

$$\mathcal{R}_{\text{sphere}} = \frac{1}{r^2}. \tag{3.21}$$

It can also be shown that, for S^2, the Ricci curvature scalar is equal to $2/r^2$.

The reason that **parallel transport** around a closed loop is used for defining the curvature of a manifold is because the Riemann curvature that one obtains is *independent* of the coordinate system being used to describe the manifold. Any other way of transporting a vector around a closed loop will depend on the coordinate system and hence will be frame dependent, thus violating the basic premise of relativity.

The Riemann curvature tensor $\mathcal{R}$ gives rise to two objects, namely the Einstein tensor $\mathcal{G}$ and the Ricci scalar; both of these quantities also describe the geometry of a curved manifold but with not as much detail as $\mathcal{R}$. For example two manifolds that are different may have the same Einstein tensor and Ricci scalar, but this is not the case for $\mathcal{R}$.

Note that for a sphere in any dimension, the Ricci curvature scalar is > 0, that is, the sphere has a *positive curvature*; generically a space with positive curvature is called a spherical space; there are two more generic types of curvatures, namely spaces with *zero curvature*, namely flat space, and spaces with *negative curvature* called hyperbolic spaces. These spaces are shown in Fig. 3.32. Intuitively, for a space having a positive scalar curvature at a point, the volume of a small sphere about the point has a smaller volume than a sphere of the same radius in flat space. In

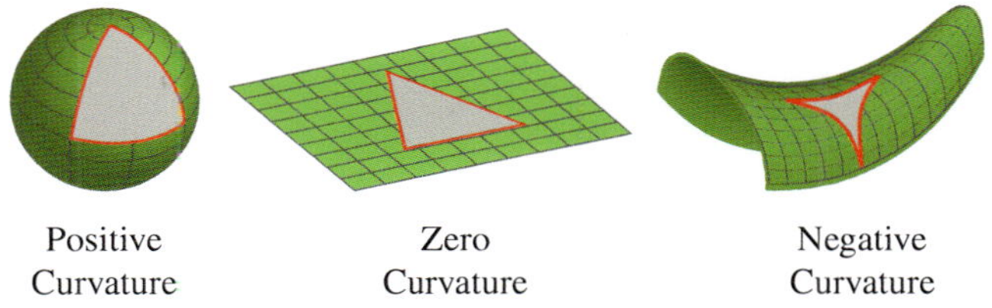

Fig. 3.32 Three types of curvature.

contrast, for a space having negative scalar curvature at a point, the volume of a small sphere is instead larger than it would be in flat space.

3.10 The Answer

"How straight is straight" led us to the concept of geodesics that generalizes the concept of a straight path to curved space, which in turn required an understanding of the parallel transport of a vector.

To describe curved geometry, we discussed, qualitatively, how does one represent a non-trivial curved space. For this purpose we discussed a few of the central ideas of geometry, namely that of parallel transport, geodesic, distance, spacetime distance and curvature.

The example of the surface of a sphere (embedded in three dimensions), namely S^2, was used extensively for simplifying the discussion on curved space since all the central ideas could be given a concrete representation. Of course, it needs to be remembered that S^2 was taken to be a similitude for four dimensional curved spacetime that arises due to the effect of mass and energy on empty spacetime.

The special theory of relativity was discussed to examine how the absolute velocity of Nature, namely the speed of light, affects the structure of empty spacetime. It was shown that spacetime is split by a light cone at every spacetime point, and causality is preserved for events having a timelike spacetime distance.

The definition of curvature was briefly discussed and hinged on the concept of parallel transport; in particular, it was shown that the change of a vector when it is parallel transported around a small closed loop defines the curvature at that point.

The concepts of this chapter are a preparation for studying Einstein's theory of gravity which contains many unexpected results — such as the expansion of the Universe as well as the existence of black holes — that we discuss in subsequent chapters.

Chapter 4

Gravity

Is gravity geometry?

4.1 The Question

On being shaken free from the tree, does the apple fall towards the Earth due to gravitational attraction? Or does the Earth fall towards the 'falling' apple?

It is commonly known that Newton saw an apple fall while he was studying under a tree. On reflecting on the falling apple, Newton came upon the idea that there is a universal attractive force of gravity; in particular, that the apple fell to the Earth because it was attracted by Earth's gravity. And Newton went on to reason that exactly in the same manner as the apple falls to the Earth, the Moon is held in orbit around the Earth because of gravitational attraction, and the Earth orbits the Sun due to the same force of gravity. Of course, unlike the apple, the Moon does not fall onto the Earth and neither does the Earth fall into the Sun: this is because the outward centrifugal force, which is due to rotation, exactly cancels the inward force of gravity.

Gravitational attraction has become so deeply embedded in our minds that we cannot imagine why it took mankind so many millennia, until the appearance of Newton, to have grasped this completely 'obvious' fact. Or so it would seem. Intuition is after all as much in-born as it is induced from childhood by society. The remarkable fact is that the Earth *does not* 'attract' the apple, or anything else for that matter. The apple does not fall to the Earth nor does the Earth fall towards the apple: the motion of both is due to the curvature of spacetime. This incredibly counter-intuitive idea is one way of speaking of Einstein's theory of gravity, also called the general theory of relativity.

We study how radical and complete was Einstein's rupture with Newtonian gravitation, and how the idea of geometry in Einstein's theory of gravity completely replaces Newton's concept of the gravitational force.

4.2 Newton's Gravity and Special Relativity

One needs to have a theory of gravity to understand the Universe at large, from planets to stars to galaxies and onwards.

Einstein proposed the special theory of relativity in 1905. Its central postulate is that there is a maximum velocity c in the Universe equal to the velocity of light in vacuum: all physical changes in the Universe propagate in space with velocity less than, or at most equal to, c. Einstein's theory combines space and time into a four dimensional spacetime in which space and time are on an equal footing.

All the theories of Physics have to be consistent with the special theory of relativity. Newton's laws of motion were modified and replaced by relativistic equations of motion. The consistency of special relativity with quantum mechanics led to relativistic quantum field theory. Maxwell's theory of electromagnetism was already consistent with special relativity and hence needed no alteration. The laws of thermodynamics and statistical mechanics also needed no changes.

The big puzzle for Einstein was how to modify **Newton's law of gravitation** and thus make it consistent with relativity. Newton's theory yields a gravitational force of attraction between masses m_1 and m_2, separated by a distance R, given by

$$F = G\frac{m_1 m_2}{R^2} \tag{4.1}$$

where G is Newton's gravitational constant with dimension $L^3M^{-1}T^{-2}$ and has a numerical value of

$$G = 6.67 \times 10^{-11}\,\mathrm{Nm^2/kg^2}. \tag{4.2}$$

Newton's laws of motion, combined with the force of gravitation given by Eq. (4.1), yield Kepler's three laws of planetary motion — and was one of the outstanding successes of Newtonian physics.

There are a number of problems with Newton's gravitational force given by Eq. (4.1). Firstly, the two masses m_1 and m_2 are observed simultaneously and found to be at a distance R apart; this statement depends on a particular inertial frame since the concept of simultaneity in the special theory of relativity depends on a frame: in another frame, moving with some non-zero relative velocity with respect to the first frame, the masses will not be simultaneously at the those points. Hence Newton's gravitational force is only valid in a particular frame — thus violating the principle of relativity that all physical laws must be the same in every inertial frame.

Another problem with Eq. (4.1) is that it is independent of time: the force between masses m_1 and m_2 is *instantaneous*. An instantaneous force violates the fact that all physical changes in the Universe can at most propagate with the velocity of light. Let the two masses be separated by a few million light years, and suppose mass m_1 moves to a new position, which is now at a distance $\tilde{R}$ from m_2; then Newton's gravitational force between the two masses is instantaneously given by $\tilde{F} = Gm_1m_2/\tilde{R}^2$; in other words, the gravitational force changes instantaneously, namely at an infinite speed, which is clearly faster than the velocity of light and hence is inconsistent with the special theory of relativity.

Einstein's greatest achievement is that, in 1916, he could go on to propose the general theory of relativity which is consistent with the special theory of relativity and provides an explanation of gravity radically different from Newton's theory.

4.3 Equivalence Principle: Accelerating Frame

Einstein's theory of gravity is a drastic and far-reaching break from our everyday intuition. To ease the transition to the new way of conceiving gravity, we follow the historical route followed by Einstein so as to motivate the introduction of essentially new ideas.

The special theory of relativity introduces the idea of an inertial frame and has been discussed in Sec. 3.7.[1] Einstein went further and established a deep connection between an accelerating *non-inertial* frame of reference and gravity. Consider the

[1]A rigorous definition of an inertial frame is quite complicated. In essence, an inertial frame of reference is defined by an observer moving at a constant velocity.

motion of a body of mass m on which gravity acts creating a force of mg, where g is the acceleration due to gravity; suppose the body is also subjected to non-gravitational forces denoted by F; Newton's second law of motion then states that

$$ma = mg + F \tag{4.3}$$

where a is the acceleration that the body is undergoing.[2]

For now we take g to be a *constant* so that we are considering motion in a Universe that has everywhere a uniform (constant) gravitational acceleration. We will later relax the constancy of g. Einstein sought an explanation as to why the 'inertial mass' multiplying the acceleration term, namely ma, is the same mass m that appears in the force due to gravity, namely mg.

This led Einstein to think that gravity was a very unique and special kind of Newtonian 'force' that is similar to the acceleration term ma on the left-hand side. Suppose we choose a non-inertial frame that is accelerating at rate a; in the new non-inertial (accelerating) frame the coordinate $\tilde{x}$ is given by

$$\tilde{x} = x - \frac{1}{2}at^2. \tag{4.4}$$

In the non-inertial frame Newton's equation of motion is given by[3]

$$m\tilde{a} = m(g - a) + F. \tag{4.5}$$

Einstein noted the 'inertial forces' carried by a, solely due to the fact that the observer constituting the non-inertial frame is accelerating, can completely cancel the effect of gravity if one chooses $a = g$; such an accelerating frame of reference is called a *freely falling frame*. In the freely falling frame gravity is absent, namely

$$m\tilde{a} = F. \tag{4.6}$$

Thus, by choosing an accelerating coordinate system, with $a = g$, one can completely remove the effects of gravity. A particle moving in the freely falling frame does not experience gravity and is completely equivalent to the particle moving in an inertial (non-accelerating) frame and experiencing the gravitational acceleration. A person in a freely falling frame will not feel his own weight: this weightlessness is what is experienced by sky divers, or by astronauts orbiting the Earth since their spaceship is in a free fall as it orbits the Earth. Figure 4.1 shows astronauts in free fall experiencing weightlessness.

The equivalence of an accelerated frame to a constant force of gravity, called the **Equivalence Principle**, was discovered by Einstein in 1907, and he later recalled that it was "the happiest thought in my life". The Equivalence Principle opened the way for subsequent developments that culminated in the general theory

[2]Acceleration is given by $a = d^2x/dt^2$.
[3]Note $\tilde{a} = d^2\tilde{x}/dt^2$.

Fig. 4.1 A freely falling frame. Weightlessness can be fun.

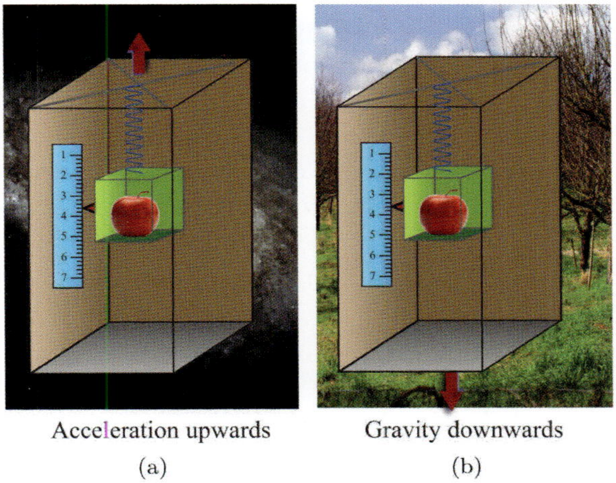

Fig. 4.2 Equivalence between (a) acceleration and (b) gravity.

of relativity. The Equivalence Principle states that *there is no experiment that can distinguish uniform acceleration from acceleration due to constant gravity.* The Equivalence Principle states that, in Nature, there is no absolute acceleration: constant acceleration caused by non-gravitational forces is exactly equal to the effect created by uniform gravity.

As shown in Fig. 4.2, consider weighing an apple in a closed lift that is accelerating at rate g; the Equivalence Principle states that inside the lift one cannot tell the difference between whether the lift is accelerating or is stationary but experiencing constant gravitational acceleration given by g: both lifts will have the same result for the weight of the apple.

It is important to note that, although the example we used to illustrate the Equivalence Principle is taken from Newton's equations of motion, it continues to hold for the case of the special theory of relativity since the Equivalence Principle depends only on the behavior of inertial and freely falling frames of reference.

A uniform gravitational field, like constant acceleration, is an idealization; in Nature gravity varies from one spacetime point to another, similar to the change in the acceleration of a car traveling on a winding road. The Equivalence Principle is really a *local statement* that holds in the close vicinity of a single spacetime point. In other words, in the neighborhood of every spacetime point one can completely replace the effects of gravity by choosing an equivalent freely falling (accelerating) frame.

Every spacetime point has an equivalent freely falling frame, and the collection of all these frames, taken together, describe the total effect of gravity. In practice the precise size of the neighborhood of a point depends on the scale on which the gravitational potential varies; for example, the gravitational potential inside an elevator is a constant to a very good approximation and hence the interior of an elevator can be considered to be a freely falling frame of reference, as in Fig. 4.2.

Einstein unsuccessfully tried for many years — using only the special theory of relativity and the Equivalence Principle — to derive other known results such as the shift of the perihelion of Mercury and so on. The reason for his failure was that one needs the full theory of gravity, complete with the field equations for the curvature of spacetime, to derive all the consequences of the relativistic theory of gravity. It was only in 1916 that the derivation of the general theory of relativity was completed by Einstein. The most spectacular *prediction* of the general theory of relativity was the bending of light near the Sun, and which was verified in 1919.

We show in the following sections that the Equivalence Principle leads one to the fact that time must slow down and space must be curved in regions where the effect of gravity is strong.

4.4 Gravity: Slowing Down Time

The rate of flow of time is fastest in empty space, and slows down in the presence of matter-energy. We study the effect of gravity on the flow of time by applying the Equivalence Principle. An accelerating spaceship is analyzed to determine how the flow of time is affected by acceleration. The Equivalence Principle is then invoked to conclude that a similar result would be obtained if acceleration is replaced by a gravitating mass creating the same acceleration.

Consider the following experiment, as shown in Fig. 4.3: a spaceship is traveling with acceleration a and has two astronauts, one at the top of the spaceship indicated by T (for top) and the other at the bottom denoted by B (for bottom). Both astronauts have identical (atomic) clocks. Astronaut T sends, to astronaut B, two light signals with a time interval of Δt_T. The question we want to answer is: what is the time interval Δt_B that is measured by astronaut B?

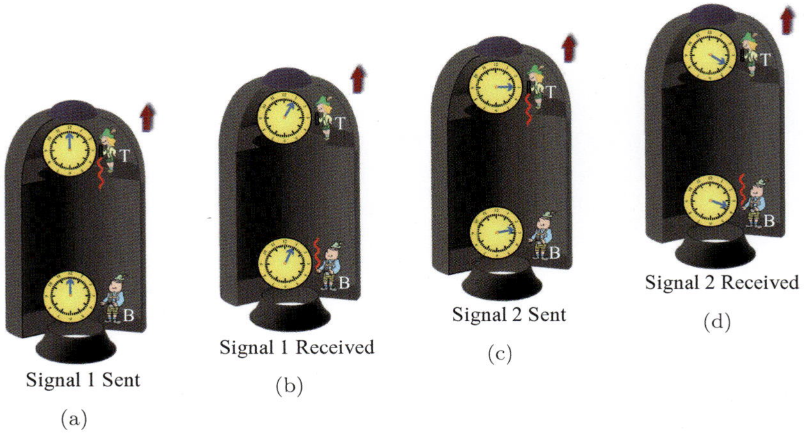

Fig. 4.3 Acceleration has an effect on clocks.

A short calculation, given below, shows that the interval Δt_B is *less* than Δt_T. In other words, the clock at position B is running *slower* than the clock at position T.[4]

We can derive the leading contribution to the slowing down of time at B using Newtonian mechanics. The calculation is carried out in an inertial frame in which the spaceship is accelerating with acceleration a. We then use the Equivalence Principle to derive how gravity affects the flow of time.

At time $t = 0$ the first light signal is sent from point T, at a height $z_T(0) = h$, to point B, which is taken to be at position $z_B(0) = 0$, as in Fig. 4.3(a). The first signal is received by B at time t and the positions of T and B at time t, as in Fig. 4.3(b), are given by

$$z_T(t) = h + \frac{1}{2}at^2$$
$$z_B(t) = \frac{1}{2}at^2.$$

The light signal has traveled for a duration of time t and hence the distance traveled by the first light signal in going from T to B is given by

$$z_T(0) - z_B(t) = h - \frac{1}{2}at^2 = ct. \tag{4.7}$$

As shown in Fig. 4.3(c), let the second light signal be sent by T at time Δt_T, when it is at position $z_T(\Delta t_T)$; the signal is received by B at time $t + \Delta t_B$ and at

[4]Note the words "slower" and "faster" don't have any absolute meaning since both the clocks at T and B are working perfectly well; instead what this manner of speaking indicates is that if one *compares* the rates at which the two clocks are running, the one at B will be seen to be running slower than the one at T.

position $z_B(t+\Delta t_B)$ as in Fig. 4.3(d). The second light signal travels for time $t+\Delta t_B - \Delta t_T$ and covers a distance of $c(t+\Delta t_B - \Delta t_T)$, which in turn is given by

$$z_T(\Delta t_T) - z_B(t+\Delta t_B) = c(t+\Delta t_B - \Delta t_T).$$

Note however that

$$z_T(\Delta t_T) = h + \frac{1}{2}a(\Delta t_T)^2$$
$$z_B(t+\Delta t_B) = \frac{1}{2}a(t+\Delta t_B)^2.$$

Keeping only terms that are linear in Δt_T, Δt_B and t/c one obtains, by eliminating t from above equations, that[5]

$$\Delta t_B = \Delta t_T\left(1 - \frac{ah}{c^2}\right) < \Delta t_T. \quad (4.8)$$

The result given in Eq. (4.8) shows that the time interval of signals sent by T are shorter when received at B; for example signals sent from T at 1 second time intervals arrive at B with time intervals of say 0.5 seconds. In other words, the flow of time at point B is *slower* than the flow of time at point T.

The derivation shows that for an accelerating spaceship, the time at the bottom of the spaceship flows slower than the time on its top. The derivation so far has been based completely on Newtonian physics. The *Equivalence Principle* allows one to go beyond Newtonian physics and carry the result to the behavior of gravity as described by Einstein: we conclude that, in the presence of gravity, time slows down — *the stronger the effect of gravity the more that time slows down.*

In 1959 Pound and Rebka placed an atomic clock at the top of a seven storied office building and compared its frequency with that of a similar clock placed in the basement. Using the so-called Mossbauer effect, they confirmed the prediction of the Equivalence Principle to within 1%. In 2004, physicists Sebastian Fray and colleagues used an atomic interferometer to perform the most accurate test ever of the Equivalence Principle at the level of atoms. They compared the acceleration of two isotopes of rubidium in the Earth's gravitational field (arXiv.org/abs/physics/0411052). As expected, the atoms accelerated at the same rate to a fractional accuracy of 1.7×10^{-7}.

[5] A careful analysis, based on the results of the special theory of relativity, shows that, for the spaceship having velocity v, terms of order $(v/c)^2$ have been ignored, which are negligible. Furthermore, keeping only terms that are linear in t/c amounts to ignoring terms of order $(h/c^2)^2$, which are also negligible. In fact, the terms retained provide sufficiently accurate results for the Global Positioning System to work effectively.

One may ask: by how much can gravity slow down the flow of time? Can gravity completely stop the flow of time? In Sec. 5 on Black Holes we will show that, as measured by observers far from the black hole, time slows down to exactly zero on the horizon of the black hole. To understand this and other remarkable properties of black holes, we need to have some understanding of the manner in which geometry is determined by the matter content of space, and in particular we need to understand how to describe the geometry of curved spacetime.

The local Equivalence Principle holds for only for small volumes of space and for short time intervals. To describe the physics of large volumes and long time intervals, one would need to put together, consistently, many local frames.

4.5 Gravity: Bending Spacetime

The Equivalence Principle applies to all physical phenomena including light. In empty space, light observed from an inertial frame moves in a straight line. We examine the behavior of light in the presence of gravity.

Consider the following experiment in empty space. An elevator is being accelerated *vertically upwards*. Let a beam of light be sent in a horizontal direction and suppose it enters from one side of the elevator and exits from the other side.

- Consider an observer sitting *outside* the elevator, namely the observer is in an inertial frame of reference, as shown in Fig. 4.4(a). The observer will see the light entering and leaving the elevator while traveling in a straight trajectory.
- Consider an observer sitting *inside* the elevator; namely, the observer is in an accelerating frame of reference, as shown in Fig. 4.4(b); for this observer, light enters into the elevator at a higher point than its exit point. Hence, to the observer inside the elevator, the trajectory of light will appear to follow a parabolic trajectory.

For a body that has mass m and is traveling under the influence of acceleration due to gravity, the parabolic trajectory follows from a straightforward calculation using Newton's second law $F = ma$; however, light has zero mass and hence one cannot use Newton's law to determine the trajectory of light. Nevertheless, the Equivalence Principle tells us that light must follow a parabolic trajectory and this pointed the way to a theory of gravity based on curved spacetime.

Furthermore, from the Equivalence Principle an accelerating elevator is equal to a stationary elevator experiencing gravitational acceleration. Hence, it follows that, as is the case for the accelerating elevator, the effect of gravity on light is to bend its trajectory into a parabolic shape, as shown in Fig. 4.5.

In empty space, in the absence of gravity, light travels in a straight line, without any bending. What does it mean to say that gravity bends the trajectory of light? Light has the highest velocity in nature and light travels between two points in the shortest time. If light bends, the simplest explanation is that it does so because

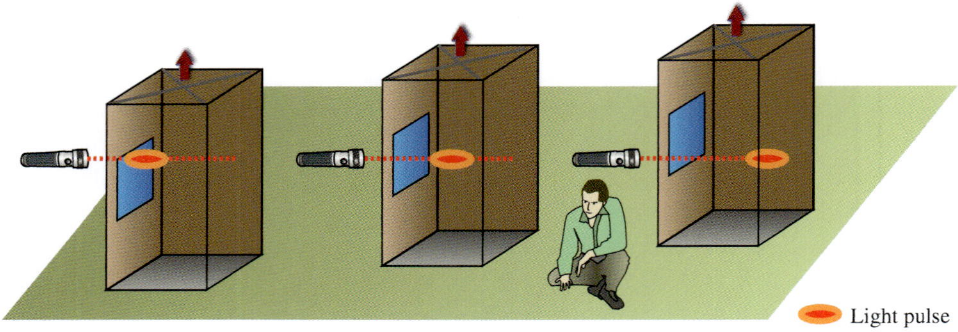

(a) Observer outside of the elevator

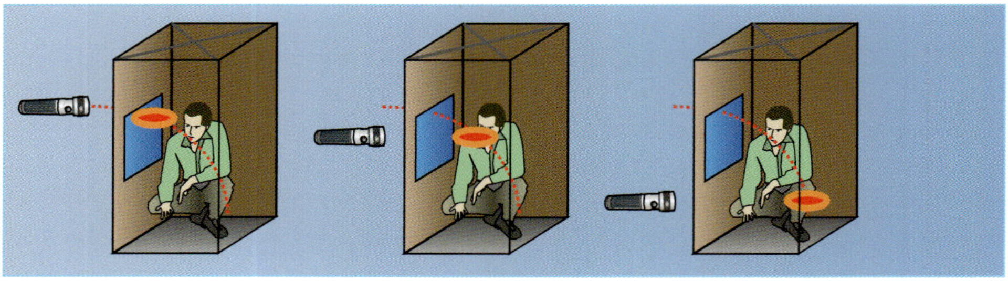

(b) Observer inside the elevator

Fig. 4.4 (a) Light viewed in an inertial frame. (b) Light observed in an accelerating frame.

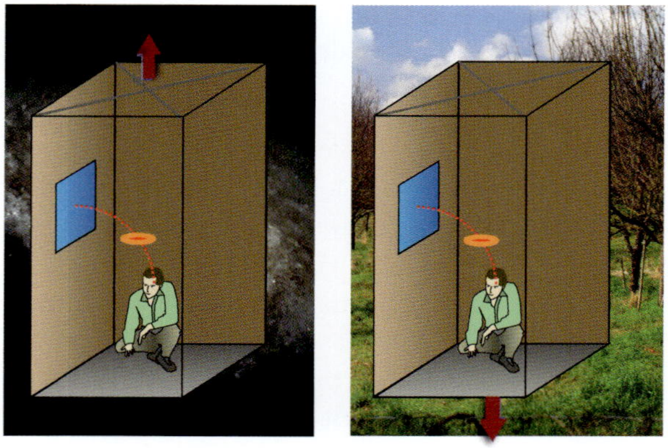

(a) (b)

Fig. 4.5 Equivalence between (a) acceleration and (b) gravity holds for light as well.

space itself is not flat, or equivalently, *spacetime is curved.* A discussion on curved spacetime has been given in Chapter 3. This is the route that Einstein followed to finally arrive at a geometrical explanation of gravity.

4.6 Geodesics and Freely Falling Frames

Einstein postulated that the geometry of spacetime is determined by the matter-energy content of the Universe, and in turn that geometry determines how matter moves in spacetime.

Objects that have only gravity acting on them move on paths that are geodesics. Geodesics diverge or approach each other as a consequence of curvature.

To concretely see how geometry determines the motion of matter consider, as shown in Fig. 4.6, the example of a large dense mass m occupying a small volume; we examine the motion of two particles in the presence of this mass.

To illustrate Einstein's formulation of gravity, suppose that — due to mass m — space is curved into the geometry of a sphere S^2. Particles in this geometry will follow the geodesics on the sphere. There are geodesics (great circles) on the sphere S^2 (that encloses the mass) and intersect at the south and north pole, as in Fig. 4.6. Suppose that the two particles start at the south pole with velocity (tangent) vectors pointing away from each other: the particles' trajectories diverge since the two geodesics diverge. At the equator the two velocity vectors become parallel. As the geodesics cross the equator they start to converge at the north pole and hence the two particles meet.

The Newtonian interpretation of the motion of the two particles is that they initially diverge from each other since their initial velocities point in different directions. As they travel the mass at the center attracts the two particles due to the *Newtonian force of gravity*; by the time they cross the equator, the effects of gravity have become large enough to change the direction of their velocities so that they start to converge, and finally meet at the north pole.

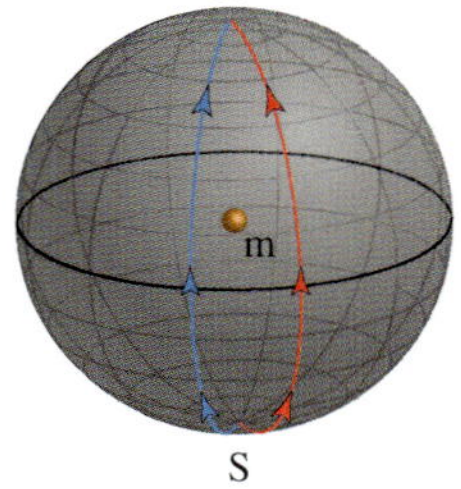

Fig. 4.6 A heavy mass is placed at the center of the sphere. Two diverging, and then converging, geodesics.

This example illustrates the fact that the effect of gravity on motion, according to Einstein, is entirely due to the curvature of spacetime — and which is incorrectly interpreted in Newtonian mechanics as a consequence of a 'gravitational force'.

Astronomical bodies approach each other since the geodesics they are traveling on are converging due to the curvature of spacetime. This is the great insight and change of one's understanding of gravity brought about by Einstein's geometric theory of gravity. In the general theory of relativity, gravity is not a 'force' but, instead, is a manifestation of the *curvature of spacetime.*

It is important to note that gravity is a manifestation not just of the curvature of space but rather, of the curvature of *spacetime.* A ball thrown upwards follows a parabolic path. This is far from being a geodesic only in space but rather is a geodesic in spacetime. Spacetime is curved by the Earth's mass; even though the ball moves a short distance in space, it moves an enormous distance in time since it is ct that appears in the distance function given by $ds^2 = -(cdt)^2 + (d\mathbf{x})^2$. A ball moving for one second has distance of ct equal to about 300,000 kilometers. This allows the slight amount of spacetime curvature due to the Earth to have a noticeable effect. Note the Earth itself is an accelerating frame following its own geodesic. The ball's trajectory appears as a parabola to us standing on the Earth and hits the Earth since the two geodesics, namely that of the ball and of the Earth, intersect, as shown in Fig. 3.15(c). On hitting the Earth, non-gravitational forces come into play and stop the ball.

The null spacetime interval was discussed in Sec. 3.8. Light travels with null distance function $ds^2 = 0$ and the path of light is called a *null geodesic*, which is a special kind of geodesic reflecting the fact that the of speed light is a constant. A null geodesic is the *shortest distance* between two spacetime points, as shown in Fig. 3.15(b), and is the generalization to curved manifolds of the concept of a straight line.

Freely falling frames of reference travel along geodesics. The Equivalence Principle shows that the velocity (tangent) vector of a freely falling frame is a constant and hence undergoes parallel transport. Each point on the geodesic is equal to a freely falling frame in which the frame experiences no gravity.

All objects in spacetime that are solely under the influence of gravity are in a state of free fall and hence move on geodesics. All astrophysical objects such as planets, stars, galaxies and clusters of galaxies are moving along geodesics; this statement is based on ignoring the effect of non-gravitational entities such as radiation, dark matter and so on.

If an object's trajectory *deviates* from a geodesic, it is due to non-gravitational interactions analogous to the case in flat spacetime, where deviating from constant velocity means that a force is acting on it. Note that objects on the Earth's surface like us are not moving on a geodesic; the reason being that we would like to be in free fall, but the surface of the Earth exerts a force on us to break the free fall; hence we feel the force of gravity, namely our weight, because we are being accelerated by the floor — by non-gravitational forces — to stay stationary. In particular, for the case of the elevator that is accelerating, as in Fig. 4.2, it is the non-gravitational forces arising from the material of the floor that acts to create an acceleration — for observers inside the elevator — that is equivalent to the acceleration due to gravity.

We hence conclude that gravity is not an attractive force: in fact it is not a force at all! Rather, gravity is another name for the geometry of spacetime.

4.7 Geodesics and Cosmological Time

A transparent and important exemplar of the significance and the role of geodesics is in Big Bang cosmology, and which is discussed in some detail in Chapter 6. In the Big Bang model of cosmology, the Universe starts ('is born') at a finite time in the past and is considered to be 13.78 billion years old. The following question naturally arises: according to whose clock, or more accurately, according to which inertial frame, is the age of the Universe being measured? This question can be addressed by studying the role of geodesics in cosmology.

At the cosmic scale, all the galaxies are considered to be equivalent and constitute a non-interacting gas of 'particles' (galaxies) that all recede away from each other. In most models of cosmology, it is assumed that space is homogeneous and isotropic; namely the geometry of space at every point of space, and in all directions, is identical, respectively; in particular, this implies that all the galaxies are identical. In the actual Universe of course all galaxies are different; but when one considers cosmological distances and time scales, all the galaxies are reduced to point masses, with their differences being negligible.

It is further assumed in Big Bang cosmology — called the **Weyl postulate** — that the geometry of the Universe consists of a family of *non-intersecting* geodesics which constitute a system of **co-moving frames**. Figure 4.7 shows a family of geodesics emerging from the origin of time, placed at the occurrence of the Big Bang, and which are all diverging from each other. A co-moving frame is very special in the sense that it is only in this frame that the expanding Universe will appear homogeneous and isotropic.

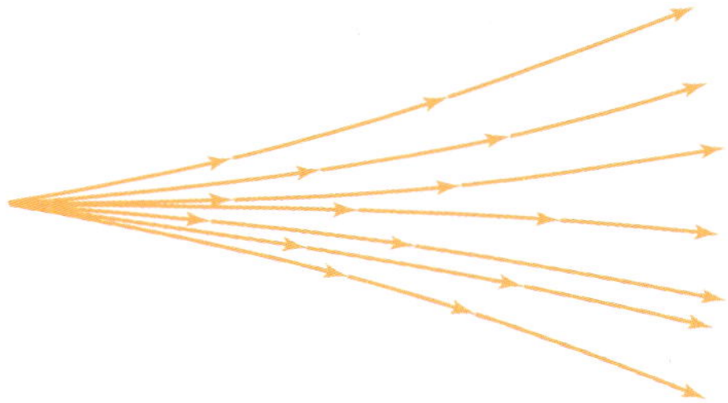

Fig. 4.7 Non-interesecting geodesics emerging from the origin of time, where the Big Bang takes place.

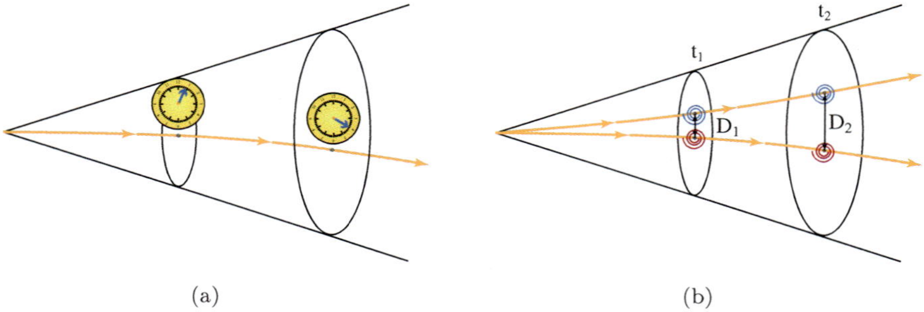

Fig. 4.8 The Universe is represented by a disk. (a) Each geodesic carries its own clock and measuring rods. (b) The distance between galaxies is increasing.

Every galaxy moves on a geodesic and has its own observers with clocks and measuring rods; namely, each galaxy has its own frame of reference moving with the galaxy (that is why it is called a co-moving frame). Figure 4.8 is a representation of a co-moving frame. Figure 4.8(a) shows that the clock for a given galaxy is running at same rate as the Universe evolves in time: one second today, as measured on the Earth, is equal to one second at the start of the Universe. So to answer the question about the age of the Universe, it is being measured by the clocks in each galaxy. In particular, clocks are running at the same rate in all geodesics (galaxies). Figure 4.8(b) shows that the distance measured between different galaxies is increasing as the Universe evolves.

Noteworthy (optional content) 4.1: Friedmann–Robertson–Walker Cosmology

The **Friedmann–Robertson–Walker metric** provides a specific model of the geometry of the Universe at the cosmological scale and is also a realization of the Weyl postulate. Recall from Eq. 3.14 that the Friedmann–Robertson–Walker metric is given by

$$ds^2 = -c^2dt^2 + R^2(t)[dx^2 + dy^2 + dz^2]$$

where $R(t)$ is an increasing function of time. As discussed in Sec. 3.7, the Friedmann–Robertson–Walker metric describes the geometry of a four dimensional spacetime that consists of a *flat* three dimensional space, but with the distance between any two points increasing as the Universe evolves in time.

The way one measures time in curved spacetime is to fix one's position, namely set $dx = dy = dz = 0$, For the Friedmann–Robertson–Walker metric, this reduces ds^2 to $ds^2 = -c^2dt$, and which in turn determines how fast the clock at that point is running. Hence, the flow of time is encoded in the term c^2dt^2, which for the Friedmann–Robertson–Walker metric is identical to this term in the distance function for empty space, given in Eq. (3.12). This is the reason that clocks today run at the same rate as clocks at the starting of the Universe as shown in Fig. 4.8(a) — because the term c^2dt^2 does not depend on the evolution of the Universe.

The space part of the distance function ds^2 is being increased by the function $R(t)$. What this means is that, as the Universe evolves, all distances between different points increase, as shown in Fig. 4.8(b); in particular, for $t_2 > t_1$, the distance between the points has increased since $D_2 > D_1$; and in fact $D_2 = [R(t_2)/R(t_1)]D_1$, as discussed in Noteworthy 6.1.

4.8 The Pattern of Spacetime and Gravity

The Equivalence Principle shows that gravity is intimately linked to the shape of space and to the flow of time. Gravity is the manifestation of the geometry of spacetime and, in particular, the workings of geometry are reflected in the warping and bending of space and in the changing of the flow of time from point to point of spacetime.

To construct a description of gravity one can proceed in two steps. Firstly, one can ascertain — using the local Equivalence Principle and freely falling frames — how light bends and time flows at different points of spacetime and, in effect, determine the lengths that rods measure and times that clocks record. Secondly, one then needs to put together all the spacetime points in such a manner that the distances that rods measure and times that clocks record are consistently accounted for.[6] In other words, one needs to consistently put together, to 'glue' together, all the spacetime points in such a manner so that, as one moves from point to point, one smoothly reproduces the measuring rods and clocks at every point.

The pattern in which the spacetime points are arranged encodes all the information provided by the local (accelerating) freely falling frames and yields the curved geometry of the spacetime continuum.

The 'gluing' together of all the spacetime points to form one composite system is analogous to how one draws a map of the world. The local information given by the freely falling frames is equivalent to knowing the distance between any two cities; to draw the map one needs to lay out all the cities in such a manner so that the distance between any two cities is equal to the measured distance. Knowing no better we can try and lay out the cities on a flat piece of paper; we will soon find out that, as illustrated in Fig. 4.9(a), we are left with gaps in the paper; in particular, the island of Mauritius incorrectly appears at two distinct points in the map, quantified by the angle of deficit.

[6] The units of length and time at different points are all being compared to a fixed standard measuring rod and clock, and varies in relation to these standards.

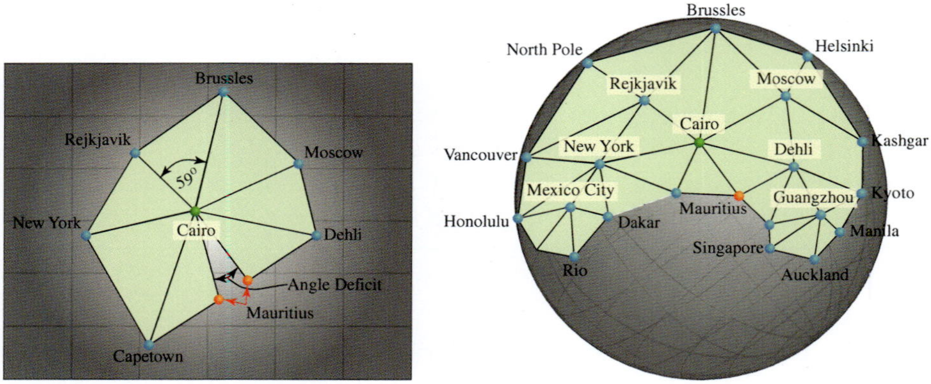

(a) Cities on a Plane

(b) Cities on a Sphere

Fig. 4.9 Cities are mapped on a plane and on a sphere. On the plane, there is an angle deficiency which does not occur on the sphere. (Maps not drawn to scale.)

Since we know that there are no 'gaps' on the surface of the Earth and each city has only one location, we will realize, after a while, that the gaps represent the curvature of the Earth's surface. After a lot of trial and error we will discover, as shown in Fig. 4.9(b), that the only way of consistently laying out all the cities — so that there are no gaps — is on the surface of a sphere. In other words, the distances between the various cities encode information about the spherical geometry of the Earth's surface.

There is another way of drawing the map of the world. One first postulates that the Earth has the shape of a sphere and *then* places all the cities on the globe, one by one, according to their distance from each other. Being able to consistently place all the cities provides a verification of the postulate. The distances between the cities follow from the fact that they are all placed at their location on the globe and one again obtains Fig. 4.9(b).

Remarkably enough, Einstein took the second route in defining the geometry of spacetime. Instead of putting rods and clocks with changing units of measurement at every point of spacetime and then consistently gluing them together to form a smooth spacetime continuum, Einstein made a bold and dramatic statement about the global nature of gravity in the general theory of relativity.

4.9 Curvature and Matter

Until the advent of Einstein's theory of gravity, space and time were considered to be an empty 'stage' given *a priori*, a precursor to material phenomena. In fact Newton

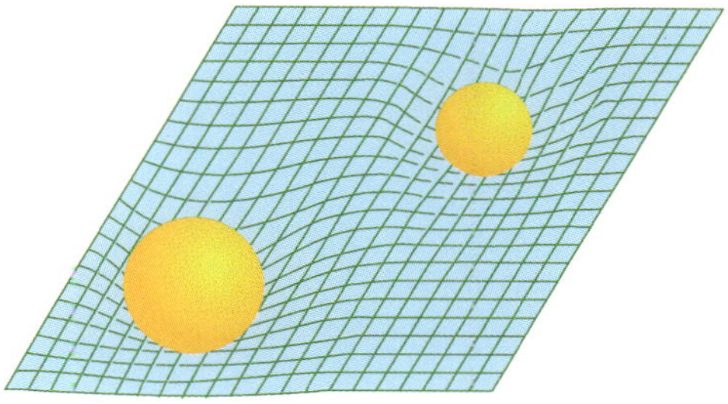

Fig. 4.10 The curvature of space due to two masses.

thought of space as an absolute and unchanging structure — with an inertial frame being in uniform motion relative to an absolute and immovable space. Newton also considered time to be identical for all observers, flowing at the same rate everywhere in the Universe. With the advent of Einstein all that changed.

Einstein, after a long period of intense effort from 1905 to 1915, completed his theory of gravity in 1916 and concluded that matter-energy curves spacetime and matter moves on geodesics that are determined by the curvature of spacetime. In our earlier discussion, we found that geodesics are the shortest paths between any two points on a curved space; a body moving on a geodesic is in free fall and does not feel the effects of curvature.

We now address the question: how does matter determine the curvature of spacetime? A simple analogy is to consider two steel balls placed on an elastic membrane, as shown in Fig. 4.10. One can think of the steel ball as the analog of matter and the elastic membrane as the analog of spacetime; the 'weight' of the steel ball deforms and curves an otherwise flat membrane into a curved surface, as shown in Fig. 4.10. If one were to release a tiny steel ball on the surface of the deformed membrane, one would see that the ball would move on a geodesic due to the curvature of the surface, which is analogous to the motion of particles in a spacetime having curvature.

The mass of a steel ball is the analog to the matter-energy of a star or a galaxy. There is a quantity, denoted by $\mathcal{T}$ and called the **energy-momentum tensor**, that mathematically represents the total matter-energy-momentum content of a star and must at least contain the energy-momentum four dimensional vector. Furthermore, there are stresses and strains in a medium that contain energy, and these are also included in $\mathcal{T}$.

A major breakthrough made by Einstein in understanding the nature of gravity was to connect his intuitive ideas of curved spacetime with the mathematics of curved manifolds discovered by Riemann, and briefly discussed in Sec. 3.9. Einstein could connect the purely mathematical and formal framework of Riemann geometry with the physical content of spacetime, and which led him to putting down the following fundamental equation relating geometry to matter

$$\mathcal{G} = \frac{8\pi G}{c^4}\mathcal{T} \tag{4.9}$$

$$\text{Einstein curvature tensor} = \frac{8\pi G}{c^4} \times \text{Energy-momentum stress tensor}$$

where G is Newton's gravitational constant. The quantity $\mathcal{G}$, called the **Einstein curvature tensor**, is purely geometrical and is constructed from the **Riemann tensor**; the energy-momentum tensor $\mathcal{T}$ carries all the non-geometrical matter-energy content of spacetime. By equating these two quantities, Einstein successfully geometrized the description and explanation of gravity.

Einstein's field equations, as Eq. (4.9) are referred to, determine the geometry of spacetime that results from the matter-energy content of the Universe. They are the crowning glory of Einstein's theory of gravity, in which $\mathcal{G}$ — carrying the geometry of spacetime — 'comes alive', changing and responding to the changing material content of the Universe — encoded in the energy-momentum tensor $\mathcal{T}$.

Noteworthy (optional content) 4.2: Dimensions of Einstein's Equations

---

The Einstein tensor $\mathcal{G}$ is proportional to the Riemann curvature tensor $\mathcal{R}$, which has been shown in Eq. (3.20) to have the dimension of $1/L^2$, where L has the unit of length. Hence

$$[\mathcal{G}] = 1/L^2.$$

The unit of the stress tensor $\mathcal{T}$ is energy per unit volume, which yields

$$[\mathcal{T}] = ML^{-1}T^{-2}$$

where M is a mass scale and T is a time scale; G and c have units of

$$[G] = L^3M^{-1}T^{-2}; \quad [c] = LT^{-1} \Rightarrow [G/c^4] = T^2M^{-1}L^{-1}.$$

Hence

$$[G/c^4][\mathcal{T}] = 1/L^2$$

and, as expected, the right-hand side of Eq. (4.9) also has the dimension of $1/L^2$.

Note that, due to Einstein's equations given in Eq. (4.9), $8\pi G/c^4$ is the amount of curvature $\mathcal{G}$ that is created by a unit of the stress tensor $\mathcal{T}$. The numerical value

$$\frac{8\pi G}{c^4} = 2.077 \times 10^{-43}\,\mathrm{s^2kg^{-1}m^{-1}}$$

is a very small number and is the reason that the amount of curvature created by any ordinary piece of matter is vanishingly small. In other words, spacetime is extremely rigid and very hard to bend. This is the reason that the spacetime around us looks perfectly flat. It is only with large conglomerations of matter that curvature of spacetime becomes non-trivial.

The Einstein tensor $\mathcal{G}$ is derived from the metric tensor $g_{\mu\nu}$, discussed in Noteworthy 3.4. Einstein's field equations given in Eq. (4.9) also fixes $g_{\mu\nu}$. And once the metric tensor $g_{\mu\nu}$ is known, the geometry of spacetime is completely determined.

Knowing $g_{\mu\nu}$ allows us to determine the measuring rods and clocks for all observers — at every point of spacetime. The fact that we can consistently 'glue' together the varying clocks and measuring rods of the observers at different points of spacetime is guaranteed by the existence of the metric tensor.

As is the case for all fields, the gravitational field, which recall is simply another name for $g_{\mu\nu}$, carries energy and momentum, and in turn, affects the geometry of spacetime. Hence, Einstein's theory of gravity forms a nonlinear system, with the gravitational field reacting back on itself. The nonlinear nature of gravity is the root cause of its mathematical complexity as well as of all the counter-intuitive and unexpected results that it produces.

4.10 Gravitational Radiation

A major prediction of Einstein's theory of gravity is the existence of gravitational waves, which necessarily follows from the speed of light being the maximum speed of physical change in the Universe.

Consider moving a large mass from one position to another; the geometry of spacetime must change as a result of the move; however, due to the limit set by the speed of light, the change can only propagate out at the speed of light. There are no gravitational waves in Newtonian gravity as all changes in gravitational attraction are instantaneous, propagating at infinite speed.

In general, an accelerating mass should emit gravitational radiation, similar to an accelerating electric charge emitting electromagnetic radiation. However, there are some difference between gravitational and electromagnetic radiation due to the fact that there are no opposite charges in gravity. For example, a star that is pulsating in a perfectly spherically symmetrical manner will *not* radiate. Similarly, a spinning ring or a disk will *not* radiate due to the conservation of angular momentum; this case has important ramifications since it is for this reason that a spinning black hole does not radiate. Although a spinning black hole is rotating and hence accelerating, it is not losing energy to gravitational radiation and the black hole geometry has an exact stationary solution, called the Kerr solution and discussed in Sec. 5.10.

Any accelerating asymmetric mass, technically one that has a non-zero third time derivative of its quadrupole moment, will emit gravitational radiation. For

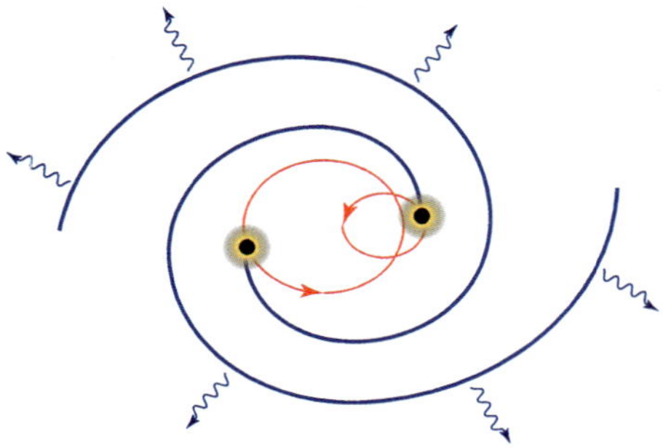

Fig. 4.11 Two pulsars, rotating about a common center of mass, and emitting gravitational waves causing them to slowing down. The solid lines indicate the wave front of gravitational radiation. (Orbits for a mass ratio of 2 and eccentricity of 0.5.)

example, the Earth loses a negligible 200 joules per second to gravitational radiation — the energy used by a light bulb — as it orbits the Sun.

Gravitational waves have not yet been directly detected. A pulsar is a spinning neutron star and is discussed in Sec. 9.13. The slowing down of the orbiting period of a **binary pulsar** — due to loss of energy radiated off in the form of gravitational waves — is predicted by Einstein's field equations and provides an *indirect* test for the existence of gravitational waves. A binary pulsar, made of two solar mass for each neutron star, and spinning at a distance of 2×10^8 m, emits 1.4×10^{28} joules per second in the form of gravitational radiation — and which is about 100 times greater than the electromagnetic radiation emitted by the Sun. The radiation from a binary pulsar is illustrated in Fig. 4.11.

The observed behavior of the 'Hulse–Taylor binary pulsar' — a pair of stars both of which are pulsars (spinning neutron stars) — agrees with the prediction of Einstein's theory and led to Joseph Taylor and Russell Hulse being awarded the 1993 Nobel Prize in Physics; their results have provided impressive indirect experimental evidence for the emission of gravitational radiation.

The Equivalence Principle, discussed in Sec. 4.3, states that gravity is the manifestation of a particle moving on a geodesic; the particle is assumed to have a negligible mass and hence does not affect the geometry of spacetime. It should be noted that the Equivalence Principle cannot account for gravitational waves. In contrast, Einstein's field equations are independent of the results of the Equivalence Principle and provide a complete description of gravity that is applicable for large masses that react back on the geometry of spacetime as they move.

4.11 The Answer

In our daily experience of space and time, our sensory perceptions mislead us into thinking that space and time are rigid and unchanging entities, *a priori* structures that constitute an empty stage in which events take place. Newton's theory of gravity is based on these perceptions of space and time. In contrast, Einstein's formulation of gravity *unifies* space and time into a single entity — based on the speed of light being a universal constant. Furthermore, the geometry of spacetime emerges as a dynamic and ever changing entity, being shaped by the mass-energy content of the Universe, and in turn determining the motion of material entities.

The key to unlocking the salient quality of gravity lies in the equivalence between acceleration and gravity. An accelerating frame is locally equivalent to gravity — this is the intuitive formulation of gravity. A point to be noted is that, unlike the case of electric charges, one can never create a device that can 'shield' the effects of gravity; since gravity is equivalent to an accelerating frame, there is nothing to shield because an accelerating frame is a way of viewing Nature and not something intrinsic to what is being viewed.

Einstein's theory of gravity explains all the successes of Newton's formulation of gravity and, in addition, predicts many completely unexpected and novel phenomena. Some of the major new phenomena that are revealed are gravitational waves, the bending of space and warping of time, and geometric singularities that occur in various types of black holes.

The fact that $8\pi G/c^4$ is so small results in gravity being a very weak force, when compared to the other interactions in Nature that determine the structures of atoms and molecules. The weakness of gravity has a rather dramatic consequence: it allows enormous conglomerations of matter to aggregate without undergoing **gravitational collapse**. The weakness of gravity is the reason that large objects like stars and planets can form with great separation, that gigantic primordial gas clouds can exist and subsequently condense into vast and stable galaxies and that cosmology can be laid out on a truly cosmic scale. In fact, the weaker gravity is, the more colossal is the Universe that it generates. And conversely, if gravity was stronger, the Universe would not be so large and neither would it have such a long lifetime.

The intuitive ideas of Einstein are expressed by the mathematics of curved manifolds and lead to a quantitative and exact description of gravity. Einstein's theory of gravitation, encoded in the general theory of relativity, is one of the most magnificent constructions of the human mind and has stood the test of a variety of experiments for over 100 years.

The success of Einstein's theory, however, does not mean that it is the final statement on gravity. To the contrary, one of the biggest challenge facing theoretical physics is to reconcile Einstein's gravity, which is essentially a classical field theory, with the principles of quantum mechanics. This exciting and as yet unsolved challenge has led to superstring theory, discussed in Chapters 14 and 15.

Chapter 5

Black Holes

Where does time end?

5.1 The Question

When we are immersed in the ongoing flow of time, we can easily imagine the end of time as eternity. And hence we think it will take an eternity to reach the end of time. But is this really the case though?

Black holes have tantalized the imagination of the public for many years. From science fiction accounts of their terrible destructive powers to their ubiquitous presence in scientific research, the term "black hole" is widely known.

Black holes are named as such since no object, not even light, can escape from their gravitational pull, once it strays too close to it. All objects that are captured by a black hole, when they hit the center of the black hole, are thought to be crushed into a single point.

Do these bizarre geometrical objects really exist in our Universe? If so how can we detect them? Are black holes really 'black' or can matter inside the black hole escape from it?

To answer our questions about black holes we need to understand, in qualitative terms, Einstein's theory of gravitation, a beginning for which was made in Chapter 4. Armed with a theoretical understanding of geometry, we discuss the properties of black holes and how these strange physical objects can create regions of spacetime where time itself comes to a complete halt!

We discuss the semi-classical synthesis of Einstein's theory of gravity with quantum mechanics — with gravity providing a classical 'background' in which quantum processes take place — and show how this leads one to the rather startling but compelling conclusion that black holes, in fact, radiate. It can be shown that black holes emit black body radiation, and in doing so slowly evaporate and finally completely disappear.

5.2 Brief History of Black Holes

The possible existence of objects similar to black holes has been known for many centuries. In 1783, based on the corpuscular theory of light, John Michell showed that for sufficiently massive stars Newtonian gravity predicted that not even light could escape from these stars, which he called "dark stars". Michell's ideas were published in 1795 by Pierre Simon Laplace in his first edition of an astronomy guide. Laplace subsequently dropped the concept of a "dark star" in the third edition (1808) since Thomas Young's experiment in 1803, on the interference of light, seemed to conclusively rule out the corpuscular theory of light.

Einstein's theory of general relativity was published in 1916 and the black hole solution to Einstein's theory was derived by Karl Schwarzschild in the same year. Einstein was not very pleased with the solution found by Schwarzschild since it implies that spacetime at the center of a black hole has infinite spacetime curvature, giving rise to a geometric singularity. For this reason black holes at that time were known as **Schwarzschild singularities**. As late as in 1939 Einstein wrote a technical paper in which he concluded that "The essential result of this investigation is a clear understanding why the 'Schwarzschild singularities' do not exist in physical reality." Both Einstein and British astronomer Arthur Eddington vigorously opposed the idea that Schwarzschild singularities could occur in Nature and provided many arguments to substantiate their views, all of which were later seen to be incorrect.

An important indication that black holes might be astrophysical objects was in a 1939 theoretical paper by J. Robert Oppenheimer and Hastland Snyder, which states that: "When all thermonuclear sources of energy are exhausted, a sufficiently heavy star will collapse. Unless (...) reduces the star's mass to the order of that of the Sun, this contraction will continue indefinitely." It was speculated by Oppenheimer that a sufficiently heavy star would undergo endless contraction and collapse into a black hole as its final state.

It was only almost half a century after Schwarzschild's derivation and following Einstein's demise in 1955 that a systematic study of these solutions were taken up. The term *black hole* itself is of a fairly recent origin, having been coined by black hole theorist John Wheeler in 1967. By the late 1960's, astronomical data and theoretical arguments provided increasing evidence that not only do black holes occur in Nature, but also that black holes are the necessary and inevitable result of the stellar evolution of very massive stars.

Black holes are amongst the most outrageous and remarkable predictions of Einstein's theory of gravitation. The concept of black holes has evolved into one of the bedrocks of modern astrophysics and astronomy — explaining a wide variety of astrophysical phenomena.

5.3 Laplace's Dark Star

The idea of Michell and Laplace is really quite simple. The corpuscular theory of light postulates that light is made up of small particles (corpuscles) that travel at the universal speed of light. If one considers light being made up of small particles, then one can imagine a star dense enough so that even light cannot escape from it. A star from which light cannot escape would be dark, and hence the name **dark star**. This leads us to the following question: what are the conditions required to permanently trap light?

A stone thrown upwards soon falls back to the surface of the Earth. Can one throw the stone with a high enough velocity that it escapes from the Earth's gravity and travels into outer space? The answer is yes: for a particle to escape from the Earth's gravitational pull the particle needs a certain minimum velocity. One may further ask a similar question: can a particle in principle escape from a very massive star? The answer depends on the density of the star, and we now address this question in some detail.

Consider a particle in a bound state on the surface of a star and hence with net negative energy $E < 0$. Since the particle is held to the surface by the star's gravitational potential U, one way to escape from the bound state is by acquiring enough *kinetic* energy T so that the particle's total energy $E = T + U$ becomes at least equal to zero.

We can find the minimum velocity required for the particle to escape from the star. Let the mass of the star be M and its radius R. The energy E of the particle of mass m moving with vertical velocity v is given by

$$E = \frac{1}{2}mv^2 - G\frac{Mm}{R}.$$

For the particle in a bound state, it's net energy is negative and hence

$$E < 0 : \text{Bound state.}$$

The *escape velocity* v_{escape} of the particle is achieved when $E = 0$. Hence

$$\begin{aligned} E = 0 &= \frac{1}{2}mv_{\text{escape}}^2 - G\frac{Mm}{R} \\ \Rightarrow v_{\text{escape}} &= \sqrt{\frac{2GM}{R}}. \end{aligned} \tag{5.1}$$

Example 5.1. Escape velocity from the Earth. The mass and radius of the Earth are $M_E = 5.97 \times 10^{24}$ kg and $R_E = 6.37 \times 10^6$ m respectively. The resultant escape velocity is given by

$$\begin{aligned} v_{\text{escape}} &= \sqrt{2\frac{(6.67 \times 10^{-11}\,\text{Nm}^2/\text{kg}^2)(5.97 \times 10^{24}\,\text{kg})}{6.37 \times 10^6\,\text{m}}} \\ &= 11.2\ \text{km/s} \simeq 25{,}000\ \text{miles/hour}. \end{aligned}$$

From Eq. (5.2), we see that the escape velocity depends *inversely* on the radius of the star R. What does this mean? It means that if we hold the mass constant and reduce the radius R, then the escape velocity would increase. Note we can reduce the star's radius by squeezing the mass of the star into a smaller volume, in other words, by increasing the *density* of the star. Suppose we compress the mass M of the star to a new radius $\tilde{R} < R$; given that the volume of a sphere is $\frac{4}{3}\pi\tilde{R}^3$, the new density of the star is given by

$$\tilde{\rho} = \frac{\text{Mass}}{\text{Volume}} = \frac{3M}{4\pi\tilde{R}^3} > \rho$$

where ρ is the density of the star before compression. The escape velocity $\tilde{v}_{\text{escape}}$ from a body with the star's mass but smaller radius $\tilde{R}$ is given, from Eq. (5.2), by

$$\tilde{v}_{\text{escape}} = \sqrt{\frac{2GM}{\tilde{R}}} > \sqrt{\frac{2GM}{R}} \quad \text{since } \tilde{R} < R.$$

The intuitive reason for this result is the following: the higher the density of star, the smaller is its radius, the closer is the surface to the center of the star, and hence the stronger is the force of gravity at the surface.

We can keep on increasing the value of the escape velocity v_{escape} by squeezing the mass of the star into a smaller and smaller volume and hence reducing the value of R. We know from the special theory of relativity that the maximum velocity that any object in the universe can have is the velocity of light, namely

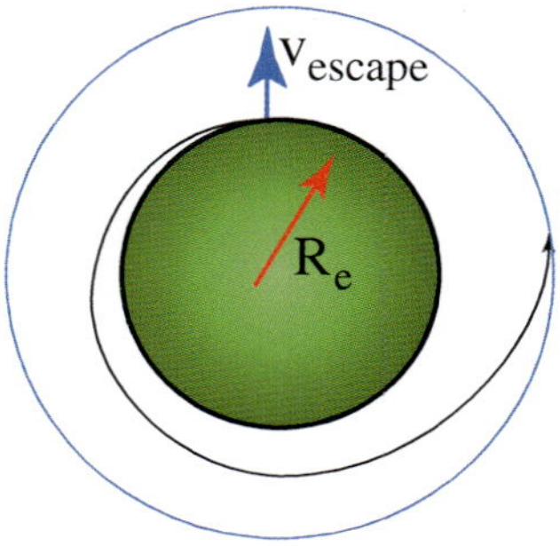

Fig. 5.1 Escape velocity.

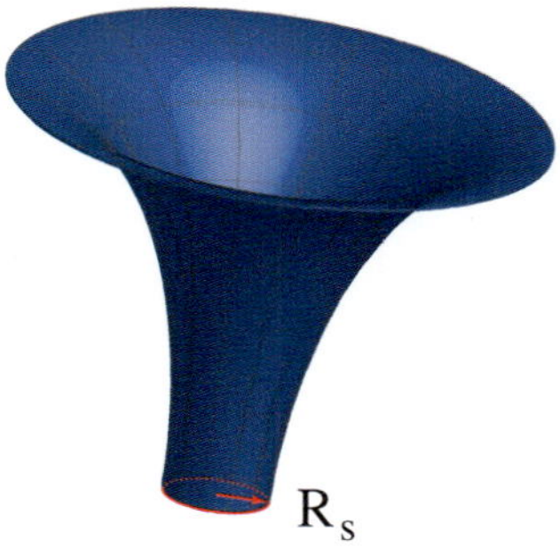

Fig. 5.2 Schwarzschild radius.

$c = 2.99 \times 10^8$ m/s. When the escape velocity v_{escape} approaches c, this is an indication that nothing, not even light, can escape from the gravitational pull of a highly dense star.

Let the radius into which the mass M is squeezed be such that the escape velocity becomes equal to the velocity of light; the radius is then called the **Schwarzschild radius**, denoted by R_S (see also Fig. 5.2), and is given by

$$v_{\max} = c = \sqrt{\frac{2GM}{R_S}}$$
$$\Rightarrow R_S = \frac{2GM}{c^2}. \tag{5.2}$$

The object is then said to form a **black hole**.

Surprisingly, exactly the same result for the Schwarzschild radius as given in Eq. (5.3) is obtained by a much more complicated calculation based on the general theory of relativity.

The Earth's Schwarzschild radius is given by

$$\begin{aligned} R_S(\text{Earth}) &= \frac{2GM_E}{c^2} \\ &= 2 \times \frac{(6.67 \times 10^{-11}\,\text{Nm}^2/\text{kg}^2)(5.97 \times 10^{24}\,\text{kg})}{(2.99 \times 10^8\,\text{ms}^{-1})^2} \\ &= 8.9 \times 10^{-3}\ \text{m} \simeq 1\,\text{cm}. \end{aligned}$$

Thus we see that if the entire mass of the Earth is compressed into a sphere with a radius a bit less than 1 cm, it will form a black hole. For the Sun, the Schwarzschild radius is about 3 km. It is known from astrophysics that stars with a final mass greater than 3 times the mass of our Sun end their stellar evolution by undergoing gravitational collapse and forming a black hole.

Laplace showed that a 'dark star' having the density of the Sun, but with a radius 250 times larger, would prevent light from escaping.[1]

There are however profound differences between the concept of a 'dark star' and a black hole: a 'dark star' is ordinary Newtonian matter, only very massive, whereas a black hole is a new structure, a new phenomenon that is not reducible in any sense to a Newtonian object. To understand how and why a black hole is a unique theoretical and physical object, it needs to be studied from first principles which is the focus of the remaining sections of this chapter.

5.4 The Schwarzschild Geometry

The most elementary object in Newtonian gravity is a point mass m, which generates a gravitational potential equal to $-Gm/r$ at a distance r. We would, similarly, like to start our study of Einstein's theory of gravity from the geometry generated by a point mass. The closest that we can come to studying the geometry of a point mass is the Schwarzschild geometry, which we now address.

The **Schwarzschild geometry** is generated by matter that is static and spherically symmetric; an example of such a form of matter is a static and perfectly spherical solid ball. Suppose the static solid ball, with total mass M and radius R_M, is located at the origin of our coordinate system; hence, for distances greater than R_M space is empty and we have, from Einstein's field equation given in Eq. (4.10), that

$$\mathcal{G} = 0; \quad r > R_M. \tag{5.3}$$

[1] "A luminous star, of the same density as the Earth, and whose diameter should be two hundred and fifty times larger than that of the Sun, would not, in consequence of its attraction, allow any of its rays to arrive at us; it is therefore possible that the largest luminous bodies in the universe may, through this cause, be invisible." (Pierre Laplace, 1798, *The System of the World*, Book 5, Chapter VI.)

On solving the equation above one obtains the geometry of a spherically symmetric and static spacetime that, at spacetime point (t, r, θ, ϕ) (in spherical coordinates), is completely specified by the following distance between neighboring points[2]

$$ds^2 = -c^2\left(1 - \frac{2GM}{c^2 r}\right)dt^2 + \frac{dr^2}{\left(1 - \frac{2GM}{c^2 r}\right)} + r^2[d\theta^2 + \sin^2\theta d\phi^2];\ r > R_M. \quad (5.4)$$

The Schwarzschild geometry has curvature in both space and time directions and the geometry changes in the time and radial directions of the metric. It is worth noting that unlike the case for the Friedmann–Robertson–Walker metric, discussed in Noteworthy 4.1, the flow of time for the black hole geometry given in Eq. (5.4) depends on how far one is from the black hole; this distortion of time by the black hole is discussed in Sec. 5.6.

Due to spherical symmetry of the problem, the dependence on the angles θ, ϕ is the same as the flat space sphere S^2, namely $d\theta^2 + \sin^2\theta d\phi^2$. The matter content of this geometry is located in the region with $r < R_M$. Note that far from the origin, that is for $r \to \infty$, the Schwarzschild geometry reduces to flat Minkowski space, as in Eq. (3.13), with metric given by

$$\lim_{r\to\infty} ds^2 = -c^2 dt^2 + dr^2 + r^2[d\theta^2 + \sin^2\theta d\phi^2]. \quad (5.5)$$

From Eq. (5.5) above we have the important result that the Schwarzschild coordinates (t, r, θ, ϕ) are constant (time-independent) and are the ones that a distant observer — having a flat spacetime geometry — would be using to study the properties of a star or a planet.

The Schwarzschild geometry has the following physical realizations.

- Consider a planet or a star that can be treated as being approximately static and spherical. For example, our Sun rotates at its equator once in about 27 days and is a perfect sphere to an accuracy of one part in 100,000. Hence, for many problems, considering the Sun as being static and spherical is an excellent approximation. For such cases, the Schwarzschild geometry describes the spacetime *external* to the planet or the star. Based on the properties of matter that composes the planet or a star, *another* metric for $r < R_M$ is obtained and is matched to the Schwarzschild metric at $r = R_M$.
- *Black hole* is a term for a region of spacetime from which gravity prevents all objects, including light, from escaping. The Schwarzschild geometry is a means of representing the geometry of a black hole.
- The extreme limit of $R_M \to 0$ is the case of a solid ball, with fixed mass M occupying a single *point*, namely all the mass M is concentrated at the origin

[2] Deriving the metric is far too complex for a non-technical text.

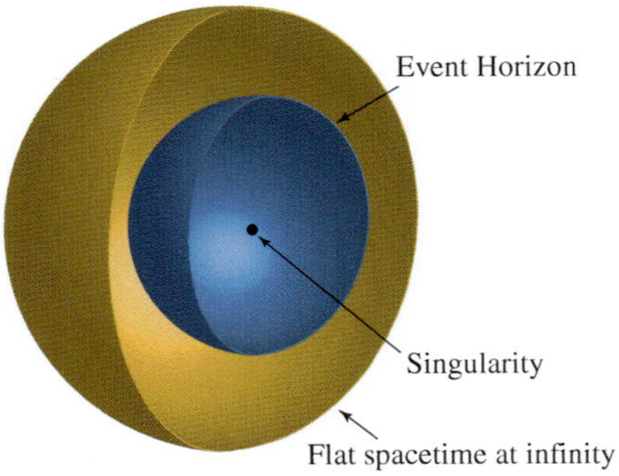

Fig. 5.3 Main features of the Schwarzschild black hole: the singularity at the center of the black hole; the event horizon enveloping the singularity and flat spacetime far from the black hole.

$r = 0$. The metric given in Eq. (5.4) is the geometry of this point mass, valid for all $r > 0$. Note that in Einstein's theory of gravity every point mass, which is localized at a point, yields a black hole geometry — although for non-astrophysical objects the Schwarzschild radius would be extremely small.

- There is an apparent divergence of the metric given in Eq. (5.4) since, at the Schwarzschild radius of $R_S = 2GM/c^2$, the denominator of the dr^2 goes to zero; this divergence is spurious and can be circumvented by a change of coordinates. We can hence conclude that for the black hole the Schwarzschild geometry is valid for all $r > 0$.
- Spacetime points at a distance of R_S from the center of the black hole span a surface of a perfect sphere, namely S^2, and is called the *event horizon*.

Figure 5.3 shows the main features of Schwarzschild black hole.

Noteworthy (optional content) 5.1: Embedding a Black Hole in Three Dimensions

The curvature of a black hole can be visualized by embedding it in three dimensional space $\Re^3$; note the **embedding space** is not a physical space, but rather a way of visualizing the curvature of a manifold. Since the black hole is static, we take slices for $t =$ constant, that is, for $dt = 0$, as they have the same geometry. Since the black hole is spherically symmetric we consider the 'equatorial' slice with $\theta = \pi/2$, and which yields the Schwarzschild metric to be

$$ds^2 = \frac{dr^2}{1 - \frac{2GM}{c^2 r}} + r^2 d\phi^2; \quad 0 \leq r \leq \infty; \quad -\pi \leq \phi \leq \pi.$$

This metric can be embedded in three dimensional space and is shown in Fig. 5.4; note the embedding in $\Re^3$ is only possible for $r > 2GM/r^2$.

The embedding yields a surface that corresponds to the Schwarzschild geometry, with points that are not on the surface having no significance. One way of thinking of

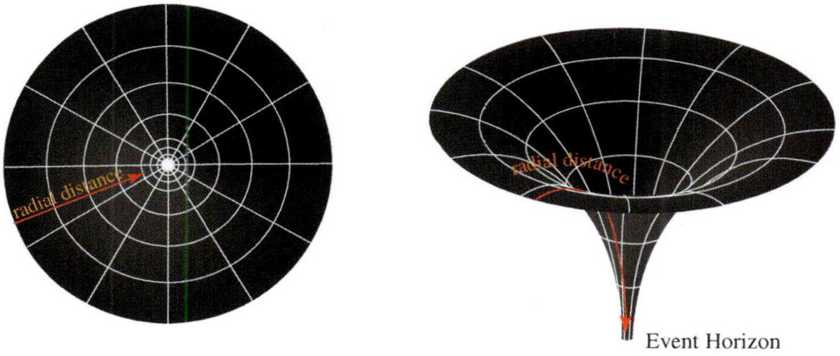

Fig. 5.4 The curvature of space increases as one approaches the black hole.

the embedding is to consider the z-direction as indicating the degree of curvature. For instance a two dimensional space with no curvature would be a flat surface, independent of z as one moves towards the origin of r. For a black hole, the embedded surface warping downwards indicates curvature: the larger the change in the value of z the higher the curvature of the black hole.

5.5 Astrophysical Black Holes

One of the most spectacular predictions of Einstein's theory of gravity is that black holes can exist in Nature. As discussed earlier, although Einstein himself did not believe black holes could physically exist, theoretical and experimental evidence for them has been mounting inexorably. From the 1960's onwards, theoretical and astronomical evidence has been obtained for the formation of black holes that result from stellar evolution. Supermassive galactic black holes are discussed in Sec. 8.8 in the context of galaxy formation. Black holes are discussed in Sec. 9.14 as one of the possible end points of stellar evolution.

Very massive black holes formed by the collapse of great clusters of stars at the center of galaxies have been detected as well as galactic supermassive black holes, which were produced during the early stages of galaxy formation. More speculatively, it is conjectured that primordial black holes, including tiny ones with small masses, may have been formed immediately following the Big Bang due to the prevailing high densities of matter.

Mathematically speaking, a black hole is a particular geometry of spacetime that, generically, has infinite curvature at its center. A star producing the Schwarzschild spacetime does not rotate. In 1916 this static model had been found by Schwarzschild only a few weeks after Einstein proposed his theory of gravity, so it was considered fairly easy to set it spinning (due to angular momentum conservation, actual black holes are expected to be spinning). However, many years

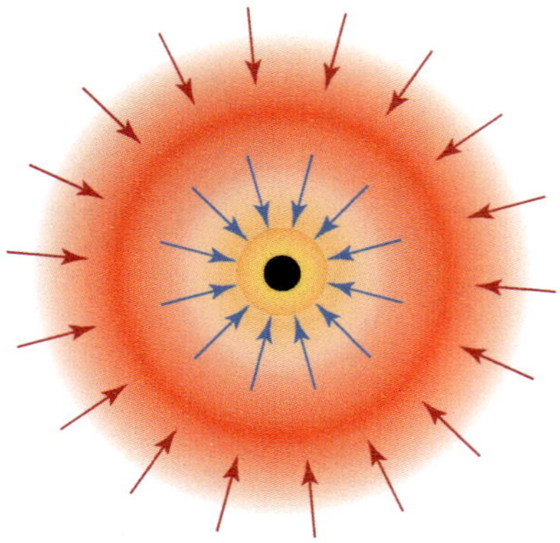

Fig. 5.5 A black hole being formed by the radial collapse of a star.

passed without any success. In 1963 physicist Roy Kerr adopted a shrewd strategy by tackling the problem indirectly and obtained the long-sought-after metric. The spherically symmetric and static black hole is given by the **Schwarzschild metric** while the cylindrically symmetric spinning black hole is given by the **Kerr metric**.

Consider the end point of stellar evolution of a very massive star. Collapsing stars have a wide variety of physical properties and one would think they should produce a large diversity of black holes. However, a series of theoretical results — principally the work due to Werner Israel, Brandon Carter, Stephen Hawking, and David Robinson — leads to the remarkable conclusion that a collapsing star loses all of its individual characteristics and ends up in a final state that is uniquely determined by only its mass and angular momentum (rate of rotation). Thus a (rapidly) spinning **Kerr black hole**, discussed later in Sec. 5.10, is the black hole which occurs in Nature.

Consider the special case for which a star has no angular momentum and collapses radially into a static and spherically symmetric black hole having mass M, as shown in Fig. 5.5. The essential feature of the black hole's horizon and infinite curvature are present in this example. The actual process for the **radial collapse** of a non-rotating star is very complicated, but the final result is very simple: the star forms a Schwarzschild black hole that is completely determined only by the mass M.

5.6 Black Holes: Dilating Time

We discussed in Sec. 4.4, using the Equivalence Principle, that time is dilated and clocks that are near gravitating masses run slower when compared to distant clocks. We had also raised the question whether the flow of time could be brought to a complete halt. We now examine how time flows in the Schwarzschild geometry. A distant observer has a clock that measures the flow of time specified by the

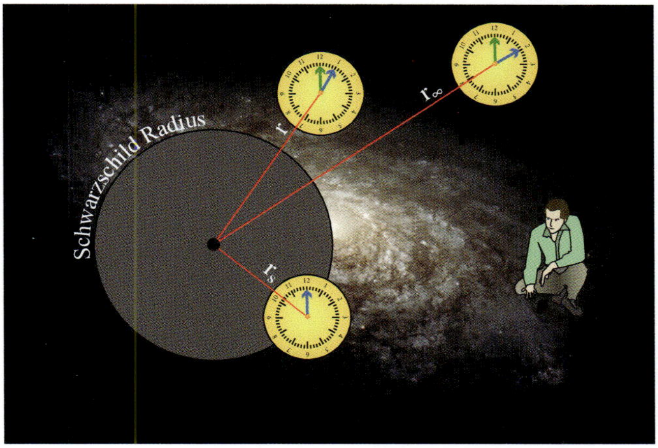

Fig. 5.6 To a distant observer at r_∞, time appears to slow down at a position r near black hole and comes to a complete halt at R_S, which is the event horizon of the black hole.

variable t. We ask, how does the flow of t compare with the flow of time near the black hole?

Consider an object stationary at some fixed value of r, θ, ϕ; since the object's position is fixed, we have $dr = d\theta = d\phi = 0$ and the spacetime interval ds^2 now refers entirely to the interval of time at fixed position (r, θ, ϕ) for a small change of distant time dt. **Proper time** is the time measured by an observer who is at a *fixed location* in space (in other words, proper time is the time as measured by a clock in a frame in which the clock is at rest). When we talk of time dilating and slowing down, one is *comparing* the proper time, measured by the observer who is at fixed position in space, to the time measured by an observer at a far distance from the black hole, as illustrated in Fig. 5.6. Note that both observers are in the same frame of reference.

To remind ourselves that we are measuring time at a fixed location, we label ds^2 as $-c^2 d\tau^2$, where τ is the proper time at (r, θ, ϕ). From Eq. (5.4), proper time is given by

$$ds^2 \equiv -c^2 d\tau^2 = -c^2 \left(1 - \frac{2GM}{c^2 r}\right) dt^2$$

$$\frac{d\tau}{dt} = \sqrt{1 - \frac{2GM}{c^2 r}} \quad \Rightarrow \quad \tau - \tau_0 = (t - t_0)\sqrt{1 - \frac{2GM}{c^2 r}}. \tag{5.6}$$

From Eq. (5.6), we see that time measured near the horizon, namely $\tau - \tau_0$, is dilated when compared to a distant clock, since

$$\tau - \tau_0 \leq t - t_0 \quad \text{for} \quad r \geq \frac{2GM}{c^2}.$$

In other words, the interval of time that has elapsed near the black hole, equal to $\tau - \tau_0$, is *less* than the time recorded by a clock at infinity, which shows time elapsed to be equal to $t - t_0$; see Fig. 5.6. *Time is dilated and slows down as one approaches the black hole.*

The black hole does something even more unexpected. Let the fixed distance, at which we are measuring proper time, approach the horizon of the black hole, namely

$$r \to 2GM/c^2 = R_S \ : \ \text{Schwarzschild radius.} \tag{5.7}$$

We have, from Eq. (5.6), the following

$$\tau - \tau_0 = 0; \quad r = R_S. \tag{5.8}$$

In other words, the flow of time at the radial distance of R_S from the black hole — as measured by a distant observer — comes to a complete halt and has stopped to flow. Recall that the spherical surface of radius R_S enclosing the center of the black hole is the event horizon of the black hole. All black holes have an **event horizon** with the remarkable property that the flow of time comes to a complete halt at the event horizon. No external observer can view any events that lie inside the event horizon.

Consider a star collapsing to form a black hole. A distant observer will never see the star's collapsing surface cross the Schwarzschild radius R_S. Instead, the observer will see the surface enclosing the star's core approaching the radius R_S more and more closely, with the radiation from the collapsing star being more and more redshifted as the surface approaches radius R_S; for radiation of wavelength λ emitted from the star and observed at a distance R, the observer will receive redshifted radiation with wavelength $\lambda + \Delta\lambda$, with

$$\Delta\lambda = \left[1 - \frac{R_S}{R}\right]^{-1/2} - 1.$$

Hence, as $R \to R_S$, we have infinite redshift with $\Delta\lambda \to \infty$. The distant observer can never observe the surface of the star crossing the Schwarzschild radius R_S.

The luminosity of the collapsing star, at time t after the collapse, rapidly goes to zero and is given by

$$L(t) \propto \exp\left(-\frac{t}{\tau}\right)$$

where, for a star with $M = 10M_\odot$, we have $\tau = 10^{-4}\,\text{s}$ (the subscript $\odot$ indicates that it is the value for the Sun, *i.e.* $M_\odot$ is the mass of the Sun). The collapsing star rapidly fades into darkness.

5.7 Black Holes: Bending Space

A measure of how space is curved is given by the **Riemann tensor** $\mathcal{R}$ discussed in Sec. 3.9. For the Schwarzschild black hole it can be shown that at the spacetime point (t, r, θ, ϕ) the curvature of the geometry given by the 'square' of the Riemann tensor, yields the following[3]

$$\mathcal{R}^2(t, r, \theta, \phi) = 12\left(\frac{GM}{c^2}\right)^2 \frac{1}{r^6}. \tag{5.9}$$

As expected, since the geometry is static and spherically symmetric, the curvature depends only on r — the distance from the center of the black hole — and does not depend on t, θ, ϕ.

A few important conclusions can be made. Firstly, the curvature has a singularity (infinity) at $r = 0$ where the mass M is located. Since the mass occupies zero volume, being point-like, it creates a singular geometry; similar to curvature, the energy density at the singularity is also infinite. Secondly, there is no curvature singularity at the event horizon, given by $r = 2GM/c^2 = R_S$, although there is a spurious singularity in the distance function ds^2, as seen in Eq. (5.4). However, the spurious singularity — called a **coordinate singularity** — does indeed have an important physical significance in that it reflects the existence of the event horizon.

The bending of space is given by the fact that, compared to the distance dr measured by a distant observer who has an ambient flat geometry, an observer located at distance r from the black hole measures a *larger* distance given by $dr/\sqrt{1 - R_S/r} > dr$. This is the reason that in Fig. 5.4 the embedded surface dips below the $z = 0$ flat plane. The distance function at $r = R_S$ cannot be discussed in the coordinate system that we have chosen and one has to use the so-called Eddington–Finkelstein **time-dependent coordinates** for which there is no singularity at $r = R_S$.

One can think of the geometry of the black hole as a series of concentric two dimensional surfaces — at distances labeled by r — enclosing the singularity, but with increasing curvature given by $\mathcal{R}^2 \simeq 1/r^6$. Note that the curvature of the two dimensional surfaces that enclose the Schwarzschild singularity are higher than S^2 since, from Eq. (3.21), for a sphere $\mathcal{R}^2_{\text{sphere}} \simeq 1/r^4$. The curvature of a sphere diverges as $r \to 0$; that is, a sphere of very small radius has very high curvature since the surface must wrap up very tightly. One can think of the black hole singularity as being equivalent to a very tightly wrapped surface of vanishingly small radius leading to a curvature that is infinite.

[3]Note that the dimension of $\mathcal{R}$ is $1/L^2$, where L is a length scale; similar to the analysis done in Noteworthy 4.2, it can be shown that the dimension of the right-hand side of Eq. (5.9) is $1/L^4$, as expected.

The black hole shows the effects of extreme spacetime curvature — a phenomenon unique to Einstein's theory that has no counter part in Newton's theory of gravity.

5.8 Event Horizon: Black Hole's Boundary

The boundary of the black hole is the event horizon; recall that at the event horizon the flow of time — as measured by observers far from the black hole — comes to a halt.

The best way to examine the nature of the event horizon is to send an astronaut *into* the black hole and follow his trajectory. We know that the astronaut will follow a geodesic as he travels in the curved geometry of the black hole. Let us consider the special case of an astronaut having fixed θ, ϕ and falling in radially. Unlike Newtonian mechanics where one needs to only specify the astronaut's position as a function of Newtonian time, for motion in curved spacetime the astronaut occupies both a position in space and carries his own clock as well. The geodesic equations fix both the proper time of the astronaut and the spacetime trajectory that the astronaut follows.

Consider an astronaut starting at initial time t_0 from position r_0 and with zero velocity. Let r_0, r — the astronaut's initial and current position respectively — both be near the event horizon R_S. It can be shown that

$$r - R_S = (r_0 - R_S)\exp\{-c^3(t - t_0)/(2GM)\}; \quad r_0 > R_S.$$

Hence, in the frame of the distant observer, the astronaut takes infinite amount of time to reach R_S since

$$t \to \infty \quad \Rightarrow \quad r \to R_S.$$

In the frame of the distant observer the astronaut is forever approaching the event horizon, getting closer and closer, as shown in Fig. 5.7. In other words, for the distant observer the astronaut never crosses the event horizon.

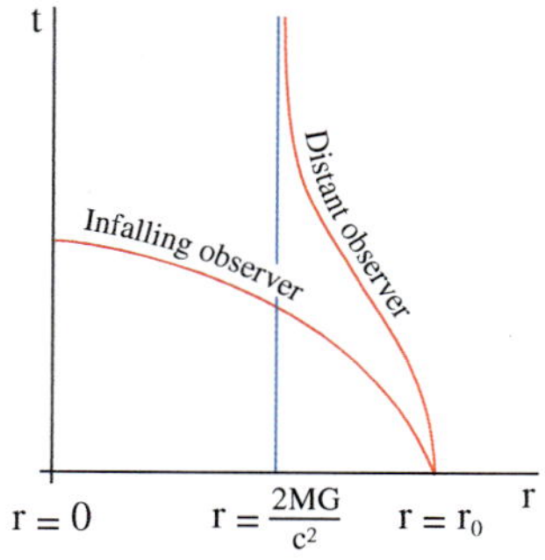

Fig. 5.7 The time recorded by two different observers of an astronaut falling into the black hole, one observer who is stationary and far from the black hole and the other observer is the astronaut who is falling into the black hole.

Let us examine what is happening in the frame of the astronaut. We need to compute the proper time for the case of the astronaut changing his position, in going from position r_0, at proper time τ_0, to position r at proper time τ. It can be shown that, if the astronaut falls along a geodesic, proper time is given by

$$\tau - \tau_0 = \frac{2}{3\sqrt{2GM}}\left(\sqrt{r_0^3} - \sqrt{r^3}\right). \tag{5.10}$$

The proper time interval $\tau - \tau_0$ is the time measured by the astronaut's watch (frame of reference) as he moves from postion r_0 to position r and the trajectory is shown in Fig. 5.7.

The formula for proper time given in Eq. (5.10) is valid inside the horizon, and all the way to the singularity at $r = 0$. In the frame of the astronaut who is falling in, he crosses the event horizon and hits the black hole's singularity at the origin $r = 0$ in finite proper time $\tau_{\text{singularity}}$; from Eq. (5.10), we have

$$\tau_{\text{singularity}} = \tau_0 + \frac{2}{3\sqrt{2GM}}\sqrt{r_0^3}.$$

In other words, according to the time kept by the astronaut's watch he hits the singularity in finite time $\tau_{\text{singularity}}$.

For a black hole with $M = 3M_\odot$, the Schwarzschild radius $R_S = 8.5\,\text{km}$. If the astronaut starts at a distance of $r_0 = 30\,\text{km}$ and at $\tau_0 = 0$, he will hit the singularity in $\tau_{\text{singularity}} = 0.124$ ms.

Once an object crosses the event horizon there is no turning back — the object inevitably hits the singularity. Consider the flow of a river heading towards a waterfall, as shown in Fig. 5.8. The analog of the astronaut heading for the black hole is a fish that is swimming towards the waterfall, as shown in Fig. 5.9. At the point of no return, the river has a perfectly normal flow — analogous to the curvature of spacetime being finite at the event horison. On crossing the point of no return there is no turning back for the fish — similar to the astronaut being unable to escape the singularity once he crosses the event horizon — and the fish going over and hitting the waterfall's bottom is analogous to the astronaut hitting the singularity.

5.8.1 *Stationary observer*

Note that outside the horizon we can have a stationary observer with $dr = d\theta = d\phi = 0$; to stay stationary the observer needs to exert a radial force against the inward acceleration which, for a particle with mass m, is given by

$$f_r = \frac{GMm}{r^2\sqrt{1 - \frac{2GM}{c^2 r}}}. \tag{5.11}$$

Note as $r \to 2GM/c^2$ the observer approaches the horizon and the force required increases, becoming infinite at the horizon. Inside the horizon there can be no

Fig. 5.8 A waterfall is analogous to the event horizon of a black hole.

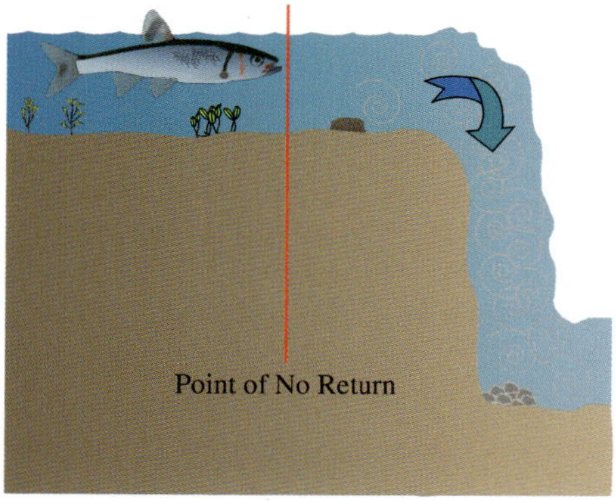

Fig. 5.9 If the fish gets too close to the edge of the waterfall, it will no longer be able to return.

stationary observers but rather all objects, including light, are heading towards the singularity. Hence, to describe the black hole inside the horizon, one needs to use a time-dependent in-falling coordinate system instead of the Schwarschild coordinates (t, r, θ, ϕ).

5.9 Permanent Trapping of Light

Once it has crossed the event horizon, no object, not even light, can escape from the interior of a black hole. The analysis of the motion of light is slightly different

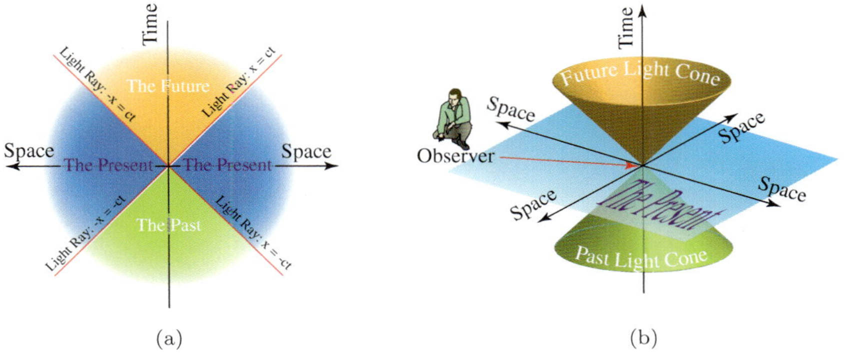

Fig. 5.10 Light cones, the present, the future and the past.

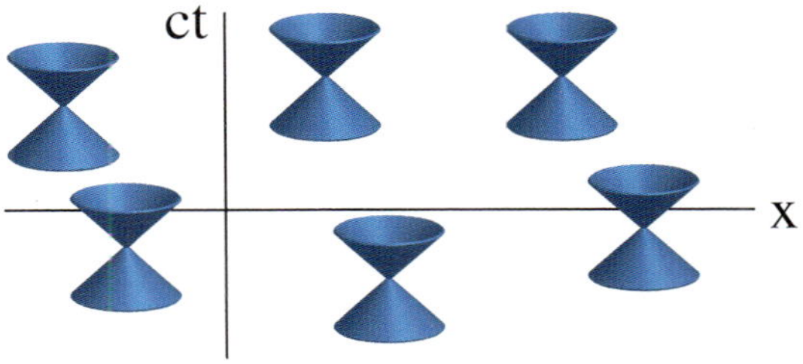

Fig. 5.11 Light cones in Minkowski space.

from an object with mass since light is (locally) traveling at the speed of light and hence travels on a null geodesic.

Since the speed of light is fixed to be c, the motion of light can be represented by light cones. For simplicity consider one space and one time dimension. Since light can travel either to the left or to the right, the two possible trajectories are given by $x = \pm ct$ as shown in Fig. 5.10(a). In higher dimensions light travels on a light cone, specified by $\sqrt{x^2 + y^2 + z^2} = \pm ct$, as shown in Fig. 5.10(b).

In flat Minkowski spacetime light cones at all points of spacetime are identical, as shown in Fig. 5.11; in contrast, curved spacetime manifests itself by the tilting of the direction of light cone and by its squashing or opening from point to point.

Since the black hole has perfect spherical symmetry it is sufficient to consider light that is traveling radially — either towards the black hole or away from it. As shown in Fig. 5.3, there are three distinct zones of the black hole's geometry, namely Region II consisting of the event horizon and Regions I and III that are outside and inside the event horizon, respectively.

The behavior of light near a black hole depends on how it is observed. Two frames of reference are particularly important, namely a stationary frame (observer) at infinity with coordinates (t, x, y, z) and an in-falling frame (of the astronaut). These two frames will observe two very different manifestations of the same black hole.

5.9.1 *Light cones: Stationary observer*

We can ask the following question, namely: what is the behavior of light as seen by a stationary observer, who is at a great distance from the black hole, and who never crosses the horizon? Figure 5.12(a) shows the light cones of a black hole in Region I, which is outside the horizon, and for Region II, which is the horizon. The stationary observer cannot receive any signal from Region III, namely from inside the horizon, and hence no light cones are shown inside the horizon.

Noteworthy (optional content) 5.2: Null Geodesics and Speed of Light

---

Light travels on null geodesics for which

$$ds^2 = 0.$$

This equation is valid for all observers and hence is invariant (unchanged) under coordinate transformations. To determine the light cones that are seen by an observer at different distances from the black hole, we need to choose a coordinate system in which the observer can be described. Consider the fixed (time-independent) coordinate system (t, r, θ, ϕ) that is used for writing the Schwarzschild metric in Eq. (5.4); in this coordinate system, the distant observer can be placed at a large value of r. Since the metric is spherically symmetric, consider geodesics that are radial, namely, for which θ, ϕ are fixed and hence $d\theta = d\phi = 0$; we then

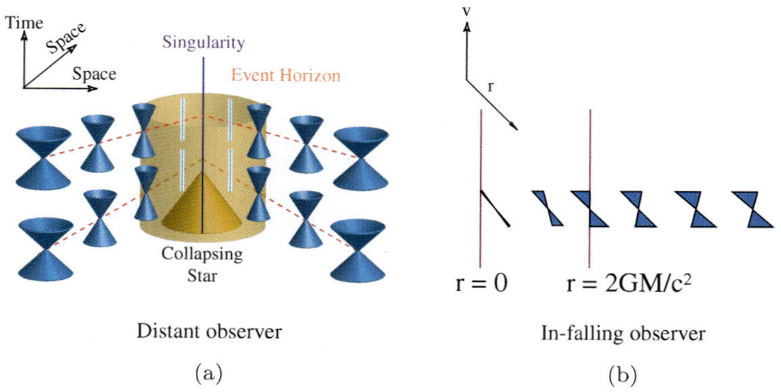

Fig. 5.12 (a) Light cones near a black hole seen by a stationary and distant observer. Light cones collapse to a line at the horizon since light is stationary at the horizon. (b) Light cones as seen by an observer falling into a black hole, drawn for the oblique coordinate system (v, r). Inside the horizon, light traveling in all directions is heading for the singularity.

have, from Eq. (5.4)

$$0 = ds_F^2 = -c^2\left(1 - \frac{2GM}{c^2 r}\right)dt^2 + \frac{dr^2}{\left(1 - \frac{2GM}{c^2 r}\right)}$$
$$\Rightarrow \frac{dr}{dt} = \pm c\left(1 - \frac{2GM}{c^2 r}\right) \quad : \text{ velocity of light at position } r. \tag{5.12}$$

dr/dt is the velocity of light since a change in the position dr of the light ray — on the *null geodesic* $ds^2 = 0$ — corresponds to a change in time dt. Note dr/dt is the velocity of light as observed by the *stationary frame*.

Equation (5.12) yields, in Region I and at a point r, the expression for the velocity of light dr/dt; for light traveling *away* from the black hole, the velocity is positive and the velocity of light heading *towards* the black hole is negative. Far from the black hole, the light cones are obtained by letting r become large so that Eq. (5.12) reduces to $dr/dt = \pm c$; this means that far from the black hole, spacetime is flat and light travels on trajectories $r = \pm ct$. However as one approaches the black hole the speed of light, as observed by the distant observer, slows down as given in Eq. (5.12) and the light cone starts to narrow; these light cones are shown in Fig. 5.12(a).

Region II is the horizon for which $r = R_S$ and we have

$$\frac{dr}{dt} \to 0 \ \text{ for } \ r \to R_S = \frac{2GM}{c^2}$$

and the velocity of light has come to *zero* — as observed by a distant frame of reference. The distant observer cannot see light crossing the horizon and hence, as expected, concludes that light comes to a halt at the event horizon. In other words, light becomes trapped on the event horizon of the black hole. The light cones *collapse* to a *single line* since light is *stationary* at the horizon — and is shown in Fig. 5.12(a).

In Region I, the distant observer sees light slowing down as it approaches the horizon and finally coming to a complete halt at Region II, for which $r = R_S$.[4] This is the reason that the flow of time also stops since there is no movement of any kind. Light is permanently trapped on the event horizon. The distant stationary observer cannot see any of the spacetime points inside Region III — which appear as a *hole* in space, with nothing going in and nothing coming out. Hence, the name of *black hole* to describe this phenomenon.

In summary, the term *black hole* exemplifies two cardinal features of the Schwarzschild singularity: firstly, spacetime beyond the horizon is simply absent

[4]The slowing down of light, as seen by a distant oberver, is given by Eq. (5.12).

to an external observer, that is, there is a hole in space and secondly, the term black states that nothing emerges from this hole in space.

One may wonder how can we talk of the velocity of light as being anything but equal to the constant c. Recall that in curved spacetime the velocity of light is only defined *locally*, namely at a spacetime point. Hence, the velocity of light at the location of the distant stationary observer is indeed equal to c. The distant stationary observer has a complete frame of reference in which he can receive information about the time and length scales at *every* point of spacetime. It is in this frame of reference, in which the observer has a fixed position, that the velocity of light changes from point to point and is a function of distance from black hole r — being zero at the horizon of the black hole.

5.9.2 *Light cones: In-falling observer*

The discussion in Sec. 5.9.1 was confined to Regions I and II — outside and at the horizon — but could go no further: time-independent coordinates were used and these cannot describe events inside the horizon.

The Regions I, II and III can be observed by an in-falling frame of reference (observer) who crosses the horizon. For an in-falling observer who is crossing the horizon, a *time-dependent* coordinate system must be used. In a time-dependent in-falling coordinate system, there is no event horizon. The in-falling frame (observer) will not experience the horizon and will continue through, and hit the black hole singularity in *finite* proper time.

We choose the in-falling **Eddington–Finkelstein** coordinate system, which only describes observers free-falling into the black hole.[5] The coordinates (t, r) used in Eq. (5.4) are transformed to the Eddington–Finkelstein coordinate system given by (v, r); the radial coordinate r is the same as in (t, r) and the time coordinate is replaced by a new timelike coordinate v, which is related to coordinates (t, r) by the following transformation

$$v = t + r + \frac{2GM}{c^2}\log\left|\frac{c^2}{2GM}r - 1\right|; \quad r = r.$$

Note that lines of constant v are at an angle of 45° to lines of constant r, and the oblique coordinates are drawn as such in Fig. 5.12(b).

The coordinate system (v, r) is the appropriate one to describe an in-falling observer: there is no coordinate singularity at the event horizon, unlike the case of the coordinate system (t, r) used in Eq. (5.4), and we can describe the light cones seen by an in-falling observer all the way to the singularity. An in-falling observer, heading for the black hole, will observe the light traveling along v = constant

[5]The out-falling Eddington–Finkelstein coordinate system only describes observers who are moving *away* from the black hole and is only defined for observers located *outside* the horizon.

trajectories; the light cones shown using the in-falling Eddington–Finkelstein (v, r) coordinates are given in Fig. 5.12(b) and discussed below.

- For Region I, outside the event horizon, the light cones point 'upwards', with one direction in which outgoing light can move *away* from the black hole and another direction for ingoing light that is heading towards the black hole. As shown in Fig. 5.12(b), as the light cones approach the horizon, the light cone is tilted towards the horizon.
- Region II is the event horizon. At the horizon, one axis of the light cone is *aligned* along the horizon: the ingoing light can travel towards the singularity but the outgoing light now remains *fixed* at position of $R_S = 2GM/c^2$ — as shown in Fig. 5.12(b). This is the reason that, at the horizon, the in-falling observer loses contact with Region I outside the horizon since even light moving radially outwards from the black hole center is trapped on the event horizon.
- For Region III, inside the event horizon, both the 'ingoing' and 'outgoing' light rays are heading towards the black hole singularity. The light cone keeps tilting and being squashed in such a manner that, at the origin of the black hole, the light cone is squashed into a *line* — as shown in Fig. 5.12(b). In other words, all light, and all other objects as well, that are inside the horizon inevitably hit the singularity.

In summary, the observer who is falling into the black hole has a frame of reference that can, in principle, observe all points of the spacetime. In the in-falling observer's frame, the light cones outside the horizon will tilt; at the horizon, the outgoing light gets aligned with the horizon — and light becomes trapped on the horizon. Inside the horizon, the light cone tilts further so that all light rays are heading towards the singularity. Figure 5.12(b) shows the light cones on the horizon, as well as outside and inside the horizon.

The event horizon is a one-way 'membrane' through which one can always cross over into the interior of the black hole but one can never go the other way and leave the interior of the black hole. Only those objects that cross the horizon — and are destined to hit the singularity — can witness the singularity. The singularity inside the event horizon can never be observed directly by the observers outside the horizon: space only up to the horizon can be observed by them. In fact there is a famous **cosmic censorship** hypothesis that states a 'naked singularity', which is a singularity without an event horizon, cannot exist in Nature.

5.10 Spinning Black Holes

All realistic cases of a gravitational collapse of stars — due to the conservation of angular momentum — must lead to a black hole that is *spinning.* A major unanswered question for many years was whether angular momentum would cause instabilities in the event horizon of a black hole — leading to the absence of the gravitational singularity; in effect, it was not known if a black hole that was spinning could exist. It was only in 1963 that Roy Kerr exactly solved Einstein's field equations and obtained a solution for a spinning black hole.

Most spinning galactic black holes, due to their enormous masses, are only weakly coupled to matter outside the black hole and, to a good approximation, may be considered to be stationary and isolated.

Nobel laureate astrophysicist and black hole theorist **S. Chandrasekhar** has the following to say about the Kerr solution. "To the extent they (black holes) may be considered to be stationary and isolated, to that extent, they are all, every single one of them, described *exactly* by the **Kerr solution**. This is the only instance that we have an exact description of a macroscopic object. (...) the only elements in the construction of black holes are our basic concepts of space and time; they are, thus, almost by definition, the most perfect macroscopic objects there are in the Universe. And since the general theory of relativity provides a unique two-parameter (...) description, they are the simplest objects as well."

No other macroscopic object has an exact description such as the Kerr black hole; this is because all macroscopic objects that we encounter in Physics are governed by a variety of forces that are approximated by a diverse collection of theories; in particular, as long as the microscopic atomic and subatomic constituents are important, one needs to have virtually infinitely many variables for an exact description of a macroscopic object.

Figure 5.13(a) is an artistic representation of the swirling of spacetime caused by a spinning mass; dark lines indicate the trajectory of particles in the swirling geometry. The effect of spinning on a spherical body makes it into an ellipsoid, and is schematically shown in Fig. 5.13(b).

A **Kerr black hole** is completely fixed by only two parameters, namely its mass M and angular momentum J. In Newton's theory of gravitation a spinning spherical mass has the same gravitational potential as a static one. In contrast, Einstein's theory predicts that a spinning mass creates a rotating gravitational field that causes the very fabric of spacetime to spin and results in the 'dragging of the

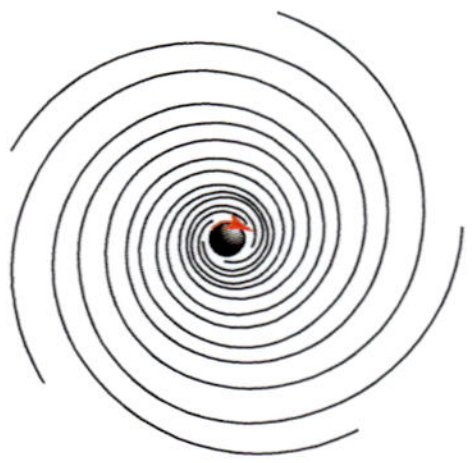

(a)

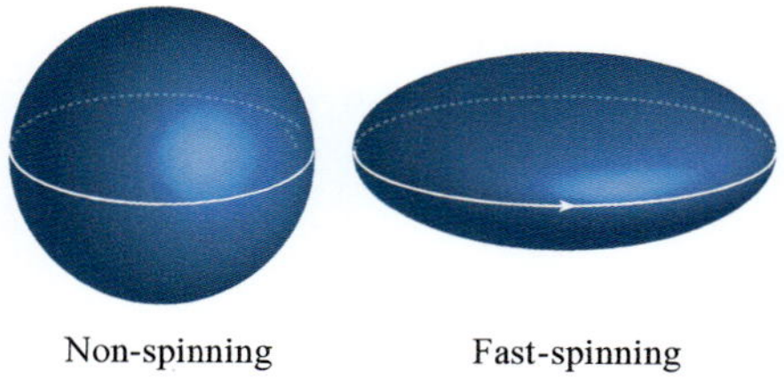

(b)

Fig. 5.13 (a) Spacetime swirls near a spinning mass. (b) A spinning sphere becomes an ellipsoid.

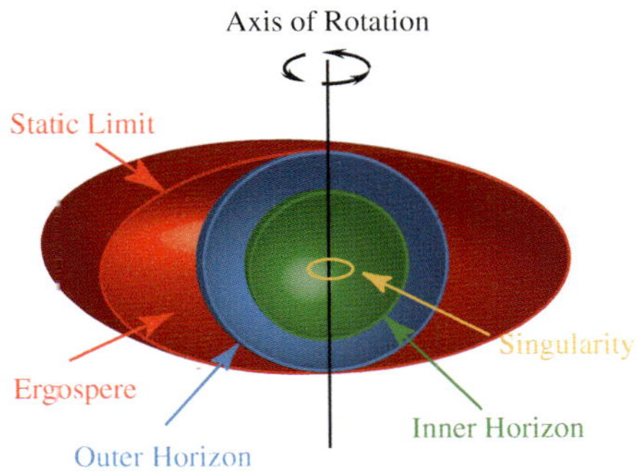

Fig. 5.14 Spinning Kerr black hole, with its three distinctive surfaces, and the domain of the ergosphere. The geometric singularity is located at the spinning ring.

inertial frame' along its angular direction of motion: the curvature of spacetime due to a spinning mass will cause any object that comes close enough to rotate, even though there was no external torque applied to it.

5.10.1 *Kerr black hole*

Consider a total mass M in the shape of a perfect circle (ring) with a radius of $a = J/(cM)$ that is spinning with angular momentum J.[6] The Kerr black hole provides a description of the spinning ring, which generates a spacetime geometry with angular momentum J. The mass M is spinning in the equatorial plane, as shown in Fig. 5.14, and at a radius a from the center.

Both the Kerr and Schwarzschild black holes are stationary, meaning that the geometry of the black holes does not change with time. The Schwarzschild black hole is *static*, meaning that there is no movement at all; in contrast, the Kerr black hole is dynamical since the mass generating the geometry of the black hole is spinning with angular momentum J.

As discussed in Sec. 4.10, an accelerating mass that has perfect cylindrical symmetry does not emit gravitational radiation; for example, a mass in the form of a disk, if made to spin, will not radiate. Therefore, although spinning involves acceleration, the Kerr black hole does not radiate gravitational waves because the black hole is generated by the spinning of a ring that has perfect cylindrical symmetry.[7] Hence, since no gravitational radiation is generated by the spinning ring, we expect to find that Kerr geometry is stationary, namely does not explicitly depend on time.

---

[6]The parametrization of a is chosen for later convenience.

[7]In Noteworthy 5.7, a perfectly circular spinning ring with charge is discussed. For this case as well, there is also no gravitational radiation due to perfect cylindrical symmetry of the charged spinning ring.

The geometry of the Kerr black hole is shown in Fig. 5.14 and is more complicated than the Schwarzschild case. Unlike the single event horizon of the Schwarzschild black hole, the Kerr black hole has the following *three different surfaces* and a domain of spacetime called an **ergosphere**.

- A stationary surface, also called the **static limit** for reasons to be explained in Sec. 5.11, that is outside the outer horizon.
- Two exactly spherical event horizons — called the inner and outer event horizons. The stationary surface has an 'equatorial bulge' and just touches the outer horizon at the north and south poles.
- Between the stationary surface and the outer event horizon is the ergosphere. The ergosphere is discussed in Sec. 5.11.

The ellipsoidal shape of the stationary surface, which is the boundary of the ergosphere, is intuitively obvious since the spinning ring causes an equatorial bulge; but what is a bit more mysterious is that the inner and outer horizons are *perfectly spherical* S^2 surfaces, even though the spinning ring does not have spherical symmetry.

The spacetime structure of the Kerr black hole is quite unlike the Schwarzschild case. On crossing the outer Kerr event horizon the role of space and time is switched and is again switched on crossing the inner Kerr event horizon — thus restoring the original role of space and time that holds outside the outer horizon. Hence, inside the inner horizon, the role of space and time is the same as it is far from the black hole.

Unlike the Schwarzschild singularity, which is localized at the *point* where the mass M is located, the curvature singularity for the Kerr metric is located on the *circle* occupied by the mass M, which is inside the inner event horizon and in the equatorial plane at the fixed radius of $a = J/(cM)$, as shown in Fig. 5.14. One can think of the point singularity of the Schwarzschild black hole smoothly transforming into a circle for a spinning black hole. The mass is located on the circular singularity; since the black hole spins about the vertical axis the equatorial plane is unchanging and the circular singularity is stationary. The center of the Kerr black hole is a regular spacetime point.

The Kerr geometry smoothly reduces to the Schwarzschild case for $J = 0$; taking the limit of $J \to 0$ leads to the inner event horizon collapsing to the origin and the static surface and outer event horizon coalescing to form the Schwarzschild event horizon; the Kerr singularity concentrated on the circle smoothly deforms to the Schwarzschild point singularity.

Noteworthy (optional content) 5.3: Black Hole Singularities

For the Schwarzschild black hole, time t outside the event horizon has the distinctive property that one cannot go back in time but must always go forward; in contrast an object can go back and forth in the r direction. Inside the event horizon, due to the singularity at the origin an object must always move to smaller and smaller values of r. The singularity is called **spacelike** since this singularity cannot be avoided. Every geodesic inside the horizon, including the null geodesics, must end up at the singularity.

The singularity of the Kerr black hole is called a **timelike** singularity; null geodesics do not necessarily have to terminate at the timelike singularity. In general, there are geodesics inside the horizon that can completely avoid the singularity of rotating black holes.

The timelike singularity of the Kerr geometry implies that an observer inside the inner horizon can avoid the singularity by navigating to avoid it.[8] However, the observer cannot return to the Universe he came from due to the event horizons; there is a view that the observer inside the inner horizon can completely escape from the black hole but would end up in a different Universe connected to our Universe by the Kerr black hole. In this sense, one can think of the Kerr black hole as a wormhole connecting two different Universes.

The distinctive feature of spinning spacetime is a phenomenon called the **Lense–Thirring effect**. A gyroscope is similar to a spinning flywheel — see Fig. 5.15. In a spinning spacetime the Lense–Thirring effect predicts the precise amount by which the gyroscope must precess. Precession is the change in the direction of angular velocity keeping its magnitude fixed. The Lense–Thirring effect is very small, one part in a trillion, and hence one needs great precision to be able to measure it.

The Earth is a spinning sphere that completes one revolution per day and its external geometry is described by the Kerr metric. The precession of a gyroscope

Fig. 5.15 Spinning gyroscope.

[8] Recall that since the Schwarzschild singularity is spacelike, all observers who fall through the event horizon must hit the singularity in a finite amount of (proper) time.

due to its orbiting the Earth is 6.5″/year with a further Lense–Thirring contribution of 0.042″/year due to the dragging of inertial frames caused by the spinning of the Earth. The Gravity Probe B, launched to measure the Earth's Lense–Thirring effect, completed its mission in 2005 and verified Einstein's predictions to an accuracy better than 1%.

Noteworthy (optional content) 5.4: The Kerr Metric

For completeness, the Kerr metric is given below. In the **Boyer–Lindquist coordinates** the geometry of spacetime generated by a mass M in the shape of a perfect circle, spinning about the z-axis with angular momentum J, is given by

$$\begin{aligned} ds^2 &= -c^2dt^2 + \frac{\rho^2}{\Delta}dr^2 + \rho^2 d\theta^2 + (r^2+a^2)\sin^2\theta d\phi^2 \\ &\quad + \frac{2GMr}{c^2\rho^2}(a\sin^2\theta d\phi - cdt)^2 \\ a &= \frac{J}{cM}; \quad \rho^2 = r^2 + a^2\cos^2\theta; \quad \Delta = r^2 - 2\frac{GM}{c^2}r + a^2. \end{aligned}$$

Note that r is *not* the radial coordinate of the spherical polar coordinates. The Cartesian coordinates are given by

$$\begin{aligned} x &= \sqrt{r^2+a^2}\sin\theta\cos\phi \\ y &= \sqrt{r^2+a^2}\sin\theta\sin\phi \\ z &= r\cos\theta. \end{aligned}$$

The two spherical event horizons have radii $r_0^{\pm}$ given by the two solutions of

$$\Delta(r_0) = 0 \quad \Rightarrow \quad r_0 = r_0^{\pm}. \tag{5.13}$$

Far from the black hole $r \to \infty$ and the asymtotic flat spacetime, given by $ds^2 \to -c^2dt^2 + dr^2 + r^2d\theta^2 + r^2\sin^2\theta d\phi^2$, is the Minkowski metric in spherical polar coordinates given by t, r, θ, ϕ, as given in Eq. (3.13).

5.11 Extremal Kerr Black Hole and Ergosphere

The angular momentum of a Kerr black hole satisfies the inequality

$$J \le \frac{GM^2}{c}.$$

For $J > GM^2/c$ the **Kerr geometry** has a singularity without any event horizon, namely a naked singularity, which is forbidden by the cosmic censorship hypothesis. Almost all Kerr black holes in astrophysics are **extremal**, namely with angular momentum given by

$$J_{\max} = GM^2/c \ : \ \text{extremal Kerr black hole}$$

for the following reason: matter keeps spiraling into the black hole increasing its angular momentum until the extreme value is reached. For a black hole of one solar mass, with a circumference of 18.5 km, the extremal black hole has the maximum spin rate of 0.000062 s per revolution.[9] This gives a tangential speed at the outer horizon of about 299,800 km/s, close to the speed of light.

To simplify our discussion we henceforth discuss only the extremal Kerr black hole: the two event horizons coalesce to a *single* event horizon, as shown in Fig. 5.16, and is perfectly spherical with the radius given by

$$r_0^{\pm} = GM/c^2 = r_0 \ : \ \text{extremal Kerr black hole.} \tag{5.14}$$

The space and time coordinates in the interior of the extremal Kerr black hole have the same significance as outside the event horizon and the circular singularity is timelike. The effect of a rotating black hole is most clearly seen by examining its effect on a stationary observer. An observer far from a rotating black hole can be stationary using the power of, say, a rocket. For the Schwarzschild case, a stationary observer is possible all the way up to the horizon of the black hole. In contrast, stationary observers are forbidden inside the *ergosphere* — which defines the region of space between the stationary surface and the outer event horizon.

Inside the ergosphere the observer is dragged by the spinning of spacetime and the observer *cannot* be stationary, but, rather, *must* rotate with the black hole. This is the reason that the stationary surface is also called the static limit, to indicate that it is the boundary of the region of Kerr spacetime that allows for stationary observers who carry a static coordinate system.

The impossibility of having a stationary observer inside the ergosphere is a reflection of the singular nature of the Kerr black hole due to its spinning, whereas the horizons are a reflection of the singular nature of the black hole for observers approaching the black hole radially.

The ergosphere for an extremal Kerr black hole has a radius of GM/c^2 at the poles and a radius of $2GM/c^2$ at the equator, as shown in Fig. 5.16. For

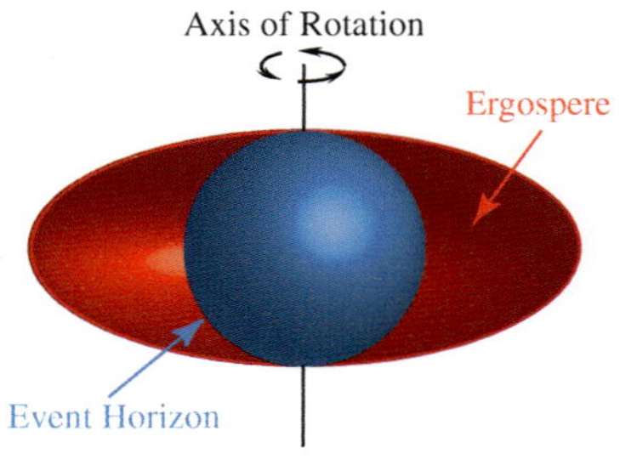

Fig. 5.16 Single event horizon and ergosphere of an extremal spinning (Kerr) black hole.

[9]Note in one second the black hole has 16,126 revolutions!

a rotating observer inside the ergosphere, at fixed r, θ, there is a minimum and maximum allowed angular velocity. The frame-dragging becomes more extreme as one approaches the event horizon until, at the horizon, all observers orbit with an angular velocity of the horizon, which is equal to $Jc^4/(4G^2M^2)$.

5.12 Energy from a Kerr Black Hole

The radius of the horizon of an extremal spinning Kerr black hole, given in Eq. (5.14), is equal to *half* the radius of a static black hole with the same mass, as given in Eq. (5.7). As discussed in Sec. 5.16, the entropy of a (classical) black hole must always increase. The entropy of a static black hole, as given in Eq. (5.16), is proportional to the area of the horizon. A Kerr black hole **increases** its entropy as its spinning slows down by increasing the radius of its event horizon. Consequently, due to energy conservation, the energy carried by the angular motion of a black hole can, in principle, be extracted from a spinning black hole by slowing it down.

Mass-energy can be converted to radiation using the gravitational effect of a black hole in a manner similar to how radiation is released by nuclear fusion. Consider fusing two nuclei with masses M_1 and M_2 into a new nucleus with mass M; due to nuclear binding energy, conservation requires that fusion must release radiation-energy equal to $(M - M_1 - M_2)c^2$. In the same manner one can send an approximately free particle into a close orbit (outside the horizon) around a black hole. The particle becomes strongly bound to the black hole and has negative gravitational binding energy, which in turn requires that an equal amount of radiation-energy be released as shown in Fig. 5.17.

A consideration of black hole geometry shows that for a particle with mass m placed in a stable innermost orbit around the *Schwarzschild* black hole, about 6% of its mass-energy, namely $0.06\,mc^2$, can be extracted as radiation. Furthermore, for a *Kerr* black hole, up to 42% of the mass-energy of the particle, that is, $0.42\,mc^2$, can be extracted using this process.

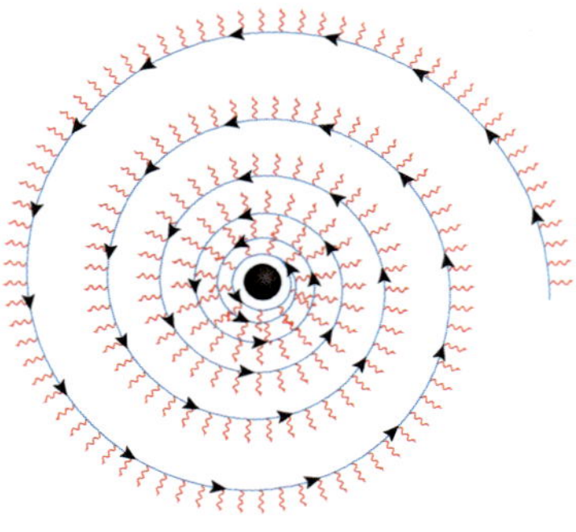

Fig. 5.17 A particle falling into a black hole radiates off energy.

Compared to nuclear fusion, which at best converts about 1% of the particle's mass-energy into radiation, black holes provide a far more efficient and powerful means of converting the mass-energy of a particle into radiation.[10]

Another form of energy that can be extracted from a black hole is the rotational energy of the extremal Kerr black hole that lies entirely in the region of (spinning) spacetime that is *outside* the (single coalesced) event horizon. It is estimated that an extremal Kerr black hole can have 29% of its energy in the form of rotational energy. Energy is extracted by reducing the black hole's angular momentum. The amount of energy that can be extracted from the Kerr black hole is given by $(M - M_0)c^2$, where M_0 is given by

$$M^2 = M_0^2 + \frac{c^2 J^2}{4G^2 M_0^2}.$$

The Penrose process extracts rotational spin energy of the Kerr black hole using the motion of particles; in contrast, the Blandford–Znajek mechanism does so using electromagnetic effects. The tremendous outpouring of energy from quasars is thought to be due to the extraction of spin energy from supermassive Kerr black holes, at the center of active galaxies, via the Blandford–Znajek mechanism.

Extracting all the rotational energy of a Kerr black hole yields a static Schwarzschild black hole with mass M_S and zero angular momentum.

5.13 Reissner–Nordstrom Charged Black Hole

The Kerr spinning black hole is of great astrophysical importance since most black holes in the Universe have some angular momentum. Charged black holes have not yet been observed but are nevertheless worth studying to understand how gravity is affected by charged particles. Einstein's field equations, which are non-linear partial differential equations, are notoriously difficult to solve exactly and, in general, need to be studied numerically; however, it is often only the exact solutions that allow one to understand new qualitative features of the solution that are almost impossible to see in numerical solutions or analytical approximations. For these reasons, the Reissner–Nordstrom charged black hole, which is an exact (static) solution of Einstein's field equation — for a black hole that has a charge Q in addition to the mass M — is of great theoretical interest.

---

[10]The burning of stars is powered by the fundamental nuclear fusion reaction of four protons into a helium nucleus and is given by

$$4\,{}^1\mathrm{H} \rightarrow {}^4\mathrm{He} + 26.72\ \mathrm{MeV\ in\ radiation}.$$

The mass of a proton is 938 MeV; hence the fraction of mass-energy released in fusion is $26.72/(4 \times 938) \simeq 0.01$, which is about 1%.

Noteworthy (optional content) 5.5: Reissner–Nordstrom Metric

To determine the gravitational effects of a point charge Q, the Maxwell field A_μ describing a charged particle has to be coupled to the Einstein gravitational field $g_{\mu\nu}$. For completeness, the Reissner–Nordstrom metric is given below. Let spacetime have the spherical polar coordinates given by (t, r, θ, ϕ); the geometry of spacetime generated by a mass M that carries charge Q is given by the metric

$$ds^2 = -\left(1 - \frac{r_S}{r} + \frac{r_Q^2}{r^2}\right) c^2 dt^2 + \frac{1}{1 - \frac{r_S}{r} + \frac{r_Q^2}{r^2}} dr^2 + r^2\left(d\theta^2 + \sin^2\theta d\phi^2\right)$$

$$r_S = \frac{2G}{c^2} M; \quad r_Q^2 = \frac{G}{4\pi\epsilon_0 c^4} Q^2.$$

Note that G is Newton's gravitational constant and ϵ_0 is the permittivity of empty space. The coordinates (t, r, θ, ϕ) are fixed spherical coordinates that are measured by a stationary observer at an infinite distance from the black hole's center and are given by

$$x = r \sin\theta \cos\phi$$
$$y = r \sin\theta \sin\phi$$
$$z = r \cos\theta.$$

Far from the black hole $r \to \infty$ and the asymptotic flat spacetime, given by $ds^2 \to -c^2dt^2 + dr^2 + r^2(d\theta^2 + \sin^2\theta d\phi^2)$, is the Minkowski metric in spherical polar coordinates, given in Eq. (3.13).

Only the radial component of the Maxwell field is non-zero and is given by

$$A_r = \frac{Q}{r}.$$

In contrast to the Schwarzschild black hole, the solution has a number of new features. In particular, there are now *two* event horizons (null surfaces) on which light is trapped. The two event horizons for the Reissner–Nordstrom case are given for values of $r = r_H$ that are determined by the following equation

$$1 - \frac{r_S}{r_H} + \frac{r_Q^2}{r_H^2} = 0$$

$$\Rightarrow r_{H\pm} = \frac{1}{2}\left(r_S \pm \sqrt{r_S^2 - 4r_Q^2}\right).$$

The Reissner–Nordstrom geometry has three distinct cases.

- $r_S > 2r_Q$; there are two event horizons as shown in Figs. 5.18 and 5.19. The greater the charge Q of the black hole, the closer are the two horizons.

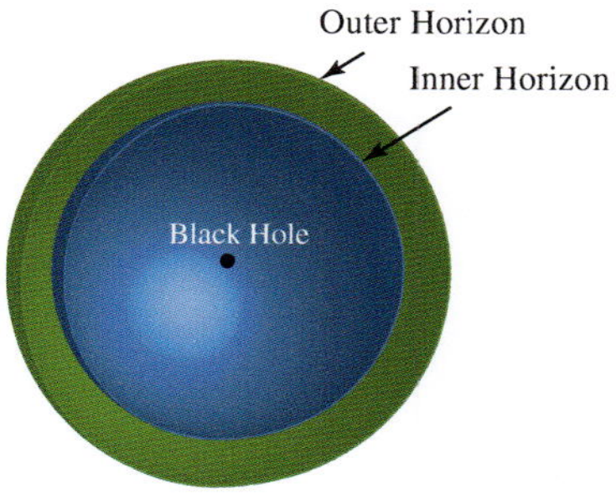

Fig. 5.18 The outer and inner event horizons of the Reissner–Nordstrom black hole.

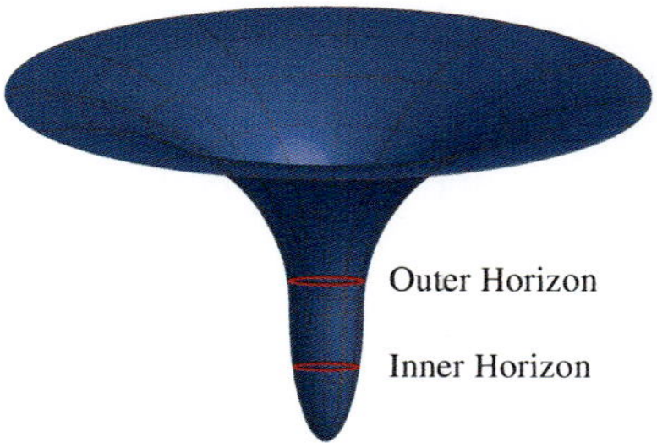

Fig. 5.19 The embedded representation of the Reissner–Nordstrom black hole showing the outer and inner event horizons.

- *Extremal Reissner–Nordstrom* being given by $r_S = 2r_Q$. The charge has increased so that the two event horizons collapse into a single horizon.
- The case of $r_S < 2r_Q$; there is no longer any horizon. This case is considered to be unphysical since it would yield a *naked singularity*, that is, the curvature singularity at $r = 0$ would not be enveloped by a horizon; arguments based on supersymmetry also rule out this case.

The two event horizons split spacetime into three regions, as shown in Fig. 5.18 and 5.19, namely

- A region with $r > r_{H+}$
- A region given by $r_{H-} < r < r_{H+}$
- A region the case of $r < r_{H-}$

A complete study of this spacetime geometry is beyond the scope of our discussions. Suffice to say that an in-falling observer inside the inner horizon has the same space and time coordinates as outside the outer horizon. What this means is that it is possible to avoid the singularity at the origin since it is timelike.

5.14 Black Hole Entropy

Entropy is essentially a statistical mechanical concept that describes the classical randomness arising from a system being at temperature T. Entropy is defined by

$$S = k_B \ln \Gamma \tag{5.15}$$

where Γ is the number of *microstates* that correspond to a given macroscopic state of the system. The macroscopic state of a system is specified by its bulk properties such as pressure and volume and so on, whereas its microstates correspond to *all* the possible states of the constituents that make up the macroscopic system, be it particles, atoms, molecules or fields.

A black hole's macroscopic state is defined by its mass M, charge Q and angular momentum J. The entropy of a black hole is determined by its allowed microstates.

If one computes the entropy of a classical black hole, one finds that its entropy is infinite. A way of understanding this result is that all objects that cross the horizon are lost to the outside observer, but with no change in the macrostates of the black hole; hence, the number of microstates of the black hole needs to be infinite since adding a finite number of microstates brings no change to its macrostates; an infinite number of microstates gives rise to infinite entropy. The black hole having infinite entropy can lead to many inconsistencies in Physics; in particular, it can lead to the loss of information leading to the inability to predict — even probabilistically — the time evolution of a system.

In general, quantum mechanics leads to the 'freezing out' of many classical microstates since the allowed microstates in quantum mechanics are, for example, *constrained* by the Heisenberg Uncertainty Principle — and hence for a physical system there are fewer quantum microstates than the allowed microstates in classical physics. Quantum mechanics in general leads to a 'granularity' of allowed states and removes all the inconsistencies arising from classical mechanics. In particular, the quantum principle has to be taken into account for correctly counting the microstates of the black hole.[11]

Consider the example of an ideal gas with N particles. The macroscopic variables for an ideal gas can be taken to be its volume and pressure, with temperature being an intensive variable, defined as the average kinetic energy per particle; the microstates of an ideal gas are all the possible positions and velocities of all N particles. In deriving the entropy for an ideal gas, a situation similar to the case of the black hole arises, in that the number of classical microstates yields an inconsistent result — namely, that the entropy $S(T, V)$ for an ideal gas is not extensive.[12]

---

[11]The microstates of the black hole will be discussed in Sec. 15.6.

[12]One expects all thermodynamic properties to be extensive in the sense that, for the case of the ideal gas, if one doubles the volume of the ideal gas, all the macroscopic quantities should also double.

Entropy is rendered extensive for an ideal gas — using the principles of quantum mechanics — by *reducing* the number of classical microstates by a factor of $N!$[13]; this leads to the following modification of entropy

$$\exp\{S(T,V)/k_B\} = V^N\zeta^N \quad : \text{ Incorrect}$$

$$\Rightarrow \exp\{S(T,V)/k_B\} = \frac{V^N\zeta^N}{N!} \quad : \text{ Correct}$$

$$\zeta = \left(\frac{2\pi m k_B T}{h^2}\right)^{3/2} .$$

Dividing by $N!$ is due to quantum mechanics since the N identical particles whose microstates are being counted are *indistinguishable*; hence, a permutation of the particles does not lead to a new microstate, which, in turn, implies that $N!$ classical microstates correspond to a single quantum microstate. In other words, it is the indistinguishability of the quantum microstates that leads to a reduction of the allowed microstates by a factor of $N!$.

Similar to the case of the ideal gas, we expect that the infinite entropy of an classical black hole is incorrect because one is over counting the number of the black hole's classical microstates Γ. A classical black hole, defined by Einstein's theory, does not take any quantum effects into account. A classical black hole violates the Second Law of Thermodynamics, which states that the entropy of an isolated system never decreases. Consider, for example, throwing a bottle of hot water into a Schwarzschild black hole. Once the bottle crosses the horizon all contact is lost with the bottle and the entropy that it is carrying is removed from the Universe; in effect a classical black hole *reduces* the entropy of the Universe, in clear violation of the Second Law of Thermodynamics. Whatever crosses the event horizon — carrying any amount of entropy — is lost forever to the Universe outside the horizon.

Since the classical black hole has infinite entropy, any amount of entropy that is carried by an in-falling body into the black hole does not increase the entropy of the black hole. This poses serious problems of consistency since it is known that black holes are present in all galaxies. In particular, supermassive spinning black holes — at the center of active galaxies — interact with the accretion disk around it and generate tremendous amounts of energy. Violating the Second Law would have major observable consequences.

5.14.1 *Bekenstein–Hawking entropy*

The microstates of a black hole were obtained by Hawking in 1974. He studied the effect of the black hole on the vacuum state of photons and electrons and found

[13] $N! = 1 \cdot 2 \cdot 3 \cdots N$.

that, due to quantum effects, the black hole is not completely black and in fact emits radiation. Black hole radiation in turn provides a measure of its entropy, which in fact turns out to be finite.

Hawking calculated the free energy F of the black hole at temperature T — in presence of the electron–photon quantum fields — using techniques of quantum field theory. Although there has been a wide spread recognition of Hawking's results on black hole temperature and related topics, there has as yet been any experimental evidence to support the correctness of these results.

The energy of a black hole with mass M can be shown to be $E = Mc^2$ — it behaves like a particle with energy given by the expression well known from special theory of relativity. A remarkable result of black hole theory, found by Hawking, is that for all known black holes, the entropy is, in suitable units, simply *one-fourth* of the area of the *outermost event horizon* A_H, namely[14]

$$S = \left(\frac{k_B c^3}{\hbar G}\right)\frac{A_H}{4}. \tag{5.16}$$

Note the important fact that the black hole's entropy is proportional to $1/\hbar$; in the classical limit of $\hbar \to 0$, the temperature of the black hole goes to zero and its entropy becomes infinite — reflecting the fact that the classical black hole has an infinite number of microstates. The (leading order) quantum effect, as reflected in the appearance of $1/\hbar$, reduces the number of microstates from infinity to a finite number and, as expected, the 'granularity' of quantum mechanics yields a finite value for the black hole's entropy.

The unit of area in which the area of the black hole's horizon is measured is given by $\ell_P^2 = \hbar G/c^3$ and yields

$$S = \left(\frac{k_B c^3}{\hbar G}\right)\frac{A_H}{4} = k_B \frac{A_H}{4\ell_P^2}; \quad \ell_P^2 = \frac{\hbar G}{c^3}.$$

The fundamental length is given by Planck's length $\ell_P = 10^{-35}\,\text{m}$. The entropy of a black hole is the only place (we know so far) where the fundamental constants of Nature, namely the Newton's gravitational constant G, speed of light c and quantum mechanical Planck's constant $\hbar$, appear together — and give us the first glimpse into the world of microscopic **quantum gravity**.

For a Schwarzschild black hole with radius r_S, the area of the event horizon is given by $A_H = 4\pi r_S^2 = 16\pi G^2 M^2/c^4$; hence the **Bekenstein–Hawking entropy**

[14]Bekenstein had concluded, on general grounds, that for the Second Law to hold the entropy of a black hole must be proportional to the area of the event horizon. He, however, could not compute the constant of proportionality, since it requires a microscopic theory that couples black holes to quantum fields.

for the *Schwarzschild black hole* is given by

$$S_S = \left(\frac{k_B c^3}{\hbar G}\right)\frac{A_H}{4} = \frac{4\pi k_B G}{c\hbar}M^2 \tag{5.17}$$

$$\Rightarrow S_S \simeq 1.05 \times 10^{77}\left(\frac{M}{M_\odot}\right)^2 k_B. \tag{5.18}$$

For the *Kerr black hole* the entropy is given by

$$S_K = \frac{\pi k_B G M^2}{c\hbar}\left(\left[1+\sqrt{1-\left(\frac{cJ}{GM^2}\right)^2}\,\right]^2 + \left(\frac{cJ}{GM^2}\right)^2\right). \tag{5.19}$$

As discussed in Sec. 5.12, as one slows down the Kerr black hole and drives its angular momentum to zero, namely $J \to 0$, the entropy of the Kerr black hole increases until it becomes equal to the entropy of the Schwarzschild black hole; namely the limit of $J \to 0$ yields $S_K \to S_S$.

The entropy of a *Reissner–Nordstrom black hole* is given by

$$\begin{aligned} S_{RN} &= \frac{\pi k_B G M^2}{c\hbar}\left(1+\sqrt{1-\frac{\hat{Q}^2}{GM^2}}\right)^2 \\ \hat{Q} &= \frac{1}{\sqrt{4\pi\epsilon_0}}Q. \end{aligned} \tag{5.20}$$

For an *extremal Reissner–Nordstrom black hole* $GM^2 = \hat{Q}^2$ and hence the entropy is given by

$$S_{RN,\text{Extremal}} = \frac{\pi k_B}{c\hbar}\hat{Q}^2. \tag{5.21}$$

Figure 5.20 shows the entropy as a function of the mass of the black hole for three different cases.

Noteworthy (optional content) 5.6: Holography Hypothesis

Since black hole entropy is proportional to the surface area of the horizon, it apparently seems that all the information contained inside the volume of the black hole, enclosed by the event horizon, resides on the *surface* of the black hole.

This feature of the black hole has led to the 'holography' hypothesis, namely that in a full theory of quantum gravity, the surface of a gravitating system contains all the information contained inside the volume.

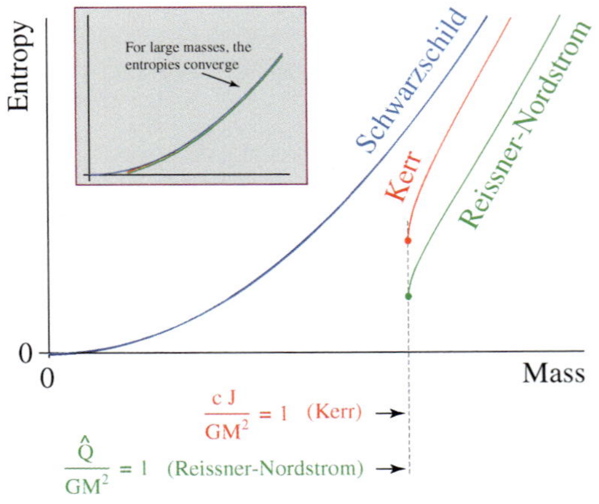

Fig. 5.20 Entropy dependence on mass for the Schwarzschild, Reissner–Nordstrom and Kerr black holes — for fixed J, Q. The inset shows that the curves converge for large masses.

5.15 Black Hole Temperature

The qualitative reasoning leading to the hypothesis that a black hole must have finite entropy led to the idea of black holes having temperatures since thermodynamics requires that any object that has entropy must also have a temperature. The puzzle of what is the black hole's temperature was solved by Hawking, who derived an expression for it, called Hawking temperature, and denoted by T_S. For a Schwarzschild black hole, it is given by

$$\begin{aligned} T_S &= \frac{c^3 \hbar}{8\pi k_B G M} \\ &= 6.2 \times 10^{-8} \left(\frac{M_\odot}{M}\right) \mathrm{K}. \end{aligned} \tag{5.22}$$

For large stars the temperature is negligible; a black hole of one solar mass has a temperature of only 60×10^{-9} K. Only black holes with mass less than 10^{11} kg, about the mass of a mountain on the Earth, would have temperature above the background 2.7 K and would be hot enough to radiate. The Schwarzschild black hole's mass becomes smaller as the temperature rises and Fig. 5.21 shows how temperature and mass change as the black hole radiates off its energy.

The Hawking temperature of a Reissner–Nordstrom black hole is given by

$$T_{RN} = \frac{c^3 \hbar}{2\pi k_B G M} \frac{\sqrt{1 - (\hat{Q}^2/GM^2)}}{\left(1 + \sqrt{1 - (\hat{Q}^2/GM^2)}\right)^2} \tag{5.23}$$

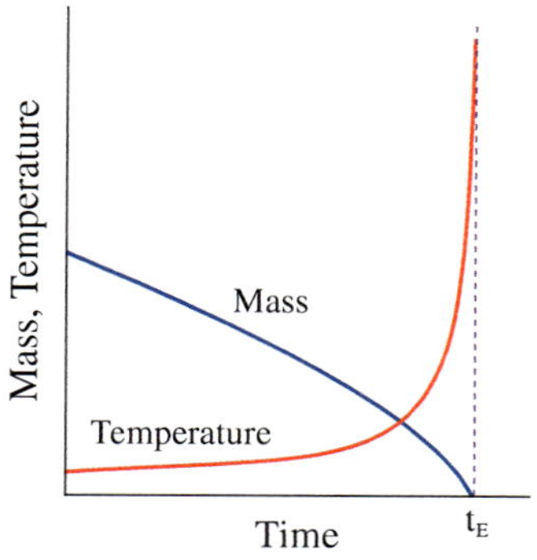

Fig. 5.21 As temperature rises, the Schwarzschild black hole's mass decreases and it evaporates.

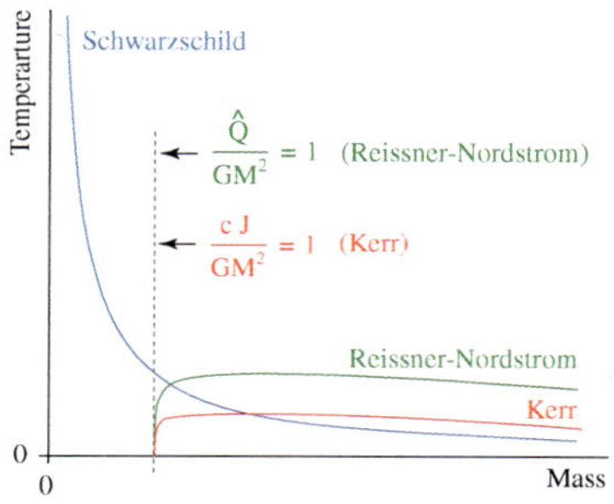

Fig. 5.22 Temperature dependence on mass for the Schwarzschild, Reissner–Nordstrom and Kerr black holes.

and for a Kerr black hole is given by

$$T_K = \frac{c^3\hbar}{4\pi k_B GM}\frac{\sqrt{1-(cJ/GM^2)^2}}{1+\sqrt{1-(cJ/GM^2)^2}}. \tag{5.24}$$

Note that as the black hole radiates, it loses mass; in contrast to the Schwarzschild black hole that loses all of its mass as the temperature increases, as given in Eq. (5.22), for both the Reissner–Nordstrom and Kerr black holes, the temperature increases until it reaches a turning point after which the temperature decreases, as shown in Fig. 5.22. The mass decreases until both, the Reissner–Nordstrom and Kerr black holes, become extremal and the Hawking *temperature falls to zero*, resulting in the black holes no longer radiating. In other words

$$T_K \to 0 \text{ as } GM^2 \to cJ$$
$$T_{RN} \to 0 \text{ as } GM^2 \to \hat{Q}^2.$$

The reason that the Hawking temperature goes to zero for the extremal Reissner–Nordstrom and Kerr black holes is because, due to existence of charge and angular momentum respectively, the mass of the black hole has to be greater or equal to a minimum value for the event horizon to exist. For the Reissner–Nordstrom black hole, from Eq. (5.23), the minimum mass is $GM \geq \hat{Q}^2$ and for the Kerr black hole, from Eq. (5.24), it is $M \geq \sqrt{Jc/G}$.

If the mass is less than the minimum value, the black hole would have a naked singularity, something forbidden by the *cosmic censorship* hypothesis.

Noteworthy (optional content) 5.7: Kerr–Newman Metric

All the three types of black holes that we have discussed, namely, the Schwarzschild, Reissner–Nordstrom and Kerr black holes, are special cases of the Kerr–Newman black hole, which is a charged spinning black hole that is generated by a perfect spinning ring that has total mass M, angular momentum J and net charge Q. The reason that the Kerr–Newman exact solution can be found is due to a remarkable factorization of its geometry. In the Kerr–Schild coordinates, the metric tensor of the Kerr–Newman geometry can be written as follows

$$
\begin{aligned}
ds^2 &= \sum_{\mu,\nu} dx^\mu dx^\nu g_{\mu\nu} \\
g_{\mu\nu} &= \eta_{\mu\nu} + \xi_\mu \xi_\nu .
\end{aligned}
\tag{5.25}
$$

Note that $g_{\mu\nu}$ is the metric tensor that completely determines the geometry of spacetime and $\eta_{\mu\nu}$ is the flat metric of empty Lorentz spacetime. The entire Kerr–Newman geometry is contained in the four vector ξ_μ, which consists of only four independent functions of spacetime as opposed to the general expression for (the symmetric) $g_{\mu\nu}$, which consists of 10 independent functions of spacetime. It is this very special and simple form of the metric tensor $g_{\mu\nu}$ given in Eq. (5.25) that is at the root of the exact solvability of the problem.

The Hawking temperature and entropy of the Kerr–Newman black hole are given by the following:

$$
T_{KN} = \frac{c^3\hbar}{2\pi k_B GM} \frac{\sqrt{1 - \frac{\hat{Q}^2}{GM^2} - \left(\frac{cJ}{GM^2}\right)^2}}{\left(1 + \sqrt{1 - \frac{\hat{Q}^2}{GM^2} - \left(\frac{cJ}{GM^2}\right)^2}\right)^2 + \left(\frac{cJ}{GM^2}\right)^2}
$$

$$
S_{KN} = \frac{\pi k_B GM^2}{c\hbar} \left(\left[1 + \sqrt{1 - \frac{\hat{Q}^2}{GM^2} - \left(\frac{cJ}{GM^2}\right)^2} \right]^2 + \left(\frac{cJ}{GM^2}\right)^2 \right).
$$

The Schwarzschild black hole is obtained by taking $J = 0 = \hat{Q}$, whereas the Reissner–Nordstrom and Kerr black holes are obtained by taking $J = 0$ and $\hat{Q} = 0$, respectively.

5.16 Black Hole Thermodynamics

As the black hole accretes in-falling matter it gains energy and entropy; one can show that a black hole obeys the First and Second Law of Thermodynamics. For simplicity, consider the case of the Schwarzschild black hole; its energy $E = Mc^2$ and entropy obey the following

$c^2 dM = TdS$: First Law of Thermodynamics (Energy conservation)

$\Delta S \geq 0$: Second Law of Thermodynamics (Entropy never decreases)

Absolute zero temperature is unattainable : Third law of thermodynamics.

Another statement of the Third Law of Thermodynamics is that it takes an infinite number of steps to reach absolute zero temperature. In black hole physics, the Third Law has an interesting realization; the temperature of a extremal Kerr black hole is exactly zero. It has been observed, based on astronomical measurements, that the Kerr black holes that occur in Nature never become exactly extremal, with their angular momentum being at most $J/Mc \approx 0.998GM^2/c^2$ and never exactly equal to its maximal value; hence the black hole temperature never reaches absolute zero, as one would expect from the Third Law.

The Second Law states that if a black hole accretes mass, then its entropy — proportional to the area of its horizon — must also increase; detailed calculations show that in fact this is the case. To return to our example, throwing in a bottle of hot water into the black hole adds to mass of the black hole by the amount of the mass of the hot water bottle and the entropy of the black hole also increases; the final entropy of the black hole will be larger than the sum of the initial entropy of the black hole and the entropy of the hot water bottle — hence maintaining the validity of the Second Law. If two black holes fuse, as shown in Fig. 5.23, then the area of the resultant black hole can be shown to be greater or at least equal to the sum of the areas of the previous two black holes, as is required by the Second Law.

Furthermore, if the black hole radiates, its mass decreases leading to a decrease in its entropy; however the total entropy of the black hole and the emitted radiation increases, showing again that the Second Law is valid.

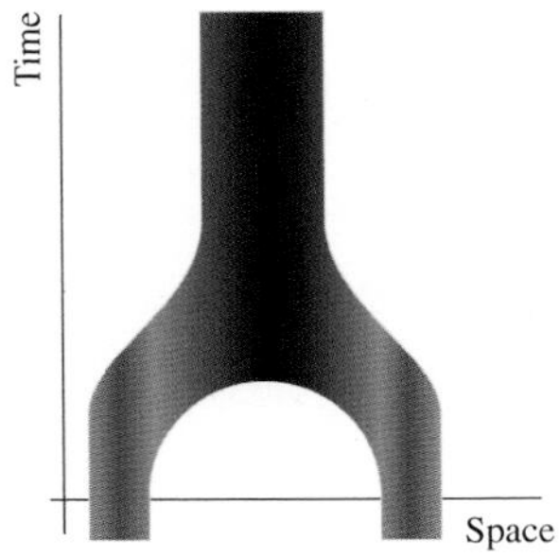

Fig. 5.23 Colliding black holes obey the Second Law: the resulting black hole has a surface area greater than the sum of the surface areas of the colliding holes.

5.17 Hawking Radiation

What is **Hawking radiation**? Does Hawking radiation have anything to do with the entropy of a black hole? We briefly discuss Hawking's analysis of black hole entropy and radiation.

The concept of the vacuum state of a quantum field was discussed in Sec. 2.10; the quantum vacuum, unlike classical vacuum, is far from being inert, but rather is alive with all sorts of vacuum fluctuations. The Casimir force, illustrated in Fig. 2.13, is a measurable effect of the fluctuations of the vacuum. In particular, we saw in Sec. 2.10, that if one turns on a sufficiently strong classical electric field, then — out of the quantum vacuum — physically observed electron–positron pairs are produced at a steady rate. The black hole is similar to the electric field and provides a classical background source for the production of particles and of radiation.

Recall that the *total energy* of the Schwarzschild black hole, which is equal to the energy in the curvature of spacetime plus the energy carried by the matter concentrated at the singularity, is equal to Mc^2.

Consider a stationary observer outside the horizon. The stationary observer has to accelerate, as given in Eq. (5.11), in order to avoid falling into the black hole. The quantum field occupies all of space (and time), and exists both inside and outside the horizon. But the observer in the accelerating frame can only observe the states of the field *outside* the horizon, with the states inside the horizon being inaccessible. Hence, to describe observations outside the horizon, one needs to *sum* over all the states of the quantum field that lie inside the horizon, thus losing information about the interior of the black hole. The loss of information leads to *classical randomness* and is described by the principles of classical thermodynamics. The result of classical randomness is that the accelerating observer detects a steady stream of (black body) radiation, called *Hawking radiation*, coming out of the horizon of the black hole. Note that when a black body is heated, it is in a state of classical uncertainty, which for the case of Hawking radiation arises due to the ignorance of the quantum states that are in interior of the black hole.

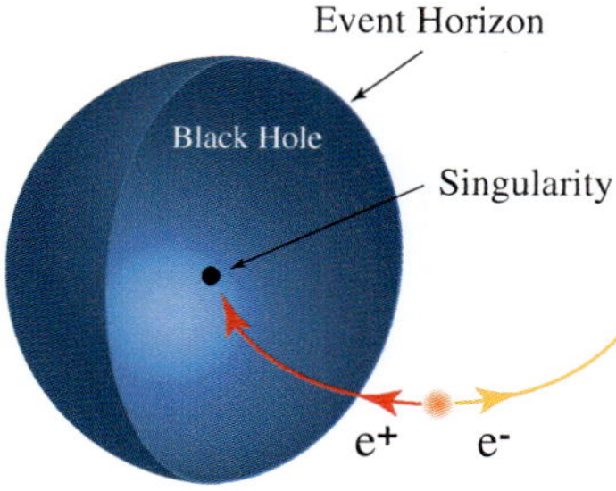

Fig. 5.24 Pair creation near the event horizon.

A metaphorical description of Hawking radiation is to consider a vacuum fluctuation near the horizon in which a photon disintegrates into an electron–positron pair, as discussed in Sec. 2.10 and shown in Fig. 2.12. All vacuum fluctuations are configurations of the vacuum, which have zero enery, and hence the electron–positron pair has net zero energy. Suppose the positron crosses the horizon, as shown in Fig. 5.24. Since the positron is inside the horizon it cannot escape and hence the remaining electron, due to momentum conservation, is ejected away from the black hole.

Since the emitted electron has positive energy, it is stated rather loosely that, due to energy conservation, the in-falling positron has *negative energy.* Hence when the positron is absorbed by the black hole, it causes the black hole's mass to *decrease* and thus results in the black hole *losing* energy by emitting an electron. This explanation invoking the concept of *negative energy* of the in-falling particle is used widely in popular books on black holes, including by Hawking as well. It should be noted, however, that the concept of negative energy inside the horizon is not correct; furthermore, the concept of negative energy being attributed to a particle does not have any mathematical basis in quantum field theory.

Vacuum fluctuations near the event horizon can result in a steady production of physical (as opposed to virtual) electrons, positrons, photons and so on, and is shown schematically in Fig. 5.25. The process of black hole radiation entails energy being transferred from the black hole to the region of space outside the horizon.

As discussed in the earlier section, the Schwarzschild black hole has a finite temperature as well as a finite entropy. Hawking has shown that the black hole, as seen by a distant observer, is emitting radiation that follows the black body spectrum. A black hole with mass M loses energy at a steady rate given by

$$\frac{dM}{dt} = -\frac{1}{15{,}360\pi}\frac{c^4\hbar}{G^2M^2}$$

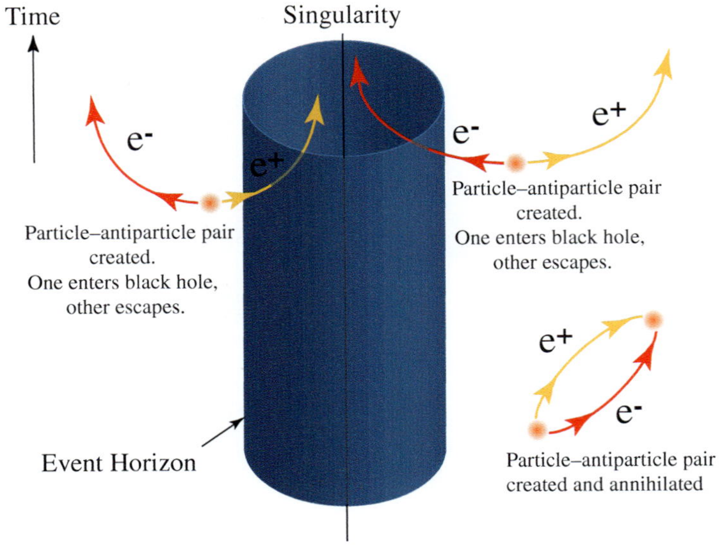

Fig. 5.25 Hawking radiation.

and fully evaporates in time t_E given by

$$t_E = 5120\pi \frac{G^2 M^3}{c^4 \hbar}. \tag{5.26}$$

Black holes with mass of our Sun or larger have extremely low temperatures and would need aeons more than the present age of our Universe to evaporate. On the other hand, tiny black holes with mass of about 2×10^{11} kg — the mass of a typical mountain — would last about 14 billion years and, if formed at the Big Bang, would be exploding today.

5.18 The Answer

Einstein's theory of gravity leads to black hole solutions that, in general, have geometric singularities enveloped by event horizons. The physical significance of the black hole was probed by studying the behavior of stationary and in-falling observers. In particular, for observers outside the event horizon, time itself ceases to flow at the event horizon.

The Schwarzschild and Kerr black holes were studied in some detail as these have distinctive properties. The Schwarzschild black hole was studied to understand the nature of the black hole's singularity. The Kerr black holes are far more complex, and we studied the extremal case to understand the role of the ergosphere and how to extract energy from the Kerr black hole.

Black holes after all are not that black, but in fact emit black body radiation. The fact that a black hole is not completely black also renders finite the entropy of the black hole. Hawking's result on black hole entropy and black hole temperature was studied for a variety of black holes.

The black hole's entropy is proportional to the surface area of its event horizon and *all the information* (entropy) of the black hole is carried on its event horizon. This is intuitively sensible since a distant observer *cannot*, in principle, observe the black hole beyond its event horizon. Hence, if the Universe is not to lose information due to objects falling into the black hole, all the information of the black hole should reside on its event horizon. The holography hypothesis is an expression of this intuitive idea.

Black holes are compact objects; a black hole is invisible to all observers who are external to its horizon — and which comprises most of the Universe. The only way that we can know of the existence of a black hole is by indirect methods, namely by its effects on ordinary matter. Black holes require a mathematical description, like the one provided by the general theory of relativity, for us to conclude that what is being observed is, in fact, pointing to an object such as a black hole.

Black holes are a prime exemplar of invisible objects that populate our Universe. The existence of these objects can be inferred from experimental data provided we have a theory that provides a mathematical model of these objects.

Chapter 6

Cosmology

How did it all begin?

6.1 The Question

Human beings have looked at the sky since time immemorial and felt their own insignificance when confronted by the immensity of the Universe. Stars that appear on the horizon are at unimaginable distances away and have burned for aeons. There seems to be no end to the space that surrounds us. Questions that used to be in the realm of mythology and religion can now be addressed by modern day cosmology.

Cosmology is the subject that addresses questions about the Universe at the cosmic scale — its birth, evolution and its ultimate fate. In particular, cosmology addresses questions such as:

- How did it all begin?
- How large is the Universe?
- What are the structures and processes of the Universe at the cosmic scale?
- Will the Universe ever come to an end? If so, then how and when will it all come to an end?

6.2 Introduction

Modern cosmology has three primary empirical results that form its experimental bedrock. Our discussion will revolve around various models that address and explain the following observational results:

- The expansion of the Universe, encoded in Hubble's law.
- The existence of the 2.7 K cosmic microwave background radiation.
- The relative abundance of the light elements.

In this chapter we focus on some of the key ideas that explain the cosmic structure of the Universe. Einstein's formulation of gravity is mathematically too complicated to analyze in any simple way; hence, we present the essential ideas of modern cosmology by explaining, instead of deriving, the results that emerge from Einstein's theory.

At the cosmic scale, a galaxy is the smallest cosmic body. Two assumptions, both discussed in Sec. 4.7, are made to simplify the study of cosmology, namely

- It is assumed that all galaxies are moving on a family of geodesics that have a common starting point in the past; these geodesics fill the Universe and do not intersect until at some distant future time, if at all.
- The Universe is assumed to be homogeneous and isotropic, namely that every position and all directions in the Universe are identical.

It is an observational result that the Universe is almost exactly flat; and for this reason, one can think of the expanding universe as an ideal (dilute and non-interacting) gas made out of identical galaxies, each having mass m, and with the expansion of the Universe crudely represented by the expansion of an ideal gas of galaxies.

We take advantage of a rather surprising result to simplify our discussion. It turns out that Einstein's result for an expanding Universe of a dilute gas of galaxies can be directly derived using *Newtonian* gravity. Intuitively, the reason this is possible is because Newtonian gravity is the limiting case for Einstein's theory when the gravitational force is small, and this is the case for the regime of cosmology that exists in Nature.

6.3 Expanding Universe: Newtonian Cosmology

The Newtonian approximation, although simple, allows for interesting insights with relatively simple mathematics. Furthermore, a study of the full Einstein's field equations for gravity in the presence of a uniform (homogenous) distribution (density) of matter can be shown to yield results that are exactly equivalent to those obtained from **Newtonian cosmology**.

Suppose two galaxies of masses m_1 and m_2 are separated by a distance r, as shown in Fig. 6.1. The gravitational potential energy of the two galaxies is given by

$$U_{\text{gravity}}(r) = -G\frac{m_1 m_2}{r} \quad : \text{ Gravitational potential energy} \tag{6.1}$$

where G is Newton's gravitational constant.

All the galaxies are moving apart from each other. Since the gas of galaxies is expanding, the galaxies have both kinetic and potential energy. We compute the kinetic and potential energy of a galaxy to find out how the total energy of a galaxy depends on the mass-energy density of the Universe.

As shown in Fig. 6.2, consider a galaxy with mass m that, at time t, is at a distance $r(t)$ from the origin of the chosen coordinate system. Suppose that $r(t)$ is

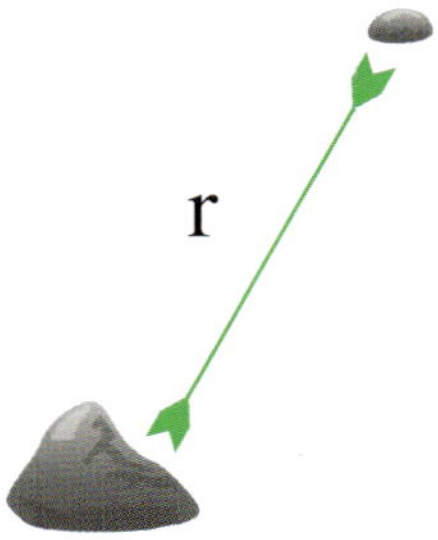

Fig. 6.1 Potential energy between two masses.

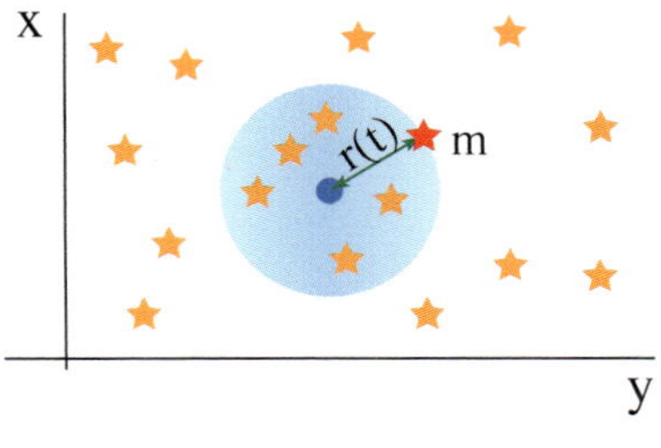

Fig. 6.2 A galaxy at distance $r(t)$ from the origin of the coordinate system.

much greater than the distance between two galaxies and much smaller than the size of the Universe. The position of the galaxy is given by

$$r(t) = R(t)r_0$$

with r_0 being the distance of the galaxy from the origin at time t_0, with $R(t_0) = 1$, where t_0 is the time at present.

We represent the **expanding Universe**, in which all distances are increasing with the expansion, by a *dimensionless* scale parameter $R(t)$. The increase in the distance between any two points is realized by $R(t)$ being an *increasing* function of time t. In Einstein's theory of gravitation, the scale parameter $R(t)$ is obtained from the metric tensor that describes the geometry of an expanding Universe, and is given in Eq. (6.15).

The velocity of the galaxy is given by

$$v(t) = \frac{dr}{dt} = \frac{dR}{dt}r_0 = \left(\frac{dR}{dt}\right)\frac{1}{R} \cdot Rr_0$$

$$v(t) \equiv H(t)r. \tag{6.2}$$

To simplify the notation, we denote $R(t) = R$ and write out the dependence on time only if needed. The Hubble parameter is defined by

$$H(t) = \frac{1}{R}\left(\frac{dR}{dt}\right) \quad \text{dimension: (time)}^{-1}. \tag{6.3}$$

The cosmological significance of the Hubble parameter will be discussed later. Since $R(t_0) = 1$ we have Hubble's constant given today by

$$H(t_0) = \left[\frac{dR}{dt}\right]_{t=t_0} \tag{6.4}$$

and which yields, from Eq. (6.2), the following

$$v = H(t_0)r. \tag{6.5}$$

Equation (6.5) is **Hubble's law** and which states that the velocity of a sufficiently distant galaxy is linearly related to r, its distance from Earth. The galaxy's velocity of recession is experimentally determined by the redshift of its radiation spectrum and is discussed in Sec. 7.3. Figure 6.3 shows the experimental evidence in support of Hubble's law.

The galaxy has kinetic energy due to the expansion of the ideal gas. The (Newtonian) kinetic energy of the galaxy, having a mass m, is given by

$$T = \frac{1}{2}m\left(\frac{dr}{dt}\right)^2 = \frac{1}{2}mr_0^2\left(\frac{dR}{dt}\right)^2. \tag{6.6}$$

Let M be the total mass enclosed by the sphere of radius $r(t)$ as in Fig. 6.2. The mass M is spherically symmetrically distributed to reflect that the Universe is isotropic and homogeneous. Due to Gauss's law, as shown in Fig. 6.2, only the

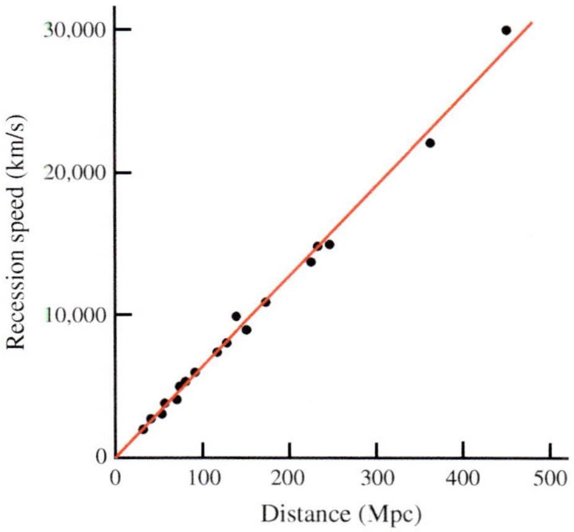

Fig. 6.3 Hubble's law: expanding Universe. The data points are the recession velocities of various galaxies plotted against their distances from Earth.

mass inside the sphere contributes to the potential energy of the galaxy with mass m and hence

$$U(R) = -\frac{GMm}{r_0 R}. \tag{6.7}$$

Let $\rho =$ the energy density of the Universe due to its mass content, called mass-energy density to differentiate it from radiation-energy density. The mass per unit volume of the Universe is given by ρ/c^2 and hence M is given by[1]

$$M(R) = \frac{4}{3}\pi r(t)^3 \frac{\rho}{c^2} = \frac{4\pi R^3 r_0^3}{3c^2}\rho$$

and which yields, from Eq. (6.7), the following

$$U(R) = -\frac{4\pi m G r_0^2 R^2}{3c^2}\rho. \tag{6.8}$$

As the Universe expands the *total* mass of all the galaxies taken together does not change but the mass per volume decreases. In particular, the mass-energy density $\rho(t)$ decreases since $\rho(t) \times$ volume $= \rho(t) \times \frac{4}{3}\pi R^3(t) r_0^3 =$ constant and yields

$$\rho(t) \propto \frac{1}{R^3(t)}. \tag{6.9}$$

[1] From the relation $E = Mc^2$, mass can be expressed in terms of energy.

The total energy, at time t, of the galaxy at position $r(t)$ is given by

$$\begin{aligned} E &= T + U \\ &= \frac{1}{2}mr_0^2\left(\frac{dR}{dt}\right)^2 - \frac{4\pi mGr_0^2\rho}{3c^2}R^2. \end{aligned} \tag{6.10}$$

Since the total energy of the galaxy is constant, we write it as

$$E = -\frac{kmr_0^2}{2} \quad : \text{Energy at } t = t_0 \tag{6.11}$$

where k, r_0 are constants.

The parameter k is a constant due to energy conservation, and has units of $(\text{time})^{-1}$. k is called the **curvature parameter** and in **Newtonian cosmology** specifies the total energy of the Universe. Hence, from Eqs. (6.10) and (6.11)

$$k = -\left(\frac{dR}{dt}\right)^2 + \frac{8\pi G\rho}{3c^2}R^2. \tag{6.12}$$

Note that the energy of a galaxy is made from two components, a negative piece U coming from gravitational attraction due to the Universe's intrinsic energy density, and a positive piece T due to kinetic motion. It is the trade off of these two forms of energy that ultimately determines whether the Universe will expand forever, yielding what is called an **open universe**, or whether it will collapse and end up at a single point, yielding what is called a **closed Universe**.

6.4 Friedmann Equation

As discussed in Sec. 4.7, a fundamental assumption in the study of cosmology is that every point of space, at a cosmic scale (distances greater than 100 million light years), is essentially equivalent to any other point of space. In other words, there is no special location in the Universe. This means that, at a cosmic scale, the Universe is *homogeneous*, that is, has the same mass-energy content everywhere, and is *isotropic*, namely all directions are equivalent.

One can consider the structure of space to change with time, but in such a manner so that at any given time t, space *continues* to be homogeneous and isotropic. One of the simplest such models for cosmology is where only the distance between two fixed points in space changes with time, and this is the Friedmann model for the cosmological evolution of the Universe.

Einstein's field equations for a homogeneous and isotropic Universe, relating the curvature of spacetime to the energy density ρ, were exactly solved by Friedmann in 1922 and independently by Lemaître in 1927. It should be noted that the energy density ρ includes both matter and radiation, and with matter and radiation having different time dependences as the Universe expands.

From Eq. (6.12), we have

$$\left(\frac{dR}{dt}\right)^2 = \frac{8\pi G\rho}{3c^2}R^2 - k : \textbf{Friedmann equation}. \tag{6.13}$$

We also have from above

$$H^2 \equiv \frac{1}{R^2}\left(\frac{dR}{dt}\right)^2 = \frac{8\pi G\rho}{3c^2} - \frac{k}{R^2}. \tag{6.14}$$

Note that $H(t) > 0$ means that the scale parameter $R(t)$ is increasing with time, and hence the distance between any two points is increasing. This is an example of a cosmological model for an expanding Universe.

The Friedmann equation yields a fundamental relation between the expansion parameter H — that measures the rate of expansion — and the density of the Universe given by ρ. The evolution equation for the density ρ requires one to study the relation of density and pressure of the Universe; we will not need the time dependence of ρ in our analysis.

6.4.1 *Hubble's law revisited*

The Friedmann equation provides an explanation of Hubble's law. From the definition of the scale parameter $r = R(t)r_0$, we have from Eq. (6.2) that $v = H(t)r$. The value of the Hubble parameter today is given by $H_0 = H(t_0) > 0$; we hence have, from Eq. (6.5), that

$$v = H_0 r \;:\; \text{Hubble's law.}$$

Hubble's law states that if we were to observe the entire Universe at time t_0, we would find that all the galaxies are receding from us with a velocity that is proportional to their distance from us: the further away is the galaxy the higher is its velocity of recession. The law is modified once the velocity v approaches the speed of light c.

If we view the Universe today, the more distant the galaxy is, the further back it is in the past since light from the galaxy took a finite amount of time to reach us. Hence, if Hubble's constant depends on time, as indeed we will find that it in fact does, then the recession of all galaxies cannot be simply be explained by the formula $v = H_0 r$, but rather H_0 needs to be replaced by a Hubble's constant that depends on time, namely $H(t)$, and is given by Eq. (6.14). The fact that the rate of expansion of the Universe is *accelerating* is discussed in Chapter 7.

Noteworthy (optional content) 6.1: The Friedmann–Robertson–Walker Metric

---

Newtonian cosmology is understood in terms of gravitational potential and kinetic energy. In Einstein's theory of gravitation, Newton's gravitational potential is replaced by the gravitational field. On solving Einstein's field equations for a Universe with uniform density of matter, one obtains an equation identical to the

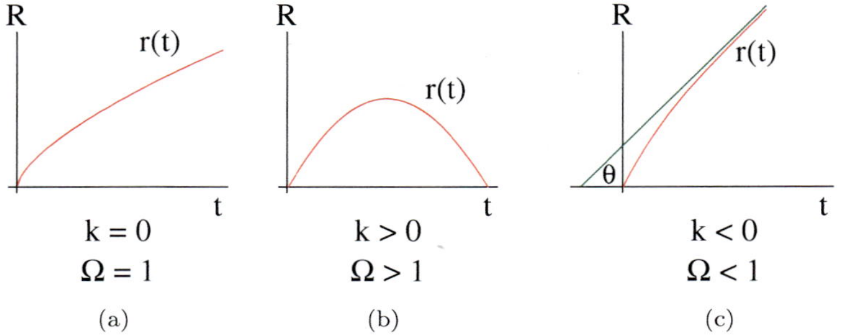

Fig. 6.4 Three types of Universes. (a) $k = 0$: a flat Universe, (b) $k > 0$: a re-collapsing Universe and (c) $k < 0$: a Universe expanding forever. The density parameter Ω is defined in Sec. 6.8.

Friedmann equation given in Eq. (6.13). A new piece of information is that in Einstein's cosmology the constant k has a geometrical interpretation. The case of $k = 0$ corresponds to a flat spacetime, $k < 0$ describes a hyperbolic geometry similar to the surface of a saddle, and $k > 0$ describes a closed geometry similar to a sphere; the three geometries are shown in Fig. 3.32.

For the more advanced reader, the geometry for the Friedmann Universe is given by the so called Friedmann–Robertson–Walker metric

$$ds^2 = -c^2dt^2 + R^2(t)\left[\frac{dr^2}{1-kr^2} + r^2(d\theta^2 + \sin^2\theta d\phi^2)\right]$$

where $R(t)$ is an increasing function of time t and is called the scale parameter. See Fig. 6.4.

For $k = 0$ the spatial metric of the Universe becomes flat, as given in Eq. (3.14), since the spatial part of the metric, in Euclidean coordinates (x, y, z), becomes

$$\begin{aligned} ds^2 &= -c^2dt^2 + R^2(t)\big[dr^2 + r^2(d\theta^2 + \sin^2\theta d\phi^2)\big] \\ &= -c^2dt^2 + R^2(t)\big[dx^2 + dy^2 + dz^2\big]. \end{aligned} \tag{6.15}$$

The expansion of the Universe means that, as time evolves, the distance between any two points of flat space is increased by the factor $R(t) > 0$.

Suppose we measure the distance between two points at time t_0 to be $R(t_0)L$; then the distance between the same two points at a later time t will be $R(t)L = [R(t)/R(t_0)] \times R(t_0)L$. In other words, the distance between the points has increased by a factor of $R(t)/R(t_0) > 1$. The Universe is not expanding into some pre-existing empty space, like a balloon that expands into an ambient space surrounding it, but rather, the *intrinsic distance* between any two points is getting larger and larger.

Cosmological Constant

Einstein's theory allows the introduction of a new parameter in the theory, namely the cosmological constant Λ, and which yields the following modification of the Friedmann equation

$$H^2 = \frac{1}{R^2}\left(\frac{dR}{dt}\right)^2 = \frac{8\pi G\rho}{3c^2} - \frac{k}{R^2} + \frac{\Lambda}{3}. \tag{6.16}$$

The parameter Λ is a measure of how much energy is contained in empty space, and is a model for dark energy discussed in Chapter 7. Dark energy provides a universal *repulsive* force between all matter; it also yields a negative pressure term, and hence the expansion of space entails doing work on the dark energy of 'empty' space! If one lets Λ depend on time, one has a theory with quintessence, which is a term that denotes a time dependence for dark energy.

6.5 The Curvature Parameter k

We study three cases of the Friedmann equation, given in Eq. (6.13), for open, critical and closed Universes, fixed by three domains for the value of the parameter k. The Universe is open (expanding forever) if $k < 0$ yielding $E > 0$, and is closed if $k > 0$ yielding $E < 0$. The Universe is in a critical state for $k = 0$, and continues to expand forever with the rate of expansion slowing down; note that $k = 0$ implies $E = 0$.

6.5.1 $k = 0$: *Critical universe*

For the $k = 0$ case, the total energy $E = 0$ and this is called the Einstein–de Sitter Universe. The value of $k = 0$ implies that the Universe is spatially flat, as can be seen from Eq. (6.15). The feature of $E = 0$ turns out to have far-reaching consequences, as will be seen when we discuss the inflationary model for the Big Bang in Sec. 6.11. From Eq. (6.13), for $k = 0$, we have

$$\left(\frac{dR}{dt}\right)^2 = \left(\frac{8\pi G\rho}{3c^2}\right) R^2. \tag{6.17}$$

Since mass M is contained within the radius $r(t) = R(t)r_0$, we have from Eqs. (6.9) and (6.17), for constant ρ_c defined for the flat Universe, namely $\rho_c = \rho(t_0)|_{k=0}$ and is called the critical density (of today),

$$\rho(t) = \frac{\rho_c}{R^3(t)} \Rightarrow \frac{dR}{dt} = \sqrt{\frac{8\pi G\rho_c}{3c^2}} R^{-1/2}. \tag{6.18}$$

Integrating the above equation yields

$$R = \left(\frac{3}{2}\right)^{2/3} \left(\frac{8\pi G\rho_c}{3c^2}\right)^{1/3} t^{2/3}. \tag{6.19}$$

The expansion parameter R is shown in Fig. 6.4(a) for $k = 0$.

By setting $t = t_0$ in Eq. (6.17), the value of ρ_c can be written in terms of the Hubble parameter H_0. Since $R(t_0) = 1$, we have from Eq. (6.18) that

$$\rho_c = \left(\frac{3c^2}{8\pi G}\right) \frac{1}{R^2} \left(\frac{dR}{dt}\right)^2 \Bigg|_{t=t_0} = \frac{3c^2 H_0^2}{8\pi G}. \tag{6.20}$$

If the density of the Universe is less than or equal to ρ_c then the Universe expands forever and if it is greater than ρ_c then gravitational attraction will overcome the expansion and the Universe will collapse back to a point.

Using the present day estimate of $1/H_0 = 14.4 \times 10^9$ years yields $\rho_c = 4$ Gev/m^3. The experimentally determined value of the current energy density of the Universe is given by 2 Gev/m^3 $< \rho(t_0) <$ 10 Gev/m^3. Recall $\rho(t_0)$ is the empirically observed value while ρ_c is the energy density for the case of $k = 0$.

6.5.2 *k> 0: Closed universe*

In this case $E < 0$, the universe expands until it reaches a maximum for $R(t) = R_m$ given by $dR/dt = 0$ and Eq. (6.13) yields

$$R_m = \sqrt{\frac{3kc^2}{8\pi G\rho(t_m)}}. \tag{6.21}$$

For $t > t_m$, $dR/dt < 0$ and the Universe contracts and is shown in Fig. 6.4(b).

6.5.3 *k< 0: Open universe*

In this case $E > 0$ and the Universe expands forever. It can be shown that

$$R(t) \simeq \begin{cases} t^{2/3}, & t \sim 0 \\ t, & t \to \infty \end{cases} \tag{6.22}$$

which is shown in Fig. 6.4(c).

6.6 The Cosmological Constant Λ

The cosmological term Λ in Einstein's theory of gravitation leads to empty space — devoid of all matter and radiation — having a constant energy density throughout the Universe; the energy of empty space is also called the vacuum energy. The vacuum energy density is positive if Λ is positive, and negative if it is negative. Observations support a positive cosmological constant Λ that leads to a positive vacuum energy. Ordinary matter and positive vacuum energy have a gravitational repulsion for each other. The vacuum energy thus has negative pressure opposing the attraction of ordinary gravitation and thus has the correct properties for being a candidate for dark energy discussed in Sec. 7.5.

The cosmological constant has an interesting history. Einstein had initially introduced Λ into his model for cosmology in order to obtain a static Universe. On later learning of Hubble's discovery of an expanding Universe and of the solution of Einstein's theory of gravity that yields an expanding Universe, Einstein considered the introduction of the cosmological constant his "greatest blunder".[2] It is an irony of fate that Einstein may indeed have been right after all... the only difference being that the numerical value that he assigned to Λ was not quite correct.

The current view is that the Universe has a cosmological constant, in appropriate units, with the value of $\Lambda \simeq 10^{-122}$! This incredibly small number is consistent with the flatness of the Universe. It also gives the correct age for the Universe; if one takes $\Lambda = 0$, then the age of the Universe turns out to be about 10 billion years, which is less than the age of the oldest observed stars in the Universe. The current best estimate of the age of the Universe of $\simeq$ 13.78 billion years is consistent with the age of the oldest observed stars.

Noteworthy (optional content) 6.2: Time and Space in Cosmology

In discussing the age and size of the Universe, the question naturally arises as to what is the frame in which these time scales and distances are being measured. As discussed in Sec. 4.7, Friedmann's model of cosmology is best described in a co-moving frame. In this frame of reference, each galaxy has its own clock and has a set of measuring rods laid out through space. The co-moving frame for an expanding Universe is shown in Figs. 4.7, 4.8 and 6.5.

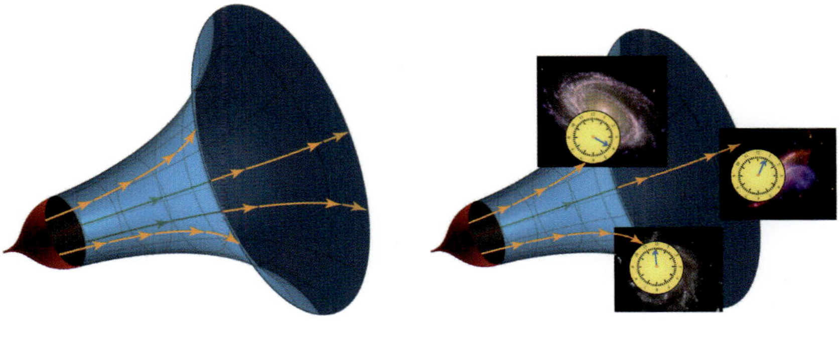

Fig. 6.5 Diverging geodesics in an expanding Universe. Each galaxy moves on a geodesic and carries its own clock and measuring rods.

[2]G. Gamow, 1970, *My World Line* (Viking, New York).

All the galaxies diverge from each other and the clocks move with the galaxy. As discussed in Sec. 4.7 and Noteworthy 4.1, a specific feature of the Friedmann cosmology — which can be seen from the expression for the spacetime interval for a flat expanding Universe given in Eq. (6.15) — is that the *flow* of time does not change with the expansion of the Universe. Hence, the rate at which time flows at different instants of the expansion of the Universe is the same. In other words, one second after the Big Bang is the same as one second today.

In contrast to the flow of time, the *distance* between points of space increases with the expansion of the Universe and is quantified by the scale factor $R(t)$. As shown in Fig. 4.8(b), if a galaxy measures the distance D_1 and D_2 to a neighboring galaxy at two instants t_1 and t_2, respectively, then the Friedmann Universe states the following

$$D_2 = \left[\frac{R(t_2)}{R(t_1)}\right] D_1 > D_1; \quad t_2 > t_1.$$

The result that $D_2 > D_1$ is due to the expansion of space itself. The expansion of space leads to an increasing distance between any two galaxies, as measured by any galaxy. There is no 'center' from which everything is receding; rather, all distances are increasing and hence the expansion.

The description of the cosmological behavior of space and time is based on the Friedmann model of the Universe. One does not expect the model to hold all the way to the occurrence of the Big Bang since other processes come into play to create a far more complicated environment. Nevertheless, the Friedmann model is a good guide as what to qualitatively expect and we will base our discussions on this model to study the main features of the Big Bang.

6.7 Age of the Universe

The age of the Universe depends on the model. For $k = 0$ we have $R \simeq t^{2/3}$ and

$$\begin{aligned} H(t) &= \left(\frac{1}{R}\frac{dR}{dt}\right) = \frac{\frac{2}{3}t^{-1/3}}{t^{2/3}} = \frac{2}{3}t^{-1} \\ \Rightarrow t_0 &= \frac{2}{3}H_0^{-1} \approx 9.6 \times 10^9 \text{ years} \end{aligned} \tag{6.23}$$

where $1/H_0 = 14.4$ billion years. The best estimate of the age of the Universe is $t_0 = 13.798 \pm 0.037$ billion years and hence the critical Friedmann model underestimates the age of the Universe.

In general, we have $v = r_0(dR/dt)|_{t=t_0}$, and from Hubble's law $v = H_0 r$. From Fig. 6.4(c) $\tan\theta = dR/dt = v/r_0 = R(t_0)/t_u$ where t_u is the upper bound (maximum) lifetime of the Universe. In other words, if the Universe has been expanding with velocity v since its beginning, its age would be t_u. Since $r = r_0 R(t)$

we have

$$t_u = \frac{r_0 R(t)}{v} = \frac{1}{H_0} \;:\; \text{expansion time.} \tag{6.24}$$

Note that for the Einstein–de Sitter universe

$$t_0 = \text{lifetime of Universe} = \frac{2}{3}H_0^{-1} < H_0^{-1}.$$

The lifetime, as expected, is less that $1/H_0$ since the Universe had been expanding at a faster rate in the past and is now slowing down due to gravitational attraction.

6.7.1 *Critical density: $k = 0$*

The critical mass-energy density ρ_c is the theoretical value for which the energy of the Universe is exactly zero, and from Eq. (6.20), is given by

$$\rho_c = \frac{3c^2 H_0^2}{8\pi G}.$$

Hence, the theoretical critical density today is given by

$$\begin{aligned} \rho_c/c^2 &= \text{critical mass density} = \frac{3H_0^2}{8\pi G} = 0.9 \times 10^{-26}\,\text{kgm}^{-3} \\ \Rightarrow \rho_c/c^2 &= 3\,\text{H atoms/m}^3. \end{aligned} \tag{6.25}$$

Also

$$\rho_c = 4\ \text{GeV/m}^3.$$

Compared to the density of water, which is $10^3\,\text{kgm}^{-3}$, the critical density today looks very small. However, converting the value of ρ_c/c^2 to the density per Mpc yields the following

$$\rho_c/c^2 = 1.37 \times 10^{11} M_\odot/(\text{Mpc})^3$$

where $M_\odot$ is the mass of our Sun. Since $10^{11} M_\odot$ is the mass of a typical galaxy, and that galaxies are typically separated by 1 Mpc, the density of the Universe can be seen to have the same order of magnitude value as the critical density today. This estimate shows that the Universe is at least approximately at the critical density.

Using Eq. (6.25) the curvature parameter k follows from Eq. (6.14) as

$$k = \frac{8\pi G}{3c^2}(\rho(t_0) - \rho_c). \tag{6.26}$$

The question of whether the Universe is open or closed is determined by the density ρ of the Universe. If $\rho(t_0) > \rho_c$ we have $k > 0$ and gravitational attraction overpowers expansion leading to contraction, and a closed universe. If $\rho(t_0) \leq \rho_c$ then $k \leq 0$ and gravity is not strong enough to stop the expansion.

6.8 Energy of the Universe

Taking into account experimental uncertainties, the critical mass-energy density of the Universe today, as mentioned in Sec. 6.5.1, is given by

$$2\,\text{GeV/m}^3 < \rho_c < 10\,\text{GeV/m}^3. \tag{6.27}$$

Is $\rho < \rho_c$ or $\rho > \rho_c$? Do we live in an open or closed Universe? This is going to be determined by $\rho(t_0)$.

To start with, there is an error in the determination of ρ_c due to the imprecision in the experimental value of H_0. From the latest measurement (2013), the best value is given by

$$1/H_0 = (13.798 \pm 0.037) \times 10^9\,\text{years} = (4.354 \pm 0.012) \times 10^{17}\,\text{seconds}.$$

Define the density parameter by

$$\Omega_0 = \frac{\rho(t_0)}{\rho_c}. \tag{6.28}$$

If we take only luminous matter into account

$$0.005 < \Omega_L < 0.04. \tag{6.29}$$

The observed bounds on Ω_0 including dark matter is

$$0.1 < \Omega_0 < 0.4. \tag{6.30}$$

The value of Ω_0 in Eq. (6.30) has a lot of uncertainties due to crude estimates of the amount of dark matter. Since $\Omega_0 < 1$, we can conclude that, so far, the Universe is open. However, if one takes into account of the cosmological constant Λ, which gives another contribution to the density parameter, we obtain $\Omega = \Omega_0 + \Omega_\Lambda$; observations lead to the result

$$\Omega = 1 \pm 0.2. \tag{6.31}$$

Based on the standard model of cosmology, the total mass-energy of the Universe contains 4.9% ordinary matter (particles of the Standard Model), 26.8% dark matter and 68.3% dark energy, as discussed in Sec. 7.5. Hence, the vacuum energy (dark energy) as expressed in Ω_Λ accounts for 68.3% of the mass-energy of the Universe. Dark energy is thought to be responsible for the apparent acceleration of the distant galaxies that was first observed in 1998.

Observations strongly indicate that the large scale geometry of the Universe is almost exactly flat with $k = 0$ and with the value of Ω being uncannily close to 1. This result is all the more remarkable since a detailed study of the Friedmann equation shows that, if at any time in the past the Universe was nearly flat, then both radiation and matter would drive it exponentially fast to a curved universe. In other words, a flat geometry is unstable to small perturbations under the evolution of the Universe. The result that $\Omega \simeq 1$ implies that, at the Big Bang, Ω must have been close to 1 to one part in 10^{30} to have its present day value. This requires a '**fine-tuning**' of the initial state of the Universe to one part in 30 and has bothered physicists as being an *ad hoc* solution; the concept of inflation was partly developed to address this conundrum.

A flat Universe implies that the total energy of the Universe is zero, with the mass-energy and radiation-energy of the galaxies being canceled by the negative gravitational binding energy. The flatness of the Universe finds an explanation in the concept of a Universe that undergoes exponential expansion moments after its birth, called inflation and is discussed in Sec. 6.11.

6.8.1 *Mass-energy and radiation density*

The total energy density ρ is given by the sum of mass-energy carried by particles ρ_m and radiation energy ρ_r carried primarily by electromagnetic radiation (photons) and neutrinos. Hence, the energy density of the Universe is given by

$$\rho = \rho_m + \rho_r. \tag{6.32}$$

Note that the energy density ρ depends on temperature T that, in turn, is changing with time t, namely $T = T(t)$. Hence, we will use both the notations of $\rho = \rho(T)$ and $\rho = \rho(t)$, depending on the phenomena being analyzed.

From Eq. (6.9), as the Universe expands, the volume of the Universe increases and hence the energy density of matter falls and yields

$$\rho_m(t) = \frac{\rho_m(t_0)}{R^3(t)}. \tag{6.33}$$

How much energy density is there in radiation? The energy of radiation at temperature T carried by the different wavelengths λ is given by the black body radiation and is shown in Fig. 6.6. It obeys the Planck distribution and yields

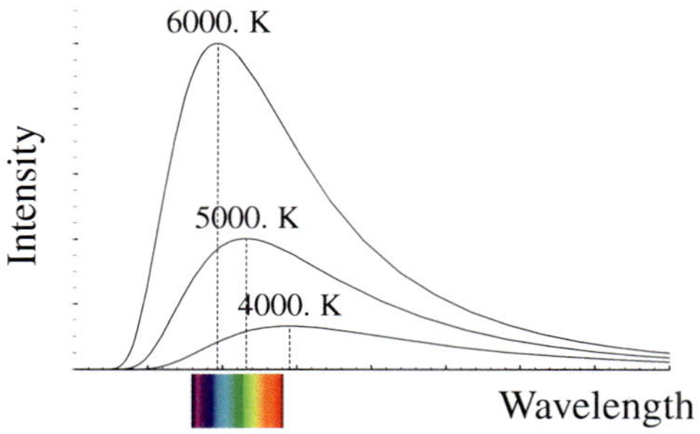

Fig. 6.6 Black body radiation. The maximum of the intensity changes with temperature.

Stefan–Boltzmann's law

$$\rho_r(T) = aT^4(t);\ a = \frac{\pi^2 k_B^4}{15\hbar^3 c^3} = 5.670373 \times 10^{-8}\ \mathrm{Wm^{-2}K^{-4}}\ :\ \text{constant}. \qquad (6.34)$$

The $T = 2.7\,\mathrm{K}$ background radiation yields energy due to photons equal to $0.25\,\mathrm{MeV/m^3}$. Adding the energy due to neutrinos to the radiant energy density yields the total energy density of radiation as given by

$$\rho_r(2.7) = 0.4\,\mathrm{MeV/m^3}.$$

As the Universe expands, it also cools. The maximum amount of radiant energy is carried by the wavelength, denoted by λ_m, for which the intensity is a maximum, and which is shown in Fig. 6.6; its relation to temperature is given by **Wien's law**

$$\lambda_m T = \textbf{constant : Wien's Law}. \qquad (6.35)$$

As the Universe expands, the wavelength λ_m increases since R increases and

$$\lambda_m \propto R.$$

Hence from Eqs. (6.34) and (6.35)

$$T \propto \frac{1}{R} \quad \Rightarrow \quad \rho_r(t) = \frac{\rho(t_0)}{R^4(t)}. \qquad (6.36)$$

In the early Universe temperature T was very high and $R(t)$ was small; hence

$$\rho \simeq \rho_r \propto \frac{1}{R^4(t)} \quad : \text{Radiation dominated} \qquad (6.37)$$

whereas at present the temperature has cooled and hence

$$\rho \simeq \rho_m \propto \frac{1}{R^3(t)} \quad : \text{Matter dominated}. \qquad (6.38)$$

To ascertain for what temperature T_0 matter becomes dominant, note that observations show that today $T = 2.7\,\mathrm{K}$, and which yields

$$\rho_r(2.7) = 0.4\,\mathrm{MeV/m^3} \quad \text{and} \quad \rho_m(2.7) = 0.5\,\mathrm{GeV/m^3}.$$

Hence

$$\rho_r(T) = \rho_r(2.7)\left(\frac{T}{2.7}\right)^4 \quad \text{and} \quad \rho_m(T) = \rho_m(2.7)\left(\frac{T}{2.7}\right)^3.$$

The temperature at which matter and radiant energy become equal is given by

$$\rho_r(T_0) = \rho_m(T_0) \Rightarrow T_0 \simeq 3000\,\mathrm{K}.$$

Once the Universe cools to below 3000 K, electrons and nuclei can combine to form neutral atoms; at higher temperatures atoms would be ionized, with electrons being stripped off the nuclei. Cosmologists call this period of the Universe the

recombination epoch. Photons can no longer be absorbed by the ionized atoms and neither are the photons scattered by free charges. Matter *decouples* from radiation and the Universe becomes transparent to photons. The cosmic background radiation that we detect today arises from the photons that could propagate due to the decoupling.

From the recombination epoch onwards, matter dominates over radiation in the energy density of the Universe. Note that at present $\rho_m \simeq 10^3 \rho_r$, and hence the present Universe is matter dominated.

6.9 The Very Early Universe

Extensive observations show that the Universe started with a **Big Bang** — namely, a hot explosive beginning. Einstein's theory of gravity cannot explain how space and time come into existence.

At the moment of the **Big Bang**, the volume of the Universe is infinitesimal and with infinite curvature; one needs a quantum theory of gravity to address the occurrence of the Big Bang. Hence in Einstein's gravity, "What was before the Big Bang?" is not a meaningful question.[3] Chapters 11 and 12 on particles and forces may need to be read for following the discussions in this section.

6.9.1 *Planck scale*

The **Big Bang** occurs at zero time and not much is known about this state. According to superstring theory all forms of energy at the occurrence of the Big Bang are the states of the superstring. It is speculated that after 10^{-43} s of the Big Bang, the Universe 'cools' to a temperature of about 10^{32} K, with the radius of the Universe being 10^{-35} m, called the Planck length; gravity is unified with the other three forces.

6.9.2 *Grand Unification scale*

After 10^{-38} s, gravity has decoupled from the other forces. Of the infinitely many states of the superstrings, only the quarks, leptons and gauge bosons are present; all the higher Planck mass states of the superstrings are locked out of the physical spectrum of states. Once gravity has decoupled from the other forces, it is described by the Friedmann metric.

[3] The question of what existed before the Big Bang can be addressed in superstring theory. Superstring theory — a leading candidate for the unification of all particles and forces as well as a theory of quantum gravity — has a description of cosmology that provides a more general framework than Einstein's, and is discussed in Sec. 15.7. In particular, in superstring theory the Big Bang is explained by a collision — taking place once in about a trillion years — of two 'E3-branes'.

The three forces, namely electromagnetic, weak and strong, remain unified as one single force and with a single coupling constant upto a temperature given by

$$T \simeq 10^{29}\,\mathrm{K}. \tag{6.39}$$

Quarks and leptons exist in a plasma ('free') state at energy scale of $\sim 10^{16}$ GeV.

6.9.3 *Electroweak scale*

10^{-12} s after the beginning, strong interactions decouple from electromagnetic and weak interactions. The Higgs field condenses and acquires a spontaneous vacuum expectation value, discussed in Sec. 13.8. The typical energy of particles in the Universe is given by the electroweak scale, namely

$$E_{\mathrm{EW}} \sim 1\ \mathrm{TeV} = 10^3\,\mathrm{GeV} \tag{6.40}$$

and yields the **Higgs phase transition** temperature, given from Eq. (6.39) by

$$T_H \simeq 10^{29}\,\mathrm{K} \times \frac{10^3}{10^{16}} = 10^{16}\,\mathrm{K}. \tag{6.41}$$

The strong force separates from the electromagnetic and weak interactions, which are still unified into a single force. The condensation of the Higgs field endows masses to the quarks, leptons and weak bosons, and is discussed in Sec. 13.8.

6.9.4 *Condensation of quarks*

When the energy density has reached

$$E_{\mathrm{hadrons}} \sim 1\ \mathrm{GeV} \tag{6.42}$$

which is 10^{-6} s after the Big Bang, quarks condense to form hadrons.

The matter/antimatter symmetry is also broken, with matter becoming dominant.

6.9.5 *Formation of light nuclei*

10^2 s $\simeq$ 3 min after the Big Bang, the light nuclei, namely ^{2}H, ^{3}He, ^{4}He and ^{7}Li, are formed.

6.9.6 *Formation of atoms*

About 10^{13} s $\simeq$ 380,000 years after the Big Bang, the radiation-dominated era ends with the Universe cooling to

$$T \simeq 3000\,\mathrm{K}.$$

At this temperature, atoms form, and the absence of free charge makes the Universe transparent to radiation which *decouples* from matter.

6.9.7 *Formation of galaxies*

After 1–2 billion years galaxies are formed.

6.9.8 *Present*

13.78 billion years after the Big Bang, galaxies have formed which are receding away (Hubble's law) and the 3000 K radiation has further cooled to the 2.7 K background radiation.

6.10 Big Bang Nucleosynthesis

In addition to Hubble's law and the 2.7 K background radiation, the third pillar of Big Bang cosmology is the **primeval abundance** of the light elements, namely hydrogen, deuterium, helium and lithium, or H, D $= {}^2$H, ^{4}He and ^{7}Li. From nuclear astrophysics, it can be shown that the light elements ^{4}He and ^{7}Li could not have been produced — in the observed quantities — inside stars; hence their formation and relative abundance must be explained by conditions existing soon after the Big Bang.

Quarks condense into hadrons — almost solely protons (p) and neutrons (n) — 10^{-6} s after the Big Bang. As long as the temperature is high enough, there are an equal number of neutrons and protons, which are in equilibrium via the following reactions.

$$\mathrm{n} + \nu_e \to \mathrm{p} + \mathrm{e}^-; \quad \mathrm{p} + \mathrm{e}^- \to \mathrm{n} + \nu_e$$
$$\mathrm{n} + \mathrm{e}^+ \to \mathrm{p} + \bar{\nu}_e; \quad \mathrm{p} + \bar{\nu}_e \to \mathrm{n} + \mathrm{e}^+.$$

Note that k_BT has the units of energy, and hence a convenient way of measuring temperature T is to specify the value of k_BT in energy units of MeV and use the energy of the Universe as a shorthand way of referring to its temperature.

Once the temperature of the Universe cools below 0.8 MeV in about $10^2\,\mathrm{s} \simeq 3$ minutes after the Big Bang, the reactions above cease and the number of protons and neutrons are 'frozen', fixed by the relative masses of the proton and neutron, and by the temperature of 0.8 MeV.

Since the protons and neutrons are in thermal equilibrium, their relative abundance is given by Boltzmann's formula which states that the probability of finding a particle with energy $E = mc^2$ is proportional to $\exp\{-E/k_BT\}$; and hence

$$\frac{N_n}{N_p} = \frac{\text{Number of neutrons}}{\text{Number of protons}} \simeq \exp\left[-\frac{(m_n - m_p)c^2}{k_BT}\right] \simeq \frac{1}{5}$$
$$k_BT = 0.8\,\mathrm{MeV}$$

where m_n, m_p are the mass of the neutron and proton respectively.

The proton's mass is slightly smaller than that of the neutron and is the reason that protons are more numerous.

Half-life is defined to be the time for which half the number of unstable particles have disintegrated. Namely $N(t) = N_0(1/2)^{t/t_0}$, where t_0 is the half-life. Neutrons inside a nucleus are stable; however, isolated neutrons are unstable and disintegrate into protons with a half-life of 641 sec.

The only way neutrons can survive after the Universe cools below 0.8 MeV is by being bound inside nuclei, in particular, in the form of light nuclei. However, protons and neutrons can bind together to form nuclei that are stable only for temperatures less than 0.1 MeV — disintegrating at higher temperatures. A rather remarkable fact is that the Universe cools from 0.8 MeV to 0.1 MeV in 440 sec — short enough for there to be a sizable number of neutrons even after some neutrons have disintegrated into protons. From the definition of half-life, the abundance of neutrons is given by

$$\left.\frac{N_n}{N_p}\right|_{k_BT = 0.1\ \text{MeV}} = \frac{1}{5} \times \left(\frac{1}{2}\right)^{(440/641)} = \frac{1}{5} \times \exp\left[-\frac{440 \times \ln 2}{641}\right] \simeq \frac{1}{8}$$

In the early Universe, the two elements that are copiously produced are the hydrogen nucleus, which consists of all the protons that are in excess of neutrons, and the helium nucleus, which is produced by the following three-step reaction

$$\text{p} + \text{n} \to \text{D};\ \text{D} + \text{p} \to {}^3\text{He};\quad \text{D} + \text{D} \to {}^4\text{He}.$$

Each helium nucleus ^{4}He consists of two neutrons and two protons and, to first approximation, all neutrons end up inside a helium nucleus (this estimate will be further refined to account for the production of other light nuclei). The total number of He nuclei is hence equal to $N_n/2$: half the total number of neutrons present at temperature 0.1 MeV. Ignoring the small mass difference between protons and neutrons, the mass of each ^{4}He nucleus is approximately given by $4m_n$, and the total mass of the Universe is approximately given by $m_n(N_p + N_n)$. The fraction of the total *mass* of the Universe in the form of Helium ^{4}He, denoted by Y_4, is hence given by

$$Y_4 = \frac{4m_n(N_n/2)}{m_n(N_p + N_n)} = \frac{2}{1 + N_p/N_n} = \frac{2}{9} \simeq 0.22.$$

Based on our simple but robust reasoning, we have arrived at a remarkable conclusion — that the relative abundance (by mass) of ^{4}He in our Universe is about 22%.

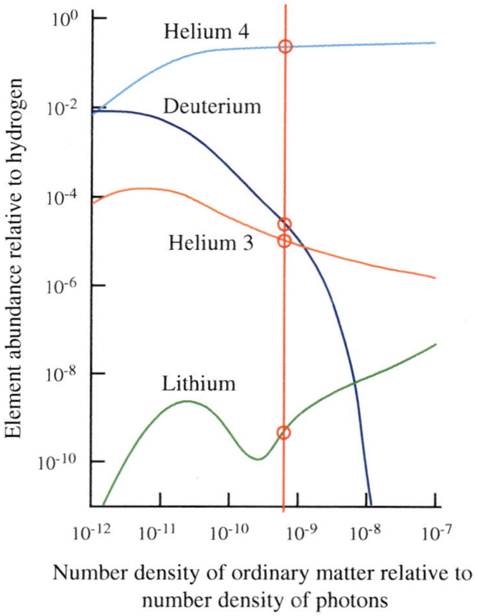

Fig. 6.7 The relative abundance of the light elements as a function of the density of baryonic matter. The vertical red line is the *observed* number density of baryonic matter and the circles are the theoretical predictions of the abundance of the light elements.

A more careful analysis of the early Universe leads to the abundance of ^{4}He being slightly higher, between 23% and 24%, with the relative mass abundance of D, ^{3}He and ^{7}Li being about $10^{-4}, 10^{-5}$ and 10^{-10} respectively.

The prediction of the relative **abundance of light elements** — ranging over ten orders of magnitude — is an extremely stringent test of the Big Bang model of cosmology. As shown in Fig. 6.7, the predictions of the model are confirmed to a high degree of accuracy by experimental data; in particular, the abundance of deuterium varies sharply with the density of matter and provides a sensitive test for the models of cosmology.

The relative abundance depends crucially on the energy density of baryon matter ρ_{baryon} since it contributes about 4.4% of the energy density required for a closed Universe. The observed pattern of the abundance of the light elements is consistent with the Big Bang provided

$$1.5 \times 10^{-31}\,\text{gcm}^{-3}/c^2 < \rho_{\text{baryon}} < 4.5 \times 10^{-31}\,\text{gcm}^{-3}/c^2.$$

The observed energy density is $\rho_{\text{baryon}} = 4.19 \times 10^{-31}\,\text{gcm}^{-3}/c^2$, which is within the range required by Big Bang nucleosynthesis.

6.11 Inflationary Universe

The hot Big Bang model of cosmology has many successes, some of which have been discussed above. There are however a number of logical and observational problems that need to be resolved. The three outstanding problems that the **inflationary Universe** solves are the following:

Flatness Problem

Recall from Eq. (6.31)

$$\Omega = 1 \pm 0.2.$$

The value of Ω close to 1 indicates that the large scale geometry of the Universe is very nearly flat. Why should the Universe be almost flat? Is this simply a coincidence or a reflection of some deeper underlying principle?

Horizon Problem

Suppose one, located at galaxy G, observes the Universe in opposite directions, say two quasars at positions A and B; if the two quasars are receding with some high enough redshift then their past causal light cones, as in Fig. 6.8, *do not seem* to have any overlap since the Universe has only a finite lifetime of 13.78 billion years. What this means is that the two quasars, or what gave rise to them, do not seem to have been in causal contact in the past.

In general, it seems as if distant parts of the present day Universe were not in causal contact with each other. This brings us into conflict with the isotropy of the background radiation which, to one part in 10^5, shows that the background temperature is uniformly 2.725 K. How can the temperature of all parts of the Universe have almost the same (but not quite as we will see shortly) temperature if they had not been in causal contact?

For the Universe to reach thermal equilibrium, all the different regions of the Universe had to be in thermal (causal) contact before the decoupling of radiation and matter took place. The problem is even more subtle: the Universe cannot be

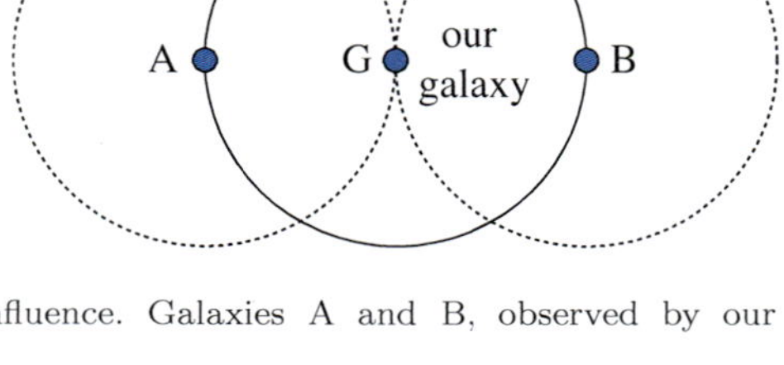

Fig. 6.8 Spheres of influence. Galaxies A and B, observed by our galaxy G, are causally disconnected.

exactly isotropic since minute amounts of anisotropy (irregularity) is required for the emergence of stars and galaxies.

Exotic Relics Problem

The theory of electroweak interactions and quantum chromodynamics, and their incorporation into Grand Unified Theories predict that magnetic monopoles and other exotic particles should be produced copiously during the hot Big Bang. All attempts to detect these exotic particles have failed, leading one to speculate that the appearances of these exotic forms of matter is forbidden for some fundamental reason.

6.11.1 *Inflaton field*

The inflationary theory postulates a period of rapid inflation of the volume of the Universe by a factor of almost 10^{30} that lasted from 10^{-35} seconds to about 10^{-32} seconds after the Big Bang.[4]

The inflationary Universe contains all the leptons and quarks and interactions of the Standard Model, discussed in Chapter 13; in addition, it is postulated that the Universe has another *scalar field* ϕ called the inflaton field; this field has not yet been experimentally observed.

The Universe starts off in a phase that has non-trivial vacuum state in which the inflaton field is in an unstable vacuum configuration. The inflaton field consequently undergoes a *phase transition* to its stable vacuum configuration. The concept of a phase transition is briefly discussed in Sec. 13.8.1 and also in Sec. 13.10. For example, the Standard Model, discussed in Chapter 13, hinges on the Higgs field having a non-trivial vacuum state in which the Higgs field has a non-zero expectation value.

The initial vacuum state of the inflaton field is inherently unstable. The analogy of an unstable state is a rock sitting on a mountain peak; any small move away from the peak will cause the piece of rock to plummet down and end up in the valley; this is shown in Fig. 6.9.

The inflaton field can have an expectation value, denoted by $v = E[\phi]$, and is similar to the Higgs field having a non-zero expectation value, as discussed in Sec. 13.8.3. The inflaton field is part of the geometry of spacetime, and the expectation value of the inflaton field determines the measure of distance between two points of spacetime. The potential $V(\phi)$ driving the inflaton field is shown in Fig. 6.10.

The initial 'false vacuum', before inflation, shown in Fig. 6.10, has $v = 0 = E[\phi]$. The inflaton field slowly 'rolls' down the potential hill, as shown in Fig. 6.10,

[4]The scale of inflation varies from 10^{25} to 10^{80}, depending on the model that one uses. We use the notional value of 10^{30} for the scale of the expansion to illustrate the general results of the inflationary model.

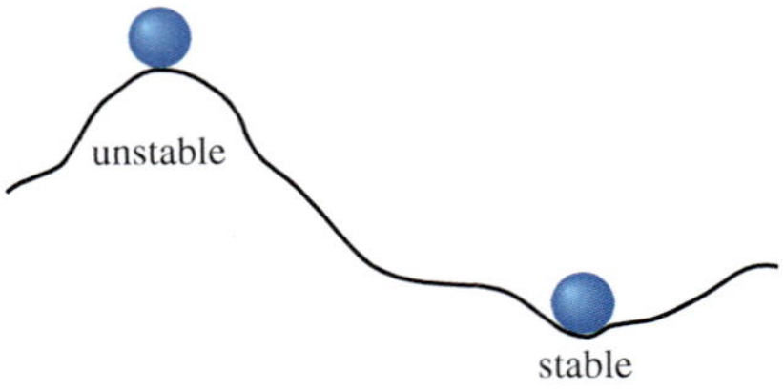

Fig. 6.9 Stable and unstable states.

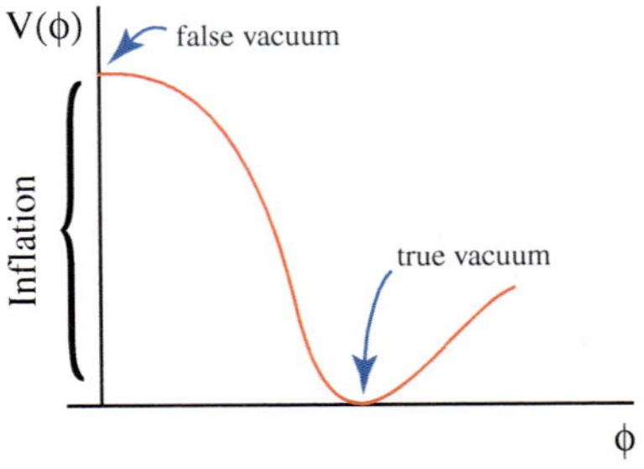

Fig. 6.10 True versus false vacuum.

to its true vacuum with $v = E[\phi] \neq 0$. The inflaton field occupies all of space, and on tumbling from its unstable value its non-zero expectation value spreads exponentially fast throughout space.

The analogy for the exponential spreading of the inflaton field is the freezing of water into ice, called a first order phase transition. If water is supercooled very carefully and without any external disturbance below its freezing temperature of 0°C, it can remain in its liquid state, but is unstable. The smallest disturbance delivered to its container will immediately cause the supercooled water to turn to ice; on being perturbed, any minute amount of impurity or piece of dirt can form a center for the nucleation of ice, and the formation of ice then spreads at an *exponential* speed through the water, turning it into ice. The liquid state of the supercooled water is analogous to the false vacuum of the system and the ice state is analogous to the true stable vacuum.

The inflaton field 'condenses' (spreads exponentially) from its unstable false vacuum to its true stable vacuum, and this exponential spreading inflates space exponentially. As mentioned earlier, the inflationary phase of the Universe takes place from 10^{-35} sec to 10^{-32} sec, with the volume of the Universe increasing by a factor of about 10^{30}. During the inflationary period the scale factor $R(t)$ increases exponentially fast and is given by

$$R(t) = e^{at}; \quad 10^{-35}\,\text{sec} \leq t \leq 10^{-32}\,\text{sec}.$$

After inflation is over, the Universe settles down to the usual expansion. Figures 6.11 and 6.12 show the effect of inflation and the subsequent expansion on the structure of the Universe.

On reaching its stable vacuum, the inflaton field decouples from the rest of the fields and has no further effect on the Universe. There is an immense amount of 'latent heat' of energy present in the inflaton's false vacuum, and which is released when the inflaton field 'rolls' slowly to its true and stable vacuum. The energy released, similar to the release of latent heat of fusion for the water–ice transition, is converted to ordinary matter, that is to quarks, leptons and force fields. The sudden

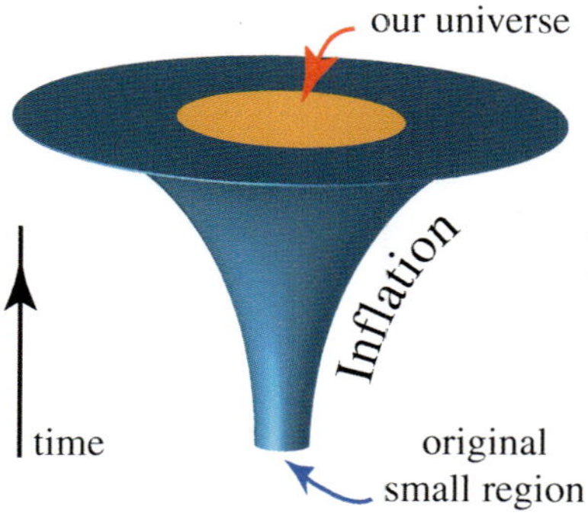

Fig. 6.11 Inflation.

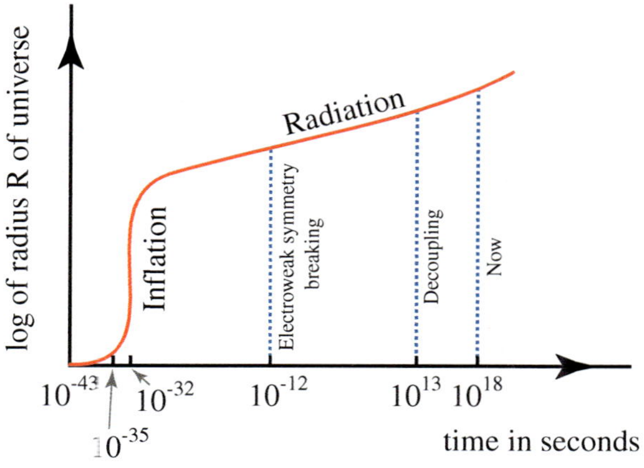

Fig. 6.12 Timeline of the Universe.

release of enormous amounts of energy is manifested as an intense high temperature cosmic explosion and takes the form of a fireball — a *second explosion* resulting in the *re-heating* of the Universe due to inflation.

6.11.2 *Flatness problem*

The Universe that results from inflation is naturally a flat Universe. The reason being that our Universe starts its inflation from an infinitesimal volume of space and with zero energy. The positive energy in the inflaton field is exactly equal to the negative energy due to gravitational attraction. Hence the Universe that results from inflation has zero energy, that is $E = 0$. In the Friedmann equation a Universe with $E = 0$ leads to a flat spacetime.

The energy of the inflaton field has some unusual properties, having negative pressure and giving rise to a repulsive gravity. Some of the energy from the inflaton may contribute to dark energy, which is thought to account for about 68.3% of the mass-energy of the Universe.

The inflationary phase causes exponential stretching and leaves the Universe as being flat to one part in 10^4; the inflationary model also correctly predicts that it is the quantum fluctuations in the early Universe that precisely generate the level of anisotropy required for galaxy and star formation.

6.11.3 *Horizon problem*

At the start of inflation, the size of the Universe was about 10^{-30} m and after inflation its size was about 1 m. Hence, inflation causes 'stretching out' of space by a fantastic factor of 10^{30} and irons out any inhomogeneity, asymmetry and anisotropy that different parts of space may have had, making the present day Universe isotropic and homogeneous.

The inflationary scenario changes the conception we have of the causal past of our Universe. Figure 6.8 shows that distant quasars do not apparently have a causal connection, and this is shown in Fig. 6.13(a): galaxies A and B observed by galaxy G are causally disconnected. In Big Bang without inflation, events u and v that evolve into galaxies A and B, were not in causal contact as shown in Fig. 6.13(a). In the Big Bang *with* inflation, as shown in Fig. 6.13(b), events u and v were in causal contact at spacetime point C when inflation took place, as shown by the red line in Fig. 6.13(b). Since they emerge from the same region around point C of spacetime that underwent inflation, galaxies A and B were in thermal equilibrium before inflation — and continue to be in thermal equilibrium even after inflation is over.

6.11.4 *Exotic relics*

Due to the immense expansion of the volume of space during the inflation period, the density of exotic particles produced becomes negligibly small, and hence explains their absence in today's Universe.

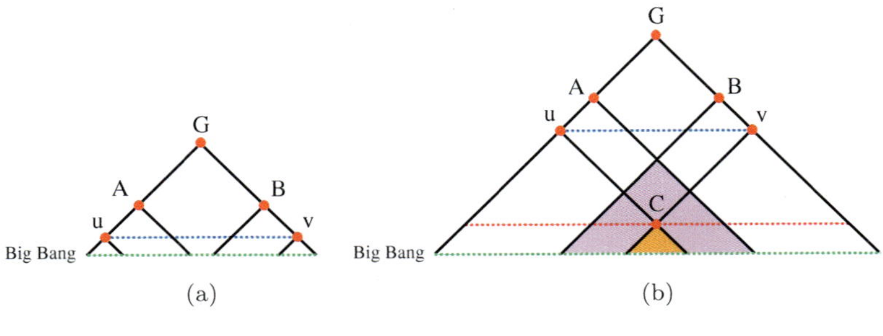

Fig. 6.13 Light cones illustrating the causal structure of the early Universe. (a) Events u and v are not in causal contact and consequently, when observed by galaxy G, the galaxies A and B have no causal connection. (b) Due to inflation, events u and v emerge from the same point C and hence are in causal contact; consequently A and B are causally connected.

In addition to resolving the apparent contradictions of hot Big Bang cosmology, a rather surprising conclusion of the inflationary model is that, quite literally, the entire Universe emerged out of an infinitesimal volume of space and from zero energy!

6.12 Cosmic Background Radiation

After inflation, as discussed in Sec. 6.11, the Universe was hot and filled with radiation as well as free protons and electrons. The radiation and free charges were all in a state of a rapidly evolving **thermal equilibrium**. This is the reason that the radiation of the Universe has a black body spectrum.[5]

As the Universe evolves, it cools and expands. In the early life of the Universe, for thermal energy $kT \gg 1\,\text{eV}$, radiation ionized any atoms that formed, leading to radiation being coupled to free electrons and protons. Once the Universe has cooled to 3000 K, the energy of radiation falls below 13.6 eV that is required for ionizing the hydrogen atom, and hence leading to the formation of atoms and radiation becoming decoupled from matter. Since then the Universe has expanded by a factor of 1000 and hence photons from the early Universe now have a temperature of about 3 K. The cosmic photons have number density $= 4 \times 10^8/\text{m}^3$ and energy density $= 2.5 \times 10^5\,\text{eV}/\text{m}^3$. These photons are in equilibrium with a black body spectrum experimentally determined to be at a temperature of $2.725 \pm 0.001\,\text{K}$ and shown in Fig. 6.14.

The remarkable fact of the cosmic black body spectrum is that it has the same intensity to one part in 10^5 in all directions showing that the cosmic background is almost perfectly isotropic, as shown in Fig. 6.14. Minuscule deviations from isotropy

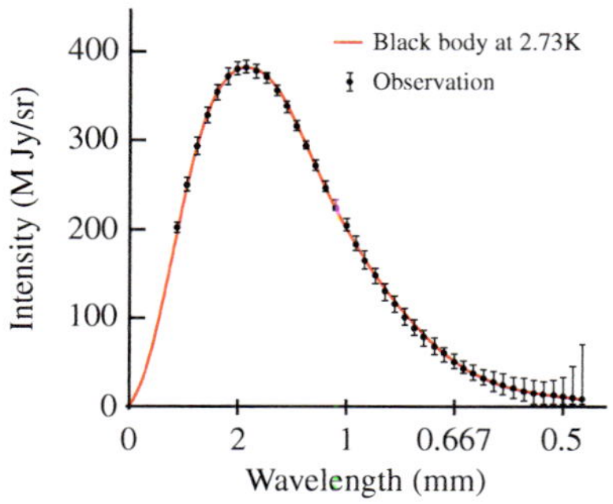

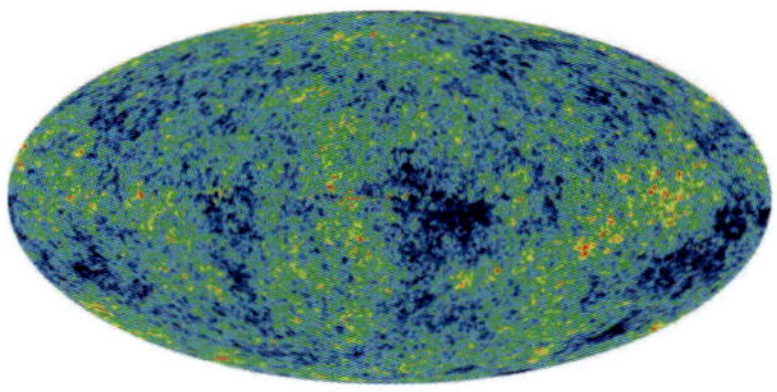

Fig. 6.14 The cosmic microwave background. The figure on the left is the observed black body cosmic background. The figure on the right shows the deviations of the cosmic background from homogeneity.

---

[5]Black body radiation is the spectrum of radiation emitted by a body that is in thermal equilibrium with radiation at some temperature T.

provide a clue of how and why galaxies are formed. The cosmic background radiation together with Hubble's law and the relative abundance of the light nuclei form the experimental cornerstone of the Big Bang model of cosmology.

6.13 Primordial Microscopic Black Holes

In our discussion of Hawking radiation we had found that black holes evaporate by radiating off their energy and finally *explode* since their temperature goes as the inverse of their mass. The word explosion is used to describe the end point of black hole evaporation; the reason being, as shown in Fig. 5.21, the rate of evaporation increases rapidly, going to infinity as the mass approaches zero.

The lifetime of a black hole is given, from Eq. (5.26), by

$$t_E = 5120\pi \frac{G^2 M^3}{c^4 \hbar} = 8.4 \times 10^{-17} \left[\frac{M}{\text{kg}}\right]^3 \text{ s} \tag{6.43}$$

where M is measured in the units of kilograms. Black holes with solar mass or larger will evaporate on time scales running into trillions of years. The role of supermassive primordial black holes is briefly discussed in Sec. 8.8. In some models of the Big Bang, due to very high densities and temperatures, primordial microscopic black holes, with very light mass, could have been formed. Microscopic black holes, due to their small mass, evaporate much faster.

The present age of the Universe is 13.78 billion years. Figure 5.21 shows that a primordial black hole with mass of about 10^{14} g produced at the Big Bang would explode during the present epoch, releasing distinctive high energy gamma rays that could be detected in the tiny variations of the background cosmic microwave radiation. Black holes with masses much smaller should have all exploded by now and those with larger masses should do so in the future. Hence, if one can observe exploding black holes, this would provide remarkable confirmation for the existence of primordial black holes as well as of quantum effects taking place in gravity. There has so far been no experimental confirmation of these expectations.

6.14 The Dark Age of the Universe

When the Universe was about 380,000 years old, its temperature had cooled to 3000 K — the Sun's surface temperature; the Universe's temperature was cool enough to allow for the protons and electrons to combine and form hydrogen atoms. When protons and electrons are freely moving, they can scatter radiation by a process called Thomson scattering and are thus opaque to all wavelengths of radiation. However, once a proton and an electron combine to form a (neutral) hydrogen atom, they no longer scatter radiation; instead, the hydrogen atoms can only absorb incident radiation with specific wavelengths, and hence become transparent to radiation for the rest of wavelengths.

The Universe was filled with the uniform red glow of cosmic black body radiation during the formation of large clouds of hydrogen gas formed by the combination of protons and electrons that are almost completely transparent to light. As the Universe expanded and the temperature fell further, the cosmic background radiation shifted to the infrared and appears completely dark to the human eye. The Universe does not emit any visible light for roughly the next 200 million years until the formation of stars gives the first light that is not a part of the cosmic background radiation. This period of the Universe is called the Dark Age, and extends from about 380,000 years after the Big Bang to about 200 million years after the Big Bang.[6] The Dark Age comes to an end with the reionization of the hydrogen atom as well as by the appearance of the 'normal' visible stars, which are organized into small dwarf galaxies containing no more than a million stars.

Astronomers have been searching for clues, based on the laws of Physics and indirect observations, as to what transpired during the Dark Age. During this period the Universe emitted radiation that, due to the expansion of the Universe, would appear to consist of radio waves, microwave and infrared radiation. Special telescopes are being built to detect the so-called 21 cm line of the hydrogen atom, as it is expected that this line contains rich information about what occurred during the Dark Age.

The main stages of the Dark Age are thought to be the following.

(1) Approximately 380,000 years after the Big Bang, the Universe cools to about 3000 K leading to the formation of hydrogen atoms that, in turn, assemble into large transparent clouds. Since the Universe becomes transparent, the Universe's radiation, having a black body spectrum corresponding to the temperature of 3000 K, can propagate freely. This is the radiation that we observe today as the cosmic microwave background — redshifted due to the Universe's expansion.
(2) Dark matter, accounting for about 84.5% of all matter, is spread unevenly throughout the Universe. Regions with larger concentrations of dark matter attract hydrogen and helium atoms to form denser clouds. These in turn attract more gaseous matter leading the hydrogen cloud to coalesce into gas dense enough to start thermonuclear fusion and hence leading to the birth of the first stars.
(3) The earliest stars were about 25 to 250 times more massive than our Sun and were formed approximately 100 million years after the Big Bang. Given their large mass, these stars burned furiously for only a few million years

[6]The occurrence of the Dark Age is not very precise, with estimates in the range of 100 to 800 million years. We use the approximate time of 200 million years to represent this range.

before exploding into supernovae and spewing out heavy elements into the Universe. The remnants of the exploding mega-stars were mostly black holes. The radiation and wind of particles from these mega-stars were so intense that they dispersed the neighboring clouds of hydrogen and helium and thus prevented the nearby formation of other stars.

(4) To ionize a hydrogen atom one needs 13.6 eV of energy, which is the energy carried by an ultraviolet photon. The radiation from the mega-stars, or maybe intense X-rays from black holes, *re-ionized* the hydrogen atoms, breaking these up into protons and electrons that could scatter light. Hence, the Universe continues to be dark and opaque to light.

(5) The Universe at this stage consisted of mega-stars and clouds of hydrogen and helium with only dwarf galaxies (cluster of stars) being allowed to form. The exploding mega-stars seeded the formation of regular stars and the creation of the first dwarf galaxies. The heavier elements such as oxygen and carbon created by the exploding mega-stars changed the chemistry of the gaseous clouds allowing for the coalescence of smaller amounts of gas into smaller stars.

(6) Due to the enormous expansion of the Universe, the density of protons and electrons became low and hence the Universe continued to be transparent. The radiation of the newly formed normal stars created visible light that propagated freely and thus brought the Dark Age to a closure, about 200 million years after the Big Bang.

6.15 Black Holes: Entropy of the Universe

The entropy of our Universe is the sum of the entropy of our visible Universe plus entropy of dark matter plus entropy of black holes, that is

$$S_{\text{Universe}} = S_{\text{visible}} + S_{\text{black holes}} + S_{\text{dark matter}}.$$

The entropy of the visible part of the Universe is due to the entropy of volume filling radiation (contributions from stars is negligible). To illustrate this consider a spherical volume of the Universe that is contained in a ball of radius 13.78 billion light years. Currently the Universe's entropy density due to radiation is $S_0 = 2970(T/2.75\,\text{K})^3\,\text{cm}^3$; hence, the visible entropy of the Universe due to background radiation at $T = 2.75\,\text{K}$ is given by

$$S_{\text{visible}} \simeq 2.35 \times 10^{87}\, k_B.$$

From Eq. (5.18), we see that even a few massive black holes with $M = 10^5 M_\odot$ have entropy larger than the entropy of the entire visible Universe! Note that almost 30% of all galaxies have supermassive black holes at their core with masses given by $10^6 M_\odot \le M \le 10^9 M_\odot$; there are about 10^{11} galaxies and, from Eq. (5.18), the entropy of black holes, hidden behind the event horizons, is estimated to be

$$S_{\text{black holes}} \simeq 10^{101} k_B \Rightarrow \frac{S_{\text{visible}}}{S_{\text{black holes}}} \simeq 10^{-14}.$$

Hence, the entropy of the visible Universe is completely negligible when compared to the entropy carried by black holes, being less than one part in 10^{14}. One explanation is that most of the supermassive black holes were formed in (a) the protogalactic cores, from almost perfectly spherical gravitational collapse of large amounts of matter or (b) were produced at the onset of galaxy formation. This would account for the almost neglible release of visible entropy into the Universe during the formation of the supermassive black holes. The production of supermassive black holes by the merger of smaller black holes can be ruled out by entropy considerations, since such a merger process would release far more visible entropy than the observed amount.

6.16 The Answer

The timeline for the history of the Universe is shown in Fig. 6.15.

The present day view of cosmology is that the Universe begins with a fiery explosion (the Big Bang); the remnant of the explosion is observed today in the form of the cosmic background radiation. After the explosion, the Universe goes through a very short period of inflation that explains why the observed Universe is

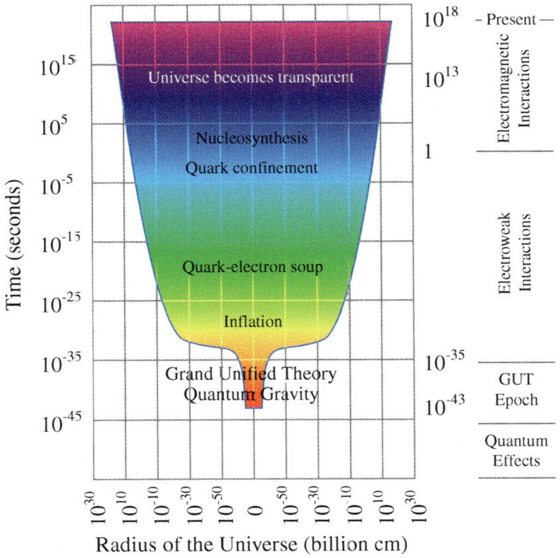

Fig. 6.15 Evolution of the Universe.

almost exactly flat. The Universe after inflation starts the expansion that is observed today in which all galaxies are receding from each other.

The quarks condensed into protons and neutrons and charged particles were in equilibrium with radiation.

As the Universe cools further, the light nuclei are formed 3 minutes after the Big Bang. The primordial cosmic soup produces the light nuclei that make up 98% of the observed matter in the Universe, with the remaining 2% being produced later in stellar fusion. Radiation decouples from matter about 380,000 years after the Big Bang leading to the Dark Age that lasted about 200 million years. Stars formed at the end of the Dark Age gave forth starlight, bringing the visible Universe into existence.

Big Bang Cosmology provides a mathematical description and explanation of an expanding Universe based on Einstein's theory of gravity. The size and age of the Universe are model dependent, and using the model, we could make many precise estimates of the geometrical structure of the Universe, including the value of the Hubble's constant.

Observations show that the Universe appears to be almost exactly flat, having a total energy of zero — the positive energy of the matter content of the Universe being exactly balanced by its negative gravitational binding energy. The flat Universe will continue to expand forever.

Chapter 7

Dark Universe

Why is the Universe so dark?

7.1 The Question

During the day, the sky may be blue or overcast but either way very little can be seen of the Universe. On a nice day, one can see the Sun and possibly the Moon and with some luck even a few planets. During the day, it is hard to recognize that we are living in a huge Universe consisting of an immense number of stars. The real magic happens at night when, paradoxically, it is darkness that enables us to see beyond our immediate neighborhood.

Through careful observation at night we discover that there are trillions of stars organized into hierarchical structures permeating an enormous Universe. Yet, if there are so many stars, then why isn't the night sky bright? And could it be so that the darkness of the night sky harbors more (to the naked eye) invisible mysteries? Hence, in this chapter, we would like to ask: **Why is the universe so dark?**

7.2 Dark Sky

The daily cycle of light and dark has a profound impact on life on Earth. Amongst many other things, it affects the weather, food production, photosynthesis and the organization of mankind's society. At first, the fact that the night is dark may not be very surprising. After all, when the Sun goes down, the main light source disappears from view and darkness can be expected. However, careful observation of the night sky shows that there is an enormous number of stars. For example, Fig. 7.1 shows many stars and the remnants of a supernova in the Large Magellanic Cloud. These stars are far away so just like distant light bulbs, we expect them to be dim as well. But is such an expectation valid?

An elegant argument, often referred to as **Olbers' paradox** after the German physician and astronomer **Heinrich Wilhelm Matthäus Olbers**, shows that the darkness of the night sky is far from trivial. Olbers' paradox is based on the assumptions of a static and infinite Universe filled, at large scales, with a relatively uniform distribution of stars. These assumptions were quite reasonable for early astronomers since there was no indication of the Universe evolving — hence it must be static; since each improved lens revealed more distant stars — it must be very large as stars are found in every corner of the sky and the distribution of stars is uniform.

Consider a number of shells of stars centered around Earth at relatively large distances as illustrated in Fig. 7.2. The intensity of light decreases as $1/r^2$ when moving away from its source so that stars in more distant shells will be dimmer proportional to $1/(4\pi r^2)$ (*i.e.* inversely proportional to the surface area of a sphere: $A = 4\pi r^2$). At the same time, however, the number of stars increases proportional

Fig. 7.1 Composite image of N49, a supernova remnant in the Large Magellanic Cloud. Source: NASA.

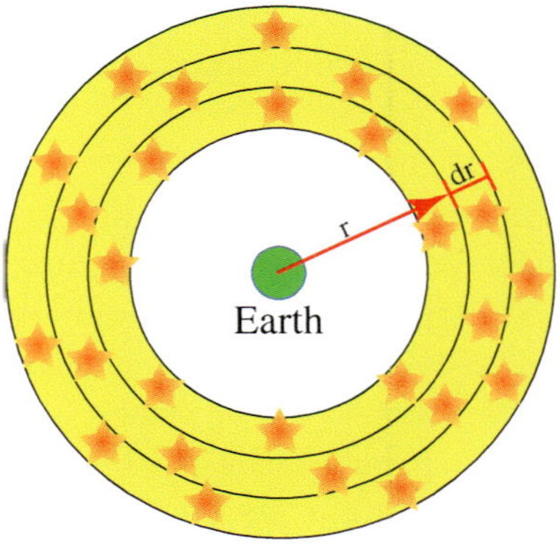

Fig. 7.2 The number of stars per unit volume is the same in each shell.

to $4\pi r^2$ for more distant shells due to the increase of the volume of the shells (the number of stars per unit volume of space is assumed to be constant). Consequently, the total amount of light reaching Earth from each shell is constant since the $1/r^2$ and r^2 terms cancel out. Now if the Universe is infinite, there is an infinite number of shells and even if the amount of light per shell is small, the sum of the light from all the shells is infinite and the night sky should be infinitely bright. Of course this is not the case, but why?

Noteworthy (optional content) 7.1: Olber's Paradox: The night sky should be bright

---

Take a shell ranging from a distance of r to $r + dr$. We can obtain its volume by calculating the volume of a sphere with radius $r + dr$ and subtracting the volume of a sphere with radius r. The general formula for the volume of a sphere is given by: $V = \frac{4}{3}\pi r^3$. Hence we have

$$V_{\text{shell}} = \frac{4}{3}\pi \left((r + dr)^3 - r^3\right) = \frac{4}{3}\pi \left(3r^2\, dr + 3r(dr)^2 + (dr)^3\right) \approx 4\pi r^2 dr.$$

The terms $3r(dr)^2 + (dr)^3$ can be neglected if we make sure that the shell is very thin in relation to the size of the sphere or in other words, we take $dr \ll r$. Given a density n of stars per unit volume and an average luminosity L per star, we then have for the total amount of light O coming from a shell at a distance r

$$O = \frac{nL \times 4\pi r^2 dr}{4\pi r^2} = ndr. \tag{7.1}$$

Integrating O from 0 to infinity gives the total amount of light O_{total}:

$$O_{\text{total}} = \int_0^\infty nLdr = [nLr]_0^\infty = \infty. \tag{7.2}$$

The fact that the night sky is not bright leads one to the inevitable conclusion that one or more of the assumptions leading to Olbers' paradox must be wrong.

Careful observation indicates the assumption of a constant density of stars at large scales is correct. Indeed, our current understanding of the Universe considers it to be both homogeneous (*i.e.* the distribution of stars at large scales is everywhere the same in the Universe) and isotropic (*i.e.* the distribution of stars at large scales is the same in all directions). Furthermore, we have no indication that distant stars are any different from nearby stars in terms of the physical laws that govern them (otherwise, distant stars could intrinsically be less luminous). Indeed, modern astronomy, just like the rest of science, is founded on the notion that the fundamental laws of Nature are 'universal' and applicable throughout the entire Universe. We must therefore infer that the Universe is not static and/or not infinite. As discussed in Chapter 6 on cosmology, the Universe is not infinite and neither is it static. We discuss this further.

The astronomer **Edwin Hubble** discovered in the 1920's that the Universe is expanding and that this expansion occurs at the same rate throughout space by analyzing the spectra of distant stars. He found that there is a linear relationship between the redshift in the spectrum of a star and its distance. This linear relationship is called **Hubble's law** and applicable to all stars sufficiently far from Earth (see also Fig. 6.3, and Secs. 6.3 and 6.4.1). Since the redshift is directly proportional to the recession speed (see below), this means that the greater the distance of a star, the greater its recession speed.

Hubble's law is not valid for nearby stars.

7.3 Origin of Redshift

The question that immediately arises is: what is the origin of the redshift? The first thing that comes to mind is probably the Doppler effect. In the **Doppler effect**, approaching waves are compressed (*i.e.* blueshifted in the case of light or giving a higher pitch in the case of sound) and receding waves are stretched (*i.e.* redshifted in the case of light or giving a lower pitch in the case of sound). Hence one might surmise that the redshift of distant stars is due them receding from us. However, one needs to be careful here. Although it is indeed true that distant stars are receding, the redshift is *not* due to the Doppler effect in the usual sense where a wave is shifted immediately upon emission due to the motion of the source. To understand this, we need to distinguish between two different ways in which stars can recede.

Perhaps the most straightforward way to do so is by imagining space as a two dimensional rubber sheet that is being stretched and stars as points on this sheet.

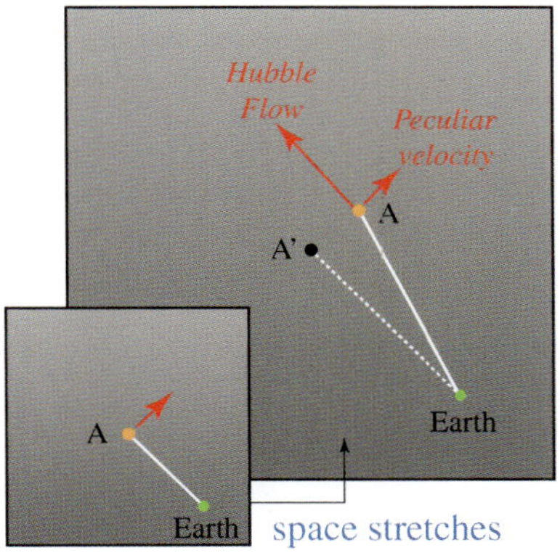

Fig. 7.3 Peculiar velocity and Hubble flow. A′ indicates where the star A would be if the peculiar velocity were 0.

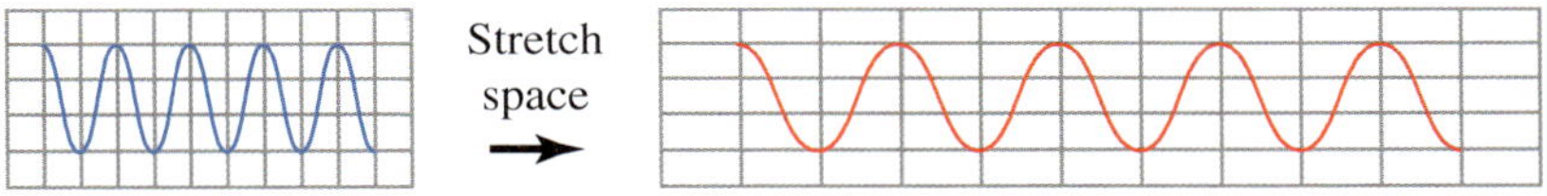

Fig. 7.4 Stretching space causes light to be redshifted.

The first kind of motion, called the **peculiar motion**, is relative to the sheet. The second kind of motion, called the **Hubble flow**, is due to the stretching of the sheet. The difference is illustrated in Fig. 7.3. It turns out that the recession of distant stars is almost entirely due to the Hubble flow. As illustrated in Fig. 7.4, the stretching of space also leads to a redshift and it is this redshift that is mostly responsible for Hubble's observations.

The next questions is: If distant stars have a greater redshift and hence are receding faster following Hubble's law, then what does this mean for the expansion of space? As illustrated in Fig. 7.5, Hubble's law can be explained by stretching space everywhere at a constant rate.

Of course, if space is stretching at a constant rate everywhere, then one can go backward in time and one must logically conclude that the Universe started in a point or perhaps in a very small region. Furthermore, the Universe is remarkably homogeneous and isotropic. This would be hard to explain if the Universe had always expanded at the current rate since its very beginning because gravity in the early Universe should have led to more clumping. This issue was addressed by the inflation hypothesis according to which the Universe expanded extremely fast initially (from the size of about an atom to the size of the Milky Way in about 10^{-32} s) as discussed in Chapter 6.

Thus it can be seen that the darkness of the night sky leads to some truly startling insights. Namely, the Universe has a finite age and is expanding and hence

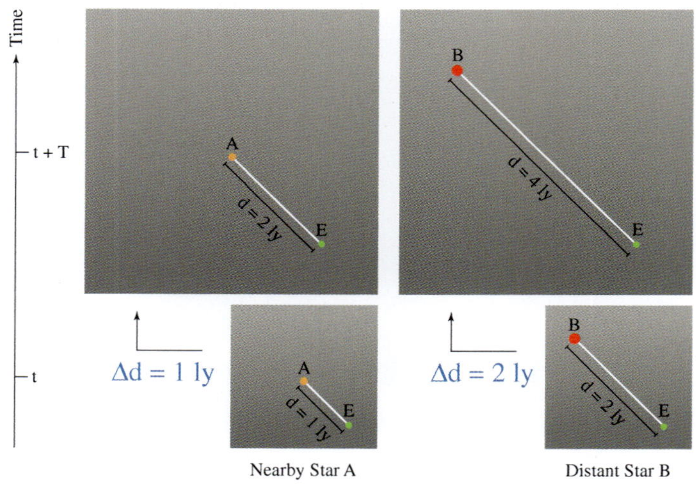

Fig. 7.5 If space is stretched at the same rate everywhere, distant stars will recede at a higher speed than nearby stars. Here, at time t, star B is at twice the distance from Earth (E) than star A. At time $t + T$ space has stretched to double the size at time t. As can be seen, in T time units star A gains a distance of 1 light year (ly) while star B gains 2 ly.

is neither static nor infinite, having begun in the Big Bang. The findings clearly show that the assumptions leading to Olbers' paradox are wrong and hence there is no need for the night sky to be bright.

There is more that the darkness of the night can reveal. Next let us look at what is commonly called "dark matter".

7.4 Dark Matter

It is likely that the luminous matter (stars) seen at night only make up a small part of all matter. Firstly, there is still a great deal of inter-galactic gas (mostly hydrogen and helium left over from the Big Bang). Indeed, there is probably close to 10 times as much matter in inter-galactic gas than in stars. Interestingly enough, however, there seems to be even more matter that cannot be seen. That is to say matter that does not interact with electromagnetic radiation (light) at all or at most only very little. Since this kind of matter is invisible it is generally called **dark matter**. As the name indicates, dark matter is thought to be matter-like, found near ordinary matter, and just like matter clumping. It is therefore of a very different nature from dark energy which evenly permeates the entire Universe as discussed in Sec. 7.5.

Below, we will discuss three key pieces of evidence for the existence of dark matter.

7.4.1 *Evidence 1: Galaxy rotation curve*

The stars that make up a galaxy do not stand still. Just like planets revolve around the Sun, stars in general revolve around the center of the galaxy. For a (stable) circular orbit around a central mass, the centripetal force F_G due to gravity needs to exactly cancel out the centrifugal force F_C. Using Newton's universal law of gravitation and the classical expression for the centrifugal force, we hence have

$$\left.\begin{aligned} F_G &= \frac{GmM}{r^2} \\ F_C &= \frac{mv^2}{r} \end{aligned}\right\} v^2 = \frac{GM}{r} \Rightarrow M = \frac{rv^2}{G} \tag{7.3}$$

where r is the distance from the center, m the mass of orbiting object, M the mass of central object and v the orbital speed of the object with mass m. With the help of Eq. (7.3) one can then plot the expected orbital velocities of stars versus their distance from the galactic center based on the relatively well known masses of the observed stars, as in Fig. 7.6 (red line).

This type of plot is generally called a **galaxy rotation curve**. Rather remarkably, when the orbital velocities are measured directly (*e.g.* with the help of the Doppler shift[1]), one finds that the curve flattens out (blue line), thus strongly deviating from the predicted curve.

One way to explain the actual curve is by assuming the existence of a large amount of invisible mass (*i.e.* dark matter) with orbital velocities indicated by the

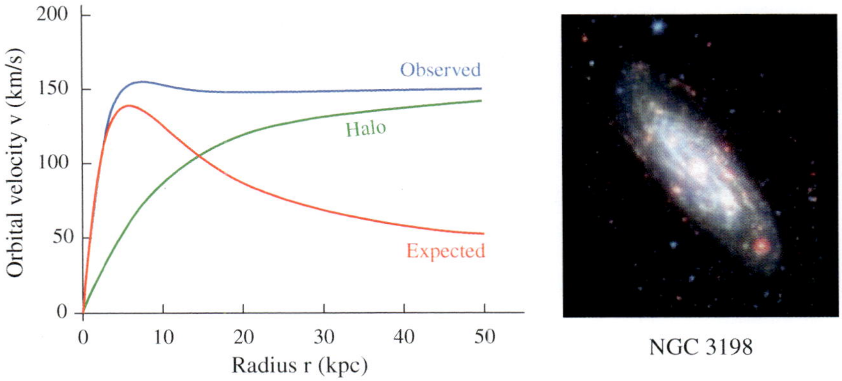

Fig. 7.6 Rotation curves for the spiral galaxy NGC3198. The red line is the expected curve calculated with Eq. (7.3). The blue line is the curve obtained by directly measuring the orbital velocities. The discrepancy can be explained by an invisible halo as indicated by the green line.

[1]Due to gravitational binding, the intra-galactic effects of the Hubble flow are very small and can generally be ignored.

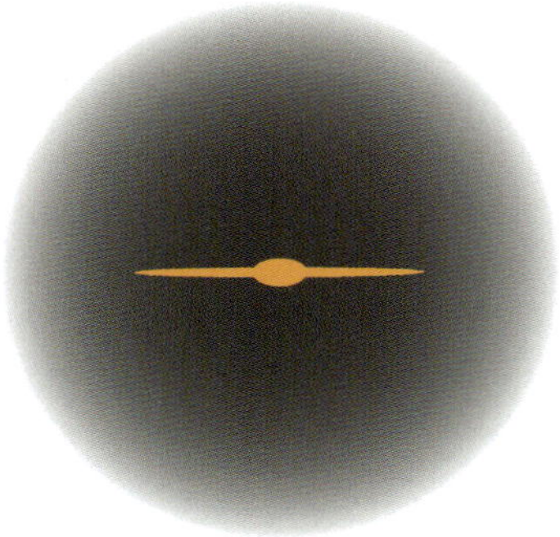

Fig. 7.7 The dark matter halo extends far beyond the visible galaxy. In the case of the Milky Way, its diameter may exceed 500,000 ly.

green line in Fig. 7.6. This dark matter then forms a halo around the visible galaxy and extends far beyond the visible galaxy as illustrated in Fig. 7.7.

The galaxy rotation curve is the first piece of evidence for the existence of dark matter.

7.4.2 *Evidence 2: Cluster mass*

The determination of the mass of astronomical phenomena is necessarily indirect. Above, Newton's laws and orbital velocities were used. Another method, employed in this section, is based on the **virial theorem** for gravitationally bound masses.

7.4.2.1 *Virial theorem*

In general, a virial theorem relates the forces between objects in a bound system due to potentials to the average kinetic energy of the objects in the system.[2] In the context of Newtonian mechanics, one has

$$E[T] = -\frac{1}{2}E[V] \tag{7.4}$$

where $E[T]$ is the average over time of the kinetic energy and $E[V]$ the average over time of the gravitational potential energy between the masses. For the theorem to be valid, two assumptions are necessary:

- The averages are well defined (this means that stars do not escape, join, go supernova, *etc.*)
- The velocities and positions of the masses are well defined and finite.

[2]The word "virial" is derived from the Latin word "vis" meaning "force" or "energy".

Before outlining the more general proof for gravitational systems, let us first consider a simple Sun–Earth-like system where a relatively small mass m moves in a circular orbit around a stationary central body with a relatively large mass M. If the small mass is at a distance r from the large mass, the small mass' gravitational potential energy is given by

$$V = -\frac{GmM}{r} \tag{7.5}$$

where G is the universal gravitational constant. The kinetic energy of the small mass can be found by looking at the orbit. If the orbit is circular and stable, the gravitational and centrifugal forces acting on the small mass must exactly cancel out (otherwise the orbit would not be stable) so that

$$\frac{GmM}{r^2} = \frac{mv^2}{r} \tag{7.6}$$

from which the kinetic energy T immediately follows as

$$T = \frac{1}{2}mv^2 = \frac{GmM}{2r} = -\frac{1}{2}V \tag{7.7}$$

where Eq. (7.5) was used and we hence find that for this case Eq. (7.4) indeed holds.

Noteworthy (optional content) 7.2: Virial Theorem: N-body case

---

For the N-body case, it is convenient to start out with the moment of inertia defined by

$$I = \sum_{i=1}^{N} m_i|\mathbf{r}_i|^2 = \sum_{i=1}^{N} m_i r_i^2 \tag{7.8}$$

where m_i and $\mathbf{r}_i$ are the mass and position of the ith body (the moment of inertia is a measure of a body's resistance to a change in its rotation and hence, for rotational systems, plays a role analogous to that of mass in linear systems). The virial G is defined as

$$G = \sum_{i=1}^{N} \mathbf{p}_i \cdot \mathbf{r}_i \tag{7.9}$$

where $\mathbf{p}_i$ is the momentum of the ith body. Since the momentum is defined as $p = mv = mdr/dt$, we hence have

$$G = \sum_{i=1}^{N} \mathbf{p}_i \cdot \mathbf{r}_i = \sum_{i=1}^{N} m_i \frac{d\mathbf{r}_i}{dt} \cdot \mathbf{r}_i = \frac{1}{2}\frac{dI}{dt}. \tag{7.10}$$

Taking the derivative of the virial G we have, using Newton's law $\frac{d\mathbf{p}}{dt} = \mathbf{F}$, that

$$\frac{dG}{dt} = \sum_{i=1}^{N} \mathbf{p}_i \cdot \frac{d\mathbf{r}_i}{dt} + \sum_{i=1}^{N} \frac{d\mathbf{p}_i}{dt} \cdot \mathbf{r}_i = \sum_{i=1}^{N} m_i \frac{d\mathbf{r}_i}{dt} \cdot \frac{d\mathbf{r}_i}{dt} + \sum_{i=1}^{N} \mathbf{F}_i \cdot \mathbf{r}_i. \tag{7.11}$$

We then obtain that

$$\frac{dG}{dt} = 2T + \sum_{i=1}^{N} \mathbf{F}_i \cdot \mathbf{r}_i \tag{7.12}$$

where T is the total kinetic energy given by $T = \frac{1}{2} \sum_{i=1}^{N} m_i \frac{d\mathbf{r}_i}{dt} \cdot \frac{d\mathbf{r}_i}{dt}$.

Now in the case of gravitation, the force F is given by Newton's universal law of gravitation

$$F = \frac{GmM}{r^2} \tag{7.13}$$

which, for a collection of bodies, means that the force $\mathbf{F}_i$ on body i is the result of all the other $i - 1$ bodies acting on it

$$\mathbf{F}_i = \sum_{j=1, j \neq i}^{N} \frac{Gm_j m_i}{|r_j - r_i|^2} \hat{r}_{ij} \tag{7.14}$$

where $\hat{r}_{ij}$ is a unit vector pointing in the direction from r_i to r_j. The gravitational potential at the position of the ith body is

$$V_i = - \sum_{j=1, j \neq i}^{N} \frac{Gm_j m_i}{|r_j - r_i|} \tag{7.15}$$

and hence, for the case at hand, we find that $\mathbf{F}_i \cdot \mathbf{r}_i = V_i$ so that

$$\frac{dG}{dt} = 2T + \sum_{i=1}^{N} V_i = 2T + V. \tag{7.16}$$

Next we take the time average on both sides, *i.e.* we integrate both sides from time 0 to time τ and then divide the result by τ, after which the limit $\tau \to \infty$ is taken. For the left-hand side we then have

$$E\left[\frac{dG}{dt}\right] = \frac{1}{\tau} \int_0^{\tau} \frac{dG}{dt} dt = \frac{1}{\tau} \int_0^{\tau} dG = \frac{G(\tau) - G(0)}{\tau}. \tag{7.17}$$

In a bound system such as a gravitational system, it is reasonable to state that both $G(\tau)$ and $G(0)$ are finite. Then the right-hand side of Eq. (7.17) will be 0 in the limit of $\tau \to \infty$ so that

$$E\left[\frac{dG}{dt}\right] = 0 = 2E[T] + E[V] \Rightarrow E[T] = -\frac{1}{2} E[V]. \tag{7.18}$$

7.4.3 *Cluster mass — continued*

With the help of the virial theorem, the mass of a cluster can be inferred as follows. By observing the stars of the cluster, one can obtain a reasonable estimate of the total kinetic energy T of the cluster which then immediately yields V with the help of the virial theorem. From V, the mass of the cluster can then be calculated. If this mass is greater than the mass of the observed stars (including possibly central black holes), then there must be dark matter to make up for the difference. Many observations show this to be the case thus indicating the existence of dark matter.

7.4.4 *Evidence 3: Gravitational lensing*

According to the general theory of relativity, matter curves spacetime and light propagates following this curvature. Consequently, a large mass located between a distant source and an observer can act as a lens as illustrated in Fig. 3.1.

The strength of the lens, in other words how much light bends, is determined solely by the mass of the object that causes the bending of the light. Therefore, if the mass is known, the amount of bending can exactly be calculated. Although this might be useful under some circumstances, what is of particular interest here is that the reverse is also true. If, as illustrated in Fig. 3.1, the known position of a distant object is approximately in the line of sight of a much closer object whose position is also known, then mass of this closer object can be calculated from how much the light is bent.

If a sufficiently bright object is directly behind a massive closer object, the light of the more distant object will be bent around the closer object for all directions. This gives the spectacular so-called Einstein ring shown in Fig. 7.8(a).

(a) (b)

Fig. 7.8 (a) Einstein ring. Source: NASA. (b) Composite image of the galaxy cluster 1E 0657-66. Pink is where regular matter is dominant and blue is where dark matter is dominant. Source: NASA.

It turns out that there are many celestial objects for which the mass calculated with the help of gravitational lensing is much greater than the mass of the visible matter. Since the general theory of relativity is well established, the only possible conclusion then is that there must be other matter besides visible matter. Or in other words, there must be dark matter.

A particularly striking example of dark matter can be found in and around the galaxy cluster 1E 0657-66 shown in Fig. 7.8(b). From gravitational lensing it is found that huge amounts of dark matter (indicated by the blue tint in the figure) exist outside of the area where most of the visible matter occurs (indicated by the pink tint). In other words, dark and visible matter are separated unlike most other known observations where dark and visible matter overlap. It is hypothesized that this remarkable phenomenon is the result of a collision between two galaxies where the visible matter was slowed down due to the drag created by crossing galaxies while the dark matter moved on since it only interacts weakly with visible matter through gravity.

7.4.5 *Possible explanations for dark matter*

As discussed above, the evidence for dark matter is overwhelming. What is less overwhelming is the array of possible explanations. Literally, as of this writing, we are in the 'dark'. No dark matter has been directly detected yet. Furthermore, there is at least one serious experimental oddity which is currently unexplained. Let us start with the latter.

Many globular clusters do not seem to show any signs of dark matter while some have relatively small amounts of dark matter in the range from 25% to 50%. This is also so for a number of elliptical galaxies such as Messier 105 shown in Fig. 7.9.

Fig. 7.9 Messier 105. Source: NASA.

Table 7.1 Possible explanations for dark matter.

Type	Example constituents
Cold dark matter	Slow particles such as WIMPs or hard to see objects such as MACHOs
Warm dark matter	Certain unknown types of neutrinos, gravitinos
Hot dark matter	Fast (relativistically) moving particles such as neutrinos or other unknown particles

This begs the question, if most of the matter in the Universe is dark, then why are there so many galaxies that don't show any or much signs of it? (Current estimates are that 85% of all matter in the Universe is dark.)

Quite a few possible explanations for dark matter have been proposed, none of which having any experimental support. Currently, three main categories are considered: cold, warm and hot dark matter as summarized in Table 7.1.

Cold dark matter is as of this writing the leading candidate for an explanation of the observed phenomena. It is relatively slow moving (usually less than $0.1c$ and hence 'cold') and two rather different kinds of cold dark matter are considered. One kind consists of so-called Weakly Interacting Massive Particles (WIMPs). WIMPs are thought to only interact through the weak and gravitational forces (otherwise they would most likely have been detected a long time ago) and therefore share many characteristics with neutrinos but have much larger masses (in the range from 10s to 1000s of proton masses). However, they do not occur in the Standard Model (an extension which would imply many more particles are required) and to date, despite a huge amount of experimental work, not one particle outside of the Standard Model has ever been found. Another kind of possible cold dark matter is the so-called Massive Astrophysical Compact Halo Object (MACHO). Examples of MACHOs are small black holes, neutron stars and white dwarfs. Therefore MACHOs are as such not an exotic type of matter, rather they are made up of regular (baryonic) matter that is just hard to see. In other words, it is not really dark, it is just too dark to see from Earth but as such it would be visible if close enough or if it blocks the light from a visible source by passing in front of it. Sky surveys have shown no evidence of a significant number of MACHOs (MACHOs clearly exist but the number must be huge to explain the observed dark matter distribution).

Warm dark matter falls between hot and cold dark matter. It would need particles whose mass is neither too small (since if the mass is small, they likely move fast and hence be hot) nor too large (since if the mass is large, they likely would move relatively slowly and hence be cold). One such speculated particle would be a 'sterile' neutrino that does not interact through the weak force like regular neutrinos (and is therefore sterile). There are no such particles in the Standard Model and no hints of these particles have been found.

Hot dark matter consists of particles moving at relativistic speeds. Neutrinos qualify as hot dark matter but the best estimates put the total mass of the neutrinos near that of the baryonic mass (neutrinos only interact through the weak and gravitational forces and are very light — less that 1/10000th the mass of an electron). Consequently neutrinos cannot explain the observed amount of dark matter. There are no other good candidates for hot dark matter so it is currently not widely seen as a possible explanation for dark matter.

7.5 Dark Energy

Hubble's and subsequent observations clearly show that the Universe is expanding, and that it is expanding at the same rate everywhere. From the Big Bang theory, we know that this rate of expansion was not always the same: shortly after the Big Bang, there was an era of inflation during which the Universe expanded extremely fast, after which the rate of expansion slowed down. The question then is, has or will the rate of expansion approach a constant, decrease or increase.

From the general theory of relativity, we know that the structure of spacetime is related to mass. On a cosmological scale, due to the attractive nature of gravity, one would therefore expect that gravity slows down the rate of expansion.

In order to characterize the expansion of the Universe, in Sec. 6.8 the density parameter $\Omega = \rho/\rho_c$ was defined. Depending on the average density of the Universe, there are then in principle three possibilities listed in Table 7.2 and illustrated in Fig. 6.4: (a) the Universe keeps on expanding forever at some finite rate, (b) the Universe expands forever but at an ever decreasing rate asymptotically approaching zero, and (c) the rate of expansion decreases and become negative (after the Universe has reached a finite maximum size, it starts to shrink and ends up in a point — this is often called the "big crunch").

The critical density of the Universe is estimated to be about 5 hydrogen atoms per cubic meter while the observed density of visible matter is only about 0.2 hydrogen atoms per cubic meter. Consequently, the observed $\Omega_L \approx 0.04$ (obtained by dividing 0.2 by 5) appears to indicate an open, forever expanding universe whose expansion rate is slightly slowed down by gravity but never approaches zero.

Table 7.2 Fate of the Universe if Ω is entirely due to matter.

Ω	Eventual fate of the Universe
$\Omega < 1$	The Universe is open and expands forever at some rate
$\Omega = 1$	The Universe is flat and expands forever at an ever decreasing rate
$\Omega > 1$	The Universe is closed. Expansion comes to a halt and eventually the Universe collapses

Measurements by the Wilkinson Microwave Anisotropy Probe show, however, that the Universe is flat to within an error margin of less than 1% which suggests that Ω should equal 1. This is another indicator that visible matter is not the only matter. However, the discrepancy is too great to be solely accounted for by dark matter and consequently something else must also be at play. Taking the independent estimate that dark matter constitutes about 85% of all matter, the density parameter Ω_M due to mass can be estimated as $\Omega_0 \approx 0.26$. If the Universe is flat so that the total density parameter equals 1, there must be another part equaling about 0.74 (see also Sec. 6.8).

Other crucial observations, in the mean time, indicate that the rate of expansion of the Universe is *accelerating* rather than slowing down and/or approaching a steady value. Since an accelerating Universe would correspond to a positive cosmological constant (usually denoted by Λ, see Eq. (6.16)) in the general theory of relativity, the associated density parameter is written as Ω_Λ, and its value would then need to be about $\Omega_\Lambda = 0.74$ and yields $\Omega = \Omega_0 + \Omega_\Lambda \simeq 1$. The source of this acceleration is an unknown form of energy called **dark energy** that permeates the entire Universe and hence is cosmological in its origin and significance. It is important to note that unlike all other contents of the Universe, dark energy is gravitationally repulsive. The composition of the Universe as determined by the Planck satellite is graphically represented in Fig. 7.11.

Let us now review some evidence for dark energy.

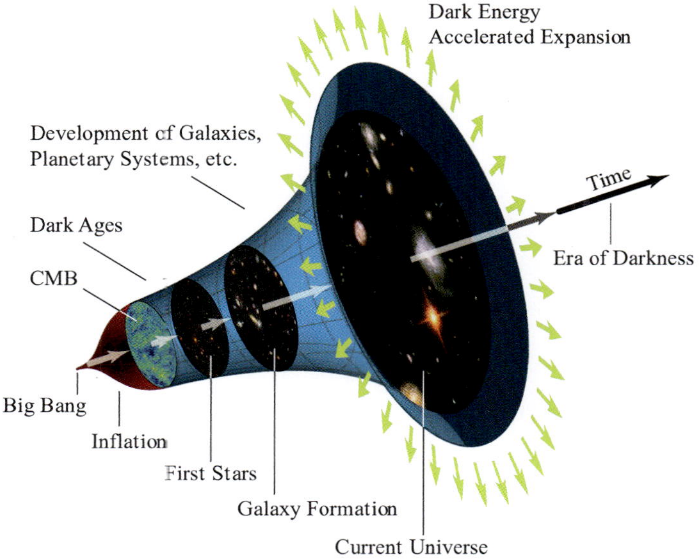

Fig. 7.10 Observations indicate that the expansion of our Universe is accelerating.

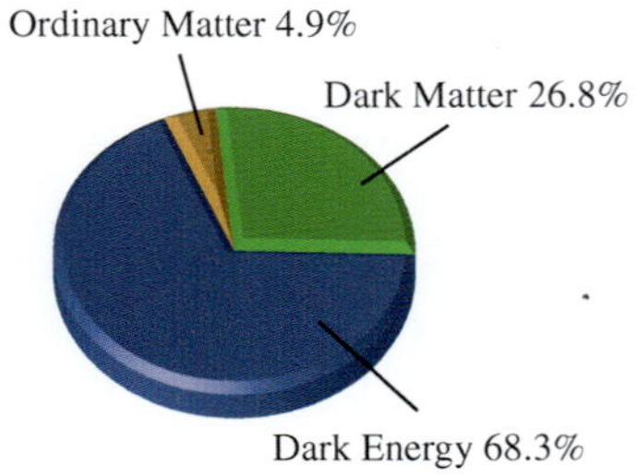

Fig. 7.11 Composition of the Universe as determined by the Planck satellite.

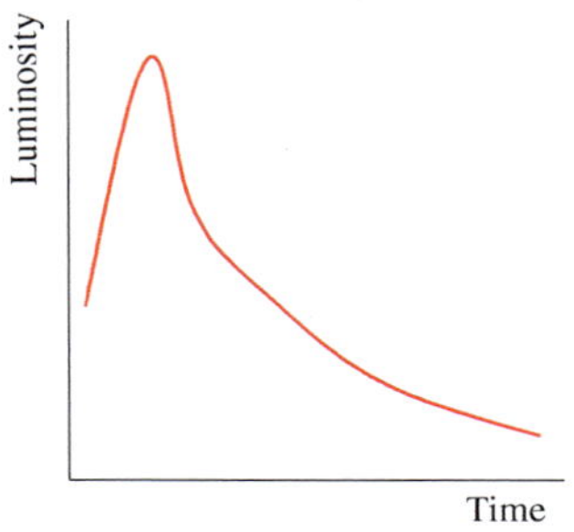

Fig. 7.12 Characteristic light curve of a type Ia supernova.

7.5.1 *Evidence 1: Accelerating expansion*

A supernova is an extremely energetic explosion of a star that releases an enormous amount of energy. There are several types of supernovae and the type of interest here is called a Type Ia supernova. A Type Ia supernova occurs when a white dwarf exceeds a certain rather well defined mass as discussed in Secs. 9.12 and 10.8. Since this type of supernova always occurs at about the same mass for a very similar composition of white dwarf, the supernovae themselves are very similar and have a characteristic light curve (illustrated in Fig. 7.12) with a very well defined peak. The peak always occurs at the same time from the beginning of the supernova and has the same luminosity. The (peak) brightness of a Type Ia supernova is about 5 billion times that of the Sun!

Now if the distance of one (or a few) nearby Type Ia supernovae can be determined accurately, then the distance of far away Type Ia supernovae can be found with the help of their apparent brightness and the inverse square law. Due to the fact that an underlying astrophysical process is used, it is close to certain that this kind of distance measurement is very good indeed.

Another method to measure the distance to far-away objects is with the help of the redshift and Hubble's law. According to Hubble's law, there is a linear relationship between the distance of a far-away star and the redshift in its spectrum. Irrespective of Hubble's law and how distant an object is, redshift is a measure of the

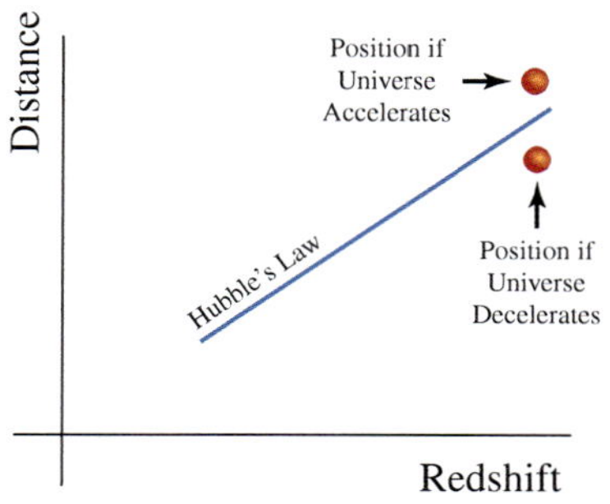

Fig. 7.13 Observed positions of distant objects depend on whether the Universe's expansion remains the same, accelerates or decelerates.

speed with which space expands. If, since the end of inflation after the Big Bang, the Universe expanded at a uniform rate, then Hubble's law must hold exactly and redshift and distance are indeed always linearly related. Since gravity is an attracting force, it may be expected that the expansion of the Universe slows down over time. If so, then a distant object whose light was emitted a long time ago when the expansion rate was higher than now will appear nearer than implied by the redshift. In a plot shown in Fig. 7.13 a very distant object is then expected to lie somewhat below the straight line representing Hubbles' law.

It was experimentally found, however, that very distant Type Ia supernovae are *further* away than expected, placing them above the line. This means that when the light was emitted, space was expanding slower than now. Or in other words that the rate of expansion of the Universe is accelerating.

7.5.2 *Evidence 2: Cosmic microwave background*

Our entire Universe is filled with radiation of a very long wavelength corresponding to the spectrum of a black body at 2.73 K. This radiation is called the **cosmic microwave background (CMB)**. Just as is the case for all other radiation, the path of a photon of the CMB is affected by the curvature of spacetime. As illustrated in Fig. 7.14, a photon passing by a large cluster of mass will gain some energy if the rate of expansion accelerates.

The energy gain can exactly be predicted if the expansion of spacetime and the mass of the supercluster is known or *vice versa*, and the rate of expansion of the Universe can be calculated. Since space has regions with above average mass (superclusters) and below average mass (supervoids), the energy of photons coming from those regions can be compared with the energy of the photons from regions of space with average mass. Doing so one finds that the rate of expansion appears to indeed be increasing.

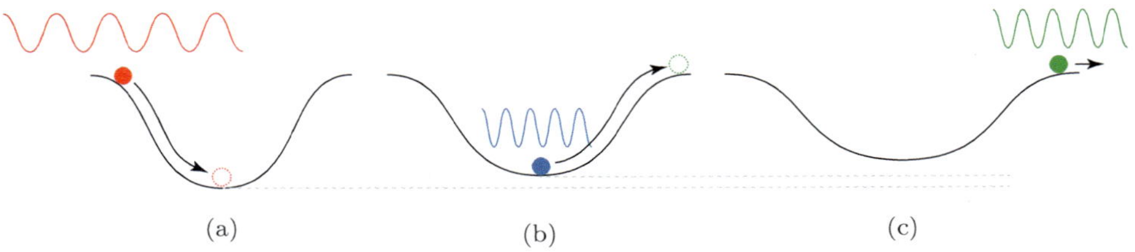

Fig. 7.14 (a) A photon enters the gravitational well of a large cluster. (b) While descending the gravitational well, it will gain some energy (and hence be blueshifted). (c) Due to gravitational binding, the size of the cluster is not significantly affected by a uniform expansion of space, and the photon will lose the energy it gained upon exiting the gravitational well. If however, space expands at an accelerated rate while the photon travels through the well, the well will become less deep and the photon will not need all the energy it gained to exit the well, leaving it with a small energy gain.

7.5.3 *Evidence 3: Flatness*

According to the general theory of relativity, matter curves space as discussed in Sec. 3.9. At very large scales, the Universe strongly appears to be the same everywhere, in other words it is both homogeneous as well as isotropic (this is usually referred to as the **Cosmological Principle**). Consequently, at the largest scales, one can calculate the shape of the Universe by only considering its average density without having to worry about the exact distribution of the Universe's contents. Doing so, one then finds that there are three possible types of shape as illustrated in Fig. 3.32. The Universe as a whole can have a positive curvature so that it is akin to a ball (in 4 dimensions) and therefore closed and of finite size. This happens when the average density of the Universe is greater than the critical density. Due to the large amount of mass, an expanding closed Universe will stop expanding at some point and then contract. The Universe as a whole can also have a negative curvature so that it is akin to a saddle. This kind of Universe can expand forever, asymptotically approaching some finite expansion rate. Finally, between the positive and negative, there is exactly one density (the critical density) for which there is no spatial curvature. This is the flat Universe. The flat Universe will also expand forever but with an ever decreasing expansion rate (the expansion rate asymptotically approaches 0). For the Universe to be flat, it needs to be exactly at the critical density giving $\Omega = 1$. See Table 7.2.

It should be noted that due to the finite age of the Universe of 13.78 billion years and due to the finite speed of light, only that part of the Universe which is within 13.78 light years from Earth can be observed (this is often called our horizon). Furthermore, the distant light arriving at Earth shows us the Universe of the past, not the Universe of the present. Therefore, the present size and shape of the Universe must be inferred and there are in principle some possibilities (such as a doughnut-shaped Universe) not listed above. The current size of the Universe is estimated to be around 93 billion light years.

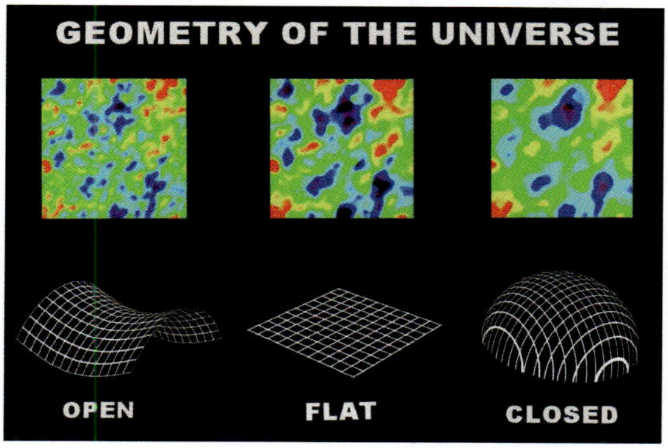

Fig. 7.15 The average size of the spots in the CMB reveals the shape of the Universe. Source: NASA.

A good way to determine the shape of the Universe is by looking at the cosmic microwave background. Even though at the largest scales, the Universe is homogeneous and isotropic, at smaller scales this is of course not so (after all, we have galaxies, stars and planets). Also, (very) small fluctuations in the temperature of the background radiation depending on the direction from which the radiation comes can be expected and are indeed observed. Hence the cosmic microwave background has some (very small) anisotropy, regions that are a bit colder and regions that are a bit hotter. Now, depending on the shape of the Universe, the average sizes of these regions will be different as illustrated in Fig. 7.15.

The Wilkinson Microwave Anisotropy Probe had determined the anisotropy of the cosmic microwave background with very high accuracy and found that the Universe is flat with an error margin of about 0.4%.

Consequently, Ω is either exactly or otherwise very close to one. However, since regular and dark matter can only account for about 26% of Ω, there must be something else to make up for the remaining 74%, namely dark energy.

7.5.4 *Possible explanations for dark energy*

There are a number of possible explanations for dark energy but none of them is supported satisfactorily by either observation or theory.

Quintessence: A dynamic field permeating all of space that may enhance (in the past) as well as oppose (currently) gravity. Quintessence is a scalar field and can be tuned to more or less match the expansion rate of the Universe. However, it is perhaps not so surprising that a cleverly constructed field can yield the observed findings (with enough patience matching a relatively simple curve is usually not all too hard). The question is whether there is a need for, or any other observable sign of, quintessence. Thus far neither is the case and consequently, this theory does not look all too promising.

Systematic Observational Errors: It is in principle still possible that the relative motion of Earth with respect to the rest of the Universe is misunderstood (knowing and correcting for this motion is important when carrying out high accuracy cosmological studies). However, considering that this relative motion is important for many studies and that it has been considered in detail, it is unlikely that there is a systematic error.

String Theory: String theory, discussed in more detail in Chapters 14 and 15, could well explain dark energy. Regrettably, as a whole, despite all its strengths, string theory has no experimental or observational evidence going for it yet.

The current best explanation is based on the cosmological constant Λ discussed in Sec. 6.6. If the vacuum has a positive energy, which in the context of field theories is not unreasonable, then a **cosmological constant** implies a negative pressure in Einstein's field equations and hence it will have an accelerating effect. Interestingly enough, the vacuum energy required for Λ to more or less match observations, is about 10^{-122} times the vacuum energy expected in field theories. Λ is very small indeed. This discrepancy is not understood. Nevertheless, the cosmological constant is currently the simplest way to explain dark energy and so it is favored by many.

The incomplete theory for tackling dark energy and matter with the cosmological constant and cold dark matter is often referred to as "Λ-CDM" or Lambda-CDM.

7.6 The Answer

Directly, the Universe is dark because it has a finite age and is expanding and thus resolving Olbers' paradox. More indirectly, however, it turns out that the Universe mostly consists of things that cannot be seen, it is invisible. There is dark energy that causes the Universe to expand faster, there is dark matter that cannot be seen and then there is the inter-galactic gas that is too faint to be seen. When all is tallied, the visible matter in stars amounts to only about 0.4% of the energy in the Universe! With so little matter being actually visible, it is no surprise that the Universe is dark. *Next time you find something dark, you might just have made a great discovery*!

Chapter 8

Galaxies, Stars and Planets

How do astronomical structures form?

8.1 The Question

How do the primordial elements that are forged in the initial fiery beginnings of the Universe combine and aggregate to form astronomical objects? How do the galaxies, stars and planets form from the apparently structureless primordial gaseous cloud that emerges from the Big Bang? We address these and related questions and in particular, study in some detail the origin and formation of our Sun and solar system.

8.2 Primordial Gas Cloud

The first stars whose starlight can reach us from their primordial beginnings were formed about 200 million years after the Big Bang, and which brought the Dark Age of the Universe to a close, as discussed in Sec. 6.14. At the end of the Dark Age, matter (atoms) was distributed uniformly throughout the Universe in the form of the *primordial* gas cloud, with only the forces of gravity and pressure working.

In the inflationary model of cosmology, minute inhomogeneities due to quantum fluctuations in the energy density are exponentially magnified by inflation and are responsible for the primordial inhomogeneity in the spatial distribution of matter. To understand the formation of galaxies, we need to know how matter organizes itself at various scales, in other words, what is the size and shape of the distribution of matter and the nature of the inhomogeneous distribution of matter. Radiation tends to disperse matter and remove inhomogeneities, and one can see that large scale structures have less likelihood of being present in the early Universe. It can be shown that radiation has the effect of dispersing all structure smaller than about fifteen kiloparsec — the distance that light travels in about 50,000 years.

The evolution of the primordial cloud and its possible breaking up into galaxies — entailing star formation — are all driven by the force of gravity; for this reason, we will ignore the other forces, unless necessary, in discussing the formation of galaxies and stars.

On the largest scale, the magnitude of any configuration away from homogeneity is down to one part in a 100,000. Gravity is a purely attractive force; hence any inhomogeneity in the distribution of matter that may emerge due to the random motion of atoms is magnified by the force of gravity. The smaller the scale, the more time is available for the magnitude of a configuration of the gas cloud to grow, and hence small clusters form first and these subsequently combine to form larger clusters.

The aggregates of gaseous matter become gravitationally bound to each other, and become approximately decoupled from the rest of the Universe. Small bound aggregates combine to form larger aggregates, and these aggregates then go on to form super-aggregates. There is hence a bottom-up growth of **clustering** that naturally emerges: the smallest scale clusters become large enough to overcome the expansion of the Universe and emerge as gravitationally bound entities.

Figure 8.1 shows the bottom-up formation of a hierarchy in the structure of the Universe. The gravitationally bound clusters finally give rise to **protogalaxies** — the predecessor of the current day galaxies; smaller protogalaxies form first, and these smaller galaxies interact with each other and combine to form larger galaxies. In other words, galaxies are formed by the merger of smaller protogalaxies.

As the Universe expands radiation becomes less important. The larger chunks of gas become sufficiently dense to continue to persist as individual clouds.

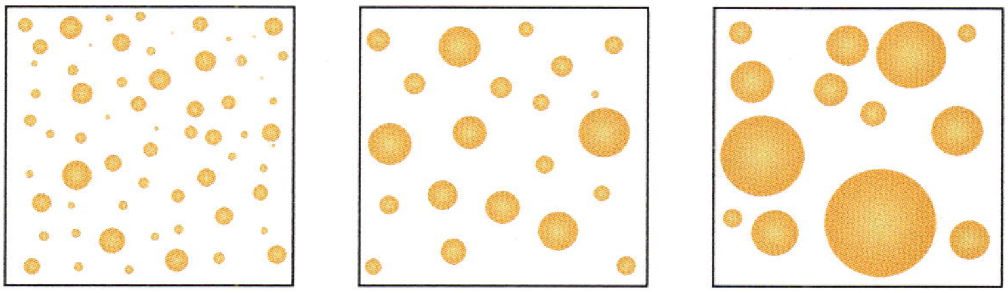

Fig. 8.1 Stages of the bottom-up clustering of the primordial gas cloud.

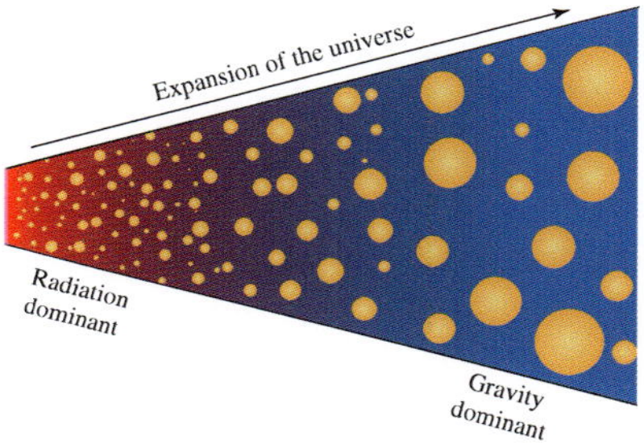

Fig. 8.2 As the Universe expands, radiation is superseded by gravity as the dominant force.

This bottom-up growth of clustering of the primordial gas cloud is schematically represented in Fig. 8.2.

The picture that we have developed is borne out by observations; the Hubble telescope, viewing vast distances and hence looking into the early Universe, shows that the young Universe had more smaller galaxies than at present.

8.3 Formation of Galaxies

Once the inhomogeneous primoridal gas cloud has broken up into gravitationally bound denser clouds, mostly composed of atoms, the question arises as to what are the conditions required for galaxy formation. All clouds do not go on to become galaxies, and we examine the properties that are required for galaxy formation.

The galaxies that are observed are fairly similar. Confining our attention to those galaxies that make a substantial contribution to the average luminosity of the Universe, we find that their properties lie in a fairly narrow range. Our Milky Way is an average size galaxy with a mass of about $10^{11}M_\odot$; most of the spiral galaxies have masses about that of our Milky Way. Only the largest of elliptical galaxies have a mass of about $10^{13}M_\odot$; furthermore no visible non-elliptical galaxy has been found with mass much larger than $10^{12}M_\odot$. Most of the galaxies have a radius from 30,000 to 100,000 light years. The gas clouds that go on to become protogalaxies and then full fledged galaxies need to yield galaxies in this range.

The mechanism, and hence the rate, at which a gas loses energy depends on its temperature. At high temperatures the loss is through radiation emitted by accelerating charges, namely protons and electrons. At low temperatures the main mechanism of loss of energy is by those electrons that are in a bound state inside an atom, and which make quantum transitions from higher to lower energy quantum states. A detailed calculation, done in 1977 by J. Silk, M. Rees and J. Ostriker, showed that if a gas cloud has a mass greater than a few trillion solar masses ($10^{12}M_{\odot}$), then the gas cannot cool fast enough to form stars, and the gas cloud stays dark without any star formation. The explanation for the size of the elliptic galaxies may be lie in the fact that in many cases, pre-existing galaxies merge to form the very large elliptic galaxies.

Clouds of gas with mass smaller than $10^{12}M_{\odot}$ and radius less than 150,000 light years, as shown in Fig. 8.3, can indeed form galaxies with star formation. The dotted red line in Fig. 8.3 is the boundary of gas clouds between those that form galaxies, and those that are too massive or too large and cannot cool off sufficiently to form galaxies. The theoretical calculations of the inability of a gas cloud with mass much larger than $10^{12}M_{\odot}$ to form a visible galaxy — based on the rate of dissipation of energy in a heated gas — seem to borne out by observations.

As the Universe expands and the hot primordial gas cloud cools, it starts to break up into disjoint pieces whose temperature depends on the cloud's mass — with the larger the mass of the gas cloud the higher its temperature. Clouds with a mass of about $10^{10}M_{\odot}$ have a temperature of 10^6 K; for a gas cloud with mass of about $10^7M_{\odot}$ the temperature is about 10^4 K. Since it is only for temperatures of about a few thousand kelvins that the gas cloud can break up further and form stars, a pre-condition for the formation of stars is that the gas cloud needs to cool down.

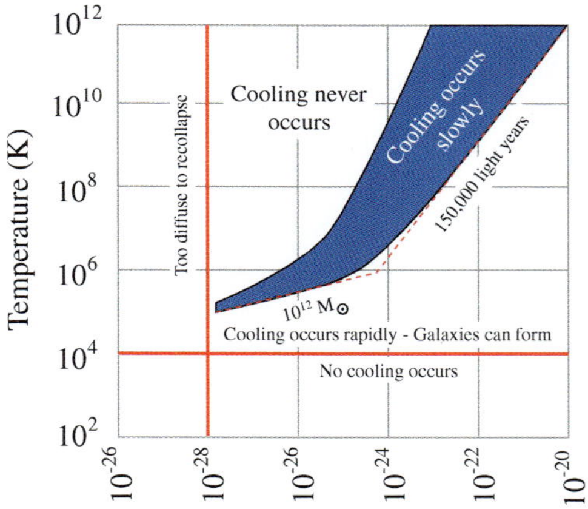

Fig. 8.3 Conditions for primordial gas clouds to give rise to the formation of galaxies. Only if cooling occurs rapidly can galaxies form.

Rotation is a universal phenomena and many galaxies are rotating. Initially, the tidal forces between neighboring galaxies create the torque (angular force) that creates rotation and counter-rotation amongst the protogalaxies, with the total angular rotation being zero. As a protogalaxy collapses and contracts to form a galaxy, the spinning becomes more rapid. As the galaxy fragments into stars, the gas distribution flattens out into a pancake shape and eventually becomes a spiral galaxy. An elliptical galaxy can form due to anisotropic random motions of stars. A spheroidal galaxy, on the other hand, can be the result of the merger of two spiral galaxies. The larger galaxies emerge from the interaction and combination of smaller galaxies as well as directly from the larger clouds of primordial gas.

In conclusion, to form galaxies, the gas cloud needs to have a mass less than $10^{12} M_{\odot}$ and a characteristic radius of about 150,000 light years. Clouds not satisfying these constraints are either too hot to cool effectively or too diffuse and are more likely to be destroyed by other galaxy forming clouds. A summary of the behavior of the primordial gas clouds is given in Fig. 8.3.

8.4 Formation of Stars

Long after the primordial gas cluster has cooled off and has given rise to galaxies — composed of first and second generation stars and what is called the **interstellar medium** — stars continue to form throughout the life of a galaxy. We discuss the generic case of the formation of stars from the interstellar medium since the formation of the first stars has been discussed in Sec. 6.14.

For star formation to be possible, in essence, large clumps of cold matter must be allowed to concentrate. Gravity compresses a gas, and in doing so raises its pressure; the increase in pressure in turn opposes this compression. A gas at higher pressure also has a higher temperature. A hot gas would disperse before it reaches the high density and temperature required for the onset of nuclear fusion, and for such a hot protogalaxy there would be no star formation, as shown in Fig. 8.4(a).

The whole formation of stars crucially hinges on the primordial gas losing energy at a fast enough rate so that it stays cool. The interstellar gas must lose internal

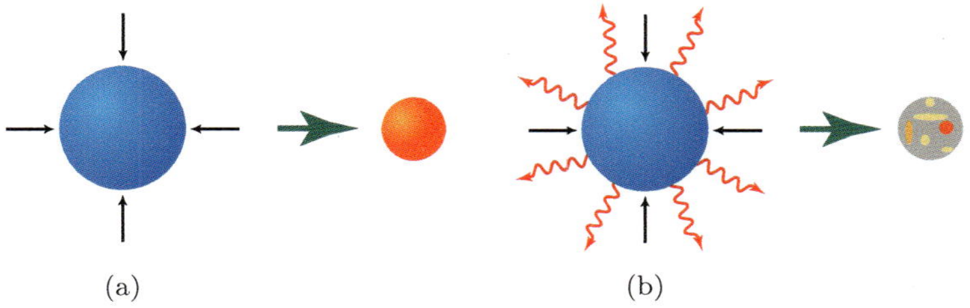

Fig. 8.4 (a) For the case that energy is slowly radiated, the gas is hot under contraction and stars are not formed. (b) The case when energy is rapidly radiated leads to the cooling of the gas and to the formation of stars.

energy fast enough so that the pressure of the compressed gas *falls* suddenly, leading to the break up of the gas into smaller parts, as shown in Fig. 8.4(b).

The formation of molecules in the primordial gas clouds is essential for the emergence of galaxies and stars. The reason being that molecules provide a mechanism for the gas cloud to rapidly dissipate energy and cool off to reach temperatures low enough for the formation of stars.

Molecules are very difficult to form in the primordial gas. It was shown in 1977 that hydrogen molecules can form in sufficient quantity required for star formation, as shown in Fig. 8.5, by an intermediate step of a hydrogen atom first capturing a free electron, and then combining with another hydrogen atom to form a molecule and ejecting the captured electron. The free electron serves as a catalyst for the formation of sufficient quantities of hydrogen molecules. The disassociation of a hydrogen molecule H_2 requires an energy of 4.5 eV per molecule and the ionization of a hydrogen atom costs 13.6 eV; the removal of 18.1 eV for these processes is a major mechanism for the primordial cloud to lose energy and hence lower its temperature.

All stars undergo a constant process of change and evolution from their birth to their death, and in this journey they change their sizes and colors. The formation of second and third generation stars proceeds a bit differently from the first generation stars. One of the main differences is that second generation stars do not need the presence of hydrogen molecules for forming stars since the interstellar medium now contains *dust* arising from the debris of earlier generation stars; dust then plays a catalytic role for the rapid dissipation of heat. As illustrated in Fig. 8.6, the existence of dust particles can accelerate the process of molecule formation.

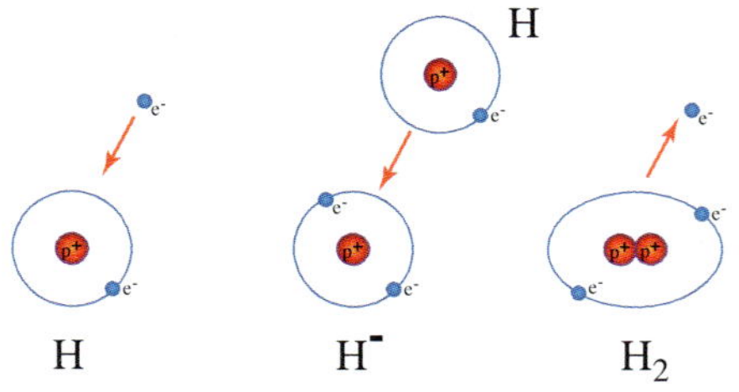

Fig. 8.5 The formation of H_2 can be sped up with the help of (temporary) electron capture.

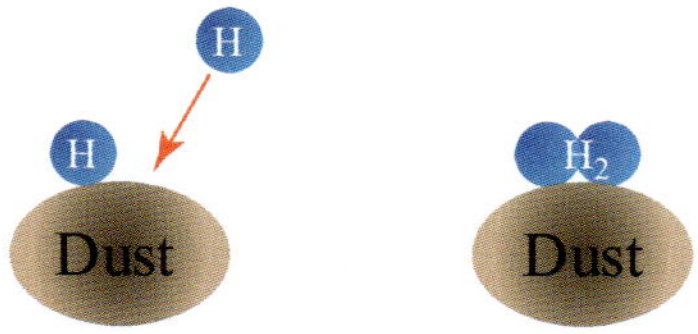

Fig. 8.6 The formation of H_2 can be sped up with the help of dust.

Interstellar dust, more appropriately called interstellar smoke, consists of solid grains — typically of radius of 0.001–0.1 micrometers — consisting of silicon, iron, carbon, water-ice and carbon dioxide-ice. The origin of interstellar dust is not well understood and is thought to be formed in the photosphere (that exists on and below the surface) of well evolved stars, and in supernovae. The dust is spewed into the interstellar medium by solar winds and by supernova explosions.

The formation of stars occurs within molecular clouds, which are one component of the interstellar medium. Molecular clouds are randomly shaped dense regions of the interstellar medium rich in molecular hydrogen H_2, having a temperature from 10 K to 100 K and with mass of about $10^2 - 10^5 M_\odot$. Due to their high density and low temperatures, the cold molecular clouds can collapse — under the force of gravity that is strong enough to overcome outward pressure — and form a dense core at high temperature that, in turn, can ignite thermonuclear fusion. The particle density inside a typical star is $10^{24}\,\text{cm}^{-3}$ whereas in a molecular cloud it is about $10^3\,\text{cm}^{-3}$. Hence some regions of the molecular cloud have to be compressed by many orders of magnitude in order to form a new star. We return to the processes leading to the formation of a new star in Sec. 8.12 when we discuss the birth of our Sun.

8.5 Stars: Hydrostatic Equilibrium

Our Sun, which is a typical star, is composed out of primordial gaseous hydrogen (73%), helium (25%) — elements that resulted from the Big Bang — and the rest (2%), formed in the core of earlier generation stars that burned out and contributed to the gaseous cloud from which the Sun was formed. Under the pull of gravity, an initial hydrogen and helium cloud concentrates into a star by the following process.

The gravitational attraction exerts an attractive force causing the gas cloud to contract. The contraction of the gas cloud increases its temperature due to energy conservation. The virial theorem given in Eq. (7.7) states that the average kinetic energy of a collection of particles is given by $E[T] = -E[V]/2$, where $E[V]$ is the average potential energy that is negative due to gravity being an attractive force. For the case of the gaseous cloud, gravitational contraction leads it to occupy a smaller volume (with higher density) and $-E[V]$ increases since the gas is more tightly bound by gravitation; hence, due to energy conservation, the average kinetic energy of the gas molecules increases. Temperature is proportional to the average kinetic energy of a particle, and hence the temperature of a gravitating gas increases as it contracts.

Stars, to a good approximation, can be considered to be an ideal gas and, from Eq. (2.2), obey the equation $PV = Nk_BT$; hence, as the temperature T of the gas increases, then so does its pressure P. As the star contracts, it occupies a smaller and smaller volume and its temperature — and hence its pressure — continues to increase. The temperature and pressure increase until

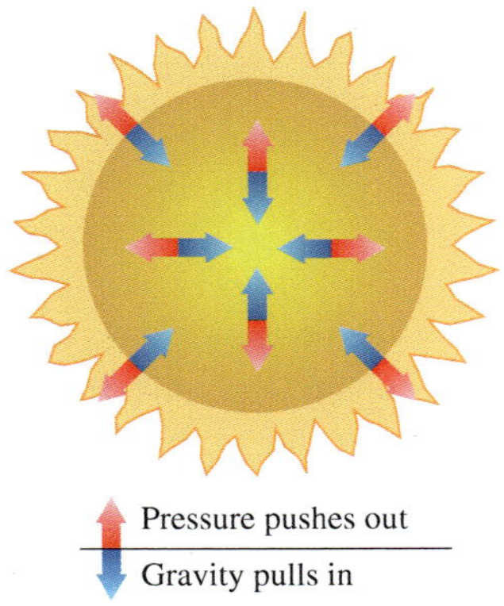

Fig. 8.7 Hydrostatic equilibrium: The outward pressure of the hot gas and the inward pull of gravity are balanced in a steadily burning star, leading to hydrostatic equilibrium.

the star ignites by nuclear fusion, which further raises the temperature of the star — thus creating a stabilizing outward pressure. The contraction stops and the star reaches **hydrostatic equilibrium** when the outward force due to pressure generated by star burning is strong enough to balance the inward force due to gravity as illustrated in Fig. 8.7. This stable situation continues until the star runs out of fuel to create the temperature required for maintaining hydrostatic equilibrium.

A detailed discussion of the formation of star such as our Sun from the gas cloud is given in Sec. 8.12.

The *maximum* and *minimum* mass for a star are seen — from observations — to lie in a fairly narrow range, with most stars having a mass from $0.1M_\odot$ to $50M_\odot$.

If the mass of a star is less than a *minimum* mass, on undergoing gravitational contraction the star generates a temperature that is less than what is required for igniting thermonuclear fusion. The lightest star, from theoretical calculations, is estimated to be about $0.08M_\odot$.

On the other hand, the *maximum* mass for a star is fixed by the condition of hydrostatic equilibrium. For a very massive star, the dominant pressure does not arise from the motion of the particles composing the gas but from radiation pressure, mostly due to photons; because the photons are relativistic, this affects the energy balance in the star — which in turn leads to a small binding energy for the star.

Noteworthy (optional content) 8.1: Stability of Stars

For a gas that is thermally isolated and does not exchange energy with its environment, its pressure P and volume V are related by

$$PV^\gamma = \text{constant}$$

where γ is the adiabatic index, and depends on whether the gas is made out of non-relativistic particles or relativistic photons.

Let E_I be the internal energy due to thermal motion and radiation, and E_G be the gravitational potential energy. The total energy of the star is given by

$$E_{\text{star}} = E_I + E_G = -(3\gamma - 4)E_I.$$

For stable stars, the internal energy E_I is dominated by the kinetic energy of the non-relativistic particles that compose the star. For the non-relativistic particles, we have that $\gamma = 5/3$ and this yields

$$E_{\text{star}} = -E_I \ : \text{Non-relativistic.}$$

Since the total energy is negative, that is, $E_{\text{star}} < 0$, these stars are stable, and remain so as long as γ is less that 4/3.

However, for very massive stars, due to thermonuclear fusion generating large amounts of radiation, radiation pressure dominates the internal energy E_I and drives $\gamma \to 4/3$; the total energy E_{star} is then given by

$$E_{\text{star}} = E_I + E_G \approx 0 \ : \text{Relativistic.}$$

The hydrostatic equilibrium is stable when E_{star} is less than zero. Once $E_{\text{star}} > 0$ — due to the absorption of energy — the stellar gas is no longer in a bound state and the star will disintegrate.

In particular, the hydrostatic equilibrium of a very massive star becomes unstable for the following reason. Since $\gamma \approx 4/3$, any small change in the total energy of the star leads to a large change in its internal energy E_I; furthermore, since the total energy is almost zero, a large change in E_I requires a corresponding change in the gravitational energy E_G. For example, a 1% decrease in the total energy E_{star} leads to a 25% increase in the internal energy and a 26% decrease in the gravitational energy.

In conclusion, any small energy loss or gain by a large mass star will lead to large changes in its internal and gravitational potential energy, thus making the hydrostatic equilibrium of the star unstable and leading to its eventual disintegration.

The nuclear processes that power the burning of stars — required to generate the pressure that balances the inward pull of gravity — are discussed in detail in Chapter 9. When a star has exhausted all its fuel, it has — depending on its mass at the end point of its evolution — three possible final states, which are discussed in Sec. 8.7. In general, the final state of a star is a neutron star or a black hole if its *initial* mass is greater than $8M_\odot$ and is a white dwarf if its initial mass is less than $8M_\odot$.

8.6 Classification of Stars

In spite of the narrow range of $0.1M_\odot$ to $50M_\odot$ for the mass of a star, there is nevertheless a bewildering variety of stars, with stars having a wide range of lifetimes, sizes, temperatures, masses, luminosity and so on. Before discussing the processes driving the formation and evolution of stars, we need to devise a scheme for classifying the different types of stars, as each type of star needs an explanation addressing its specific characteristics.

8.6.1 *Spectral classification*

Light is radiated from a star's outer shell, which is called the **photosphere**. The photosphere is the deepest region of a star from which photons can be emitted to regions outside the star. The ionization of the photosphere reveals the atomic spectra that are most important for a star. Light from the star has dark lines showing the presence of atoms that absorb certain wavelengths of light. The spectral analysis of light from a star shows the composition of the photosphere.

Most stars are currently classified using the letters O, B, A, F, G, K, and M. The O class of stars are the hottest and the letter sequence indicates successively cooler stars up to the coolest M class. The spectral classification of stars is according to the presence or absence of spectral lines in their radiation. The spectral class types are shown in Fig. 8.8, with O class of stars being the most massive ($>16M_\odot$) and M stars being the least massive ($<0.4M_\odot$). The Sun falls in the spectral G class of stars.

8.6.2 *Hertzsprung–Russell classification*

Intuitively, the *brightness* of stars seems like a natural basis for classifying stars, with more and less 'bright' stars being grouped together. However, how bright a star appears to a detector on Earth depends on how far the star is from the Earth — and consequently brightness is not an intrinsic property of a star. In astronomy, the concept of **luminosity** is used for classifying stars. Luminosity is defined as the *total amount* of electromagnetic energy emitted per unit time by a star (or other astronomical objects); hence, luminosity is an intrinsic property of a star.

Two salient properties of stars — namely the **surface temperature** of the star and how much net radiation it is emitting — form a natural basis for classifying all stars. There is a correlation between the luminosity of a star and its surface temperature, with stars that are blue being higher in temperature than those that are red. A classification scheme, proposed by Ejnar Hertzsprung in 1911 and

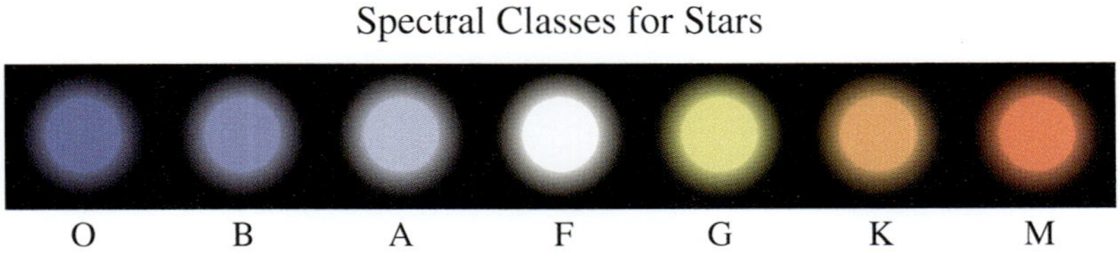

Fig. 8.8 Spectral class types of stars.

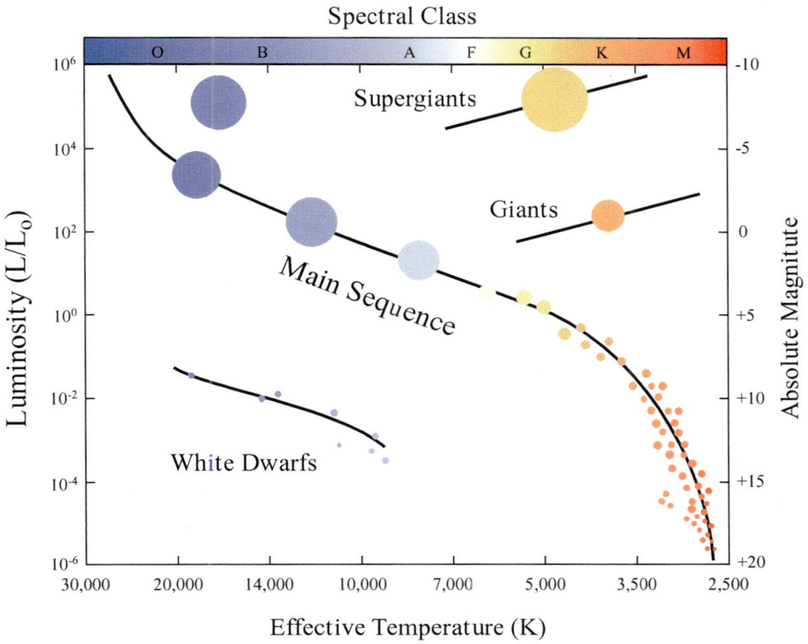

Fig. 8.9 The Hertzsprung–Russell diagram showing the luminosity of stars versus their surface temperature. Stars spend most of their lives on the so-called main sequence.

(independently) by Henry Norris Russell in 1913 — based on the luminosity and the temperature of a star — brought tremendous order to our understanding of stars. The Hertzsprung–Russell classification yields a two dimensional diagram in which a star's luminosity and temperature are the axes, and with all the stars being placed at a particular point on the Hertzsprung–Russell diagram.

The **Hertzsprung–Russell diagram** is given in Fig. 8.9; the luminosity of the Sun is set at $L_\odot$ and the luminosity of other stars are ranked accordingly. Note the counter-conventional labeling of temperature in the Hertzsprung–Russell diagram, which decreases as one moves along the x-axis.

Stars move on the Hertzsprung–Russell diagram as they evolve, and we will discuss a few cases of stellar evolution in some detail in later sections. There are three main groups of stars in the Hertzsprung–Russell diagram.

- About 80–90% of all the observed stars fall on a well-defined line giving what is called the **main sequence**.
- There are a few Red Giants, or Blue Supergiants. These stars have evolved away from the main sequence as they have consumed more than 10% of their hydrogen and become cooler Red Giants. Eventually, the Red Giants mostly become planetary nebulae, whereas Red Supergiants in a few cases become Blue Supergiants.

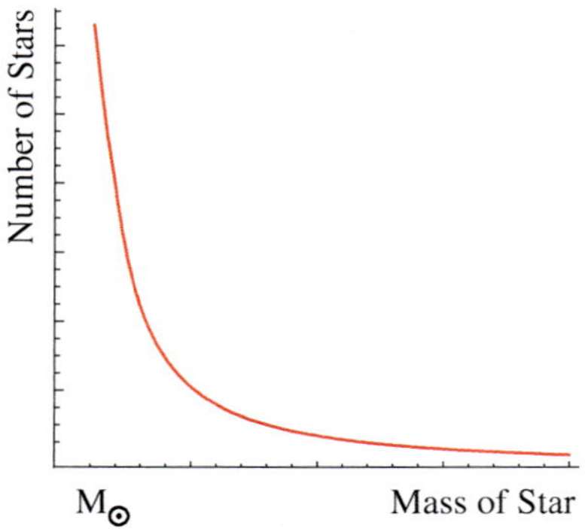

Fig. 8.10 The number of stars with a large mass is much smaller than the number of stars with a small mass.

- There are a few very faint stars near the bottom left of the diagram — these are white dwarfs, discussed in Sec. 9.8; these stars are one of three possible end points for the evolution of stars, and discussed in Sec. 8.5. No matter what is the initial mass that a star starts off with, it always ends up as a white dwarf if it has a final mass that is less than $1.4M_\odot$, called the Chandrasekhar mass.

The properties and evolution of a star are largely determined by its initial mass; in particular, a star's mass determines its temperature and luminosity during the star's main sequence lifetime. The more mass it has, the hotter and more luminous it is and the shorter is its lifetime. The number of stars with a mass of M, namely $N(M)$, is known from observations; for $M_\odot$ being the mass of our Sun, we have the following

$$N(M) \approx \left[\frac{M_\odot}{M}\right]^{2.35} \quad ; \quad M > M_\odot \tag{8.1}$$

and is shown in Fig. 8.10.[1]

One can see from Fig. 8.10, the lighter stars are more numerous; the more massive a star the more rare is its occurrence.

One expects that the hotter a star the more luminous it should be; and in fact this turns out to be correct. The relation between mass M and luminosity $L(M)$ for the main sequence of stars (normalized by the Sun's values) is given by

$$\frac{L(M)}{L(M_\odot)} = \left[\frac{M}{M_\odot}\right]^{3.5} \quad : \text{Main sequence stars.} \tag{8.2}$$

Most of the stars spend their lives as main sequence stars; during this period, their surface temperature and luminosity can vary considerably.

[1]For stars below one solar mass, $N(M) \approx M^{-\alpha}$, with (a) $\alpha = 2.3$ above half a solar mass, (b) $\alpha = 1.3$ for stars between 0.08–0.5 solar masses and (c) $\alpha = 0.3$ for stars below 0.08 solar masses.

Most stars join the main sequence of stars in Hertzsprung–Russell diagram after their initial process of formation and leave the main sequence in Hertzsprung–Russell diagram on exhausting their fuel — going on to becoming red giants, supergiants and so on. The Sun, for example, as it evolves will — at the end point of its life — *exit* from the region of the main sequence of the Hertzsprung–Russell diagram since it will increase its luminosity considerably while decreasing its surface temperature.

The lifetime $T(M)$ of main sequence stars is given by the following

$$\frac{T(M)}{T(M_\odot)} = \left[\frac{M_\odot}{M}\right]^{2.5} \quad : \text{Main sequence stars.} \tag{8.3}$$

For example the lifetime of our Sun is $T(M_\odot) \simeq 10^{10}$ years, whereas the lifetime of a star twice as massive as our Sun is $T(2M_\odot) \simeq 1.3 \times 10^9$ years.

Table 8.1 provides a summary of the mass, temperature, luminosity and lifetimes of stars, including the time they spend on the main sequence of the Hertzsprung–Russell diagram. From Table 8.1 and Eq. (8.3), it can be seen that the more massive a star the shorter is its lifetime. This result is consistent with our intuition for the following reason: the greater the mass of a star the greater is the gravitational force compressing it and hence the higher must its temperature be to generate the necessary pressure to stave off gravitational collapse. Maintaining a high temperature forces the star to rapidly consume its fuel and hence its lifetime is short.

Two questions arise from the distribution of stars on the Hertzsprung–Russell diagram, namely: why do stars appear only in certain regions of the luminosity–temperature diagram? And how and why do stars evolve from one part of the Hertzsprung–Russell diagram to another?

The first question is linked to the initial conditions of star formation that was briefly touched upon and will not be elaborated any further. The second question is related to the process of thermonuclear fusion that powers the burning of a star and is discussed under nucleosynthesis in Chapter 9.

Table 8.1 Mass, lifetime, surface temperature, luminosity of stars and time spent on the main sequence of the Hertzsprung–Russell (HR) diagram. (Recall the Sun's mass is equal to $M_\odot$ and its luminosity is equal to $L_\odot$.)

Type of star	Surface temperature	Mass (in units of $M_\odot$)	Luminosity (in units of $L_\odot$)	Years on HR main sequence
O	≥33,000 K	≥16	≥30,000	1 million
B	10,000–33,000 K	2.1–16	25–30,000	11 million
A	7500–10,000 K	1.4–2.1	5–25	440 million
F	6000–7500 K	1.04–1.4	1.5–5	3 billion
G	5200–6000 K	0.8–1.04	0.6–1.5	8 billion
K	3700–5200 K	0.4–0.8	0.08–0.6	17 billion
M	≤3700 K	≤0.4	≤0.08	56 billion

8.7 End Points of Stellar Evolution

First generation stars start their life as a gaseous mass of primordial hydrogen and helium gas clouds. Second generation stars are formed out of interstellar medium that consists of hydrogen and helium as well as the elements and dust generated by the earlier generation stars. The end point of the evolution of all stars that are large enough to ignite thermonuclear fusion is a compact object of incredibly high density, be it a white dwarf, neutron star or a black hole. *Brown dwarfs* are 'stars' that lack sufficient mass (about 65–80 Jupiter mass) to ignite the fusion of hydrogen but can fuse deuterium or lithium.

Recall from our discussion at the end of Sec. 8.5, stars with an initial mass that is greater or less than $8M_\odot$ have different end points to their evolution. When a star has exhausted all its fuel, it has — depending on the mass M that it has at the end point of its evolution called its *final mass* — the following possible final states, shown in Fig. 8.11.

- White dwarf (black dwarf): final state for stars of final mass $M < 1.44M_\odot$. A white dwarf slowly becomes a black dwarf as it radiates away all its thermal energy and cools to near zero temperature; it ends up as a completely black object. Figure 9.1 shows the evolution of a star from the main sequence to its final state as a white dwarf.
- The fate of a star with a final mass in the range $1.44M_\odot < M < 2M_\odot$ depends on the details of its evolution. It can end up either as a white dwarf or a neutron star.
- Neutron star: stars with a final mass of about $2M_\odot < M < 3M_\odot$ end up as supernovae, with a resultant neutron star at the core of the star.

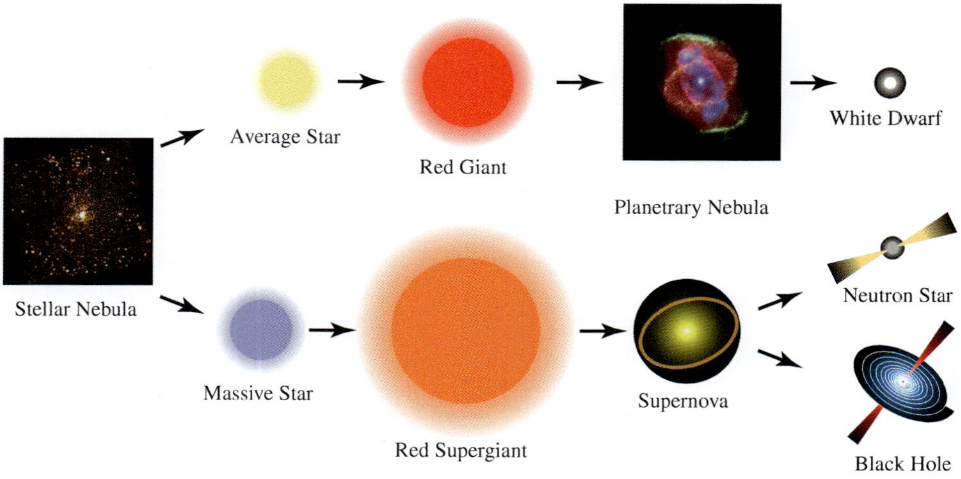

Fig. 8.11 Possible evolutionary paths of a star.

- Black holes: stars with final mass of about $M > 3M_\odot$ also end up as supernova, but with the formation of a resultant black hole due to its high final mass.

White dwarfs, neutron stars and black holes are discussed in Chapter 9, where these objects are revisited from the point of view of stellar evolution.

8.8 Normal and Active Galaxies

Normal galaxies come in the following four types and are illustrated in Fig. 8.12.

- **Elliptical Galaxies**. These range from more spherical to highly flattened galaxies. They contain old stars and have very little gas and dust. Their sizes vary greatly, from about 3 to 600,000 light years in diameter, and are mostly found in clusters of galaxies. These galaxies have masses ranging from 10^6 to $10^{13}M_\odot$. Average type of star: K.
- **Spiral Galaxies**. These are flattened systems of stars comprising a thin disk. These galaxies display both barred and unbarred spiral structures. They contain young and old stars. They have copious amounts of gas and dust and are mostly 6–160,000 light years in diameter, having a mass in the range of 10^9 to $10^{11}M_\odot$. These types of galaxies are found mostly outside of clusters. Average type of stars are: A, F, G, K.

  We inhabit the **Milky Way**, which is a barred spiral galaxy with two spiral arms; it has a radius of about 40,000 light years with an average depth of about 1000 light years and contains about 200 billion stars. Our galaxy has a supermassive black hole at its center with a mass of about 4.2 million solar masses. Our Sun is located on one of the spiral arms, about 27,000 light years from the center of the galaxy, and rotates about the galaxy's center once in every 230 million years.
- **Lenticular Galaxies**. Similar to spiral galaxies in shape and color but having no well defined spiral arms. These galaxies are flattened disk-shaped systems that are morphologically between elliptical and spiral galaxy. Star formation has mostly ceased in these galaxies, consisting mainly of aging stars.
- **Irregular Galaxies**. These galaxies are irregular in shape, with mostly young stars. These galaxies have lots of gas and dust, and have a size of about 3–300,000

(a) (b) (c) (d)

Fig. 8.12 Main types of normal galaxies. (a) Elliptical galaxy M89, (b) spiral galaxy M51, (c) lenticular galaxy NGC524, (d) irregular galaxy: The Large Magellanic Cloud.

light years in diameter, and have masses of about 10^8 to $10^{10} M_\odot$. They are found outside clusters. Average type of star: A, F.

8.8.1 *Active galaxies*

Normal galaxies are usually the older ones, with active galaxies being much younger and more active. Active galaxies represent a small fraction, only about 10%, of all the galaxies. The *total* luminosity of a normal and active galaxy is almost the same. Light from the stars of a normal galaxy is mostly in visible spectrum and comes uniformly from all parts of the galaxy. The spectrum of radiation from normal and active galaxies is quite different. Active galaxies have a compact region at their center that is a source of intense radiation — both in the visible spectrum as well as radiation with wavelength of X-rays, ultraviolet, infrared and radio waves, and is shown in Fig. 8.13.

High energy photons, due to X-rays and gamma rays, are abundant only in the active galaxies. These high energy photons free many more electrons from their bound state in atoms than is the case for normal galaxies; this provides a catalyst for the formation of hydrogen molecules and thus are an effective seeding mechanism for star formation. These high energy photons however can also break up the hydrogen molecule thus attenuating star formation. Which of these two processes plays a dominant role in the evolution of an active galaxy is still not clear and needs more detailed calculations.

All active galaxies are thought to have an *active galactic nucleus* and their classification and properties are discussed in Sec. 8.10. Figure 8.14 shows a normal and an active galaxy.

8.8.2 *Age of stars in the Milky Way*

The age of a galaxy clearly must be older than all the stars that it contains, and this rather simple fact puts a strong constraint on all theories of galaxy formation.

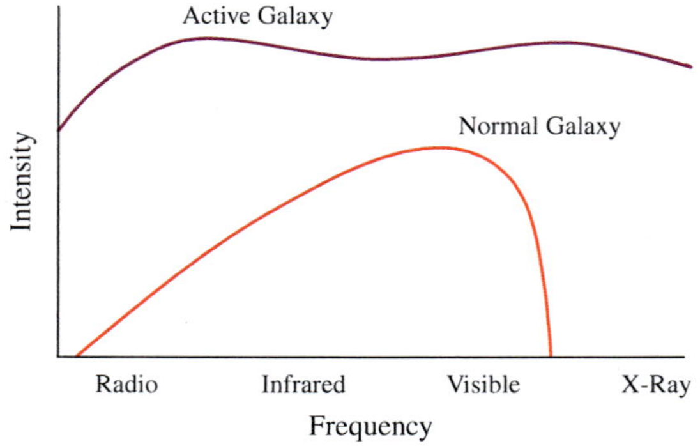

Fig. 8.13 The radiation spectrum of an active galaxy compared to a normal galaxy.

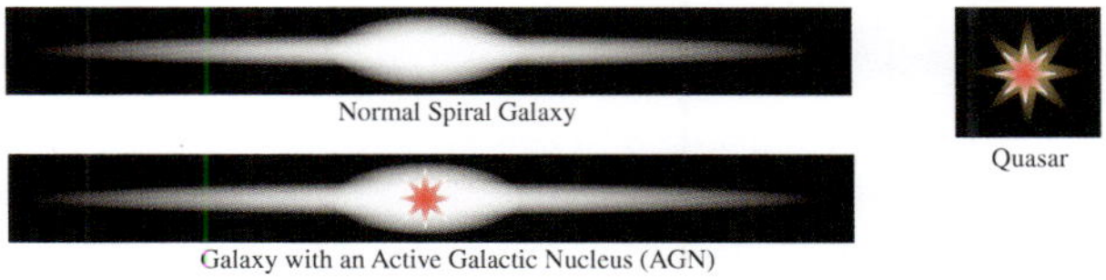

Fig. 8.14 An example of a normal galaxy compared with an active galaxy with an active galactic nucleus (AGN). Many active galaxies have a quasar at their center.

Fig. 8.15 HD-140283, also known as the Methuselah star, in the constellation Libra. Source: NASA.

Light from the stars of a nearby galaxy, or for that matter stars in our galaxy, does not need to travel billions of years to reach us; hence, no matter how the stars of nearby galaxies formed, all of them are relatively close and hence their light reaches us much faster than from distant galaxies.

The Methuselah star, shown in Fig. 8.15, is one of the oldest known stars in the Milky Way and can be found in the constellation Libra; its age is estimated to be greater than 13.6 billion years. It is only about 190 light years away from the Earth and hence the light that we are receiving has traveled for only 190 years. We can consequently conclude that our Milky Way must be at least 13.6 billion years old. The age of the Universe is estimated to be about 13.78 billion years, and hence the Milky Way must have been formed a mere 100–200 million years after the Big Bang.

Stars are being born all the time, and stars in the process of being formed can be observed. The Sun is a typical star of the Milky Way. The lifetime of a star depends on its mass, and a star with the mass of our Sun burns for about 10 billion years.

From the facts that we have obtained so far, we can conclude that — depending on the mass of the star in question — there is a possibility that some stars in our galaxy are first generation stars, and that some, like our Sun, are second generation stars. Astronomical observations bear this out, with stars, like Methuselah star, being more than 13 billion years old, and some like our Sun having an age of only 4.55 billion years.

8.9 Supermassive Black Holes

Many galaxies are thought to have a massive galactic black hole at their center with mass of about a million solar masses or higher. For normal galaxies, the black holes are quiescent since they do not have significant amounts of material falling into them. **Supermassive black holes** have masses in the range given by $10^6 M_\odot \leq M \leq 10^9 M_\odot$. Given their immense mass the logical place to look for them is the center of galaxies. Any other location in the galaxy for such an enormous concentration of mass would have created gravitational effects that would have been observed — and which is not the case. It is expected that all sufficiently large galaxies have a supermassive black hole at its center, with reliable estimates putting this number to at least 30%.

It is not clear how and at what stage of a galaxy's evolution these galactic black holes were formed, or what their role is in galaxy and star formation; there is growing evidence that galactic black holes might be crucial in seeding the formation of stars and galaxies.

The galactic black holes have a mass that is proportional to their host galaxy. It has been established, through observations, that the mass of the galactic black hole is proportional to v^n with $4 \leq n \leq 5$, where v is the average speed of the stars in the galaxy. It is not clear if the black hole fixes the mass of the galaxy or if the galaxy constrains the mass of the black hole.

Supermassive black holes have some interesting properties which distinguish them from solar mass black holes. The mass density ρ_{BH} required to form a black hole is given by dividing its total mass by its Schwarzschild volume ($R_S = 2GM/c^2$) and yields

$$\begin{aligned}\rho_{BH} &= \text{Total mass/Schwarzschild volume}\\ &= \frac{M}{\frac{4}{3}\pi R_S^3} = \frac{3c^6}{32\pi G^3}\frac{1}{M^2}.\end{aligned}$$

Hence, the density of a black hole is proportional to the inverse of its mass squared: if matter with density ρ_{BH} fills up the Schwarzschild volume, then the system will undergo gravitational collapse and form a black hole. Given the large mass of a supermassive black hole, its average density is very low, sometimes being lower than even the density of water, which is $1000\,\text{kg/m}^3$.

Recall from Eq. (5.22) that the Hawking temperature for a black hole with a large mass is very small, and that a black hole of one solar mass has a temperature of only $60 \times 10^{-9}\,\text{K}$: supermassive black holes have a Hawking temperature that is close to absolute zero. Hence, Hawking radiation from these black holes is negligible and too small to play any role in their astrophysical properties.

8.9.1 *Observing supermassive black holes*

The detection of black holes depends on the type of galaxy in which it is located. A **supermassive black hole** at the center of say a spiral galaxy creates a 'bulge' that can be seen clearly. However, we cannot directly observe the black hole, firstly because it does not emit any radiation (the hole is black) and secondly because the size of a $10^6 M_\odot$ black hole horizon is about 10^9 km — about 10 million times smaller than the bulge.

All observations of a massive black hole of necessity have to be indirect; two procedures that are of great importance are (a) the gravitational effects of black holes and (b) their properties under radio interferometry. These techniques, for example, have provided convincing evidence that our galaxy, the Milky Way, has a moderately sized $3 \times 10^6 M_\odot$ supermassive black hole at its center.

8.9.1.1 *Gravitational effects*

Newtonian gravity requires that a star's tangential velocity $V(r)$ at position r satisfies the Kepler relation

$$\frac{V^2(r)}{r} = \frac{GM(r)}{r^2}$$

where $M(r)$ is the total mass inside the volume with radius r. Measuring the velocity $V(r)$ of stars in the central region of a galaxy, in particular its radial velocity when approaching and receding from the Earth, leads to an estimation of $M(r)$. Based on the Kepler relation, a recently established empirical relationship shows a tight correlation between the mass of the supermassive nuclear dark mass, most likely a black hole, and the total mass of the galaxy bulge.

8.9.1.2 *Radio interferometry*

All supermassive spinning black holes develop an accretion disk of matter due to its strong gravitational attraction. As gaseous and other matter fall into the black hole, they heat up, radiate energy and create a rotating accretion disk. Matter is also expelled by the black hole with great energy in the form of jets of plasma, moving close to the speed of light, and radiation; the jets travel in opposite directions and can be seen at great distances from the active galaxies. The energy of the jets is finally converted into photons when it hits the ambient matter, and appears as two lobes — as shown in Fig. 8.16.

Water vapor is present in trace quantities in the accretion disk and gives rise to maser emission at radio wavelengths of 1.35 cm. These masers are bright point sources that can be detected by radio interferometry. Measurements of water masers surrounding the nucleus of nearby galaxies have revealed very fast Keplerian motion,

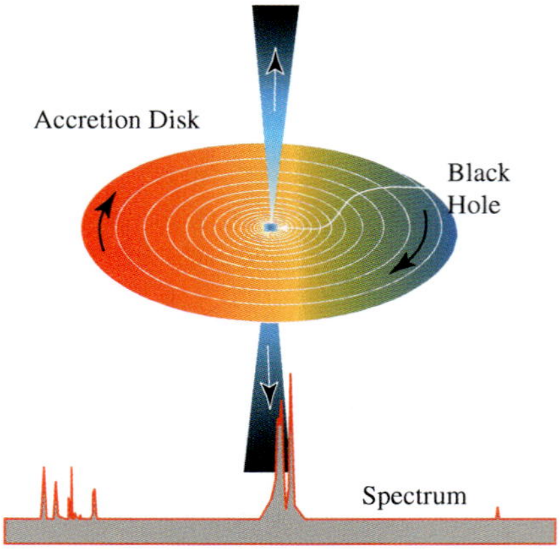

Fig. 8.16 Accretion disk of a black hole, with energetic jets beaming out from two sides.

which is only possible with a high concentration of matter in the center and leads to indirect evidence for supermassive black holes.

Figure 8.16 shows the case of NGC4258, which has a supermassive black hole of mass $39 \times 10^6 M_\odot$. The inner and outer radii are 0.46 and 0.95 ly. The bottom graph shows the radio spectrum of the maser exhibiting emission at about the velocity of the galaxy (center), redshifted emission from the receding side of the disk (left), and blue-shifted emission from the approaching side of the disk (right).

8.10 Active Galactic Nuclei (AGN)

Active galactic nuclei are *compact regions* at the center of active galaxies, which are the host to supermassive black holes having a mass of $10^8 - 10^9 M_\odot$. AGN are small objects, a few light months across, in contrast to the disk of an active galaxy that typically has a radius of about 60,000 light years. The AGN at the center of many active galaxies are thought to be the 'engine' of intense astronomical activity.

It is found that 80% of the host galaxies of AGN have early-type (bulge-dominated) morphologies, while the rest have structures characteristic of late-type (disk-dominated) galaxies. Continuous long lived AGN activity is not observed, with most AGN being observed for times when the Universe was only about one billion years old. The supermassive black holes in normal galaxies are thought to be AGN that have become quiescent due to the accretion disks having only a small trickle spiraling into the core of the AGN.

AGN are classified into Seyfert galaxies,[2] quasars and blazers; the differences between these objects are partly historical, in that they were identified by different experiments, and they are also distinguished by their optical, radio and X-ray

[2]The term Seyfert galaxy is used by some authors to denote an active galaxy, while for others it denotes a special case of an AGN.

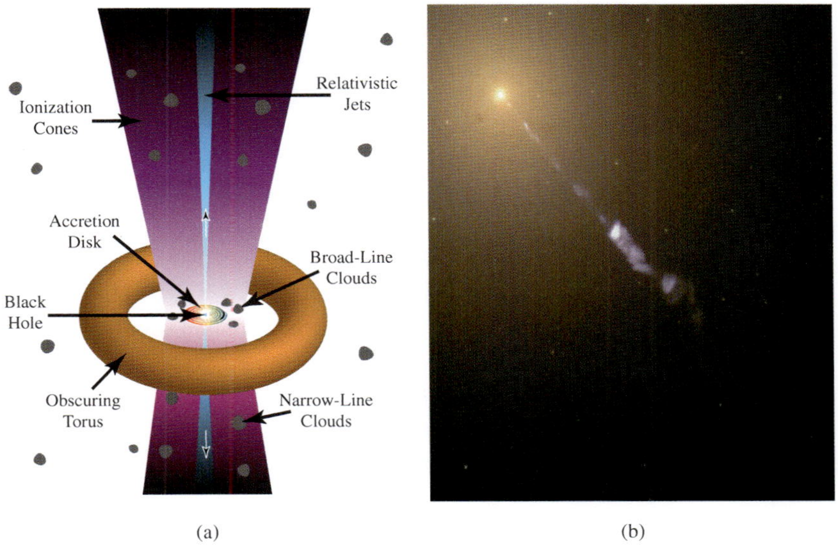

Fig. 8.17 (a) Active galactic nuclei. (b) Black hole powered jet of electrons and particle streams from the center of the M87 galaxy. Source: NASA.

spectrum of radiation. Today, most astronomers consider all the different AGN as being spinning supermassive black holes, with the differences due to the different orientations of the AGN as viewed from the Earth and due to the different rates of their accumulation and accretion. A supermassive black hole has two energetic jet beams emanating from it, as shown in Fig. 8.16. The AGN appears as a quasar when the beam is directed towards the Earth. When the beam is directed away from the Earth, the AGN appears as a Seyfert galaxy or a blazer.

The tremendous power emitted by the AGN and its other unique characteristics make it relatively easy to detect using optical, radio and X-ray telescopes. Figure 8.17(a) is a diagram of some of the key features of the active galactic nuclei with Fig. 8.17(b) showing a jet of electrons and particles streaming from the center of the M87 galaxy that is powered by a supermassive black hole.

AGN are much brighter than their host galaxies and have significant radiation, unlike starlight, over a broad spectrum from X-ray, optical to radio waves. AGN are powerful radio sources and emit highly energetic jets of out-flowing matter which are also thought to be the source of high energy cosmic rays; AGN have high intensity luminosities that are typically about 10,000 times brighter than all the stars in a typical host galaxy as illustrated in Fig. 8.13.

The source of the extraordinary power emitted by AGN points to a supermassive spinning black hole being at the core of the AGN. The AGN are powered by the gravitational binding of accreting matter that releases X-rays and other radiation, as well as being powered by the extraction of rotational energy of the spinning black hole using both the Penrose and Blandford–Znajek processes, discussed in Sec. 5.12.

8.10.1 *Quasars*

Quasars are the most luminous AGN and illustrate many of its interesting features. Quasars are very distant objects, billions of light years away; they inhabit the center of young active galaxies and appear as point sources of light, similar to stars, but with a radiation spectrum very unlike normal stars. The name "quasar" derives from the term: quasi-stellar radio source.

Quasars, similar to other AGN, occupy a *compact region* that surrounds the central supermassive black hole at the center of the galaxy. The size of the quasar is 10−10,000 times the Schwarzschild radius of the black hole. Quasars have energetic jet beams of particles and radiation that emanate from the spinning black hole and display the lighthouse effect when observed on Earth: the beam traveling towards the Earth being much brighter than the beam traveling away from the Earth.

Quasars are the most powerful sources of X-rays yet discovered, as well as sources of intense visible light. Some quasars are so luminous that they can be seen by telescopes at a distance of 12 billion light years. As is the case for the other AGN, quasars derive their energy from mass falling onto the accretion disc around the spinning black hole. Matter constantly falls into the quasar; as it falls into the black hole, matter releases energy and emits radiation, making the quasar shine with a luminosity 100 times brighter than the Milky Way; and this luminosity is produced from an extremely compact object less than 3 light years in size.

Quasars are rare objects now but were greater in number in the past, having their maximum about 10 billion years ago. The reason for this is not known although it is speculated that the population of quasars depends on galaxy formation and its relation to the central supermassive black holes.

8.11 Formation of the Solar System

This section should, logically speaking, discuss the general theory of the formation of planets from stars. A lot of progress has been made in the last few decades in detecting extrasolar planets, or exoplanets in short, with 937 having been detected by mid-2013. However, since our own solar system has been studied better than the exoplanets, we limit our discussion to it. Our solar system is primarily made up of the Sun, the planets, moons, asteroids, comets and meteors; the orbits of the various planets are illustrated in Fig. 8.18.

The process of planet formation takes place in two stages.

- The first stage is that gravitational collapse occurs in the original gaseous cloud, leading to the formation of small asteroid-like bodies, called **planetesimals**, some as large as 1/500th of the mass of the Earth. The planetesimals begin to collide and form the larger bodies of the planets.
- The second stage has to do with the fact that when planetesimals hit anything, some parts of it sticks and some portions are scattered back into space by the

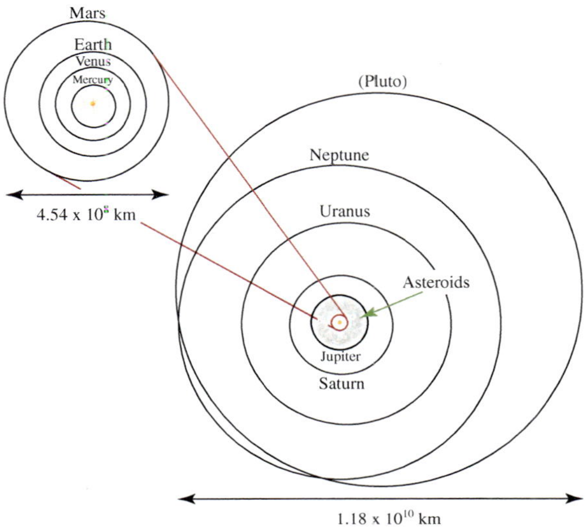

Fig. 8.18 Orbits of the planets in the solar system.

impact. The lower the density of the material, the more likely it is to escape. This leads to the heavier material coagulating and increasing the size of some of the planetesimals.

Planets can be divided into two categories, the **terrestrial planets** (Mercury, Venus, Earth, Mars) and the **Jovian planets** (Jupiter, Saturn, Uranus and Neptune). Pluto is no longer considered to be a regular planet, and has been categorized as a 'dwarf planet'; the composition of the planets is given in Table 8.2. The inner terrestrial planets are small, generally volatile-poor and are dominated by rocky material, while the Jovians contain an abundance of volatile elements.[3] There are also significant differences in the composition of Jupiter and Saturn compared to Uranus and Neptune.

Any successful theory of the formation of the solar system needs to address the following unsolved puzzles.

- The theory needs to explain the clear pattern for the composition of the planets given in Table 8.2.

[3] *Volatile* substances are those that have low melting and boiling temperatures, and the lower the sublimation temperature the more volatile is the substance; the most volatile substances are hydrogen and helium. *Refractory* materials have high melting, boiling and sublimation temperatures with the corundum Al_2O_3 having an extremely high sublimation temperature. Sublimation is a process by which a substance goes directly from a solid to a gas, and which depends on the pressure of the substance: the higher the pressure, the higher the sublimation temperature.

Table 8.2 Compositions of Earth-like and Jupiter-like planets. Note that $M_\oplus$ and $R_\oplus$ are the mass and radius of the Earth, respectively; 1 AU is the distance from the Earth to the Sun.

	Earth-like	Jupiter-like
Basic form	Rocky	Gas/liquid/ice
Mean orbital distance (AU)	0.39–1.52	5.2–30.1
Mean surface temperature (K)	200–750	75–170
Mass ($M_\oplus$)	0.055–1.0	14.5–318
Equatorial radius ($R_\oplus$)	0.38–1.0	3.88–11.2
Mean density (gcm^{-3})	3.95–5.52	0.69–1.64
Sidereal rotation period (equator)	23.9 h–243 d	9.8 h–19.2 h
Number of known moons	0–2	8–20
Ring systems	No	Yes

- A major puzzle is the origin and distribution of the solar system's angular momentum. The Sun completes one revolution in 26 days and — since its radius of rotation is very small compared to say the radius of rotation of Jupiter — it carries very little angular momentum. In fact the Sun contains about 99.8% of the mass but carries only about 1% of the angular momentum of the entire solar system. Jupiter and Saturn carry about 85% of the solar system's angular momentum; Jupiter has an angular momentum about 20 times larger than the Sun.
- A third feature that needs to be explained are the effects of the heavy bombardment endured by planets and moons during the early solar system. The current cratered surface of planets and moons as well as the tilted axis of rotation of Venus, Uranus and Pluto are thought to have been the result of these bombardments.
- And lastly any successful theory would need to explain the formation of the solar system 4.6×10^9 years ago, an obvious fact that is surprisingly hard to produce for many theories.

8.12 Solar Nebular Theory

There are many competing theories about the solar system's formation, and the consensus amongst experts is that the *solar nebular theory* is the one that accords most closely with the known observational data.[4] The solar nebular theory in essence states that a spinning nebula flattened out and condensed into the Sun and the planets and other objects in the solar system.

Since the solar system was formed during the formation of our Sun, to understand the formation of planets we need to also understand how stars in general are formed. The interstellar gas is the starting point of the solar system. As discussed

[4] Nebula is defined as a diffuse mass of interstellar dust or gas or both, visible as luminous patches or areas of darkness depending on the way the mass absorbs or reflects incident radiation.

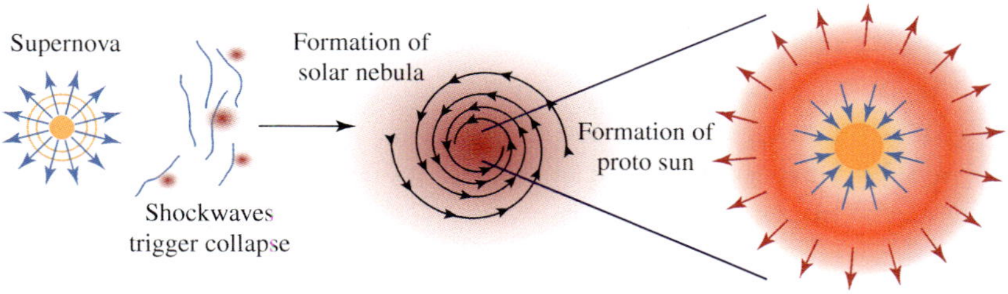

Fig. 8.19 Shockwaves from a supernova trigger the formation of a new solar system.

in Sec. 8.2, the primordial gas starts to break and collapse to form stars. The most massive portions of the gas segments evolve rapidly into massive stars on the upper main segment of the Hertzsprung–Russell diagram, while the less massive cloud segments persist in the process of collapsing.

Within a period of a few million years, the most massive stars complete their stellar evolution and explode as supernova, as shown in Fig. 8.19. Dating, based on the $^{26}_{13}$Al in chondrites found in meteorites, provides evidence that the solar system must have been formed within a few million years after the detonation of a nearby supernova.[5]

The expanding gas and debris from the supernova travel at about one-tenth the velocity of light and encounter segments of the primordial cloud that have not yet collapsed. The cloud segments are compressed by the shock wave from the high velocity supernova remnant; the compression triggers the collapse of the cloud segments, now enriched with elements synthesized in the exploded star, as shown in Fig. 8.19.

Our Sun is about 4.57 billion years old, and this age is estimated from the relative abundance of radioactive nuclei like ^{238}U and ^{235}U — that have a lifetime of over a billion years. We can empirically determine the relative abundance of ^{238}U and ^{235}U at the time the Earth was formed. Using the fact that the production ratio of ^{238}U with ^{235}U in a supernova is about 1.65 and assuming that all of the uranium currently present in the solar system was made in a single supernova, the empirical abundance of ^{238}U and ^{235}U found in the Earth points to the occurrence of a supernova explosion — estimated to have taken place about 6 billion years ago. Another evidence for the occurrence of a supernova is from the composition of ancient meteorites that contains traces of stable daughter nuclei of ^{60}Fe and of other short-lived isotopes, and which form only in exploding and massive supernova.

In fact, a more careful analysis shows that the relative abundance of ^{238}U and ^{235}U compared with other elements points to more than ten distinct sources of the material that contributed to the solar nebula. This provides the empirical basis to

[5]Chondrites contain presolar grains that were formed before the origin of our solar system.

consider the Sun to be a second generation star, which is made from the debris that was created by the first generation stars in the one and half billion years that preceded the formation of the **proto-Sun**.

The solar nebula is formed out of the collapsing molecular cloud. Assuming that the solar nebula possessed some initial angular momentum, the collapse of the solar nebula under the impact of the supernova explosion induces, due to conservation of angular momentum, a swirling motion and the flattening out of the cloud into a disk. The spinning motion of the disk leads to a proto-Sun being formed at the center of the solar nebula surrounded by a spinning disk of gas and dust. See Fig. 8.19.

The composition of the disk by mass is 73% hydrogen, 25% helium and 2% the rest. It is estimated that the *nebular disk* of gas and grains contains a mass of approximately $0.04M_{\odot}$ and is 100 AU across, and with the proto-Sun containing a mass of $1M_{\odot}$.[6]

The proto-Sun rotates rapidly and initially has a large fraction of the solar nebula's total angular momentum. At this stage the proto-Sun is still connected to the surrounding disk. Viscosity of the gas and grains in the disk causes turbulent motion to be superimposed on the smooth swirling motion of the disk. Turbulence transports parcels of gas and dust radially, with motion in the outer disk being *outwards* and the motion of parcels in the inner disk being *inwards*. See Fig. 8.19.

Turbulence results in a net transfer of angular momentum from the proto-Sun to the disk. The transfer of nebular disk mass inwards as well as outwards leads to a net loss of nebular disk mass, outwards to interstellar space and inwards to the proto-Sun, which in turn is losing mass by outflow along its axis of rotation. Taking into account the loss of disk mass and the requirement of the proto-Sun losing most of its angular momentum, some theories consider the initial nebular disk mass to be much greater than $0.04M_{\odot}$, with some estimates being about $0.15M_{\odot}$; the difference in the estimates comes from the time of the formation of the planets and their locations.

The inner disk has a temperature of about 2000 K to a distance of about 1 AU from the proto-Sun; all the volatile elements are vaporized leaving behind only the refractory materials that condense into small pockets of materials from very small to 10 cm in diameter, and are rich in calcium and aluminum. Chondrules, which are spherical objects 1–4 mm across and made from silicates, are also formed in the inner region when accreting lumps are melted by nebular flares. The shape of the solar nebula at this stage of its evolution is shown in Fig. 8.20.

The temperature of the disk falls as one moves away from the proto-Sun. Water is thought to have been abundant in the solar nebula because oxygen, among the heavy elements, is very abundant, with 940 atoms for every 10^6 atoms of hydrogen. In a hydrogen-rich gas, at all but very high temperatures, most of the oxygen

[6]The unit of distance in this section is the AU (Astronomical Unit), which is the distance of the (semimajor axis of the) Earth from the Sun, and is equal to 1.496×10^{10} m.

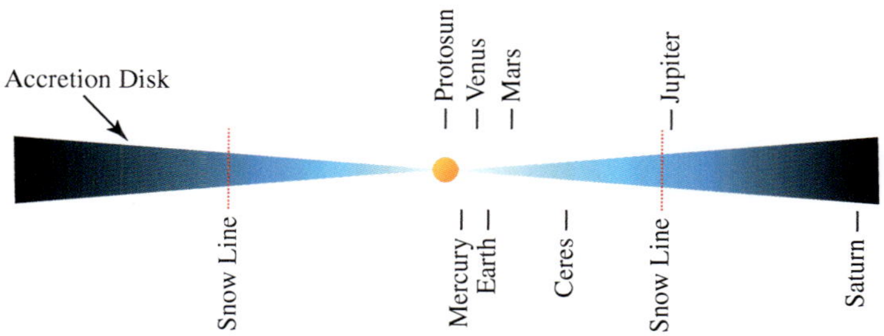

Fig. 8.20 Schematic representation of the solar nebular disk. The location of planets is where they will be once the solar system has been fully formed. Ceres is the largest asteroid. The snow line, at a distance of 5 AU, is also indicated in the figure.

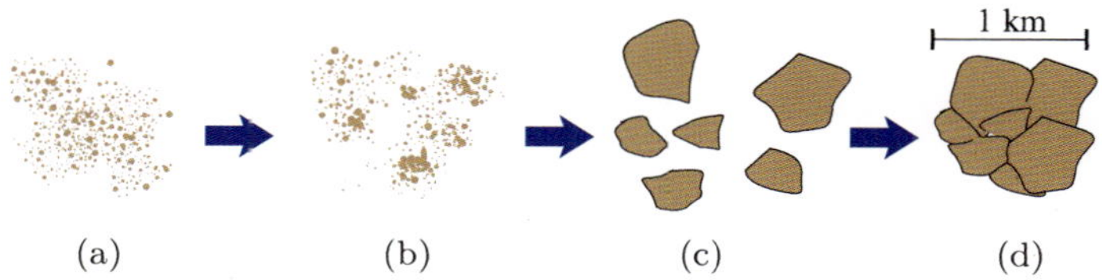

Fig. 8.21 Formation of planetesimals, from dust to granules to plantesimals.

combines with hydrogen to form water. An extremely important boundary in the nebular disk, at a distance of 5 AU from the proto-Sun, is where the temperature falls low enough for water to condense and freeze into ice; the boundary is called the *snow line*, and is shown in Fig. 8.20.

The concentration of dust into planar sheets leads to significant random collisions between grains and to the coagulation of the grains. The outcome of coagulation is to build up solids that are about 10 mm across. The process of coagulation takes thousands of years out to about 5 AU from the proto-Sun, increasing to hundreds of thousands of years at 30 AU. By the time bodies of 10 mm appear at 30 AU, the solids with distances upto 5 AU from the proto-Sun have grown to 0.1–10 km across, as shown in Fig. 8.21.

These are the planetesimals, 'little planets', being rocky at distances less than the snow line, and icy-rocky mixture beyond the snow line. At distances of 30 AU and beyond, planetesimals contain methane-ice as well. A whole hierarchy of planetesimals develops by the process of coagulation.

8.13 Formation of the Terrestrial Planets

Planetesimals, about 10 km across or larger, exert a significant gravitational attraction on other planetesimals. Planetesimals at low velocities move close to each other in the plane of the disk and — by a process of accretion — the larger planetesimals grow at the expense of the smaller ones. Accretion leads to a runaway growth of large planetesimals. Computer simulations show that for distances less than 5 AU, accretion leads to 100 planetesimals about the size of the Moon, 10 with masses comparable to Mercury's and several as large as Mars. The accretion process results in most of the planetesimals being incorporated into Venus and Earth. The time scale for the formation of terrestrial planets was of the order of 10^7-10^8 years. The freshly formed terrestrial planets and their moons were subjected to heavy bombardment by the remaining planetesimals that ended approximately 4 billion years ago, which is about 400–500 million years after the formation of the solar system started.

One would think that the composition of terrestrial planets should have more refractory materials the closer they are to the Sun. This expectation is borne out by data on the terrestrial planets, moderated by the fact that the planetesimals which were incorporated into the planets came from different regions and thus had varying composition. Terrestrial planets were formed mostly devoid of volatile material.

Asteroids are small rocky objects that condensed out of the solar nebula and were not accreted by the terrestrial planets. The largest asteroid is Ceres — being 933 kilometers across; the smallest observed asteroids are only tens of meters in size, with many being too small to detect. Many asteroids, including all of the largest asteroids, orbit the Sun between Mars and Jupiter in the asteroid belt. The Trojan asteroids share Jupiter's orbital path. Near-Earth asteroids orbit the Sun in the vicinity of the rocky terrestrial planets and pose the greatest threat to Earth.

8.13.1 *Early Earth*

The formation of the Earth started about 4.6 billion years ago and probably took a few hundred million years to be completed. That is to be compared with the time of about 4 billion years ago since the Earth has developed a solid crust.[7] About the time the Earth was formed, the gaseous proto-Sun became dense enough to ignite thermonuclear fusion. This did not happen smoothly, but likely in a sputtering way for a while. Each flaring up of the Sun spewed out particles that went streaming out from the proto-Sun. If the Earth had an atmosphere at this time, it would have been blown off, leaving the Earth as a rock with neither air nor water on its surface.

The Moon formed about 4.48 billion years ago, about $70-110$ million years after the formation of the solar system. The similar composition of the Moon and the Earth points to their common origin.

In the early stages, the Earth collected heavier material more easily, leaving lighter material such as silicon and water in orbit about the Sun. However, during

[7]The oldest continental crustal rocks on Earth have ages in the range from about 3.7 to 4.28 billion years.

the later stages of planet formation, as the Earth got bigger it could more effectively trap the lighter materials.

The rain of bodies onto the surface of the Earth generated large amounts of heat, enough to cause the heavier elements, such as iron to sink to the center. In about the first 500 million years of the formation of proto-Earth, the bombardment by the planetesimals and the heat generated by radioactive materials led to the heating of proto-Earth, and at a critical temperature, caused nickel and iron — which were distributed uniformly — to become molten. As the nickel and iron became molten, they were sufficiently dense so that the sinking globules of nickel and iron released gravitational potential energy into heat and further increased their temperature. This increase in temperature caused a *runaway migration* towards the core of the Earth, called the *iron catastrophe.*

Due to the spinning of proto-Earth, the iron catastrophe results in the formation of a rapidly spinning molten iron core, enclosed by a silicate magma. The proto-Earth's spinning iron core generates a magnetic field that results in the Earth's magnetosphere, and which deflects away the solar wind and short wavelength radiation coming from our Sun. The magnetosphere is the reason that, till today, the Earth's atmosphere has not been blown away by solar radiation and thus allows for the existence of terrestrial life.

Much of the water and ice on Earth and Mars is thought to have been delivered by comets after the formation of the planets. The Earth's atmosphere initially resulted from these comets as well as outgassing from the planet's interior, with the present day abundance of oxygen being produced later by the evolution of life.

The process forming the solar system leads to 99% of the Earth being composed out of the eight elements given in Table 8.3; the relative abundance of the elements in Earth's crust is given in Fig. 8.22. The lighter elements like hydrogen and helium escaped from the Earth since its gravitational field is too weak to hold these elements.

Table 8.3 Abundances of the eight most common elements in the Earth, which has a mass of 5.98×10^{24} kg. The Earth's *core* is composed of 88.8% iron.

Element	% weight of Earth
Iron	32.1
Oxygen	30.1
Silicon	15.1
Magnesium	13.9
Sulphur	2.9
Nickel	1.8
Calcium	1.5
Aluminium	1.4
Trace elements	1.2

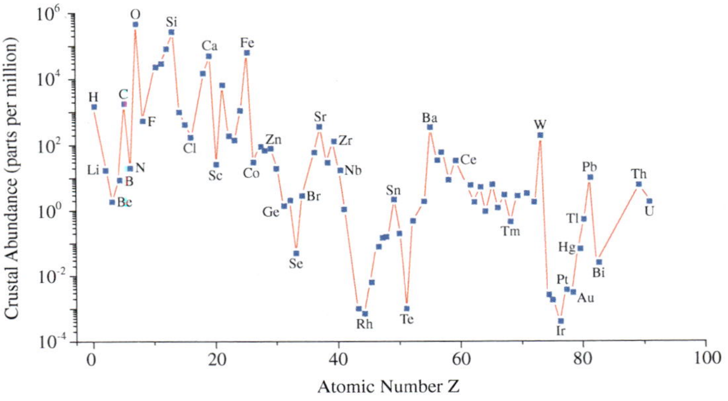

Fig. 8.22 Abundances of elements in the Earth's crust, with Z being the number of protons.

8.14 Formation of the Jovian Planets

During the period when the terrestrial planets are accreting planetesimals, the evolving proto-Sun reaches a period of instability that is caused by the initial ignition of thermonuclear fusion in its core. Thermonuclear fusion does not suddenly start in the proto-Sun; rather, it begins by fits and starts and sputters on and off for a considerable period of time, called the T Tauri stage, that is thought to last for about 10^7 years. During this stage, there is a heavy outflow of about 10% of gas from the Sun, creating a strong solar wind, as well as a high level of ultra-violet radiation. The T Tauri stage comes to a sudden end with the onset of steady hydrogen fusion in the Sun's core.

The disk material falling into the proto-Sun is reversed by the strong solar wind, and any gas and dust that has not yet been accreted by the planetesimals are driven out from the inner solar system. Water vapor, which is being blown away, on crossing the snow line at 5 AU condenses into ice and causes a blizzard, thus increasing the local density of the nebula in the snow line region. See Fig. 8.23.

In the solar nebula model, once past the snow line the planetesimals accrete into a few very large bodies called *embryos*. Figure 8.24 shows the size of these embryos as a function of distance from the proto-Sun. The mass of an embryo in the Jovian region is estimated to be about 10^{26} kg, nearly 10 times the mass of the Earth! Since there are only a few embryos past 5 AU it is unlikely that they would randomly collide.

When the large-mass embryo, which forms the kernel of Jupiter, had accreted a mass of about 10^{26} kg, its gravitational attraction was strong enough for it to sweep up surviving planetesimals, that still accounted for nearly 2% of the nebula's mass; Jupiter also absorbed the nebular gas in the region. The captured nebular gas formed an envelope with a density that increases with the mass of the embryo. The runaway capture of gas by the Jovian planets is stopped by the T Tauri phase

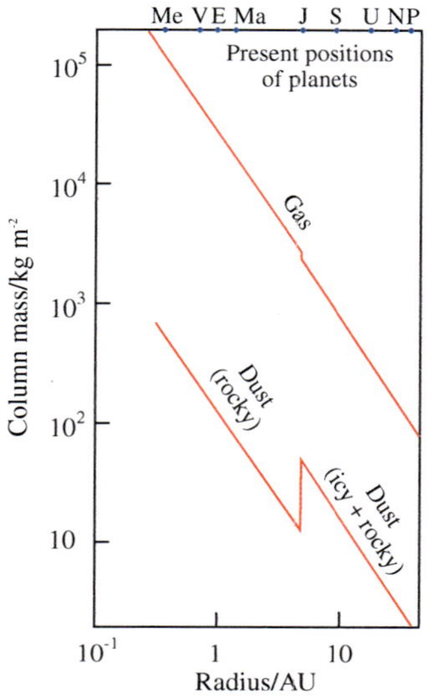

Fig. 8.23 Density of the nebula disk mass versus the distance to the Sun.

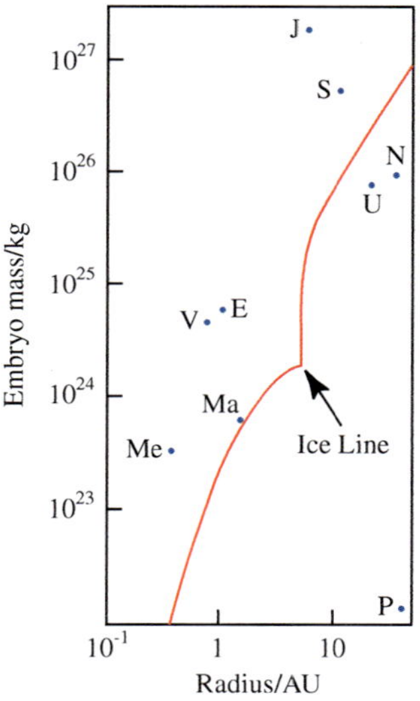

Fig. 8.24 Embryonic planetary masses versus the distance to the Sun.

of the Sun due to high radiation flux and solar winds blowing away the remaining nebular gas into interstellar space. It is estimated that the formation of Jupiter took place over about 10^6 years and Saturn a factor of two longer, halting when all the nebular gas was depleted.

The rings of Saturn, at a distance of 7000 to 80,000 kilometers from its surface, are the most extensive ring formation in our solar system. The rings are composed 99.9% of ice particles ranging from micrometer to meter in size, with a trace of rocky material. The origin of the rings is not well understood; one of the more common views is that the rings are the remnants of one of Saturn's moons that disintegrated either due to the planet's tidal forces or due to a collision of the moon with a planetesimal.

The key to understanding the differences in the masses of Jupiter and Saturn compared to Uranus and Neptune lies in the time scale at which these planets were formed. It is thought that the onset of the runaway accretion process of Jupiter and Saturn happened *before* the start of the T Tauri stage, whereas the runaway accretion for Uranus and Neptune *after* the T Tauri stage. Hence by the time Uranus and Neptune formed there was no longer any nebular gas for them to absorb, leading to their mass being lower than Jupiter's and Saturn's and their cores being much larger, proportional to their size, than Jupiter's and Saturn's.

Uranus and Neptune, by their gravitational attraction, captured a few stray planetesimals in their region. It is estimated that the formation of Uranus took about 10^7 years and that of Neptune a factor of two longer. Other planetesimals, that formed beyond Neptune, were kept out of the reach of planetary 'feeding' by their distant orbit which led to the formation of the Edgeworth–Kuiper belt; it is thought that Pluto is one of the larger planetesimals of this group, and its demotion from being considered a planet to a dwarf planet is in keeping with this view.

The formation of a massive Jupiter at 5 AU created a strong gravitational attraction that perturbed the motion of all the planetesimals close to it. In particular the orbits of planetesimals in the asteroid belt at 3 AU distance were made more and more irregular under the combined influence of the Sun and Jupiter, with some being absorbed by Jupiter and the other planets, some being sent crashing into the Sun and a few being thrown out of the solar system; the asteroid belt today contains only about 0.02% of its original planetesimal population. The gravitational effect of Jupiter also depleted, down to 3% by mass, the planetesimals in the 'feeding' zone of Mars, and is the reason for the small mass of Mars compared to Earth and Venus.

The formation of the various moons, Saturn's rings, the axis of rotation of the various planets and many other specific features of the planets can be attributed to collisions of the planets with the planetesimals. Planetesimals, due to deflections, are also expelled from the solar system. Studies of the Moon's surface leads one to conclude that bombardment by planetesimals stopped about 4 billion years ago. Collisions of planets and moons with asteroids or comets are nowadays rare.

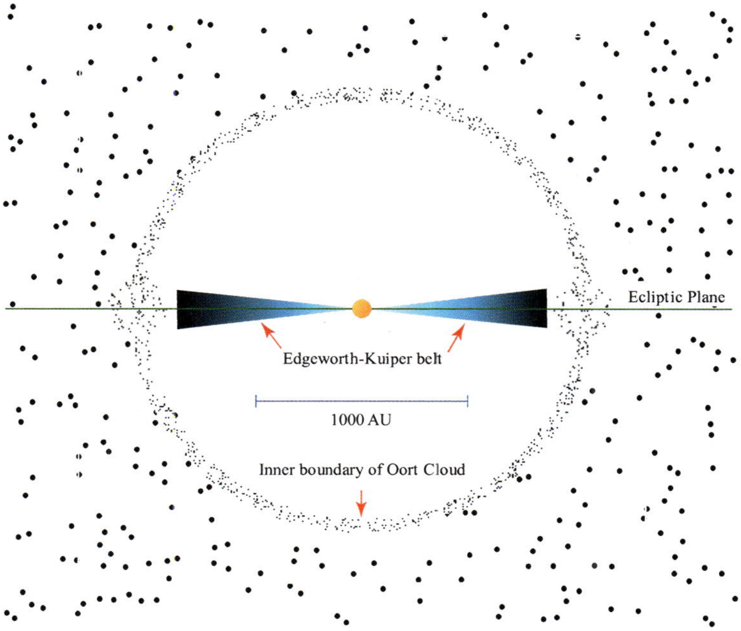

Fig. 8.25 Large scale structure of the solar system.

8.15 Large Scale Structure of Our Solar System

The large scale structure of our solar system, as shown in Fig. 8.25, has the Sun at its center, with the Oort cloud surrounding it in all directions, and the Edgeworth–Kuiper belt lying in the same elliptical plane as the planets.

Comets are objects that have very eccentric (egg-shaped) elliptical orbits that go far beyond Neptune. Comets are composed mostly of ice and dust and are hard rocky-ice objects. When a comet nucleus nears the Sun, solar radiation begins to heat the ice and vaporizes it, creating a tail that can be 10^8 m long. Comets come from two places in the solar system: the Oort cloud and the Edgeworth–Kuiper belt. The high content of volatiles like water and carbon monoxide in the cometary nucleus provides relatively direct evidence for their formation at low temperatures. Comets probably formed during the early stages of collapse and contraction of the solar nebula. The long and short period orbits of comets are shown in Fig. 8.26.

Meteors are 'shooting stars', leaving a short white trail across the sky. Meteors are made up of small 1 mm–1 cm pieces of inter-planetary dust that burn up when they enter the Earth's atmosphere at high speeds. Meteors come from two sources,

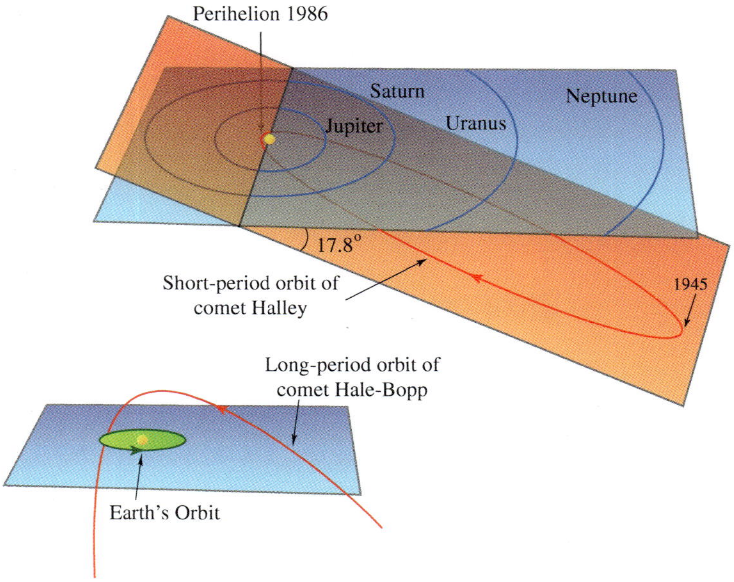

Fig. 8.26 Orbit of well-known comets.

namely (a) the remnants of the smaller planetesimals that were left over after the larger planetesimals were incorporated into planets and (b) debris from the billions of comets that have traversed through the solar system millions of times.

The **Oort cloud** is the source of the long-period comets; typically they have a period of orbit from 100,000 to 1 million years. The Oort cloud is composed of a huge collection of about 10^{12}–10^{13} comets, in orbits confined to a spherical shell centered on the Sun. The Oort cloud has an inner radius of 10^3 AU and is spread out to about 10^5 AU, almost reaching Proxima Centauri, our nearest star at a distance of 2.7×10^5 AU. In spite of the large number of comets, the mass of the Oort cloud is estimated to be 10^{25} kg, about the mass of the Earth.

It is thought that the Uranus–Neptune region is the main source for the rocky planetesimals that were ejected to form the Oort cloud. The start of thermonuclear fusion in the Sun's core created enough luminosity so that the remaining hydrogen and helium gas in the solar disk was removed by radiation pressure.

All the large objects in our solar system rotate in the counter-clockwise direction when viewed down from the North Pole, pointing to their common origin from a rotating interstellar gas cloud. About half of the comets observed from the Oort cloud have orbits that circulate the other way. This provides evidence that the Oort cloud objects are the result of random scatterings, and result in having a uniform spherical distribution.

The **Edgeworth–Kuiper (E–K) belt** is a low inclination collection of about 10^{10} comets that are a reservoir of the short-period comets, which typically have a period of less than 200 years. The E–K belt extends from about 34 AU to about 45 AU, and the inner belt is thought to supply all the short-period comets. It is thought that the E–K belt is composed mostly of icy-rock planetesimals beyond Neptune and is composed of left-overs from the formation of the solar system. The inner part of the E–K belt is close enough for its members to be visible and about 60 planetesimals from the E–K belt have been observed in the trans-Neptune region. The moon Triton is thought to have been captured by Neptune from the E–K belt. The discovery of Eris, thought to be a member of the E–K belt that was gravitationally scattered into a more distant orbit, has caused Pluto to lose its status as the ninth planet. Eris is more massive than Pluto and with a radius of 2397 km which is larger than Pluto's radius of 2306 km; both are now considered to be members of the E–K belt that have strayed into orbits with greater visibility.

8.16 The Answer

The formation of galaxies depends crucially on the cooling mechanism of the primordial gaseous clouds that emerge from the Big Bang, since high temperatures would disperse the cloud. Once a gaseous cloud can cool down and break up into smaller clusters, the formation of galaxies and its constituent stars can begin. Stars are formed out of the molecular clouds in the interstellar medium consisting of hydrogen and helium and other elements. The formation of the first generation stars takes place via processes that differ from the processes giving rise to second and later generation stars.

Stars are classified according to their **spectral classification** (surface temperature) and intrinsic luminosity and yield the Hertzsprung–Russel diagram. In general, stars with an initial mass greater than $8M_\odot$ evolve to a neutron star or a black hole, and stars with mass less than $8M_\odot$ end up as white dwarfs. A precise measure of the final state of a star is its final mass; stars with a final mass less than $1.4M_\odot$ evolve to white dwarfs whereas stars with a higher mass can end up as a neutron star or as a black hole.

The formation of planets takes place in two stages, namely the gravitational collapse of the initial gaseous cloud into small asteroids and planetesimals, and the subsequent accretion of heavier material by planetesimals — through collisions with other planetesimals, meteors and asteroids. Planets are classified into two categories, those which are within the snow line are the rocky terrestrial planets and those beyond the snow line are the Jovian planets.

The processes leading to the formation of early Earth provide an explanation for the Earth's rotating molten iron core. The Earth's magnetic field, which plays a key role in sustaining the Earth's atmosphere, is generated by the spinning iron core.

Chapter 9

The Life of Stars

What powers star burning?

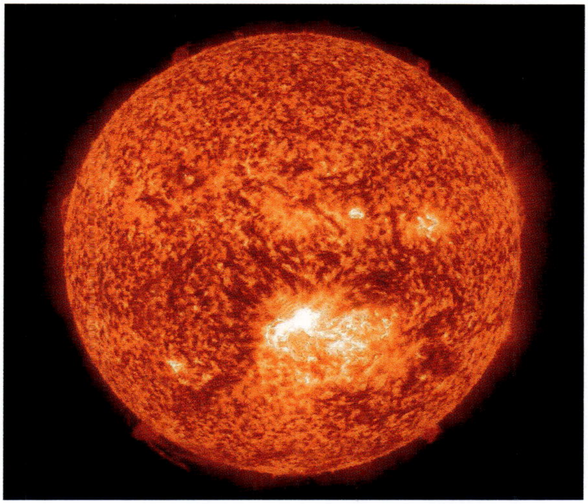

9.1 The Question

The Sun is the most important astronomical object for our planet for both inorganic and biological processes. Everyday, without fail, we see the Sun shining brightly and with no apparent sign of any change. One may wonder where does the Sun get its energy to keep on shining ceaselessly and how long will it continue to have this energy? How does it produce the sunlight that it is emanating? How was it formed? How long has it been shining and when will it stop shining?

The formation of our Sun was discussed in Sec. 8.12 and we do not pursue this aspect any further. Instead, in this chapter we focus on the Sun's source of energy and the (thermonuclear) processes that generate the Sun's radiation.

The Sun is a star, like billions of other stars in the Milky Way galaxy, and there are billions of galaxies in our Universe. Stars come in many different varieties and are at various stages in their evolution. In order to understand how our Sun burns, we study the evolution of stars in general, with the behavior of our Sun being a special case of the birth and death of a star.

9.2 Introduction

We summarize the salient points in stellar evolution that are discussed in some detail in this chapter. The term 'initial mass of a star' refers to the mass of the star when it has already evolved from its protostar phase and joined the main sequence of the Hertzsprung–Russel diagram.

- Stars in the main sequence of the Hertzsprung–Russell diagram generate energy through the burning of hydrogen at their core using thermonuclear fusion and this is discussed in Secs. 9.3 and 9.4.
- Stars in the main sequence of the Hertzsprung–Russel diagram have a lifetime given by Eq. (8.3) and spend about 90% of their lives being stationary at the point they occupy in the diagram. On exhausting their hydrogen fuel, these stars evolve to other regions in the Hertzsprung–Russel diagram.
- Stars with initial masses in the range from about $0.3M_{\odot}$ to around $8M_{\odot}$ evolve into red giants, discussed in Sec. 9.6.
- Stars with initial masses less than $2M_{\odot}$ undergo a process of helium flash as they leave the main sequence, discussed in Sec. 9.7.
- Stars with initial masses more than about $10M_{\odot}$ to $11M_{\odot}$, after exhausting their hydrogen fuel, become red supergiants, discussed in Sec. 9.9. However, very high mass stars with initial mass greater than $40M_{\odot}$ do not go through the red supergiant phase.
- The distinction between red giants and red supergiants becomes blurred for stars with initial mass in the range of $7M_{\odot}$ to $10M_{\odot}$.
- The nuclear processes that power the burning of stars with initial mass greater than $8M_{\odot}$ is discussed in Sec. 9.10. These high mass stars undergo supernova explosions, discussed in Secs. 9.11 and 9.12.
- The end point of the evolution of a low mass star is a white dwarf and that of a high mass star is either a neutron star or a black hole, discussed in Secs. 9.13 and 9.14, respectively.

To illustrate the general features of stellar evolution, consider a star in the main sequence of the Hertzsprung–Russel diagram with an initial mass of one solar mass $M_{\odot}$, and shown in Fig. 9.1. After spending 10 billion years in the main sequence, the star exhausts its hydrogen fuel and moves up along the red giant branch of the Hertzsprung–Russel diagram. There is a huge expansion of the star's volume in its red giant phase, and it spends roughly one-tenth of its life as a red giant before evolving to its final point as a white dwarf.

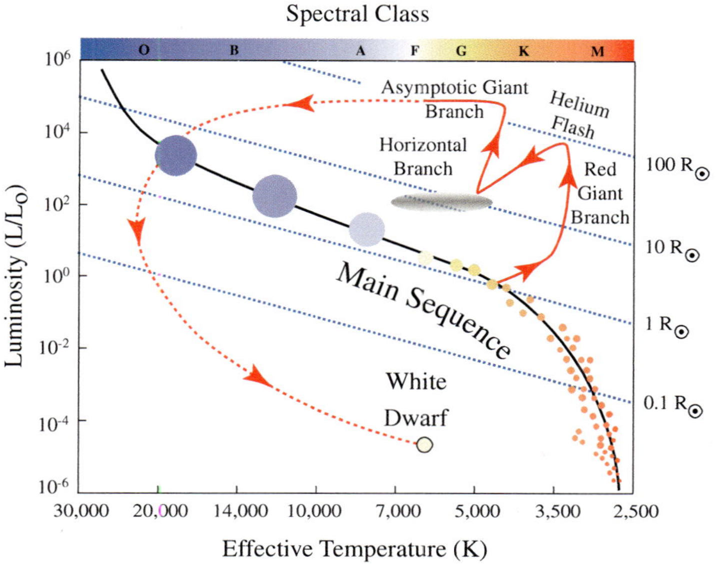

Fig. 9.1 The evolution of a star from the main sequence to its death as a white dwarf.

In the star's red giant phase, the star heats up until it starts burning helium at its core — and moves in a very short time to the left side of the horizontal branch, shown in Fig. 9.1, while still burning the hydrogen that is surrounding the star's core. The star spends only about 1% of its life on the horizontal branch, moving to the asymptotic giant branch, where helium–hydrogen burning takes place around an inert carbon–oxygen core.

During the period that the star is a red giant, radiation pressure creates strong stellar winds, which is a flow of neutral and charged gas ejected from the upper atmosphere of a star. The stellar winds blow off the star's outer mass — since it is being held by low gravity — causing the star to lose a large fraction of its initial mass. The star burns all its fuel and travels along the trajectory on the Hertzsprung–Russel diagram to its final stage of a white dwarf, where it has a low luminosity and a high surface temperature — as shown in Fig. 9.1.

The evolution of high mass stars (greater than $10M_\odot$) to their final state of being either a neutron star or a black hole cannot be shown on the Hertzsprung–Russel diagram. This is because, as discussed in Sec. 8.7, neutron stars have a high temperature but low luminosity; and black holes have almost zero temperature and zero luminosity.

9.3 Nuclear Fusion: Star Burning

The life and death of a star are determined by the interplay of gravity and pressure. All stars are in a gaseous state when they are formed; discussed in Sec. 8.5, gravity constantly compresses the gas and the compression can be stopped only if the pressure of the gas can balance the inward force of gravity. The star's temperature is raised by the compression of the gas due to the force of gravity.

The temperature is raised by compression until the onset of nuclear fusion at the core of the star — a process by which heavier elements are synthesized by fusing protons and neutrons — and which results in the release of prodigal amounts of energy that maintain the temperature of the star. Pressure is created by the temperature of the gas, which exerts an outward force that balances gravitational compression, and the star is said to be in *hydrostatic equilibrium*, discussed in Sec. 8.5. Fusion goes on until all the fuel of the star is exhausted, leading to the death of the star.

The process of thermonuclear fusion acts as a **stellar thermostat**, keeping the star's temperature — and hence the fusion rate at the core of the star — more or less constant. An increase in the temperature results in a higher fusion rate and generates a higher amount of radiation than can escape from the core of the star. This excess radiation results in the heating of the core and leads to a slight expansion of core (against the inward pressure of the outer layers) — resulting in a cooling of the core and a subsequent reduction in the fusion rate. On the other hand, if the temperature falls, the radiation generated cannot maintain hydrostatic equilibrium, leading to a slight contraction of the core and resulting in an increased fusion rate — and the subsequent heating up of the star's core.

The 'stellar thermostat' holds for stars as long as the electrons are non-degenerate; this mechanism, however, fails in the case of the helium flash, where the electron gas is degenerate, and is discussed later in Sec. 9.7.[1]

All nuclei are the bound states of protons and neutrons, which are held together inside the nucleus due to the strong interaction between protons and neutrons. Nuclear fusion combines the nuclei of lighter elements to form heavier elements. In Chapter 11, it is seen that both the proton and neutron themselves are the bound states of three quarks, and the strong interaction between the quarks, mediated by the colored gluons, holds the quarks together. The force between the proton and neutron is the so-called **nuclear force**; this force is approximately represented by the effective **Yukawa interaction** that results from the gluonic force between the quarks, and which is briefly discussed in Sec. 11.10.

Noteworthy (optional content) 9.1: Notation for Atomic Nuclei

---

In general, the nucleus of an atom is composed of protons, each of which carry a unit of electric charge, and neutrons, which are electrically neutral. Let the number of protons be Z, which is also the charge carried by the nucleus, and let N be the number of neutrons. An element is *defined* by the number of its protons, or what

[1] A degenerate and non-degenerate electron gas are discussed in Noteworthy 9.3.

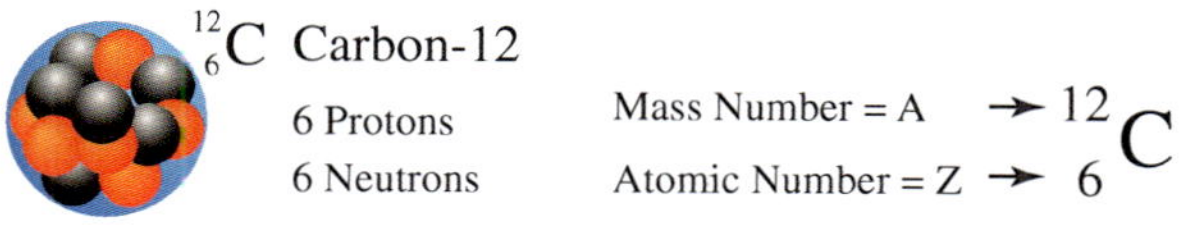

Fig. 9.2 Atomic and mass numbers for the most common isotope of carbon: carbon-12.

is the same thing, the amount of charge it has. The periodic table is organized by elements, with the protons ranging from 1 for hydrogen to 98 for californium.

For a given element (and hence a nucleus with a fixed number of protons), the number of neutrons can vary a lot and the various nuclei having different numbers of neutrons are called **isotopes** of the given element. The **mass number** A of a nucleus is given by $A = N + Z$, where Z is called the **atomic number**, which is also the number of protons. The notation for an element E is $^{A}_{Z}\mathrm{E}$ as illustrated in Fig. 9.2. A nucleus is said to have A number of *nucleons*, with both the proton and neutron being referred to as nucleons.

Hence, the hydrogen nucleus is denoted by $^{1}_{1}\mathrm{H}$, whereas the heaviest stable isotope that has been produced so far is the californium nucleus, denoted by $^{251}_{98}\mathrm{Cf}$; the deuteron d is an isotope of hydrogen that is made from one proton and one neutron and is denoted by $\mathrm{d} = {}^{2}_{1}\mathrm{H}$. Since the number of protons, namely Z, is fixed for each element, one often denotes a nucleus, which is the case for most of the discussions in this book, by only its mass number, namely, as $^{A}\mathrm{E}$; recall A is equal to the number of protons plus the number of neutrons. For example, in this notation, the helium nucleus is denoted by $^{4}\mathrm{He}$ and the most common isotope of carbon, making up about 98.89% of all the carbon of the Universe, as $^{12}\mathrm{C}$.

9.4 Stellar Thermonuclear Fusion

The generation of high temperatures by a star is powered by thermonuclear fusion, namely nuclear fusion caused by high temperatures. We examine what is the minimum temperature required for igniting and sustaining thermonuclear fusion; we further discuss this topic in Sec. 10.4.

To understand how fusion works, consider the simplest case of the fusion of two protons leading to the formation of the deuteron — which is a bound state of one neutron and one proton (here one of the protons turns into a neutron through beta decay); the formation of a deuteron is an essential step in the fusion reactions that are taking place inside a typical star. The inter-proton potential is shown in Fig. 9.3 (the sharp kink in the figure is an approximation), where the repulsive part is due to Coulomb repulsion between positively charged protons and the attractive square well is an approximate representation of the attractive strong nuclear force that acts on nucleons (protons and neutrons).

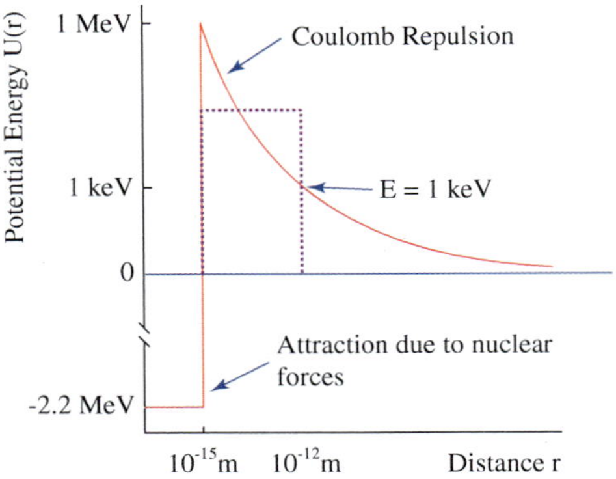

Fig. 9.3 The nuclear potential for protons is shown in the figure, which is a combination of the Coulomb repulsion between their positive charges and their nuclear attraction. The dashed line shows the potential barrier that the proton crosses for the onset of fusion. Protons with energy of 1 keV can tunnel through the Coulomb barrier.

As shown in Fig. 9.3, for two protons to fuse they have to overcome Coulomb (electrical) repulsion until they are closer than a distance of 10^{-15} m — and then the nuclear forces can bind them into a stable bound state. What, in effect, happens is that once the two protons have crossed the Coulomb barrier, strong and weak interactions can act on the two protons. One of the protons undergoes a decay via the weak interaction into a neutron, by a (rare) process called the inverse β-decay, and yields

$$\mathrm{p} \to \mathrm{n} + \mathrm{e}^+ + \nu_e$$

where e^+ is the positron (antielectron) and ν_e is the neutrino. Then, via the strong interactions, one has

$$\mathrm{p} + \mathrm{n} \to \mathrm{d} + \gamma$$

where γ is electromagnetic radiation and $\mathrm{d} = [\mathrm{p} + \mathrm{n}]$ is a deuteron — the bound state of a proton and a neutron.[2]

To cross the **Coulomb barrier** classical physics requires that the protons have (kinetic) energy higher than the Coulomb repulsive barrier. In stars, the kinetic energy is provided by thermal motion. The temperature T of the gas yields an average kinetic energy for the protons given by the formula $E[\frac{1}{2}mv^2] = \frac{3}{2}k_BT$; for protons to fuse, the Coulomb potential needs to be overcome so that we need a minimum temperature given by

$$\frac{3}{2}k_BT \simeq k_c\frac{e^2}{r_d}; \quad [r_d \simeq 10^{-15}\,\mathrm{m}]$$
$$\Rightarrow T \simeq 3 \times 10^9\ \mathrm{K}.$$

[2]Recall in the general notation of nuclear physics, we also have $\mathrm{d} = {}^2_1\mathrm{H} = {}^2\mathrm{H}$.

The kinetic energy of the protons for $T = 3 \times 10^9$ K corresponds to an energy of about 400 keV. The temperature at the center of our **Sun** is about 1.5×10^7 K and the average kinetic energy of these protons is about 2 keV — far below the energy classically required to be above the Coulomb barrier; hence, it seems that only protons with energies far above the average proton energy can undergo fusion. Such protons will be very rare so one has to wonder how fusion can take place at this temperature. The answer lies in quantum mechanics, which allows protons to **tunnel** across the Coulomb barrier even if their energy is *less* than the barrier energy. Figure 9.3 shows the effective potential barrier, indicated by the dotted line, that the protons tunnel through.

Tunneling is a quantum mechanical process in which a quantum particle crosses a potential barrier without having an energy greater than the barrier — a process that is forbidden in classical physics. In Fig. 9.3, a proton having an energy of about 1 keV tunnels across the Coulomb barrier that requires an energy of about 400 keV.

One can *never*, in principle, observe the particle *inside* the potential barrier. The tunneling process is *invisible*. Any measurement to observe the particle inside the barrier requires energy; the energy coming from the very process of measurement ends up supplying the particle with energy that makes its energy greater than the potential barrier — and hence brings it outside the barrier.

The quantum particle can tunnel through the potential barrier since the quantum particle's position is **indeterminate**, with the particle having a finite likelihood of crossing the barrier. The indeterminate state is also called a *virtual* state, to distinguish it from a physically observable state.

What quantum tunneling, in effect, means is that a star can fuse protons at the temperature of $T \simeq 10^7$ K — much *lower* than what is classically required by the Coulomb barrier. In fact, protons with kinetic energy of only 1 keV can tunnel through the Coulomb barrier, represented by the dotted line in Fig. 9.3, and fuse to form the deuteron.

In general, stars 'adjust' their temperature to be well below the Coulomb barrier temperature; fusion then proceeds at a rate fixed by the tunneling probability. Since the probability of barrier penetration by the tunneling process is very small, fusion proceeds at a very slow rate and the nuclear fuel lasts for astronomical time scales. The slowness of stellar fusion is mainly due to the fact that radiation 'leaks' out very slowly from the star's core, and is discussed further in Noteworthy 10.1.

Note for a nucleus with Z protons, the Coulomb barrier yields the temperature needed for fusion T_Z given by

$$\frac{3}{2}k_B T_Z \simeq k_c Z^2 \frac{e^2}{r_d} \quad ; \quad [r_d \simeq 10^{-15}\,\mathrm{m}].$$

To fuse protons the temperature of a star's core needs to be about 1.5×10^7 K. To fuse helium ^{4}He with $Z = 2$, we need a critical core temperature of $4 \times 1.5 \times 10^7 \simeq 10^8$ K, and for fusing carbon ^{12}C with $Z = 6$, we need a critical core temperature of about $36 \times 1.5 \times 10^7 \simeq 10^9$ K. All these temperatures are sufficient for crossing the Coulomb barrier due to quantum mechanical tunneling.

The electron volt, eV, is a frequently used unit of energy in atomic and particle physics. It is defined as the energy gained by an electron in moving through a potential difference of one volt. Using the definition of a volt, one finds, $1\,\text{eV} = 1.6 \times 10^{-19}$ J.

9.4.1 *Binding energy of a nucleus*

Why, and under what conditions, does combining two or more nuclei through fusion release energy? Consider a typical nucleus made up of Z protons and N neutrons, with masses m_P and m_N respectively. Let all the protons and neutrons be well separated with no interaction between them; we would then expect the total mass of this collection of nucleons to be the sum of the individual masses, namely

$$M_0 = Zm_P + Nm_N.$$

However, the actual mass of the nucleus when all the nucleons are brought together so that they bind into a nucleus, turns out to be M which is not equal to M_0 due to the binding energy holding the nucleons inside the nucleus. The quantity $M_0 - M$, equal to the binding energy of the nucleus, is called the 'mass deficit'; by Einstein's relation, mass m is a form of energy given by

$$E = mc^2$$

and hence the mass deficit gives the **binding energy** of the nucleus to be

$$B = (M_0 - M)c^2.$$

The nucleus has a mass M given by

$$M = M_0 - B/c^2.$$

The mass difference between nucleons that are free and the same nucleons when they are bound together to form a nucleus is shown in Fig. 9.4.

If we fuse Z protons and N neutrons to form a nucleus of mass M energy is released for $B > 0$; from energy conservation the energy *released* is compensated by the nucleus with mass M having a positive *binding energy* B equal to $B = (M_0 - M)c^2 > 0$. In other words, to separate the nucleus with mass M into its constituent protons and neutrons, one has to supply an energy equal to B.

If B is *negative*, then energy equal to $-B$, which is positive, will need to be *supplied* from outside in order to combine the nucleons to form a nucleus with mass M.

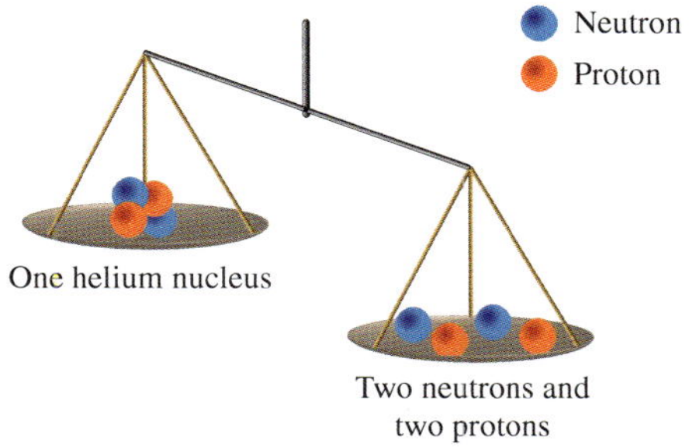

Fig. 9.4 Two protons and two neutrons have a larger mass than a helium nucleus consisting of the bound state of two neutrons and two protons.

Example 9.1. Binding Energy of a Deuteron. The atomic mass unit u is *defined* by one atom of carbon having a mass of $12u$. This yields $u = 1.66 \times 10^{-27}$ kg. The energy of an atomic mass unit of matter $= mc^2 = uc^2 = 931.494\,\text{MeV}$.

The mass of a proton and a neutron are $m_p = 1.007276u$, $m_n = 1.008665u$ respectively and the mass of a deuteron d is $m_d = 2.013553u$. The mass difference is

$$\begin{aligned}\Delta m &= (m_p + m_n) - m_d \\ &= 0.002388u.\end{aligned}$$

The binding energy of deuteron is given by

$$\begin{aligned}\text{Binding energy} = \Delta mc^2 &= 0.002388u \times \frac{931.494}{u}\,\text{MeV} \\ &= 2.22\ \text{MeV}.\end{aligned}$$

Hence in the reaction

$$\mathrm{p} + \mathrm{n} \rightarrow \mathrm{d} + \gamma$$

the energy carried by the photon γ, from energy conservation, is equal to 2.22 MeV: the binding energy of the deuteron (or about 1 million times the energy of a green light photon).

9.5 Nuclear Binding Energy: Fusion and Fission

In Fig. 9.5, the binding energies per nucleon of various nuclei are given. Note the significant fact that the binding energy per nucleon increases until $A = 56$ and $A = 62$, after which it *decreases*! There is a very shallow peak near $A = 56$ — namely for the iron ^{56}Fe and nickel ^{62}Ni atoms — and then there is a gradual decline in binding energy per nucleon for heavier elements. In terms of binding energy per nucleon, ^{62}Ni has a slightly higher binding energy than ^{56}Fe. However, since ^{62}Ni is not as rapidly produced in stellar fusion and due to other factors, ^{56}Fe is far more abundant than ^{62}Ni with about 19 atoms of iron for every atom of nickel.

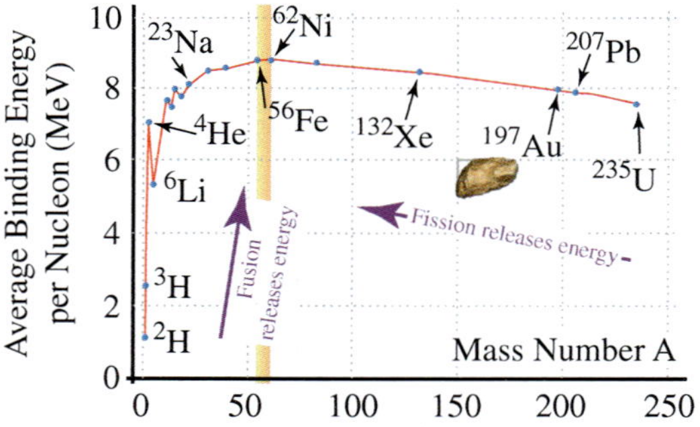

Fig. 9.5 The binding energies per nucleon in MeV of some elements are plotted versus the mass number. The curve reaches its maximum for iron (Fe).

What this means is that when nuclei lighter than iron are fused to form atoms with $A < 56$ energy is *released* since the fused nuclei are more tightly bound than when not fused. In contrast, energy has to be supplied in order to fuse nuclei to form atoms with $A > 56$.

Noteworthy (optional content) 9.2: Nuclear Binding Energy

To illustrate the fusion of nuclei consider the following. Take three nuclei with masses m_1, m_2 and m_3 with numbers of protons (atomic number) given by Z_1, Z_2, Z_3, numbers of neutrons given by N_1, N_2, N_3 and with mass numbers given by A_1, A_2, A_3 respectively; note $A = Z + N$. If the binding energy per nucleon of the nuclei is given by b_1, b_2, b_3, then the mass of each nucleus obeys the relation

$$m_i c^2 = (m_p Z_i + m_n N_i)c^2 - A_i b_i \ ; \ i = 1, 2, 3.$$

Consider the following nuclear process that fuses nucleons with masses m_1, m_2 into m_3, namely

$$m_1 + m_2 = m_3 + B/c^2.$$

Fusion releases energy only if the mass deficit B is positive. Since $Z_3 = Z_1 + Z_2$, $N_3 = N_1 + N_2$ and $A_3 = A_1 + A_2$, the energy B of fusion is given by

$$\begin{aligned} B &= m_1 c^2 + m_2 c^2 - m_3 c^2 \\ &= A_3 b_3 - A_1 b_1 - A_2 b_2 \\ &= A_1(b_3 - b_1) + A_2(b_3 - b_2). \end{aligned}$$

There are the following two cases.

- Suppose m_1 and m_2 are *nuclei lighter than iron* and m_3 is greater than m_1, m_2 but less than $m_1 + m_2$ — which holds for m_3 less than iron. Since b_1, b_2 are both less than b_3, we have that $B > 0$; hence *fusion* of light nuclei leading to an element lighter than iron *releases* energy.

- A similar analysis of *nuclei heavier than iron* can be carried out to show that the *fission* of a single nucleus heavier than iron into two or more nuclei *releases* energy.

9.6 Stellar Evolution: Formation of Red Giants

Recall from Fig. 8.9 that the Hertzsprung–Russell diagram has a group of stars called the red giants. Stars with masses in the range from about $0.3M_\odot$ to around $8M_\odot$ in the main sequence of the Hertzsprung–Russell diagram evolve into red giants.

The red giants are the result of the main sequence stars in the Hertzsprung–Russell diagram that have consumed their hydrogen fuel and have moved to the red giant sector. For the case of a star with mass similar to our Sun, its movement in the Hertzsprung–Russell diagram — over a period of billions of years from its initial state as a protostar to its position in the main sequence with a spectral classification of G star to its final state of a white dwarf — is given in Fig. 9.6. The star goes through the phase of being a red giant, and we briefly discuss the formation of the **red giants**.

Stars in the lower main sequence gradually form a core of helium as the star burns its hydrogen fuel, with the burning being fastest at the center of the star. The change in the density of helium inside the star is shown in Fig. 9.7.

Convection (movement of particles) mixes only the gas in the outer envelope of the core, with the core gradually becoming more rich in helium. A **radiative core** develops in the center of the star — from which energy is transported to the **convective envelope** by radiation rather than by convection, as shown in

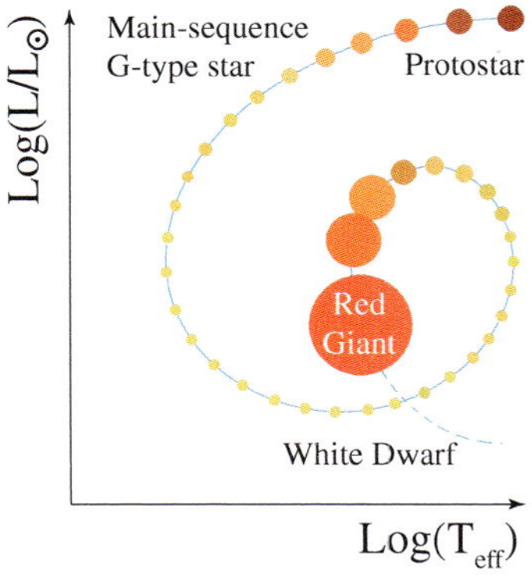

Fig. 9.6 The path of the Sun's life on Hertzsprung–Russell diagram, from a protostar to its final stage as a white dwarf.

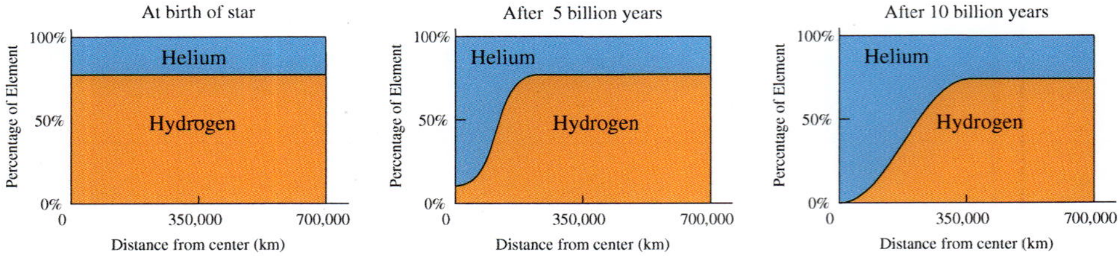

Fig. 9.7 In a Sun-like star, the ratios of hydrogen and helium are estimated to evolve as shown over its lifetime.

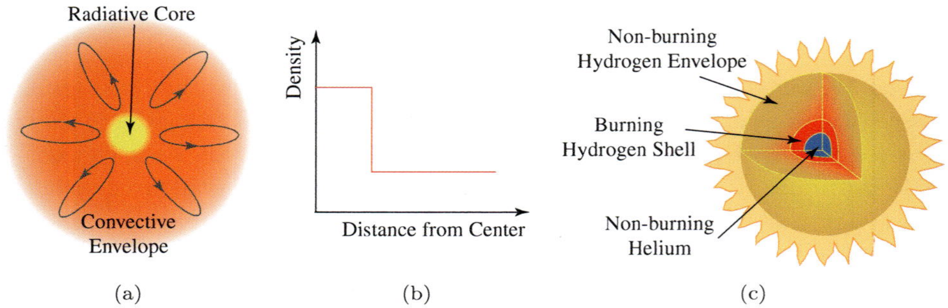

Fig. 9.8 (a) A convective envelope surrounds the radiative core. (b) Approximate density of the stellar gas. (c) Helium accumulates in the core of a star and is surrounded by a hydrogen burning shell.

Fig. 9.8(a). There is a fairly sudden change in the density of the star as one goes towards the center, as shown in Fig. 9.8(b).

A star of one solar mass, shining for about 10 billion years by fusing hydrogen at the center of the star, completely depletes its supply of hydrogen in its center — and fusion ceases at its core. The core now consists of (non-burning) helium that is referred to as helium-ash, with fusion continuing only in the convective envelope, as shown in Fig. 9.8(c). This creates a serious structural instability for the star since the temperature at the core falls due to the cessation of hydrogen burning — thus reducing the core's pressure. The helium core can no longer balance the inward pull of gravity since the pressure of the star's core is not high enough — and hence the core starts to contract.

Contraction of the helium-ash core increases the pressure on the shell of hydrogen enveloping the core and hence raises its temperature. Furthermore, the hydrogen burning shell is surrounded by a non-burning hydrogen envelope, illustrated in Fig. 9.8(c), since the pressure and temperature are too low for hydrogen burning to take place in the envelope.

The high rate of nuclear fusion in the hydrogen shell surrounding the helium core creates tremendously high pressure that pushes against and expels the non-burning outer hydrogen away from the core. Hence the core shrinks while at the same time

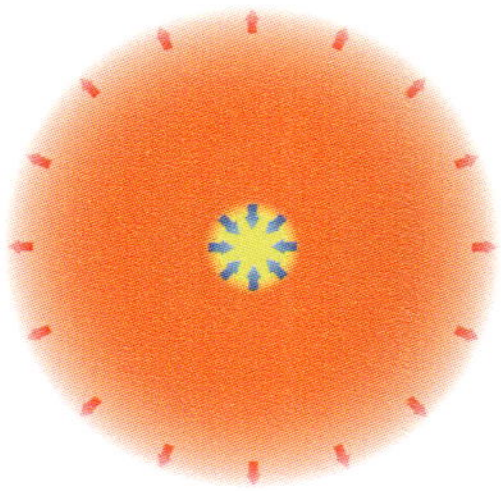

Fig. 9.9 The core shrinks while the envelope is expelled.

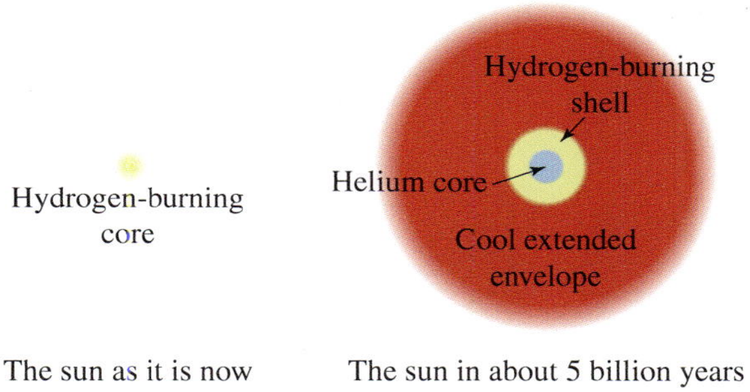

Fig. 9.10 The Sun as it is today and as it will be in about 5 billion years.

the outer portion of the star's hydrogen is expelled out to great distances, as can be seen in Fig. 9.9. The star swells to a size that is about 100 times its original size, thus creating the **red giant** star. For star like the Sun, with surface temperature of about 5700 K, its surface temperature falls by about 2000 K.

When in the future the Sun becomes a red giant, its size will reach the planet Mercury, with its helium-ash core being 1/1000th of its present size — about the size of the Earth. A red giant and its enveloping cloud is shown in Fig. 9.10.

The red giant's core contains about 25% of the star's original mass and has an immense density of 10^8 kg/m^3 — compared with the current density of the Sun of 15,000 kg/m^3; the temperature at its core, due to gravitational contraction, is now about 10^8 K. Recall that since helium has two protons (and two neutrons), the Coulomb barrier, which is proportional to the square of the proton charges, is four times higher for helium than it is for hydrogen. At the temperature of $\simeq 10^8$ K, helium now has enough thermal kinetic energy to tunnel through the Coulomb barrier at a non-negligible rate and helium nuclei can now fuse to form heavier nuclei.

The onset of helium burning happens about 100,000 years after the star has become a red giant, with helium fusion taking place by the so called **triple-alpha process** (discussed in Sec. 10.6)

$$3\ {}^4\text{He} \rightarrow\ {}^{12}\text{C} + \gamma$$

and leads to carbon as its final product.

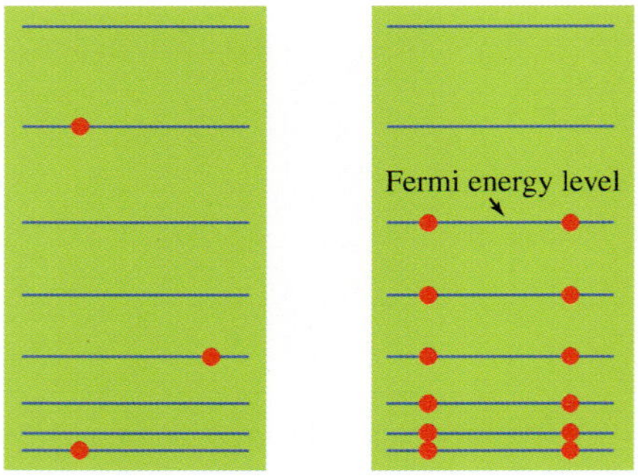

Fig. 9.11 The Pauli exclusion principle forbids more than two electrons in the same energy state. Fermi energy level is the highest occupied energy state for a degnerate electron gas.

Noteworthy (optional content) 9.3: Pauli's Exclusion Principle

---

The 100,000 years delay in the onset of helium burning is due to the **Pauli exclusion principle** that arises from quantum mechanics, which we discuss briefly.

Electrons are fermions; **fermions** have a rather special property known as the Pauli exclusion principle that states that two fermions cannot be in the same quantum state. Only one electron can occupy a point in space and has an internal state that can be either 'up' or 'down'. Only two electrons, with spins pointing in opposite directions, can occupy a particular energy level, as shown in Fig. 9.11(b).

A **non-degenerate electron gas**, as shown in Fig. 9.11(a) is one where the electron gas has a low density and can occupy one of many distinct quantum states. In contrast, as shown in Fig. 9.11(b), for a high density **degenerate electron gas**, all the electrons try and occupy the lowest available quantum states — with the highest energy state that is occupied called the Fermi level or the **Fermi energy level**, shown in the Fig. 9.11(b). The existence of the Fermi energy level creates an outward pressure: the pressure of the electron gas increases until it is high enough to provide sufficient energy to the electrons so that they have a high enough energy required for occupying the Fermi energy level.[3]

In a normal star, the density of matter, and hence of the electrons, is low so that each electron can occupy one of many states, depending on its energy. However, there are special circumstances in the evolution of a star where the Pauli exclusion principle becomes important due to the emergence of a degenerate electron gas.

---

[3]If such a high energy cannot be imparted to a fermion, as happens in the case of neutron stars and discussed in Sec. 9.12, then other processes take over.

9.7 Helium Flash

A **helium flash** occurs for stars with mass less than $2M_\odot$. In the core of a red giant with a mass less than $2M_\odot$, the density of electrons reaches such a high degree of concentration that the electrons become a *degenerate electron gas*. For a degenerate electron gas all the electrons would want to have the same energy and hence all the electrons should be in the same quantum state. However due to the Pauli exclusion principle this is forbidden, as discussed in Noteworthy 9.3; instead the electrons can only fill up the available quantum states. Starting from the lowest energy electron states, the electrons occupy higher and higher energy states due to Pauli exclusion principle as shown in Fig. 9.11(b).

The fact that electrons occupy high energy states, and resist being forced into the same quantum state, creates in turn an outward pressure in the degenerate electron gas that opposes the inward force of gravity. The outward pressure solely due to the degenerate electron gas (in the star's core) can balance the star's own gravitational self-attraction for a mass M up to a maximum value given by the Chandrasekhar limit of $M_C = 1.44M_\odot$. A white dwarf is a star for which the inward gravitational attraction is entirely balanced by the pressure solely due to the degenerate electron gas; this is the reason that all stars with a final mass less than or equal to $1.44M_\odot$ become white dwarfs.

As discussed in Sec. 9.3, thermonuclear fusion for a non-degenerate electron gas is like a self-regulating **thermostat** that maintains the star's temperature at the value required for fusion to be carried out in hydrostatic equilibrium.

In contrast, for a degenerate electron gas, the pressure counter-balancing gravity is generated by the Pauli principle and it follows that the pressure of the star is *almost independent of temperature*.[4] Hence, when the temperature rises due to fusion, there is *no expansion* of the gas and no associated appreciable fall in temperature. The energy released by helium fusion continues to raise the temperature of the core and thus increasing the rate of helium fusion. This gives rise to a thermal instability leading to an immense release of energy, namely the *helium flash*. The energy production in a helium flash exceeds the current energy production of our Sun by a factor of 10^{11}!

However, only a small fraction of the energy of the helium flash is released as radiation, with most of the energy going into creating a high temperature in the core that removes the electron degeneracy and hence causing a rapid expansion of the helium core. The star then returns to the normal and stable burning in which the gas expands as the temperature increases and thus regulating the temperature around an average value.

[4]The pressure due to a degenerate electron gas is almost independent of temperature and depends only on density. In contrast, pressure due to temperature is proportional to density times temperature and hence is strongly dependent on temperature.

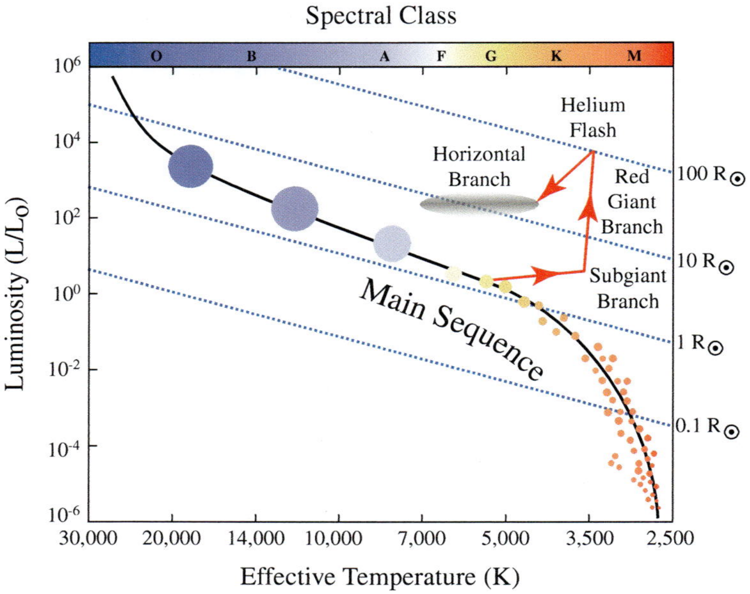

Fig. 9.12 After its time on the main sequence is up, the star's luminosity increases greatly while it moves up the red giant branch. After the helium flash, it reaches a new hydrostatic equilibrium state at the horizontal branch.

The completion of the process of helium flash (about 100,000 years after the initial onset of helium fusion) results in the helium core reverting back to having non-degenerate electrons and with the star's core reaching a new hydrostatic thermal equilibrium. The self-regulating process of thermonuclear fusion now sets in for helium burning. The evolution of a Sun-like star on the Hertzsprung–Russel diagram — undergoing helium flash — is shown in Fig. 9.12.

9.8 Formation of a White Dwarf

In a few million years, the red giant stars arising from the low and medium mass stars on the Hertzsprung–Russel diagram will have consumed all their nuclear fuel, fusing hydrogen into helium and then via the triple alpha process, fusing helium into carbon and yield a carbon–oxygen core for the star, and with the star's outer envelope having been ejected.[5] It is estimated that over 95% of all stars in the Milky Way will end up as white dwarfs.

Since its fuel is all spent, thermonuclear fusion stops in the star and the core, stripped of the enveloping cloud, appears as a small star with a white-hot surface,

[5]Stars that are more massive can end up with an oxygen–neon–magnesium core.

called a *white dwarf*. The radiation from the white dwarf is no longer due to fusion, but rather due to the heat stored in the star. Its surface temperature varies from 150,000 K to about 4000 K and its luminosity is about a hundred times less than our Sun.

A white dwarf is an incredibly dense object and is one of the possible end points of stellar evolution, as discussed earlier in Sec. 8.5. White dwarfs have a mass that is approximately equal to the Sun but with a radius that is about 100 times smaller. A teaspoonful of the white dwarf would weigh as much on Earth as an elephant, which is about 5.5 tons. For example, Sirius B is a white dwarf and has the mass of our Sun packed into a volume less than that of the Earth, with a density of about 10^{10} kg/m^3.

Figure 9.1 is a representation on the Hertzsprung–Russel diagram of how a star evolves from its initial state to its final destination as a white dwarf. The evolution of our **Sun** is expected to follow a similar path.

The white dwarf is in equilibrium with gravity not by burning nuclear fuel, but rather the star's gravitational attraction is balanced by the pressure due to the degenerate electron gas — arising from the Pauli exclusion principle. Hence the white dwarf cools off by losing energy while maintaining a fixed size for its carbon core and, in time, its temperature drops to near absolute zero and it becomes a black dwarf emitting no radiation.

9.9 Red Supergiant Stars

Stars in the upper main sequence with an initial mass greater than about $10M_\odot$ evolve into **red *super*giants** after the hydrogen in a star's core has been exhausted. These stars have surface temperatures of 3500–4500 K, and are stars with the largest volume, but not the highest mass. Betelgeuse and Antares are among the largest red supergiants.

The red supergiant stars originate from the same process that creates the red giants. On consuming their hydrogen fuel they start to expand, but because of their higher mass their core temperature is high enough to almost immediately begin to fuse helium in their core; this is the reason that they do not go through the helium flash phase. They consequently maintain their luminosity and hence move nearly horizontally across the Hertzsprung–Russell diagram to become red supergiants.

The difference between red giant and red supergiant stars is in their initial mass and the role of stellar winds. Stellar winds can cause a red giant to lose almost 20% of its mass; in contrast, for red supergiants the original star loses very little of its mass when it undergoes expansion. Hence the helium-ash core of the red supergiants has much more mass than that of ordinary red giants.

The red supergiants undergo another cycle of burning resulting in a carbon-ash core, surrounded by a helium-burning shell which in turn is surrounded by a

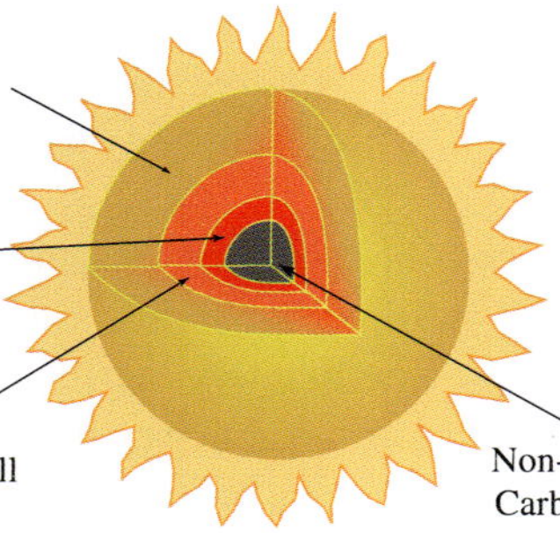

Fig. 9.13 When the pressure becomes high enough in red supergiants, the helium core begins to burn.

hydrogen-burning shell and the shells being enclosed by a non-burning envelope of hydrogen gas. See Fig. 9.13.

9.10 Evolution of High Mass Stars

High mass stars, with an initial mass greater than about $10M_{\odot}$ and constituting the upper main sequence stars in the Hertzsprung–Russel diagram, evolve much faster than the lower mass stars due to their strong gravitational attraction — which in turn leads to a ravenous consumption of their nuclear fuel. For example, for about 10 billion years our Sun will stay in the main sequence of the Hertzsprung–Russel diagram before leaving it to form a red giant whereas a star with a mass much greater than our Sun could remain in the main sequence for only 20 million years. The Hertzsprung–Russel diagram representation of the evolution of stars for high mass stars is shown in Fig. 9.14.

Stars more massive than about $40M_{\odot}$ do not go through a phase of becoming red supergiants. They rapidly burn and soon lose their outer layers and in doing so reach the blue supergiant stage, or perhaps yellow hypergiant, and finally explode to form neutron stars and black holes.

All processes that take place in the lower main sequence stars are repeated in the more massive stars. The massive star starts its journey in the upper main sequence of the Hertzsprung–Russel diagram; after consuming all its hydrogen fuel and creating a collapsing helium core, it goes on to become a red supergiant. For stars with mass greater than $1.3M_{\odot}$ the **CNO-cycle** (carbon–nitrogen–oxygen-cycle), as discussed in Sec. 10.5, is the main process of solar fusion; for the massive stars, which are burning at temperatures higher than our Sun, the CNO-cycle is a more efficient way of burning hydrogen than fusing hydrogen to make helium.

The CNO-process is strongly temperature dependent, with the temperature falling off rapidly as one moves away from the core. The temperature gradient results in the core region forming a **convection zone** that mixes the hydrogen

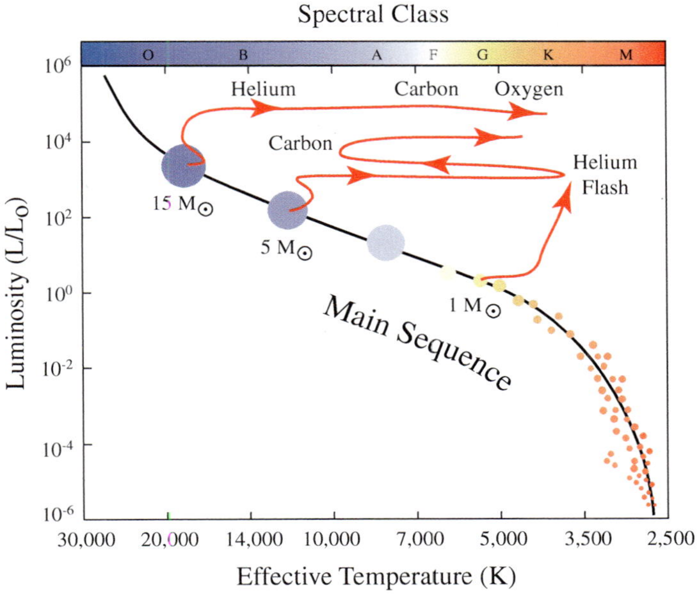

Fig. 9.14 The evolutionary tracks of stars of various masses.

undergoing fusion with the produced helium. The core convection zone of these stars is surrounded by a **radiation zone** that is in thermal equilibrium. Hence, for more massive stars, we have a convective core surrounded by a radiative envelope — the opposite case from the lower mass stars for which the radiation zone is at the core and is enveloped by the convection zone. See Fig. 9.15. Furthermore, unlike the less massive case, there is a discrete jump in the composition of the massive stars. As the core collapses the outer radiative envelope is pushed away, forming a red supergiant.

Due to its high mass core, nuclear fusion proceeds smoothly converting helium into carbon — never reaching the density required for the formation of a degenerate electron gas, and hence there is no helium flash. Furthermore since the core can reach a temperature of 600 million K, it can subsequently burn carbon into oxygen.

The high mass star continues to fuse heavier elements and, as its core contracts, the temperature rises and heavier and heavier elements are produced at an ever-increasing rate. The ash of the preceding phase becomes the fuel for the next phase of nuclear burning. Figure 9.16 shows the onion layered structure at the core of the star and Fig. 9.17 shows the location of the core within the overall structure of the red supergiant star. In the red supergiant phase, sufficiently high mass stars synthesize all elements up to iron and nickel.

The deeper a layer inside the core of the red supergiant, the *faster* it is burning. For example, for a very high mass star, hydrogen burning goes on for 10 million

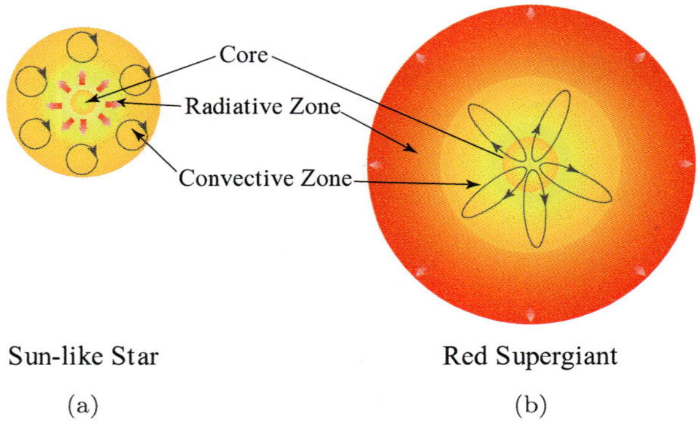

Fig. 9.15 Convective and radiative zones in (a) Sun-like stars and (b) more massive red supergiant stars of the upper main sequence. The places of the two zones are interchanged for the two cases.

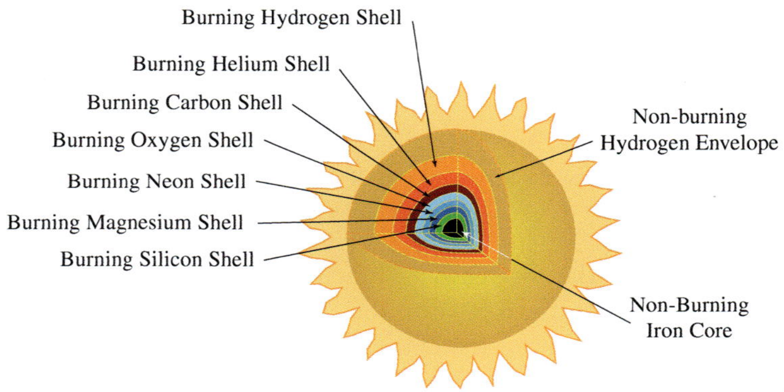

Fig. 9.16 In a highly evolved star with mass greater than our Sun, fusion of heavier and heavier elements takes place in ever thinner shells at ever higher temperatures.

years, helium burns for 1 million years, carbon burns for 1000 years, oxygen burns for only 1 year and silicon burns for merely a week. Iron is synthesized in the core of the star for less than one day!

Fusion stops at iron and nickel since the binding energy per nucleon is at a maximum and hence no energy is released by fusion that results in heavier atoms. Note gold and all the other elements heavier than iron cannot be synthesized in this phase of the star's life.

9.11 Type II Supernovae

Stars with a final end point mass greater than about $3M_{\odot}$ are thought to undergo gravitational collapse leading to an explosive supernova with a neutron star being formed at its core.

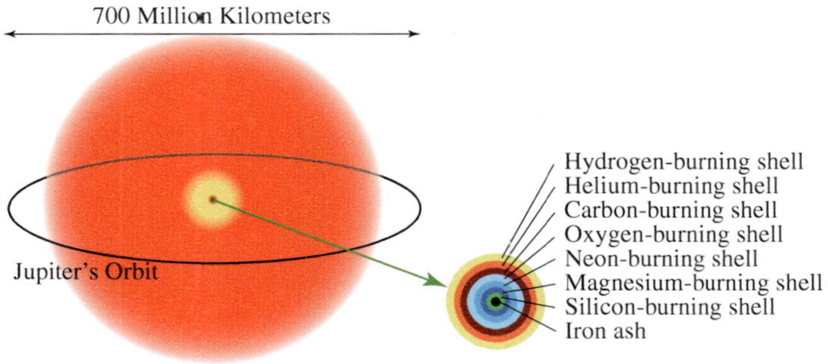

Fig. 9.17 The final stage of a red supergiant.

For stars on the main sequence of the Hertzsprung–Russell diagram, with an initial mass greater than about $10M_\odot$, all the stages of burning until silicon take place, as shown in Fig. 9.17. The production of iron in the core of the star in effect cools off the star's core since once iron starts to be synthesized, fusion of iron or nickel can no longer release energy needed for heating the star's core to balance the pull of gravity. The star is no longer able to withstand the inward pull of gravity and starts to collapse: the star's stability is destroyed forever.

Since the star's core temperature is at 10 billion K, the photons produced in fusion are energetic enough to break up the iron nucleus in the core, called photodisintegration. This is the reverse of fusion and hence the star cools rapidly due to photodisintegration since energy of the star is consumed in breaking up the iron nuclei. In less than a second, photodisintegration undoes the effect of 10 billion years of fusion and splits up all the iron nuclei in the core into protons and neutrons. As the iron nuclei are destroyed the star rapidly loses energy, which is absorbed in breaking up nuclei and the collapse of the star accelerates.

The core is now composed of only free electrons, protons and neutrons. Due to the high density of about 10^{12} kg/m^3, electrons fuse with protons yielding a core made out of only neutrons by the following inverse beta decay reaction

$$\mathrm{p} + \mathrm{e}^- \rightarrow \mathrm{n} + \nu_e + \gamma$$

where ν_e stands for an electron neutrino, which is an elementary particle and is discussed in Chapter 13.

The neutrinos escape from the core — due to the fact that they are only coupled to weak interactions — carrying away kinetic energy. This is the second mechanism that cools the collapsing star's iron core, the first being photodisintegration. The core then collapses to an incredible density of about 10^{15} kg/m^3 and the neutrons are brought into contact with each other. When neutrons are being crushed into each other, Pauli's exclusion principle — which is valid for all fermions (of which the electron is a specific case) — becomes operational for neutrons since they are also fermions.

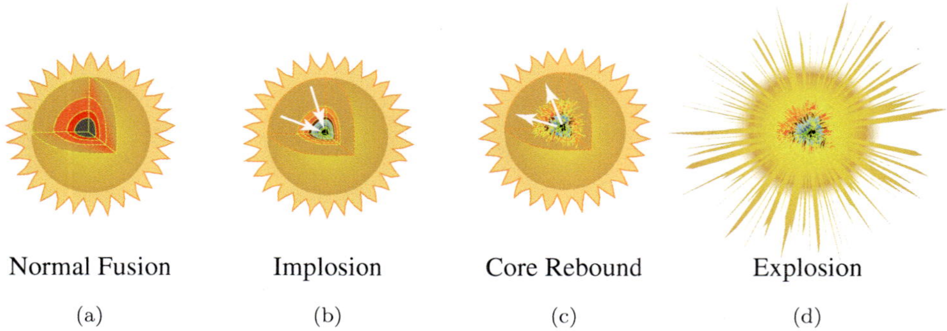

Fig. 9.18 Type II supernova: the core of a high mass star collapses, and the rebound leads to an explosion.

Similar to the case of the helium flash and white dwarfs, pressure is created by the **degenerate neutron gas** that works to *stop* the star's gravitational collapse. The collapse overshoots the resisting pressure and reaches densities of about 10^{17} kg/m^3 — the density of an ordinary nucleus: the star's former iron core has become equal to one giant nucleus!

Much like a ball bouncing off a wall, the core bounces back within about one millisecond after the start of the implosion. The bounce back generates an intense shock wave that blows away all the overlaying layers of matter surrounding the inner iron core — including all the shells of the heavy elements — into space. Stars that undergo **supernova** explosions yield as their remnants — depending on their mass and the details of the explosion – either a neutron star or a black hole.

The elements synthesized in the star are dispersed into the surrounding space and become part of the gas cloud that may later seed star and planetary formation, as discussed in Sec. 8.4.

The spectacular implosion–explosion is a *core-collapse supernova* and is called a **Type II supernova**.[6] The main stages in the occurrence of a Type II supernova (shown in Fig. 9.18) are the following.

- A high mass star generates pressure by nuclear fusion, as shown in Fig. 9.18(a).
- The star exhausts its nuclear fuel and undergoes an implosion due to its inability to stave off gravitational collapse, with the collapsing core reaching a velocity almost 25% the speed of light; this stage of a supernove is shown in Fig. 9.18(b).
- The collapse of the star is stopped by neutron degeneracy, and the core rebounds after over-shooting the collapsing forces, and is shown in Fig. 9.18(c).
- The supernova explosion takes place, leaving behind either a neutron star or a black hole, and is shown in Fig. 9.18(d).

[6]There are further subclassifications of Type II supernova that are not relevant to our discussion.

Supernovae are a key source for the production of elements heavier than iron, and in particular the supernova explosion provides the energy required for producing elements heavier than iron. Supernovae are a candidate site for the r-process, discussed in detail in Sec. 10.8, that produces highly unstable nuclei that are rich in neutrons. In fact, it is considered likely that the r-process, taking place in Type II supernovae, produces about half of all the quantity of elements beyond iron in our Universe, including plutonium and uranium.

SN 1987A is one of the most famous supernovae and was observed in 1987 at about 51.4 kiloparsecs from Earth, and was visible to the naked eye. The Japanese detector (Kamiokande-II) detected neutrinos from SN 1987A and was the first direct experimental confirmation of the immense flux of neutrinos from a supernova.

9.12 Type Ia Supernovae

A **Type Ia supernova** can occur for a white dwarf that is part of a binary star system.[7] The white dwarf's gravitational pull leads to the accretion of matter from its companion star. As it accretes, the mass of the accreting white dwarf increases until it exceeds the critical Chandrasekhar limit of $1.44M_{\odot}$.

When the mass of the accreting white dwarf exceeds $1.44M_{\odot}$, pressure due to the degenerate electron gas can no longer balance the inward pull of gravity. The white dwarf undergoes an implosion caused by gravitational attraction, leading to a tremendous increase of pressure and temperature throughout the volume of the white dwarf.

The carbon composing the white dwarf is compressed beyond the Coulomb barrier and consequently all the carbon nuclei simultaneously undergo fusion throughout the white dwarf resulting in a violent explosion. The exploding star is said to be a *carbon-detonation supernova*, also known as a *Type Ia supernova.*[8] The processes leading to a Type Ia supernova are shown in Fig. 9.19. An exploding Type Ia supernova has, for a few seconds, an intrinsic brightness that is about 5 billion times that of our Sun.

The difference between the two types of supernova can be experimentally observed. There is virtually no hydrogen or helium in the spectrum of the Type Ia supernova since it results from a carbon-core white dwarf. In contrast a Type II supernova, due to the presence of unburnt hydrogen and helium in its outer envelope that is blown away when the core explodes, shows a strong presence of hydrogen and helium in the supernova's observed spectrum.

[7] 48% of stars are single, 36% of stars are binary, 12% of stars are triple and 4% of stars are in quadruple systems.

[8] There are many subcategories of Type I supernovae and we are only interested in the Type Ia.

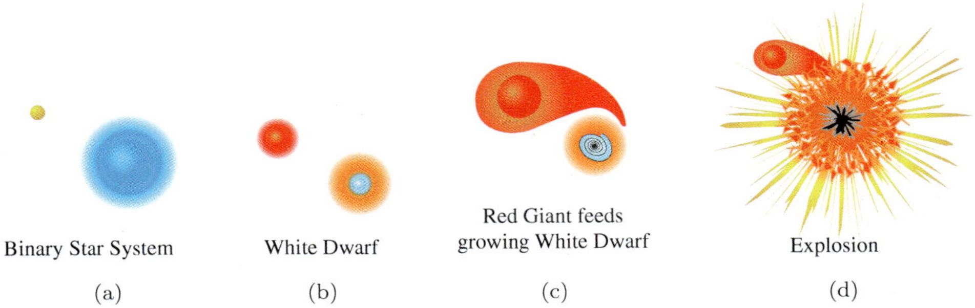

Fig. 9.19 When a carbon-rich white dwarf has pulled in sufficient matter from a nearby red giant, a Type Ia supernova occurs.

For both types of supernova, the immense energy released in the explosion is due to the conversion of gravitational potential energy of the star into energy of the explosion.

9.12.1 *Interstellar medium*

Supernovae expel heavy elements, produced in the process of supernova explosions, into the surrounding space that eventually become part of the interstellar medium (mostly molecular gas clouds) thought to be the starting point of stellar and planetary formation, as discussed in Sec. 8.12. The relative abundance of elements in the gas cloud forming a star determines the properties of the star, including the possibility of the formation of the solar system and whether life can emerge on the planets.

The kinetic energy of the gas forming the outer envelope of the star, and which is blown off by a supernova explosion, is enormous, with matter traveling at a velocity of up to $0.1c$ and creating shock waves in the neighboring gas clouds; the shock wave creates a compression wave and can nucleate the formation of stars and planets. As discussed in Sec. 8.12, the existence of short-lived radioactive isotopes in meteorites shows that a nearby supernova explosion triggered the formation and composition of our solar system some 4.5 billion years ago.

9.13 Neutron Stars and Pulsars

Neutron stars, discovered in 1968, are smaller, but heavier, than white dwarfs. The Tolman–Oppenheimer–Volkoff limit (or TOV limit) is an upper bound to the mass of stars composed of neutron-degenerate matter, that is, a neutron star. The TOV limit is analogous to the Chandrasekhar limit for white dwarf stars; the neutron star has a mass of about 1.5–$3M_\odot$, a radius of about 10–20 km, a thin solid crust of 1 km, possibly a small solid core and an interior that is a superfluid; see Fig. 9.20. The neutron star is incredibly dense, namely 3.7×10^{17} to $5.9 \times 10^{17}\,\mathrm{kg/m^3}$, which

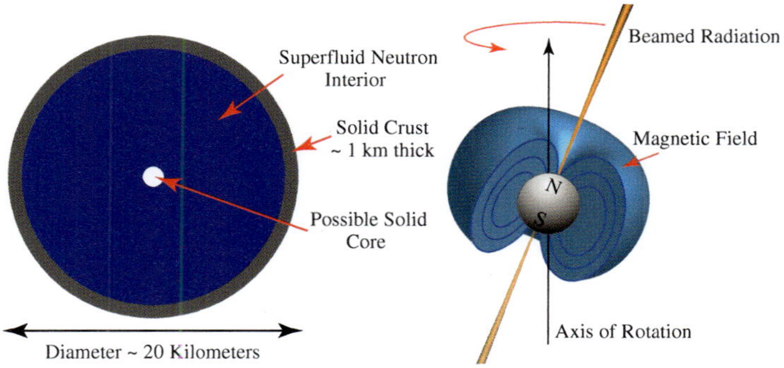

Fig. 9.20 Cross-section of a neutron star. Schematic representation of a pulsar.

is comparable to the density $3 \times 10^{17}\,\mathrm{kg/m^3}$ of an atomic nucleus. The neutron star's surface gravitational field is about 2×10^{11} times stronger than that of the Earth's — with an escape velocity of about $0.3c$.

Pulsars — which are pulsating stars — are characterized by extremely powerful, regular and short bursts of electromagnetic radiation in the radio-wave frequency range. It is now known that pulsars are rapidly rotating neutron stars, with rotation periods between about 1.4 milliseconds to 30 seconds, with the longest period observed thus far being 8.5 minutes. The radiation that one observes is coming from a plasma moving in the magnetic field of the neutron star, as shown in Fig. 9.20.

When a star undergoes gravitational collapse and forms a neutron star, almost all of its angular momentum is carried by the neutron star due to angular momentum conservation. Hence, neutron stars that occur in Nature are expected to be rotating and are pulsars when they are formed. The neutron star has a radius much less than the star from which it is formed, and this is the reason that the neutron star spins at a very high rate. Due to their high speed of rotation, many pulsars emit enormous amounts of radiation that leads to loss of energy and a slowing down of the rotation — and finally they stop radiating after 10–100 million years. For this reason, it is expected that 99% of the pulsars that formed in the 13.78 billion years of the Universe's lifetime have stopped 'pulsing'.

Noteworthy (optional content) 9.4: Velocity of Sound and Neutron Stars

---

A non-relativistic derivation for the velocity of sound yields the following

$$v = \sqrt{\frac{B}{\rho}} \tag{9.1}$$

where B is the coefficient of stiffness and ρ is the density of the medium.

We know from relativity that the maximum velocity in the Universe is the velocity of light in vacuum, namely $c = 3 \times 10^{8}\,\mathrm{ms^{-1}}$. Does sound ever reach this velocity? We know that sound velocity is highest in solids since the solid's stiffness grows faster than the density of a solid. So we have a clue that we should look at more dense material for higher **velocity of sound**. The most dense form of matter in the Universe is a neutron star that has a typical density of $\rho = 2 \times 10^{17}\,\text{kgm}^{-3}$. Due to its extreme rigidity the velocity of sound in a neutron star approaches the velocity of light when the mass of a neutron star increases to approximately $3M_\odot$.

A neutron star with a mass greater than $3M_\odot$ would have velocity of sound greater than c. Since this is forbidden by relativity, the neutron star becomes unstable and collapses into a black hole. In technical language $v \to c$ indicates that the neutron star has become unstable under infinitesimal perturbations in its pressure, and this pressure instability causes it to collapse into a black hole.

Note the important point that it is the pressure instability indicated by v exceeding c that sets the upper limit to the mass of a neutron star.

9.14 Astrophysical Black Holes

Black holes are objects that are solutions of Einstein's theory of gravity, and have been discussed from a theoretical point of view in Chapter 5. According to Einstein, gravity is not a force at all, but rather reflects the geometry of spacetime. The gravitational collapse of very massive stars, as shown in Fig. 9.21, leads to the residual mass of the star's core forming a hole in the spacetime continuum, namely a black hole.

A black hole is created when a mass M of an imploding star is compressed into a small volume having a radius less than the Schwarzschild radius R_S given by

$$R_S = \frac{2GM}{c^2}.$$

The R_S for a $M = 10M_\odot$ star is 30 km.

Fig. 9.21 An imploding star.

Spacetime curvature is infinite at the center of the black hole; it has such a strong gravitational attraction that all objects including light — which are within a distance of the Schwarzschild radius R_S from the center of the black hole — can never escape from the black hole.

Black holes come in many sizes, depending on how they were formed. We focus on three different ways, based on their mass, that black holes can be produced. Each mass scale has its own unique characteristics that can be used to detect the black holes.

- **Black Holes of a Few Solar Masses**: As discussed in Sec. 9.11, if the initial mass of the star is larger than $8M_\odot$, then there is no physical mechanism that can counteract the attractive force of gravity. Once all the nuclear fuel has been consumed, the star's structural instability leads to a supernova explosion, leaving behind a black hole at the core of the collapsing imploding star, as shown in Fig. 9.21.
- **Supermassive Black Holes**: The formation of supermassive black holes and their possible role in galaxy formation have been discussed in Sec. 8.9.
- **Primordial Microscopic Black Holes**: The production of black holes soon after the Big Bang singularity has been discussed in Sec. 6.13.

9.14.1 *Stellar size black holes*

Binary stars are quite common, accounting for about one-third of all stars. When one of the binary stars is very massive, it can undergo gravitational collapse, explode into a supernova and create a black hole of up to ten solar masses. The result of the supernova is often a binary system consisting of a black hole and a normal star. If the black hole is close enough it 'attracts' significant amounts of gaseous material from the normal star. Since both the black hole and the star are spinning, the gaseous material carries angular momentum and forms an *accretion disk*, with gaseous material orbiting in the rotational plane of the disk. See Fig. 9.22.

As discussed in Sec. 5.12, matter that spirals into a gravitational bound state around compact objects like a black hole releases a considerable amount of energy.

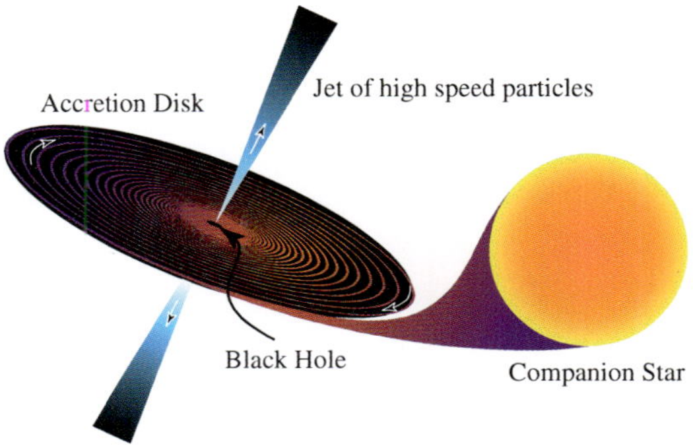

Fig. 9.22 A spinning black hole devouring a companion star. Jets of immense intensity emerge from the matter of a companion star falling into a black hole.

As the material in the disk moves towards the center, it gains velocity and energy from friction and heats up the accretion disk to high temperatures so that X-rays are produced copiously. X-ray telescopes orbiting the Earth can detect X-ray binaries since they are the brightest sources of X-rays in the sky. Optical telescopes are trained on the star that is the source of intense X-rays and its visible spectrum is observed.

Binary stars can be detected in many different ways; binaries for which both stars are visible are called visual binaries; those for which one star crosses in front of the other is called an eclipsing binary and for those binaries which are along the line of sight are called spectroscopic binaries. For astrometric binaries one of the companions is too faint to be seen, which is the case for a binary of a black hole with an ordinary star.

For **astrometric binaries**, the observed spectrum of the visible star undergoes a **Doppler shift** due to its motion around the black hole, being blue-shifted when the star approaches the Earth and being redshifted when the star recedes. From the Doppler shift one can obtain the variation in the radial velocity of the star and this in turn can be used to determine the mass of the invisible partner.[9] The *mass* of the invisible companion is the chief criterion for determining the existence of a black hole. A neutron star, which is also invisible, has a maximum mass of about three solar masses. Hence, if the invisible companion has a mass much greater than that of a neutron star, then it is taken to be a black hole.

Although a black hole does not emit radiation, it is nevertheless a source of strong spacetime curvature and exerts a powerful gravitational influence on all matter coming close to it.[10] In addition to the study of astrometric binary systems, the currently favored method for detecting black holes is to detect high energy X-rays generated by matter falling into the black hole, especially for the case where one of the stars in a binary star system has collapsed into a black hole as illustrated in Fig. 9.22.

X-rays from four binary sources have been identified as the most suitable candidates for pointing to the existence of black holes. In 1971, the binary system at **Cygnus X-1** was proposed as providing the first empirical evidence for black holes. It is near the center of the constellation Cygnus and about 6000 light years from our Sun. Cygnus X-1 is thought to be a black hole of mass $8.7M_\odot$, with a Schwarzschild radius $R_S = 26$ km and has a companion blue supergiant star having a mass of 10–$20M_\odot$, and which it orbits at a distance of 0.2 AU.

Figure 9.23 shows recent results for an X-ray binary known as Nova Muscae; the radial velocity curve of the normal star implies that the companion is a black hole of $3M_\odot$.

[9] Radial velocity is the velocity of an object along the line of sight; in our case it is the velocity of the visible star towards the Earth and away from it.

[10] Hawking radiation from black holes, which is due to quantum mechanical processes, is negligible for astrophysical black holes.

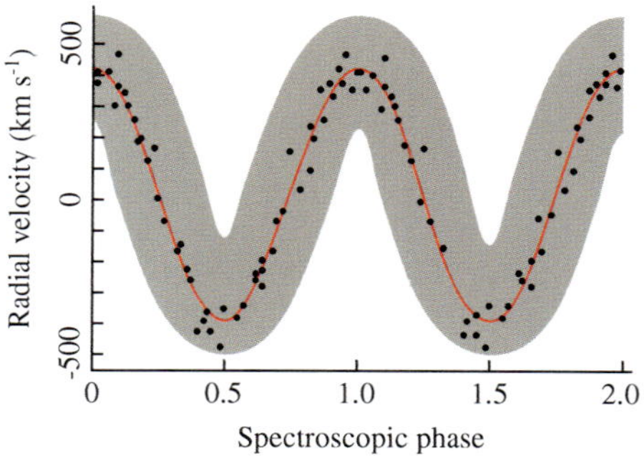

Fig. 9.23 The periodic variation of the radial velocity of a normal star in the binary system at Nova Muscae shows the presence of a companion black hole. Line = best fit. Dots = measurements. Grey band = confidence interval.

9.15 The Answer

Star burning is powered by nuclear fusion. Depending on the initial mass of the star, various pathways are open to it for its future evolution. The two main categories of stars are the upper-main-sequence and the lower-main-sequence stars on the Hertzsprung–Russell diagram, defined as stars having an initial mass greater or less than $8M_\odot$, respectively. Stars with an initial mass less than $8M_\odot$ evolve into red giants and mostly end up as white dwarfs. Stars with an initial mass greater than $10M_\odot$ evolve into red supergiants, with the final state of the stars being mostly either a neutron star or a black hole.

The three physical processes discussed so far, namely hydrostatic equilibrium, radiation transport, and energy generation via thermonuclear fusion, serve to determine the structure of a star.

We summarize in Table 9.1 our results for the various nuclear processes going on in the more massive (the upper-main-sequence stars) and less massive stars (the lower-main-sequence stars) in the Hertzsprung–Russell diagram; all the processes depend on the mass of the star.

Tunneling is at heart of stellar fusion, allowing the star to burn at a temperature much lower than otherwise — and is the reason that a solar sized star can have a

Table 9.1 Nuclear processes in stellar fusion.

Nuclear processes in main sequence of stars	Final state of star
No thermonuclear fusion	Gaseous cloud
Fusing deuterium or lithium	Brown dwarf
Hydrogen and helium burning	White dwarf
Helium and carbon burning	Neutron star
Carbon burning	Neutron star or black hole
Burning up to iron	Neutron star or black hole

lifetime of over 10 billion years. Tunneling itself is a purely quantum mechanical process, forbidden in classical physics. As discussed earlier, when the proton tunnels through the potential barrier, it is in an indeterminate and virtual state that can never be directly observed.

Tunneling is an invisible and unseen process that powers solar fusion. All the stars, and consequently the galaxies they compose, are founded on this process and points to the underlying theme of this book, namely that the visible Universe is founded on, and is a manifestation of, the invisible Universe.

The processes leading to the formation of a white dwarf, a neutron star or a black hole reveal the dynamics of the final states of a star; gravitational collapse leading to a pulsar or a black hole shows that gravity becomes an irresistible force when large enough conglomerations of mass come into play. The formation of a black hole, signifying the absence of any force capable of resisting gravity, opens up new vistas for understanding the dynamics of strong gravity. In fact, it is a testimony to the marvels of the general theory of relativity that it provides a conceptual framework for understanding phenomena that otherwise would have been close to impossible to either observe or comprehend.

Chapter 10

The Origin of the Elements

How are atoms made and why is gold rare?

10.1 The Question

Gold has been sought for by humankind for many millennia and the fact that it is a very costly and precious metal is well known. One almost never finds a piece of gold lying around like a piece of stone and all of us have a direct experience of gold's rarity. Indeed, in many gold mines one needs to go to extraordinary lengths to extract even small amounts of gold. This leads us to ask the question: **Why is gold rare?**

From the facts that gold cannot be made by chemically combining substances and that it cannot be chemically decomposed into other substances, it is clear that gold must be an element. If we want to know why an element is rare, we need to know where the elements come from in the first place. Note that the terms "element" and "atom" are the same.

Our explanation as to why gold is rare has to, at the same time, explain the *relative abundances of all the elements* found in Nature, since the same processes that create gold must also create the other elements — and in the correct proportions.

As discussed in Chapter 6 on cosmology, the early Universe is described by the behavior of elementary particles, nuclei and atoms. Therefore, the application of particle, nuclear and atomic physics to the history of the Universe allows us, in principle, to predict the abundances of all species of elementary particles, nuclei and atoms at different epochs.

The answer to the question in essence is the following: the observed abundances of the elements in the Universe at large are the result of nucleosynthesis during the Big Bang and, later, in the process of stellar evolution. The synthesis of the light elements has been discussed in Sec. 6.10 on Big Bang nucleosynthesis and will not be discussed any further. There is also a rather rare process in which elements are produced, namely in the collision of neutron stars, and is discussed in Noteworthy 10.9.

In this chapter, we analyze stellar nucleosynthesis and show how it accounts for the relative abundance of almost all the elements. Most stellar nucleosynthesis is the result of the burning of the star by thermonuclear fusion, and which in turn is determined by the mass of the star. We discussed the following in Chapter 9.

- Stars with mass above $0.08M_\odot$ burn hydrogen.
- Stars with mass greater than about $0.5M_\odot$ burn hydrogen and helium.
- Stars with mass in the range of $1-8M_\odot$ continue nucleosynthesis up till the production of carbon.
- Stars with mass greater than $10M_\odot$ synthesize all the elements up to iron and nickel.
- Elements heavier than iron are created by neutron capture via the so-called s-process and r-process, in red supergiants, helium flash and supernova.

The main focus of this chapter is on the different nuclear processes that take place in the evolution of the different mass stars.

10.2 Composition of the Universe

From chemistry, we know that gold is an atom made out of 79 protons. To assess the relative abundance of the elements, let us examine the composition of the Earth's crust.

Figure 8.22 and Table 8.3 show the abundances of the most common and interesting elements in the Earth's crust and the Earth's core; we see from the figure that gold (Au) is only 0.002 ppm, compared to iron (Fe) which is 62,200 part per million (ppm). Note that gold is more rare by a factor of 40 than silver (Ag) whose abundance is 0.08 ppm. So our intuition that gold is a rare element is well substantiated for elements in the Earth's crust and core. We need to address a wider question, namely what is the abundance of gold in the Universe, since the Earth's materials, as discussed in Sec. 8.12, predates the formation of the Earth.

Figure 10.1 shows the relative abundance of the elements in the Universe, and is compiled from data on the composition of planets, comets, inter-galactic dust and stars; note the logarithm of abundance is plotted against the various elements and shows a sharp exponential fall in abundance as we move away from hydrogen and helium; the reason is that of all the elements, hydrogen and helium

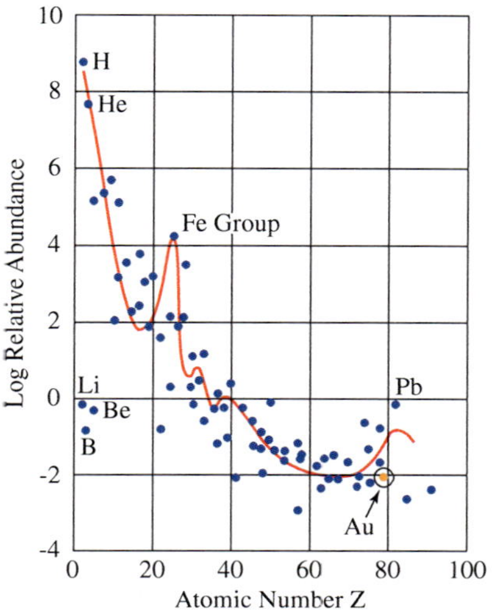

Fig. 10.1 Abundance of elements in the Universe. Gold is denoted by Au and is encircled.

compose 98% of the Universe and all the rest of the elements compose only 2% of the Universe. From Fig. 10.1, we see that, indeed, gold is also rare throughout the Universe.

From Figs. 8.22 and 10.1 we also see that the relative abundance of the elements on the Earth's crust is quite different from their abundance in the Universe; the reason is that light elements, in particular hydrogen and helium, escape readily into outer space since Earth's gravity is too weak to hold them.

10.3 Elements: Stellar Nucleosynthesis

All elements (atoms) in our Universe are the products of the manner in which the Universe started and how it has evolved to its present state. As discussed in Chapter 6 on cosmology, the current view amongst scientists is that the Universe started in the Big Bang, a violent explosion about 13.78 billion years ago, and after about 200 million years stars were formed from the debris of the Big Bang.

All elements — except for hydrogen, helium, lithium, beryllium and boron that originate in the Big Bang — are synthesized through stars. In this chapter we discuss how nuclear processes and reactions in the core of stars lead to the formation of the other elements, some of which can also arise from the collision of astronomical objects. In particular gold can also be produced in the collision of two neutron stars, discussed later.

Elements with different atomic numbers (number of protons in the nucleus) have different masses and are synthesized during different stages in the life of a star — as well as in stars with different masses. For this reason we reviewed, in Chapter 9, the evolution of stars with different masses to determine the type of star in which the various elements are synthesized and at what stage of the star's life.

10.3.1 *Main processes for nucleosynthesis*

As discussed in Chapter 9, the synthesis of the nucleus of the various elements, or nucleosynthesis in short, mostly takes place in the core of stars. In this chapter we discuss the different nuclear processes that provide the mechanism for the production of the elements and locate when and where the processes take place in the journey of a star from its birth to its death.

The entire life of a star is determined by the interplay of gravity and pressure, as discussed in Sec. 9.4. The temperature of the star is raised by compression of the gas and is further raised by the star burning its nuclear fuel. The pressure created by the temperature of the gas exerts an outward force that balances gravitational compression as shown in Fig. 8.7. This process goes on until all the fuel of the star is exhausted and leads to the death of the star.

As the star heats up the electrons become detached from the atoms, leaving behind the charged nuclei; both the electrons and the charged nuclei form a **plasma**, which is a fluid consisting of charged particles. When the temperature of the star reaches around 10^6 K we have the *onset* of stellar thermonuclear fusion, discussed in Sec. 8.14, with steady burning taking place at temperatures higher than 10^7 K.

The process of fusion stops supplying a sufficiently massive star with energy once it has formed all the elements up to the elements of the iron group. As discussed earlier in Sec. 9.5, the creation of nuclei beyond iron and nickel by fusion does not release energy. To the contrary, the synthesis of elements more massive than iron and nickel via fusion would cool the star instead of heating it further. Hence stellar fusion can only produce elements up to iron and nickel.

For forming elements heavier than iron and nickel, energy has to be *supplied* to the participating nuclei. The star's gravitational potential energy supplies this required energy in the red supergiant stars and in supernova explosions. The production of nuclei heavier than iron and nickel takes place through special processes, called the s-process, which occurs mostly in the red supergiant stars, and the r-process, that occurs in supernova explosions and in some red supergiant stars.

Stellar nuclear synthesis takes place by the following main processes (not necessarily taking place in the same star).

(1) Hydrogen burning

- The pp (proton–proton) process. We will study the pp-process in some detail as it is the process powering our Sun and hence is of particular importance for us.
- The CNO (carbon–nitrogen–oxygen) cycle

(2) Helium burning
(3) Alpha particle capture
(4) s-process and r-process

One can think of the stars as a cosmic kitchen: the process of burning the fuel required for keeping the stellar fire going in turn cooks the elements that are found in the Universe.

10.4 The pp-Process: Three-Step Hydrogen Burning

The main source of energy for most stars is hydrogen burning by the process of nuclear fusion and stars start their burning using the **pp-process**. Depending on the mass and composition of the star, the next stage in the burning can proceed in a number of ways. For low mass first generation stars having mass from 0.08–$0.5M_{\odot}$, they continue with the pp-process. Low mass second generation stars, which have carbon, nitrogen and oxygen in their original gaseous cloud, can burn hydrogen faster — at a high enough temperature — using the CNO-process (see Sec. 10.5). For stars having a mass greater than $1.3M_{\odot}$, once they have synthesized carbon, they can burn hydrogen using the CNO-process.

The three-step pp-process dominates hydrogen burning for stars that are rich in hydrogen and are in the main sequence of the Hertzsrprung–Russell diagram. As these stars consume their hydrogen fuel and convert it to heavier elements, they move off the main sequence of the Hertzsprung–Russell diagram and become red giants and supergiants.

The pp-process that fuses nucleons to liberate energy has briefly been discussed in Sec. 9.4. We now examine the pp-process in greater detail. As illustrated in Fig. 9.4, two isolated protons plus two isolated neutrons have a larger mass than a single helium nucleus that consists of the bound state of two protons and two neutrons. The mass of a proton is $m_p = 1.00728u$ while the mass of a neutron is $m_n = 1.00866u$. Consequently, the mass of two isolated protons and two isolated neutrons is $4.03188u$. However, the mass of the helium nucleus is only $4.0015u$, which is about 0.7% less than the initial incoming mass and it is this mass deficit that is converted into radiation.

The neutrons and protons can be in a bound state due to the strong nuclear force, as detailed in Chapter 11. The bound state is a lower energy state and hence due to $E = mc^2$ has a lower mass. Since E and m are related by the factor c^2, which is a huge number, converting even a small amount of mass into energy gives an enormous amount of energy. This is why the pp-process is such a good energy source.

Since its formation about 4.6 billion years ago, hydrogen burning via the pp-process has been the main source of the energy of our Sun, and will continue to power it for another 5.5 billion years. The pp-process generates 98.8% of the energy of the Sun. For all stars with a mass similar to that of our Sun, the pp-cycle takes place in three steps.

Step 1: First, two protons (hydrogen nuclei) fuse to form a deuterium nucleus. Since elements are distinguished by their atomic number, which is number of protons in an atom, deuterium is an *isotope* of hydrogen, and is denoted by $^2\mathrm{H} = [\mathrm{p} + \mathrm{n}]$; the square bracket denotes that the proton and neutron are bound together to form the deuterium nucleus. Denote by Q the energy carried by the photons γ and neutrinos ν_e that are released in the reactions; we have

$$\begin{aligned} \mathrm{p} + \mathrm{p} &\to [\mathrm{p} + \mathrm{n}] + \mathrm{e}^+ + \nu_e + \gamma \\ &\equiv {}^2\mathrm{H} + \mathrm{e}^+ + \nu_e + \gamma; \quad Q = 0.42\ \mathrm{MeV}. \end{aligned} \tag{10.1}$$

For a neutron to be created out of a proton, charge conservation demands that some other positively charged particle must appear on the right-hand side of Eq. (10.1). Indeed the positron e^+ (antiparticle of the electron) is produced, together with a neutrino. The reaction implicit in Eq. (10.1) is mediated by weak interaction, namely the process

$$\mathrm{p} \to \mathrm{n} + \mathrm{e}^+ + \nu_e \tag{10.2}$$

transforms a proton into a neutron. The neutrinos are radiated from the Sun (they carry about 2% of the Sun's energy) and about 10^{15} of them reach the Earth for every square meter per second! They do not harm us, and indeed most pass right through our bodies and the Earth, because neutrinos interact with atoms only via weak interactions.

The positrons e^+ annihilate with the electrons in the Sun's plasma, namely

$$\mathrm{e}^+ + \mathrm{e}^- = \gamma + \gamma \ ; \quad Q = 1.02\ \mathrm{MeV} \tag{10.3}$$

and contribute somewhat to the radiant energy produced by the Sun.

Step 2: In the second step, another proton combines with the deuterium nucleus to form a helium isotope and releases energy in the form of a gamma ray photon.

$${}^2\mathrm{H} + \mathrm{p} \to {}^3\mathrm{He} + \gamma \ ; \ Q = 5.49\ \mathrm{MeV}. \tag{10.4}$$

Step 3: The two steps above provide the fuel for the fusion process given by

$${}^3\mathrm{He} + {}^3\mathrm{He} \to {}^4\mathrm{He} + \mathrm{p} + \mathrm{p} + \gamma \ ; \ Q = 12.86\ \mathrm{MeV} \tag{10.5}$$

with the photon γ carrying 12.86 MeV energy. The energy carried by photons in the third step in the pp-process is the Sun's main source of energy and accounts for about half of its radiant energy.

The three-step pp-process can be summarized by the following reaction

$$4{}^1\mathrm{H} \to {}^4\mathrm{He} + 2\mathrm{e}^+ + 2\nu_e + 9\gamma \ ; \ Q = 26.72\ \mathrm{MeV} \tag{10.6}$$

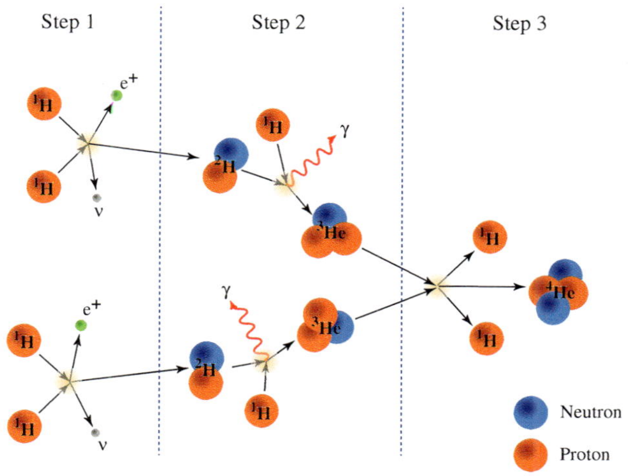

Fig. 10.2 The three steps in the pp (proton–proton) process by which the Sun currently produces almost all of its energy.

and is shown graphically in Fig. 10.2. The two e^+ are converted to energy, as in Eq. (10.3).

A total energy of $[4\,\text{mass(proton)} - \text{mass(helium)}]c^2 = 26.72\,\text{MeV}$ is released in the above reaction and this energy is carried by both γ photons (electromagnetic radiation) and neutrinos. As shown in Fig. 10.2, Step 1 and Step 2 need to occur *twice* to provide the ingredients for Step 3, leading to the energy released given by $[2 \times (0.42 + 1.02 + 5.49) + 12.86]\,\text{MeV} = 26.72\,\text{MeV}$. The neutrinos carry off about 0.5 MeV with 26.2 MeV being released as radiation. The radiation is partially radiated out in the form of sunlight and is partly absorbed in the core of the star and creates thermal energy for the star, which keeps the star's temperature high enough to stave off gravitational collapse.[1]

The rate of fusion, and consequently the rate of energy production, in the pp-process of hydrogen burning, for temperature T at the star's core, is given by

$$\text{Rate of energy production} \propto T^4 \;:\; \text{pp-process.} \tag{10.7}$$

10.4.1 *How much hydrogen does the Sun burn?*

Our Sun has a mass of $M_\odot = 2 \times 10^{30}\,\text{kg}$ and with a radiative energy output per unit time given by $dE/dt = 3.9 \times 10^{26}\,\text{W}$. We have seen that in the pp-process four protons are fused to release 26.72 MeV of energy in the form of neutrinos and radiation; 98% of the energy E released per unit mass is into radiation and is

[1]There are other side reactions that occur in addition to the above three processes, but they contribute only a minor portion to the radiation emitted by the Sun.

$dE/dm = 26.20/4\ \mathrm{MeV/m}_p = 6.68\,\mathrm{MeV/m}_p = 6.4 \times 10^{14}\,\mathrm{J/kg}$, where m_p is the mass of the proton.

To account for the observed rate at which our Sun radiates energy, note that 26.20 MeV is in radiation and hence only the fraction $26.2/26.72 = 0.98$ of the energy converted from mass into energy goes into radiation, the remaining being carried off by neutrinos. Hence, the rate at which mass must be transformed into radiation is given by

$$\frac{dm}{dt} = 0.98 \times \frac{dE/dt}{dE/dm} = 0.98 \times \frac{3.9 \times 10^{26}\ \mathrm{kg\ m^2/s^3}}{6.4 \times 10^{14}\ \mathrm{m^2/s^2}}$$
$$= 6.15 \times 10^{11}\ \mathrm{kg/s}.$$

Our Sun consequently has to fuse 615 million tons of hydrogen per second to produce its current output of energy. As mentioned in Sec. 10.4, in the pp-process about 0.7% of the mass is converted to energy, resulting in the production of 610.7 million tons per second of helium, and with the equivalent of 4.3 tons of hydrogen per second being converted into the energy of radiation.

615 million tons of hydrogen is equal to 4×10^{38} protons. There are 7×10^{56} protons in the Sun, and hence 10% of the Sun's mass will be consumed in about 10 billion years after the onset of solar fusion. This will be the end of the pp-burning stage of the Sun, as its core will contract and heat up to ignite the burning of helium. This is the next stage of the Sun's evolution, with the outer layers expanding and the Sun becoming a red giant, as discussed in Sec. 9.6.

10.5 The CNO-Cycle for Hydrogen Burning

All stars in the main sequence of the Hertzsprung–Russell diagram start their nuclear burning with the pp-process. For stars with mass greater than $1.3M_\odot$ the **CNO-cycle** is the main process of solar fusion.

Second generation stars like our Sun — that are composed of the debris of earlier generation stars — have heavier elements made by the preceding generation of stars. For such stars, the alternative process for burning hydrogen is the CNO (carbon–nitrogen–oxygen) process, in which carbon, nitrogen and oxygen act as *catalysts* for the accelerated burning of hydrogen.

There is no beginning or ending of the CNO-cycle but, for convenience, we start with the conversion of $^{12}\mathrm{C} \rightarrow {}^{13}\mathrm{N}$ by the absorption of one proton, and so on. The steps of the CNO-cycle are shown in Fig. 10.3 and given in Fig. 10.4.

$^{12}\mathrm{C}$ is a catalyst for the production of $^4\mathrm{He}$ from hydrogen, and in one cycle helium is produced and energy Q released, as given below

$$4\,{}^1\mathrm{H} \rightarrow {}^4\mathrm{He} + 2e^+ + 2\nu_e + 3\gamma\ ;\ Q = 26.72\,\mathrm{MeV}.$$

This is the same initial and final states as given in Eq. (10.6), but taking place at a much faster rate.

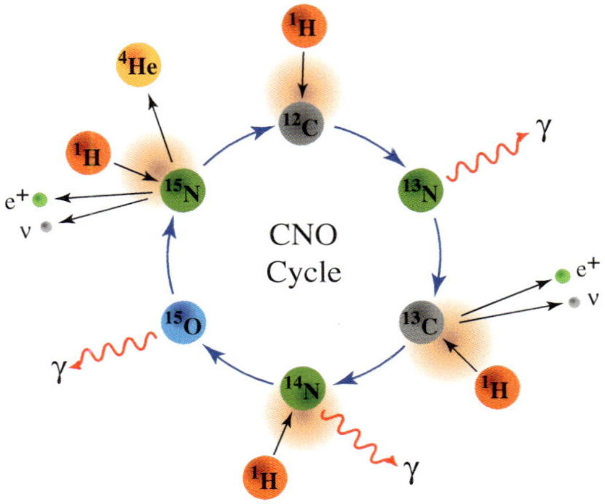

Fig. 10.3 The CNO-cycle for burning hydrogen.

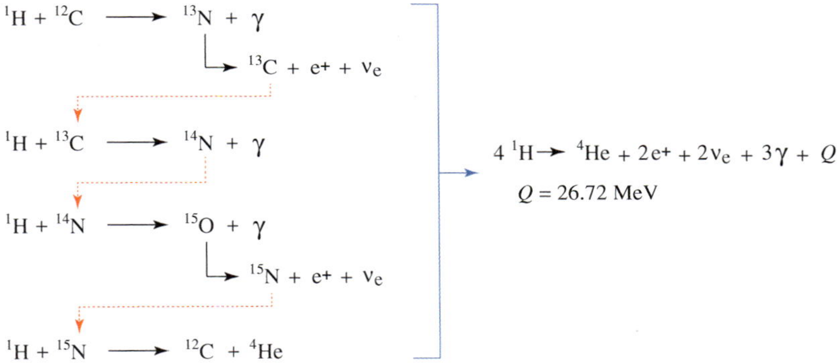

Fig. 10.4 Reactions in the CNO-cycle where carbon acts as a catalyst for the burning of hydrogen.

The *rate* at which the CNO-process burns hydrogen is determined by the slowest reaction in the cycle, which is the following

$$^{14}\mathrm{N} + {}^{1}\mathrm{H} \to {}^{15}\mathrm{O} + \gamma. \tag{10.8}$$

The density of ^{14}N at the center of our Sun is about $10^5\,\mathrm{kgm}^{-3}$, which is half that of hydrogen. At the temperature at the center of our Sun, which is about 15×10^6 K, the reaction given in Eq. (10.8) takes place about once in 5×10^8 years (per atom). It can be shown that at the Sun's temperature the pp-process is more efficient, with the CNO-process providing only about 1.7% of the Sun's energy.

Since there is greater Coulomb repulsion between a proton and the nuclei of carbon, nitrogen and oxygen, the CNO-process can only be more efficient than the pp-process for temperatures higher than the Sun's temperature. The CNO-process, for temperature T at the star's core, yields the following

$$\text{Rate of energy production} \propto T^{18} \quad : \text{ CNO-process.}$$

The rate of burning proportional to T^{18} is much greater than the fusion rate that is proportional to T^4 for the pp-process given in Eq. (10.7). For this reason, stars with a mass greater than $1.3M_\odot$ burn hydrogen by the CNO-process at higher temperatures than our Sun and at a faster rate.

For massive stars among the upper-main-sequence stars undergoing CNO-burning recall, as shown in Fig. 9.8, that the convective core, in contrast to the pp-cycle, is at *the center of the star* with the radiative envelope being on the outside. At the end of a star's burning in the main sequence, it starts to run out of hydrogen fuel and becomes a red giant and the CNO-cycle then generates most of the star's radiation.

In summary, for the higher mass stars in the main sequence of the Hertzsrprung–Russell diagram and for second generation stars with masses more than 1.3 solar masses, the CNO-cycle is more efficient than the pp-process for generating solar energy and is the main mechanism for hydrogen fusion.

Noteworthy (optional content) 10.1: Slowness of Stellar Fusion

---

Note that the process of thermonuclear fusion that fuels the burning of stars is the same process that causes the explosion in a hydrogen bomb. The explosion of a hydrogen bomb takes place in milliseconds whereas the burning of a typical star like our Sun continues for almost 10 billion years. The question naturally arises as to why is stellar fusion so slow? The answer lies partly in Step 1 of the pp-process, namely the production of deuteron from protons. The production of a deuteron requires the transformation of a proton into a neutron and as seen in Eq. (10.2) this transformation is mediated by the weak interaction.

Denoting the deuteron by $\mathrm{d} = {}^2\mathrm{H}$, the following are the processes responsible for solar fusion.

$$\mathrm{p} + \mathrm{p} \to {}^2\mathrm{H} + \mathrm{e}^+ + \nu_e + \gamma : \quad \text{Weak interaction}$$

$$\mathrm{p} + \mathrm{d} \to {}^3\mathrm{He} + \gamma : \quad \text{Strong and electromagnetic interaction}$$

$${}^3\mathrm{He} + {}^3\mathrm{He} \to {}^4\mathrm{He} + \mathrm{p} + \mathrm{p} + \gamma : \quad \text{Strong and electromagnetic interaction.}$$

The three fundamental forces of Nature are discussed in Chapter 13; all three processes above take place under the gravitational force and hence all four fundamental forces of Nature are responsible for solar fusion.

The weak interactions take place very slowly, with the process in Eq. (10.2) taking place in about 10^{-18} seconds in contrast to strong interactions that act in 10^{-23} seconds. What this means in practice is that the transformation of $\mathrm{p} + \mathrm{p} \to {}^2\mathrm{H} + \cdots$ takes place very rarely. In fact, in most of the collisions of a proton with a proton, they do not fuse but, instead, simply 'bounce off' each other; only about once in 10^{26} proton–proton collisions is a deuteron formed.

Put differently, for typical temperatures in the center of a star, on the average a proton spends about 9×10^9 years in the core of a star before it fuses with another proton and this is one of the factors that set the time scale for the astronomical lifetime of stars. In spite of the bottleneck created by the weak interactions, the high density of hydrogen and high temperature lead to the production of deuteron at the rate of 10^{12} kg/s inside a typical Sun-like star.

One may incorrectly conclude that the weak interaction is the reason for the slowness of stellar fusion. In fact, the primary reason for the slowness of stellar fusion is the slow rate at which the star *radiates off* its energy. For example, in our Sun, radiation produced by thermonuclear fusion spends roughly 50,000 years being scattered inside the Sun before it is emitted. This slow rate of energy loss is the reason why stellar fusion can produce radiation at a slow rate and still maintain the temperature required for hydrostatic equilibrium.

If stellar fusion was mediated by say electromagnetic interactions (instead of the weak interactions required for $\mathrm{p} + \mathrm{p} \to {}^2\mathrm{He} + \mathrm{e}^+ + \nu_e$), the *rate* at which radiation is generated would be the *same* as is the case now. The only difference would be in the density and temperature of the core, which for stars is fixed by the weak interaction. If solar fusion was mediated by the electromagnetic interaction (with the weak interactions not playing any role), the star would then burn at a lower temperature and density; this is because the star could slow down the rate at which the electromagnetic interaction produces radiation by reducing the density at its core, or by lowering the temperature of the core to slow down the rate of barrier penetration — discussed earlier in Sec. 9.4.

In other words, the slowness of the fusion process in the Sun is not determined by the speed at which the fundamental interactions take place, but rather by how long it takes radiation generated at the star's core to be emitted into space.

The hydrogen bomb takes a short time for completing the explosion since all the radiation escapes immediately and hence does not have any role in slowing down the fusion process.

10.6 Helium Burning: Triple-Alpha Process

The main process that sustains the star during its red giant phase is **helium burning**. A few hundred million years after a solar mass star has left the main sequence, its core density has risen to about 10^8 kg/m^3 and its core temperature has reached 10^8 K — the temperature required, using quantum mechanical *tunneling*, to overcome the Coulomb barrier for helium fusion. As mentioned in Sec. 10.3, lithium, beryllium and boron are all unstable. One would have thought that two helium atoms should combine and form a beryllium atom; but by a small margin, the beryllium isotope ^{8}Be that results from the fusion of two helium atoms — that is $^4\text{He}+^4\text{He} \rightarrow {}^8\text{Be}$ — is *unstable* with a lifetime of 2.6×10^{-16} s and decays back to two helium atoms. Nevertheless, a small quantity of ^{8}Be can exist in thermodynamic equilibrium in the core of a star; if *another* helium atom ^{4}He can be absorbed by beryllium before it decays, then this can lead to the synthesis of ^{12}C.

Carbon be formed by the following rapid two-step process

$$\begin{aligned} &^4\text{He} + {}^4\text{He} \rightarrow {}^8\text{Be} \\ &^4\text{He} + {}^8\text{Be} \rightarrow {}^{12}\text{C}^* \rightarrow {}^{12}\text{C} + 2\gamma \end{aligned} \tag{10.9}$$

as shown in Fig. 10.5 and where $^{12}\text{C}^*$ signifies an excited state of the carbon nucleus.[2] The energy levels for the second part of the process involving $^{12}\text{C}^*$ are shown in Fig. 10.6.

The process $^4\text{He}+{}^8\text{Be} \rightarrow {}^{12}\text{C}^*$ requires about 288 keV, and the net result of the two-step process, namely $3{}^4\text{He} \rightarrow {}^{12}\text{C}^*$ requires 375 keV of energy; this energy is

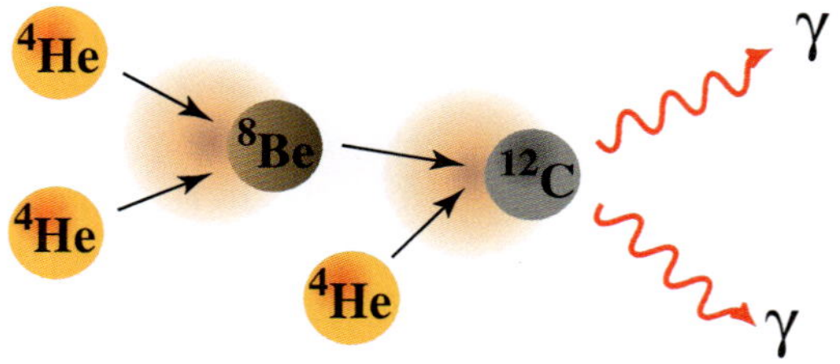

Fig. 10.5 The triple-alpha process of helium fusion.

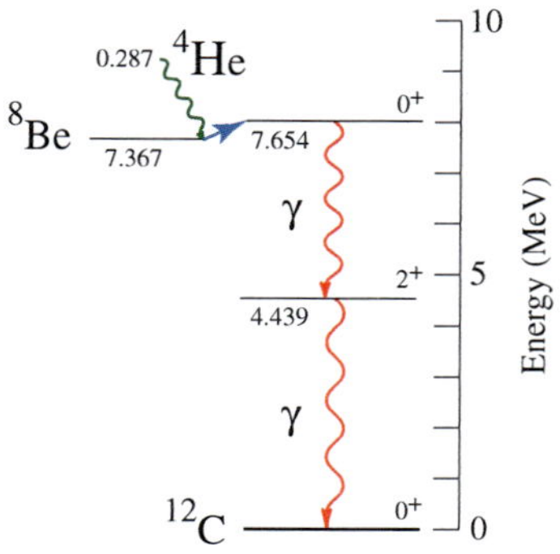

Fig. 10.6 Helium–beryllium fusion to form carbon and the energy levels of the carbon nucleus. The excited carbon state $^{12}\text{C}^*$ at 7.654 MeV decays via the emission of two photons to the ground state ^{12}C.

[2]The last equation is in general the following

$$^4\text{He} + {}^8\text{Be} \rightarrow {}^{12}\text{C}^* \rightarrow {}^{12}\text{C} + \; [2\gamma \text{ or } e^+ + e^-].$$

supplied by the kinetic energy of ^4He. The two-step rapid process is highly unlikely unless $^{12}C^*$ is a **resonant energy** level of carbon at precisely the energy required for the process given in Eq. (10.10). Resonance in this context means that there is an excited *quantum state* of the carbon nucleus that occupies an energy level at which helium fusion is taking place.

Hence, if $^{12}C^*$ is a resonant state of carbon, which in fact it is, then $^{12}C^*$ can subsequently decay to the ground state of carbon, namely leading to the production of carbon atoms ^{12}C. The energy of $^{12}C^*$ is 7.654 MeV above the energy of ^{12}C, as shown in Fig. 10.6. Hence, the net energy released in producing ^{12}C is $(7.654 - 0.375)\,\text{MeV} = 7.279\,\text{MeV}$.

In summary, the effect of the sequence of reactions given in Eq. (10.10) results in helium burning with release of energy Q in the form of radiation.

$$^4\text{He} + {}^4\text{He} + {}^4\text{He} \to {}^{12}\text{C}^* \to {}^{12}\text{C} + 2\gamma \ ; \ Q = 7.279\,\text{MeV}.$$

The helium nucleus is also called an alpha particle and the above reaction is called the **triple-alpha process**.

The process yielding carbon, together with the energy levels of the carbon nucleus, is shown in Fig. 10.6. Only about 0.04% of the excited state of carbon $^{12}C^*$ decays into the ground state of ^{12}C and releases energy in the form of two photons 2γ; the remaining $^{12}C^*$ decays back to beryllium and ^4He or directly to three ^4He — since all these nuclei are in thermal equilibrium. The small quantity of ^{12}C that is produced in this process is enough to generate the rest of the heavier elements.

Fred Hoyle predicted, in 1954, that a resonant state of the carbon atom must exist around 7.5 MeV — about 300 keV above threshold for the production of beryllium — for the carbon atom to exist in our Universe; furthermore, if the required resonance of the carbon nucleus did not exist none of the elements heavier than carbon could have be synthesized. And amazingly enough an excited energy level of the carbon nucleus was experimentally found at 7.654 MeV.[3]

10.7 Alpha Capture and Other Processes

The evolution of main sequence stars in the Hertzsprung–Russell diagram with initial mass greater than $8M_\odot$ results in red supergiants with core temperatures of up to 10^9 K. The red supergiants continue their contraction and

[3]If beryllium were stable, all the stars, including our Sun, would have spent their fuel in a few hundred million years synthesizing berryllium, thus not allowing for the time required for the emergence of life. The remarkable coincidence of the excited quantum state of carbon being exactly at the energy required to form carbon in stars, together with the instability of beryllium formed from a two-alpha process, are examples of the anthropic principle. The *anthropic principle* states that the laws of Nature must not have any bottlenecks or 'show stoppers' that would contradict the emergence of life in the Universe.

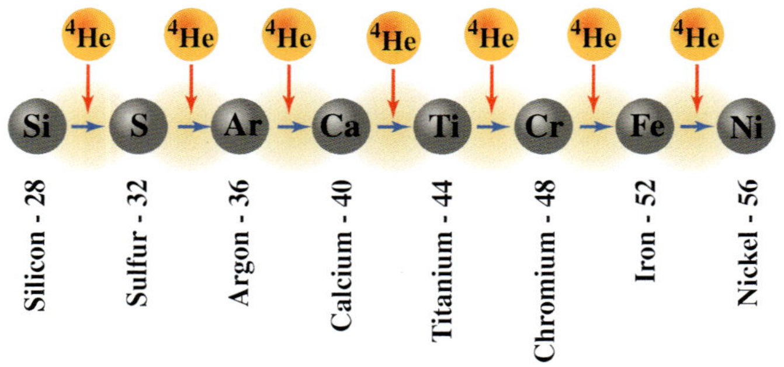

Fig. 10.7 Production of heavier elements by successive helium capture.

heating with heavier elements being formed by the successive fusion of helium nuclei — called **alpha capture**, as well as by the fusion of heavier and heavier elements.

Nuclei with atomic number $4N$ are built by successive alpha capture, where N being the number of ^{4}H accumulated by the alpha capture process.

By the alpha capture we get oxygen ^{16}O, neon ^{20}Ne, magnesium ^{24}Mg, silicon ^{28}Si, sulphur ^{32}S and so on, as shown in Fig. 10.7: the atomic mass of successive atoms is increased by four. As shown in Figs. 9.16 and 9.17, the temperature increases as one approaches the core of the red supergiant, and the different heavier elements are synthesized in *concentric* shells, the more massive elements being synthesized closer to the center of the star.

Nuclei whose atomic number is a multiple of four are produced — all the way up to iron — via the process of alpha capture. The relative abundance of elements in the Universe, as given in Fig. 10.1, shows that there are peaks near multiples of four for the atomic mass — reflecting the importance of alpha capture in synthesizing the elements.

The process of alpha capture cannot go on indefinitely since, for large enough nuclei, Coulomb repulsion will forbid the process. For atoms with masses greater than silicon and sulphur the process of fusion in red supergiants takes place in the following manner: a heavy nucleus say ^{28}Si disintegrates into its constituents, namely alpha particles, which then combine with another ^{28}Si — using a sequence of alpha particle capture similar to the one shown in Fig. 10.7 — to form nickel ^{56}Ni and that in turn disintegrates into cobalt ^{56}Co and iron ^{56}Fe.

Figure 10.8 shows the production of oxygen ^{16}O by the fusion of ^{12}C with ^{4}He and Fig. 10.9 shows the process producing magnesium. At temperatures of a billion kelvins (10^9 K) other processes can also take place. For example, as shown in Fig. 10.9, two ^{12}C nuclei can fuse to produce an atom of magnesium ^{24}Mg; similarly oxygen and neon can react with each other to form even heavier nuclei; neon can also react with carbon producing heavy elements along with neutrons, protons and alpha particles.

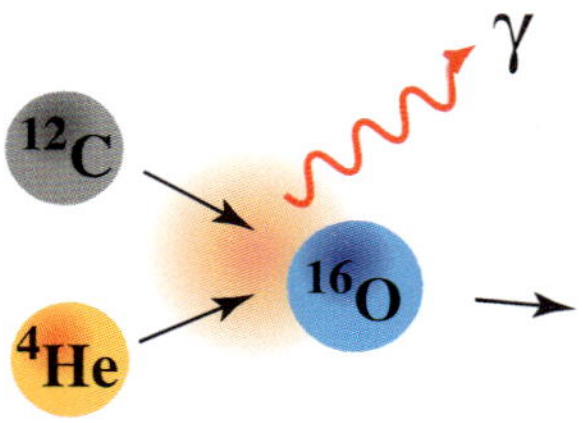

Fig. 10.8 Oxygen production by carbon–helium fusion.

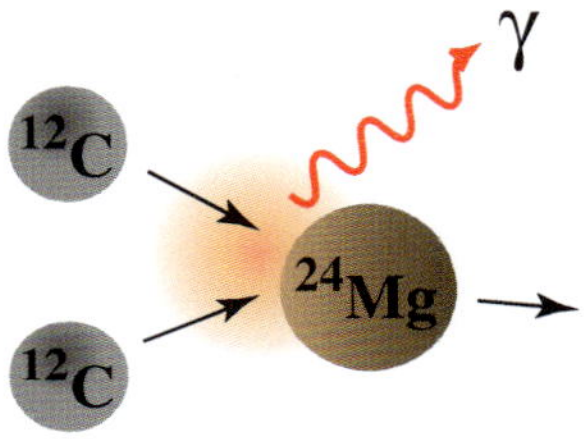

Fig. 10.9 Magnesium production by carbon fusion.

10.7.1 *Silicon melting: Photodisintegration*

Recall from Fig. 9.16 that the last and inner most shell of a red supergiant is where silicon burning takes place at temperature of 3×10^9 K, after which the core is composed of iron ash. At this temperature, for the first time thermal photons have sufficient energy to destroy the silicon nucleus by **photodisintegration**, releasing ^{4}He, protons and neutrons. These light particles combine with high mass nuclei to build more massive nuclei; in particular the released ^{4}He undergoes a series of reactions resulting in sulphur, calcium, argon and so on. To illustrate this process, consider the following photodisintegration

$$^{28}\text{Si} + \gamma \to {}^{24}\text{Mg} + {}^4\text{He}.$$

The helium nucleus released by photodisintegration can then combine with silicon to form sulphur

$$^{28}\text{Si} + {}^4\text{He} \to {}^{32}\text{S} + \gamma$$

and so on until the formation of nickel

$$^{52}\text{Fe} + {}^4\text{He} \to {}^{56}\text{Ni} + \gamma.$$

We see that silicon burning is rather special, in that due to photodisintegration silicon melts into a plasma of nuclei, alpha particles, protons and photons, all in equilibrium. And through a complex series of alpha capture and photodisintegrations, silicon burning creates all the nuclei from atomic number 30 to 56. Hence, the term **silicon melting** more accurately reflects the specific and unique features of silicon burning.

Since iron and nickel have the highest binding energy per nucleon, nuclear fusion stops with iron and nickel — fusion cannot produce any elements heavier than iron and nickel. We have to look for other processes to synthesize elements heavier than iron and nickel.

10.8 Neutron Capture: s-Process and r-Process

Beta decay is the decay of a neutron, with a lifetime of about 15 minutes when it is free, via the weak interactions into a proton, namely

$$\mathrm{n} \to \mathrm{p} + \mathrm{e}^- + \bar{\nu}_e \ : \ \beta^- \text{ decay.}$$

Neutrons are formed by protons absorbing electrons by weak interactions, namely

$$\mathrm{p} + \mathrm{e}^- \to \mathrm{n} + \nu_e \ : \ \beta^+ \text{ decay}$$

which is inverse beta decay, and is the process by which a proton transforms into a neutron as was discussed earlier in Eq. (9.1).

A nucleus can easily absorb a neutron since there is no Coulomb barrier opposing the absorption. The neutron absorbed in the nucleus, after some time, can undergo beta decay and thus transform itself into a proton. In this manner neutron capture can lead to the synthesis of the heavier elements.

The time interval between successive capture of neutrons by (heavy) nuclei ranges from one to 1000 years per nucleus. In helium burning red giants, the average 'waiting time' between successive neutron capture is about 25 years per nucleus. Hence in the slow process (**s-process**) — that takes place mostly within the core of the red supergiants — nuclei can be built up by the absorption of neutrons. Another process, namely the rapid process (**r-process**), discussed later, produces another series of elements. Figure 10.10 shows the elements that are produced by the s- and r-processes.

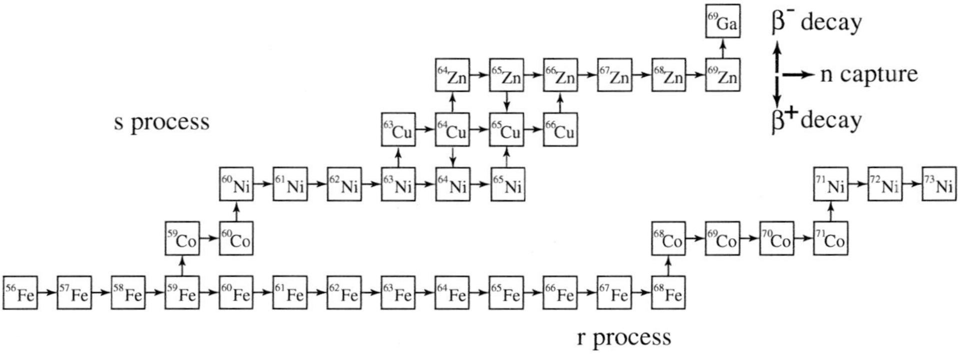

Fig. 10.10 Nucleosynthesis for $A > 60$.

The process of neutron absorption for building the heavier elements must come through a chain of beta-stable nuclei because these beta-stable nuclei do not immediately undergo beta decay and can hence absorb more neutrons and build higher and higher atomic number elements. These high-neutron nuclei in turn can undergo beta decay, with neutrons decaying into protons and forming nuclei with larger and larger numbers of protons, thus yielding the heavier elements. See Fig. 10.10.

By the s-process, iron-peak elements (iron, nickel, cobalt) with atomic mass around $A = 56$ are converted by neutron capture to elements with $A = 70$, which in turn are converted by further neutron capture to elements of atomic mass up to $A = 209$, which is an isotope of bismuth with $Z = 83$. The 'magic numbers' of neutrons in a nucleus — equal to 50, 82 and 126, as shown in Fig. 10.11 produce the peaks in the abundance of elements at atomic number $88, 138$ and 206.[4] Slow neutron capture through the s-process cannot produce polonium (^{210}Po) since it rapidly disintegrates to ^{206}Po by alpha decay. Similar to polonium, ^{232}Th thorium and ^{233}U uranium cannot be produced by slow neutron capture and require the r-process for their production.

The r-process is rapid enough to break through the instability caused by the alpha decay of ^{210}Po. The r-process occurs in Type II supernova and may also

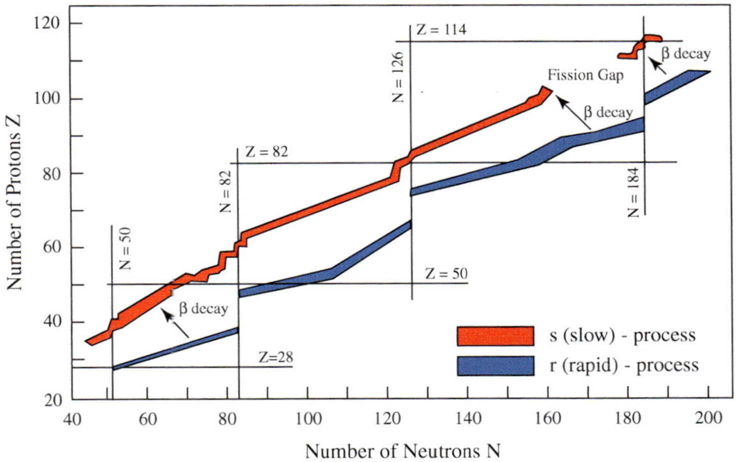

Fig. 10.11 Summary of s-process and r-process in the production of the heaviest elements.

[4]The 'magic numbers' of neutrons arise due to the closed neutron shells in nuclear physics: the stablest configurations are at the closed nuclear shells for $N = 50$, 82, and 126. Near the magic numbers the nuclei are beta stable and all decays lead to these beta-stable nuclei; hence under the r-process there is a jump in the abundance of the magic numbered nuclei since they are beta-stable.

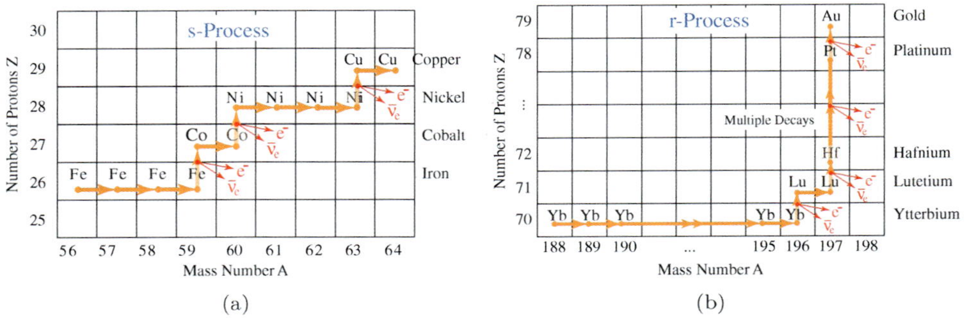

Fig. 10.12 (a) Neutron capture by the s-process. (b) Neutron capture by the r-process. Here, ytterbium-118 (^{188}Yb in figure) rapidly captures 9 neutrons and then through a series of beta decay ends up as the stable gold isotope ^{197}Au.

occur in the helium flash in red giants. Approximately half of the atomic nuclei heavier than iron are produced in the r-process.

The mechanisms of building the heavier nuclei by the s- and r-process are illustrated in Figs. 10.12(a) and 10.12(b) respectively; in the figures, unstable isotopes undergo β^--decay through the process $\mathrm{n} \rightarrow \mathrm{p} + \mathrm{e}^- + \bar{\nu}_e$. Note on the lower part of Fig. 10.12(b) that the ytterbium nucleus ^{188}Yb rapidly absorbs a series of 8 neutrons and becomes the very unstable ytterbium isotope ^{196}Yb. On the other hand, in an s-process, much fewer neutrons are captured. For example, in Fig. 10.12(a) only three neutrons are captured successively. Figure 10.12(b) also shows how by a combination of neutron capture and $\beta^{\pm}$-decays, nucleosynthesis of gold can be achieved.

In a **Type Ia supernova**, the star undergoes a rapid implosion — contracting and heating by converting gravitational energy to thermal energy. Neutrons are produced copiously and at a very fast rate. In this rapid process of neutron absorption the time between successive neutron captures is less than 1 second and is much faster than the time required for beta decay: hence elements heavier than ^{209}Po can be built from beta-unstable nuclei up to californium (^{254}Cf). Californium is the heaviest quasi-stable element and undergoes spontaneous nuclear fission in about 56 days.

Stars that become Type Ia supernovae, when they explode, can be 5 billion times brighter than the Sun, and are accompanied by the production of the heaviest known elements. The half-life of a Type Ia supernova is 55 ± 1 day, which is very close to the half-life of californium.

Type II core-collapse supernovae create a huge density of neutrons, of about $10^{18}/\mathrm{m}^3$, emerging from the imploding core. The explosion of a core-collapse supernova leads to ejection of the star's mantle, and thus to substantial enrichment of the interstellar medium with the major burning products of hydrostatic equilibrium: ^{4}He, ^{12}C, ^{16}O, ^{20}Ne, and so on. The explosion creates photons, alpha particles, and neutrons passing through the shells that enclose the exploding iron core of the star.

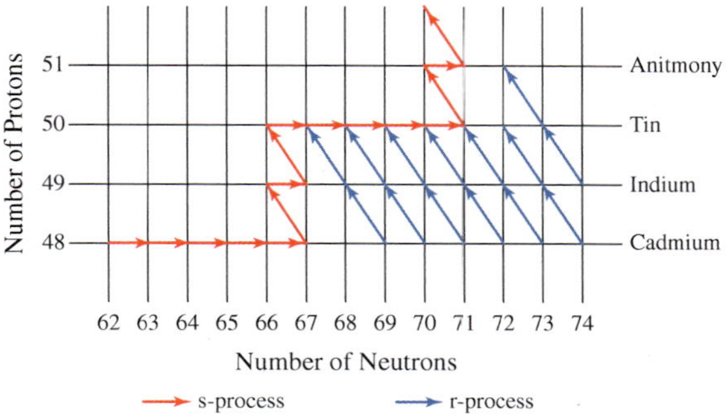

Fig. 10.13 Summary of r and s production processes.

Table 10.1 Summary of processes for nucleosynthesis in stars with different masses and at various stages of their evolution.

Nuclear process	Type of star
Hydrogen burning: pp-process	Main sequence stars
Hydrogen burning: CNO-process	Second generation stars First generation high mass stars
Helium burning (helium flash)	Main sequence stars
Alpha capture	Red giants
Slow neutron capture (s-process)	Red supergiants Type Ia and Type II supernovae
Rapid neutron capture (r-process)	Red supergiants; helium flash Type Ia and II Supernovae

In the silicon shell, significant production of iron-peak elements results: Fe, Co, Ni, Cu, Zn, with significant productions of Ca, Ti, *etc.* A similar production of elements takes place in the oxygen shell.

The somewhat lower temperature of the neon shell results in the synthesis of lighter nuclei such as Si, S, Ar, K, Ca, *etc.* The higher temperatures resulting from the supernova explosion accelerate the capture reaction leading to the production of the elements stated above. Very little explosive nuclear synthesis occurs outside the neon shell. Figure 10.13 shows how the heaviest elements of the periodic table are produced by the s- and r-processes. Figure 10.11 summarizes the heaviest elements that are produced by the s-process and r-process. The lower hatched double line shows how the heavier elements are initially synthesized and then undergo beta decay to produce elements with higher and higher atomic numbers.

Table 10.1 summarizes the nuclear processes that produce all the elements of our Universe. The various astrophysical processes that give rise to the different elements are given in Fig. 10.14.

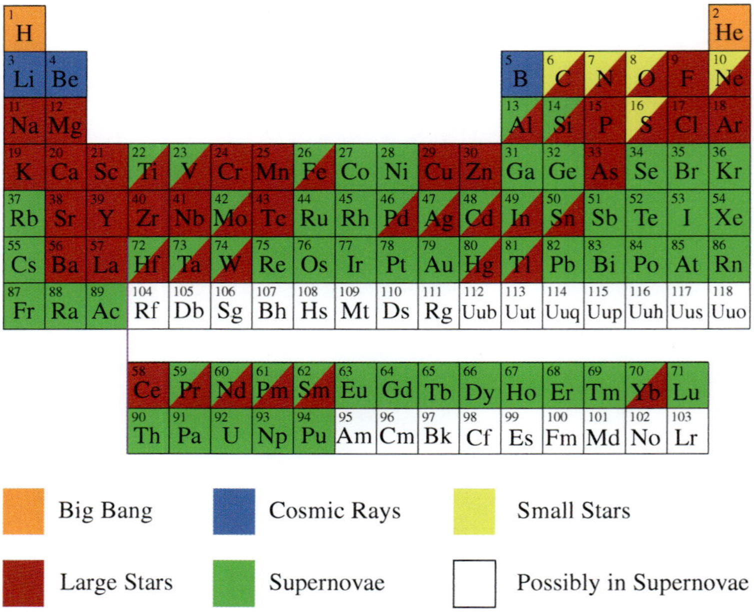

Fig. 10.14 Main origins of the elements in our Universe.

10.9 Synthesis of Gold ^{197}Au

Gold and other heavier elements cannot be synthesized in the normal burning of stars. There is a class of stars which explode in a Type Ia supernova; the explosion provides the energy required for fusion of the heavy elements, and which can yield heavy elements like gold via the r-process and shown in Fig. 10.12(b). The occurrence of supernova is very rare; for example in a typical galaxy with about 200 billion stars, it is estimated that only about one supernova occurs every 25 years and this may not be sufficient to account for all the gold in the Universe.

Recent observations indicate that, in a typical galaxy, there is a violent *collision of two neutron stars* once every 10,000 to 100,000 years, and which yields gold along with many other heavy metals. It is thought that these collisions may account for a significant amount of gold in the Universe.

10.10 Abundance of Elements in the Universe

Figure 10.1 shows the relative abundance of all the elements in the Universe; the most abundant elements, and which form the peak in the figure, are hydrogen (75%) and helium (23%); these elements are predominant due to their cosmological origins — being the result of the Big Bang. Hydrogen and helium are the most abundant of all elements in the primordial gas cloud that is the precursor to all the stars.

Note that Fig. 10.1 plots the logarithm of the relative abundance against atomic weight, and a linear decline in abundance of elements after hydrogen and up to iron means that there is an exponential drop in the abundance of the light elements. The elements from carbon to uranium account for about 2% of all the atoms in the Universe.

- Deuterium and the light elements, lithium, beryllium and boron, cannot be synthesized in the core of stars since they are unstable at high temperatures and hence they are clustered near the bottom of the figure; they do not fall on the curve since they do not originate from hydrogen by nuclear synthesis. Some of these elements may be produced in the outer layers of a star but their main source of abundance seems to be cosmological in origin — being produced soon after the Big Bang.
- There is a low occurrence of elements between helium and carbon due to difficulty in building these elements from hydrogen and helium. The secondary maximum at the iron group, shown in Fig. 10.1, is due to fusion culminating in the formation of the iron group of elements, that include iron, cobalt and nickel. The carbon–nitrogen–oxygen cycle for hydrogen burning results in the high abundance of nitrogen. In general, stellar nuclear fusion produces all the elements with $6 < Z < 28$.
- The relative abundance of elements after iron ($Z = 26$), namely for $Z > 30$, is formed by alpha particle and neutron capture, especially during the end points of stellar evolution. The elements are formed via a complicated pattern, reflecting the various processes that go in making the different elements. The elements after iron are formed by rapid and slow neutron capture, with the secondary peaks at $A = 88, 138$ and 206 due to the stability of nuclear shells against beta decay at the 'magic' neutron numbers.
- The present-day relative abundance of the heavy elements in the Universe has been gradually built up over the years by the rapid and slow processes in the core of massive stars when they have become red supergiants and also when they undergo a supernova explosion. Each new generation star contains heavy elements produced in the earlier generation stars.

Noteworthy (optional content) 10.2: Neutron-Rich Stable Isotopes

---

An element (atom) is defined by the number of protons, denoted by Z, that is in its nucleus; an isotope is defined by the number of neutrons N in the nucleus. Figure 10.15(a) shows the relative abundance of the elements and their isotopes, whereas Fig. 10.15(b) compares the number of neutrons in the stable isotopes versus the number of protons. The relative abundance of the various isotopes poses further

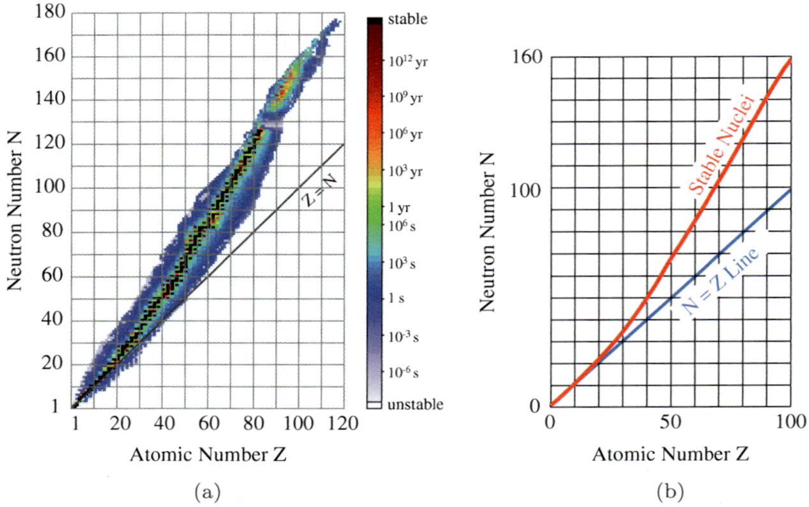

Fig. 10.15 (a) Stable and unstable nuclei with color coded half-lifes. (b) The stable nuclei have a preponderance of neutron-rich isotope.

challenges to the theory of stellar nucleosynthesis as well as providing the empirical basis for choosing the correct theory.

All elements with large Z (greater than $Z = 20$ protons) have a stable isotope that has more neutrons than protons. The reason for this preponderance of neutrons (in a stable nuclei) is due to the Pauli exclusion principle — discussed in Sec. 9.3 — when it is applied to the atomic nucleus.

Both neutrons and protons are spin 1/2 particles and hence are fermions; since they are different particles, the Pauli principle applies to both of them. An approximation that is quite accurate is to consider the protons and neutrons as separate fermion systems each having their own allowed energy levels. Figure 10.16 is a more refined representation of the nuclear potential given in Fig. 9.3; the protons, due to Coulomb repulsion, have a minimum energy denoted by E_c in Fig. 10.16, with the neutrons, since they have zero charge, having a lower energy than the protons. On filling the energy levels uptil the Fermi level, as indicated in Fig. 10.16, more neutrons than protons can be accommodated until all the energy levels are filled.

As the number of protons becomes larger and larger, the Coulomb repulsion due to their positive charge imparts higher and higher energy to the protons and pushes the proton Fermi energy level higher and higher, leading to the protons occupying higher energy levels and allowing many more neutrons to occupy energy levels less than the proton Fermi energy.

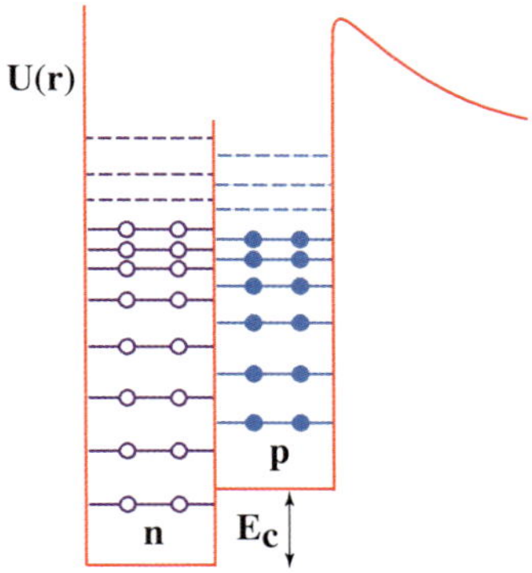

Fig. 10.16 Neutrons n and protons p in the nuclear potential well.

If there are more protons than neutrons above the filled levels, these nuclei are beta unstable, with the proton decaying to a neutron by **inverse beta decay**. Hence, in a beta-stable nuclei, and those are the ones that appear in Fig. 10.15 for the relative abundance of the elements, there are more neutrons than protons. For equilibrium, the highest energy level occupied by the protons has to be the same as that occupied by the neutrons.

Hence, for a stable isotope, in particular, one that is beta-stable, the number of neutrons in a heavy nucleus has to be greater than the number of protons; it can be shown that for a nucleus given by ${}^A_Z\mathrm{E}$ the excess of neutrons over protons grows as $(A - 2Z)^2$, namely a quadratic function of the difference between the neutrons and protons and is reflected in Fig. 10.15(b).

10.11 The Answer

The following is a summary of the synthesis of the elements in the Universe.

(1) All matter and radiation emerged around 13.78 billion years ago, when the expansion of the Universe started.
(2) Only a few light elements were formed during the initial stages of the Universe's expansion.
(3) The heavier elements, up to iron and nickel, are formed inside the core of ordinary stars. The high mass stars synthesize the heaviest elements during their red supergiant stage and when they undergo a supernova explosion.
(4) Supernovae disperse the elements, synthesized in the core of stars, into the interstellar medium.
(5) Second generation stars — like our Sun — are made up of interstellar medium composed of the remnants of the first and later generation stars.

Why is gold so rare? The answer is that all elements up to iron and nickel are synthesized in the nuclear fusion of stars, hence their relatively high abundance. An intricate web of s- and r-processes synthesize the elements heavier than iron and nickel in red giant and supergiant stars as well as in supernova explosions.

Elements with nuclear weight higher than iron, including gold, can only be synthesized in special circumstances such as very massive stars that either become red supergiants or undergo gravitational collapse and subsequently explode as supernovae. In particular, gold is synthesized in supernova explosions and in the collisions of neutron stars, both of which are very rare events. Hence the relative abundance of gold is very low.

Chapter 11

Elementary Particles

What are the elementary building blocks of matter?

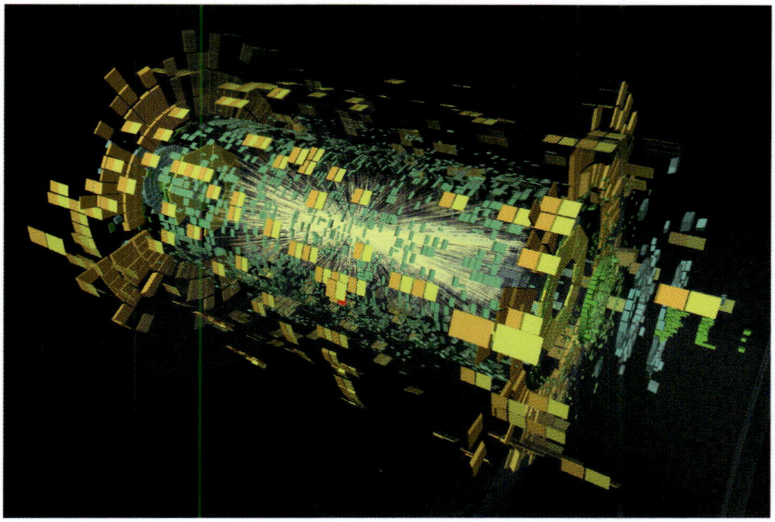

11.1 The Question

What is matter made out of has puzzled mankind for many millennia. The ancient Greek idea of an un-cuttable and irreducible 'building block' — from which the rest of matter is constituted — has been thought of by many civilizations. The notion that all materials are made out of a combination of water, earth, air and fire is similar in spirit to the idea of there being fundamental building blocks of nature.

Visible matter that composes the Universe comes in the form of various materials, all of which are now understood to be composed out of atoms and molecules. Large collections of atoms and molecules give rise to the diverse forms of materials that we see around us, including living matter.

11.2 Elementary Building Blocks

The question remains, what is matter at its most elemental, fundamental and deepest level? Are atoms the final elementary building blocks of nature? The answer is "No". We know from experiments dating back to the pioneering work by J. J. Thomson and E. Rutherford at the turn of the 20th century that an atom is a bound state of electrons and a nucleus — held together by the electromagnetic force. The nucleus was, in turn, determined by experiments to be made out of protons and neutrons. The starting point of particle physics is the study of the electron, proton and neutron.

So can we conclude that the most basic building blocks of matter are the electron, proton and neutron? The answer again is "No". Following in the footsteps of Thomson and Rutherford, accelerators were constructed to study the structure of the electron, proton and neutron. The experimental tools of high energy physics are enormous **particle accelerators** and detectors costing billions of dollars. In a typical experiment, a beam of particles, consisting of electrons or protons, is accelerated to very high energies and then made to collide with a target, or with another beam of the same or different particles, as illustrated in Fig. 11.1.

The *luminosity* of the beam is given (for a perfectly opaque target) by the number of particles crossing an area orthogonal to the beam, per unit area per time. Only if there are enough particles in the two colliding beams — hence the luminosity of the beam has to be high — can there be a sufficient number of collisions needed for the occurrence of a rare process, and which can then be studied by the particle detectors.

In experiments carried out in the 1960's a proton and an electron were collided at very high energies. The collision process is graphically represented in Fig. 11.2. It was found that, in addition to the incoming proton and electron, the collisions gave rise to a plethora of new particles that do not appear in the nucleus of any atom. These particles appear as a result of high energy collisions that create a high density of energy in a small region of spacetime; this high density of energy can lead to the 'creation', from the quantum vacuum state, of particles with larger and larger masses, something that cannot happen in ordinary circumstances due to the lack of adequate energy. The creation of particles from the quantum vacuum state

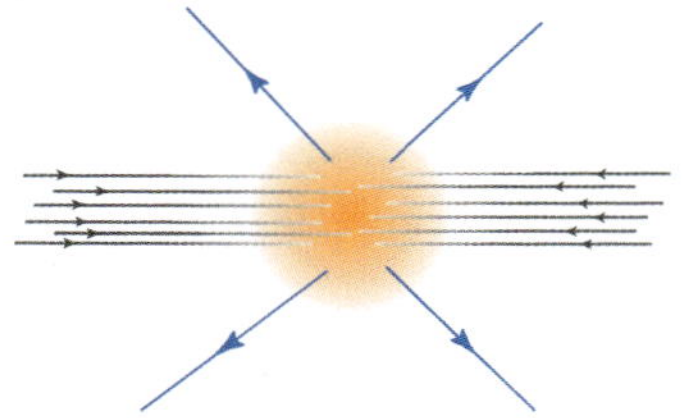

Fig. 11.1 Colliding particle beams produce many new particles.

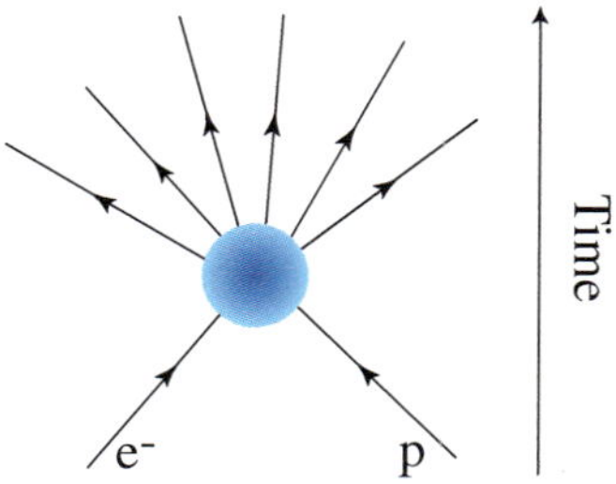

Fig. 11.2 A high energy collision between particles results in the creation, from the quantum vacuum state, of many new particles.

Fig. 11.3 Aerial view of the Stanford Linear Accelerator (SLAC).

has been discussed in Sec. 2.10. The appearance of a host of new particles resulting from high energy collisions was a puzzle to physicists. Cosmic rays also confirm the existence of particles that are not accounted for by only the proton, neutron and electron.

The grand result of the high energy experiments, carried out with the aid of larger and larger particle accelerators — such as the one shown in Fig. 11.3 — resulted, in the 1970's, in what is now called the **Standard Model**; namely, that matter is composed of a small collection of particles, called *fermions*; and interactions between these particles which are carried by another class of particles called gauge *bosons*. A leading exemplar of a gauge boson is the photon, which is the carrier of the electromagnetic interaction.

All the particles and forces are governed by the laws of quantum mechanics and the special theory of relativity. In the Standard Model, the particles and forces are structureless and point-like entities; this is the reason that the particles that constitute the Standard Model cannot be further subdivided. The result of long and arduous theoretical and experimental investigations of the Standard Model led to a series of empirical confirmations that reached a definite conclusion in 2012 with the experimental detection of the Higgs boson; this triumph of the Standard Model is discussed in Chapter 13.

In summary, the ancient question of what all of matter is composed out of has been partially answered by the Standard Model of particle physics. Many unanswered questions, however, remain: for example, what is the reason for so many 'fundamental particles', why are particles and interactions distinct, why are there so many parameters that need to be fixed by experiment and so on. In Chapter 14 on superstrings, an even more comprehensive answer than the one provided by the Standard Model is proposed to this ancient question.

In this chapter we discuss in some detail the particles that carry the matter content of Nature. In Chapter 12 we discuss the forces (interactions) of Nature and in Chapter 13 the elementary particles and fundamental forces are put together to yield the Standard Model.

11.3 Particle Accelerators and Detectors

Particle physics, or high energy physics, focuses on the study of fundamental particles that constitute all of matter and the forces that mediate the interactions between these particles. The typical size that one needs to probe is that of subnuclear distances. A nucleus has a radius of approximately 10^{-15} m, and hence one needs to have probes with a size that is less than the nuclear distances. The only probes of that size are the elementary particles *themselves*, such as the electron or the photon.

If an experiment intends to observe an elementary particle, we need to determine the energy or equivalently the momentum of the experimental probe. Quantum mechanics teaches us that all particles are described by a wave function that is characterized by the **de Broglie wavelength** λ and is illustrated in Fig. 11.4; the momentum of the particle, from de Broglie's wave concept, equals $p = h/\lambda$, with h being Planck's constant.

In analogy with optical microscopes, to probe smaller and smaller distances, we need probes of shorter and shorter wavelength in order to 'see' (observe) structures of small sizes. To study elementary particles, λ has to be less than 10^{-15} m (the size of a nucleus). As we study objects that are smaller and smaller in size, the probe's wavelength λ has to be smaller and smaller. Hence, as we probe the (subnuclear) structure of matter more and more closely we need higher and higher momentum, with the limit of $p \to \infty$ for $\lambda \to 0$. In other words, the probes needs to have a very 'high' momentum (and consequently high energy) for studying subnuclear phenomena. The term 'high' is the highest energy that is available, and in this sense, high energy physics is at the frontier of the scientific quest.

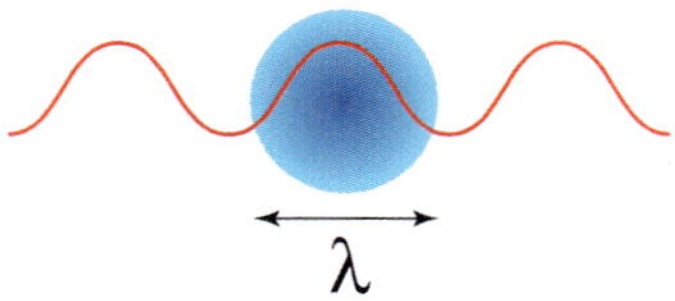

Fig. 11.4 Particles have an associated de Broglie wavelength.

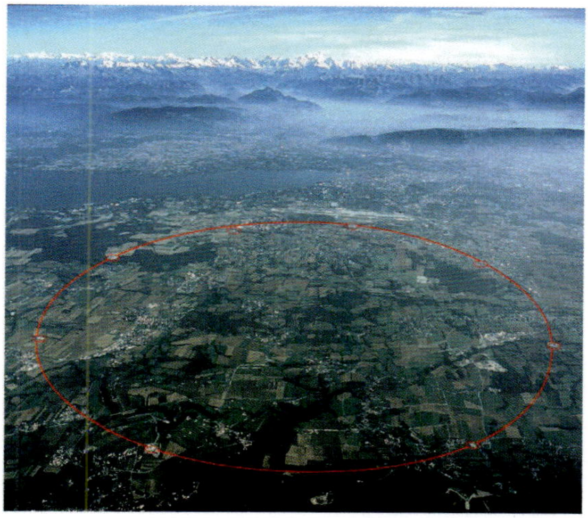

Fig. 11.5 Aerial view of the Large Hadron Collider (LHC) at CERN.

Fig. 11.6 The welding of one of the LHC magnets.

The linear collider at SLAC is shown in Fig. 11.3 and can accelerate electrons and positrons to an energy of up to 50 GeV. This energy, was however not sufficient to make the discovery in 1981 of the Z^0 particle, for which protons at the CERN accelerator with energy of about 100 GeV were collided to produce the Z^0 particles.

The highest energy in 2014 was achieved at the Large Hadron Collider (LHC) at CERN, shown in Figs. 11.5 and 11.6. The LHC consists of powerful magnets, placed around a circular hollow cylinder, that make charged particles go in a circular orbit; there are two oppositely circulating beams, one consisting of protons and the other of antiprotons — which collide at four intersection points. Each beam has an energy of up to 7 TeV and giving 14 TeV as the energy of collision[1]; the protons move at a velocity of about $0.999999991c$ at this energy. The Large Hadron Collider has a beam luminosity of up to 10^{34}–10^{35} particles per $cm^{-2}s^{-1}$.

Particle detectors identify the particles resulting from high energy collisions, and the theories of high energy physics are used for predicting and explaining the results of such high energy collisions. Experimentalists look for new particles as well as for new processes, as these yield insights into the forces and constituents of matter.

11.4 What is an Elementary Particle?

To proceed any further, we need to know: What is a fundamental, or an elementary particle? Clearly, we need to describe a particle by its *qualities* that are *permanent*, which do not change with time. For example, if a person has some identifying mark, say a scar that is permanent, then it can be used as one of the ways of identifying the person.

What are the qualities of Nature that can be used to identify a fundamental particle? Let us start with the case of classical mechanics and then go on to discuss the case of a quantum particle.

A particle in classical mechanics is specified by its mass m and the position it occupies, namely a point $x(t)$ at time t. In other words the only intrinsic quality needed to identify an 'elementary' particle in classical physics is the mass of the body; mass is a quantitative expression of the quality of inertia, the quality of matter that resists motion and acceleration. Mass occupies a position in space and follows a trajectory in time, as shown in Fig. 11.7.

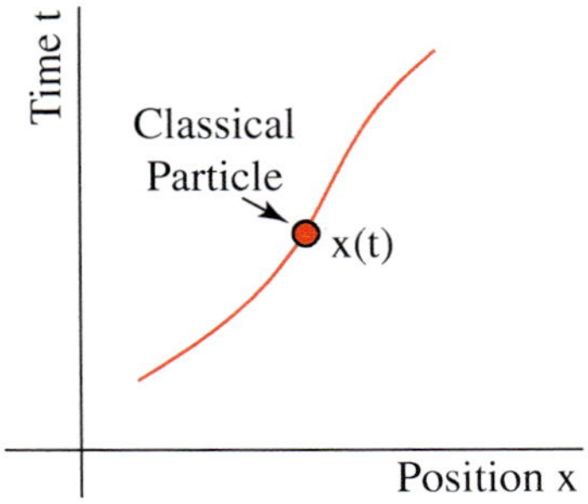

Fig. 11.7 A classical path is well defined.

[1] 1 TeV = 1000 GeV = 1 trillion eV.

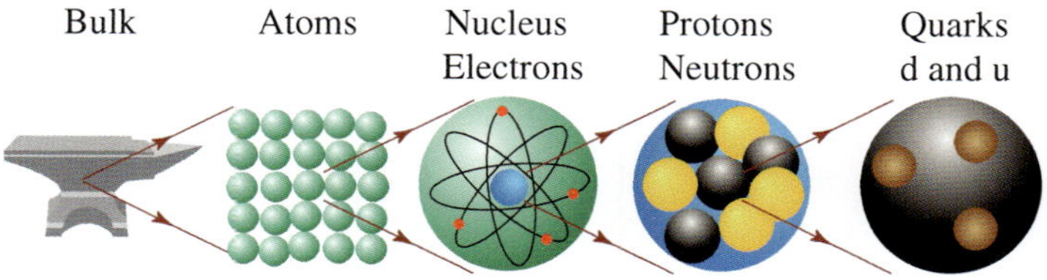

Fig. 11.8 The hierarchical organization of matter.

Before we describe an elementary particle we need to have a historical perspective as to what is meant by a fundamental or elementary particle. The concept of what is 'elementary' has changed with the increasing energy at our disposal starting from dust particles → molecules → atoms → nuclei → protons, neutrons → quarks → ⋯ as illustrated in Fig. 11.8.

Presently, the atom is fairly well understood and all atoms are composed of three 'elementary' particles, namely the electron, proton and neutron denoted by e^-, p and n respectively. An atom is composed of a **nucleus** and electrons. The nucleus, in turn, is composed out of neutrons and protons — held inside the nucleus by nuclear forces — and with the electrons being held in a bound state with the nucleus by the electromagnetic force, which is mediated by the photon. The size of a typical atom is about 10^{-10} m whereas the nucleus is a 100,000 times smaller, having a typical size of 10^{-15} m.

Unlike the case of a classical particle such as a billiard ball, elementary particles obey the laws of quantum mechanics and special relativity, and hence we need to look for a quantum and relativistic description of elementary particles.[2]

In quantum mechanics, an elementary particle at time t is described by its *state function* $\psi_t(x)$, as illustrated in Fig. 11.9. The description of an elementary particle by a state function has two aspects, namely

- Quantum properties that an elementary particle shares with other composite and non-elementary quantum objects, such as its probability distribution in space and time, its energy, momentum, angular momentum and so on.
- Other qualities and properties that are unique and intrinsic to the elementary particle.

[2]In particle accelerators, elementary particles are moving at velocities close to that of light, making the special theory of relativity of central importance. The effect of gravity is generally unimportant for elementary particles and the general theory of relativity is ignored in the Standard Model. However, as we discuss in Sec. 5.14, the interplay of general relativity and quantum physics is important in the study of black holes and this connection will be revisited in our discussion on superstrings in Chapter 15.

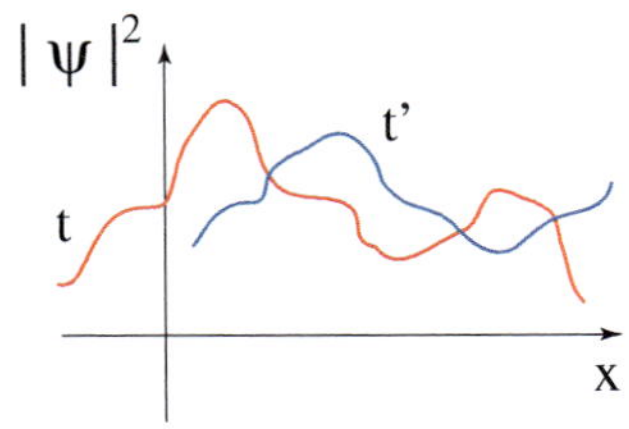

Fig. 11.9 Evolution of ψ from time t to t'.

Properties such as energy, angular momentum and momentum are not unique to elementary particles since any system including complex molecules, polymers and other composite objects are all described in terms of these quantities.

What are unique to elementary particles are certain fundamental *qualitative* properties that in turn have quantitative realizations. Similar to mass being a numerical quantity that expresses the quality of inertia in classical physics, elementary particles also have qualities that have quantitative expressions, called **quantum numbers**. In other words, quantum numbers are the *quantitative* expression of *qualitative* properties of elementary particles. The analogy of specifying an elementary particle is that of specifying a piece of food by its appearance, by its smell, by its taste and so on. Sight, smell and taste are qualitatively different from each other. Quantum numbers identify certain fundamental qualities of Nature, and assign them a quantitative description.

In summary, an elementary particle is described by a collection of quantum numbers that uniquely and exhaustively identify all the qualities of the particle.

11.5 Symmetry

Quantum numbers are permanent properties of elementary particles. To understand these permanent quantities, called 'conserved quantities', we need to understand the relationship between a **symmetry** and the **conserved quantity** that it entails. Each symmetry in quantum mechanics encodes a conservation law that, in turn, yields a conserved quantity.

Our intuitive understanding of symmetry is a highly regular pattern. A common object that is highly symmetrical is a perfect sphere. A rotation by any angle of a sphere about any axis through its center leaves the appearance of the sphere unchanged, as shown in Fig. 11.10. There are also more restrictive discrete symmetries, such as the case of a square or an isosceles triangle, as shown in Fig. 11.11, which is invariant under a few discrete reflections and rotations.

In essence a symmetry is about **invariance**; a symmetry is a property of an object that is left unchanged (invariant) by symmetry transformation. In the case of the rotation of a sphere, the property is the shape of the sphere that is left unchanged by a rotation. In Physics, it is the relevant equations that are invariant

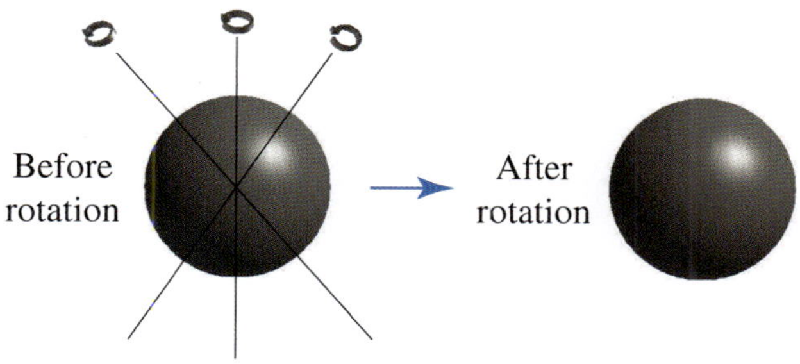

Fig. 11.10 A sphere can be rotated around any axis through its center.

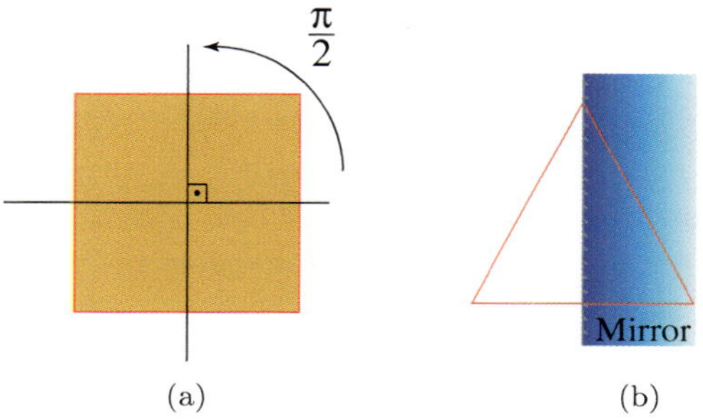

Fig. 11.11 A square can only be rotated by $\pi/2$ and an isosceles triangle only be reflected.

under the symmetry transformations. If an equation describing a phenomenon is unchanged under some symmetry transformation, the phenomenon is said to have that symmetry.

Bosons and Fermions

Bosons and **fermions** are particles that are defined by their behavior under their exchange — with bosons being symmetric and fermions being antisymmetric. Consider a two particle state function $\psi(x_1, x_2)$, with x_1, x_2 being the position of the two particles. If the two particles are bosons or fermions, they have the following property.

$$\text{Bosons:} \quad \psi_B(x_1, x_2) = \psi_B(x_2, x_1) \quad \text{symmetric}$$
$$\text{Fermions:} \quad \psi_F(x_1, x_2) = -\psi_F(x_1, x_2) \quad \text{antisymmetric.}$$

Supersymmetry unifies bosons and fermions into a single entity, called the superparticle. Under supersymmetry transformations, bosons are transformed into fermions and *vice versa*. There is an expectation among high energy theorists that at very high energies (greater than 100 TeV) supersymmetry is a symmetry of Nature.

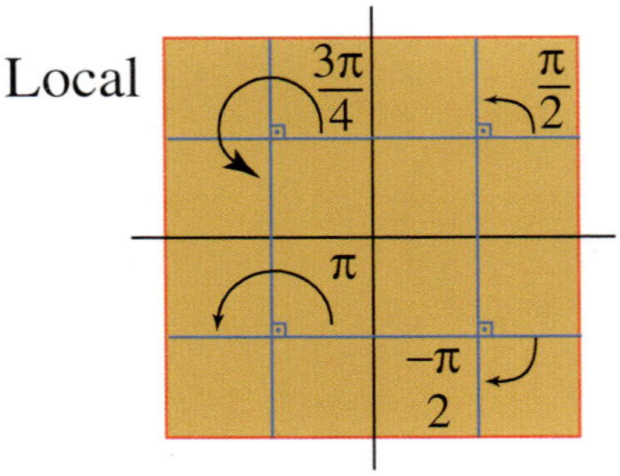

Fig. 11.12 Local versus global rotations.

Local Symmetries

The symmetries discussed so far have been global symmetries — in the sense that the same symmetry operation is carried out at all points of spacetime, that is, globally. There are also symmetries for which we can perform the symmetry transformation independently at each point of spacetime. Transformations that are performed independently at each point of spacetime are called **local symmetries**. Figure 11.12 shows a pattern where each small square can be rotated by an angle that changes from square to square — shown in the figure — without any change in the appearance of the pattern.

To illustrate the significance of a local symmetry, consider Lorentz transformations. Flat spacetime is invariant under the same Lorentz transformation throughout spacetime and Lorentz invariance is a global symmetry. However, if one demands that at *each* point of spacetime we can perform *independent* Lorentz transformations with velocity parameter $v = v(\vec{x}, t)$ that depends on the spacetime point $(\vec{x}, t)$, then for spacetime to be invariant, it can be shown that its geometry has to be determined by Einstein's general theory of relativity.

11.6 Symmetry and Conservation Law

In Physics, the symmetries that we are interested in are the symmetries of Nature. From quantum mechanics we have learnt that the evolution of the state function is determined by the Schrödinger equation. If the Schrödinger equation is invariant under some symmetry, then this symmetry persists over time and hence is a permanent property of the object that is evolving. The symmetries of elementary particles are the symmetries of their state functions.

Many of the symmetries and conserved quantities of elementary particles are hidden from the visible world, and matter has to be studied at a microscopic scale using large particle accelerators to discover these symmetries. To obtain some idea of what these symmetries could look like, we first examine some symmetries in quantum mechanics that are observed at the macroscopic scale, and the conserved quantities that follow from these symmetries.

All quantum mechanical systems, both microscopic and macroscopic, exist in spacetime and this gives rise to the following symmetries — and the concomitant quantum numbers.

- All positions in space are equivalent is a statement that the laws of Physics are **translationally invariant**. This invariance leads to the conservation of linear momentum, denoted by $\mathbf{p}$.
- The laws of Physics are the same in all directions in space leads to **rotational invariance** and results in the conservation of angular momentum, which is denoted by ℓ.
- The laws of Physics do not explicitly depend on time leads to the **conservation of energy**, which is denoted by E. For a particle at rest, its energy is given by $E = mc^2$, where m is the mass of the particle.
- The laws of Physics are unchanged when spacetime is subjected to **Lorentz transformation**; this leads to the conservation of spin (intrinsic angular momentum), and which is denoted by s.

The conserved quantities can be used to identify various physical systems. Assigning numerical values to these conserved quantities, a quantum mechanical state can be symbolically represented by a state vector denoted by $|\mathbf{p}, \ell, E, m, s\rangle$. Since these symmetries arise from the properties of spacetime, the conserved quantities also hold for elementary particles as well.

11.6.1 *Gauge invariance and gauge field*

In addition to spacetime symmetries, one of the most important symmetries of elementary particles is that of phase or **gauge invariance**. Consider an electron quantum field, denoted by $\Psi(t,x)$; the electron quantum field is a mathematical construct that represents the electron elementary particle and, in particular, can create electrons and antielectrons. Suppose one changes the value of the electron field by an arbitrary phase at *every point*, namely to $\Psi'(t,x) = e^{i\Delta(t,x)}\Psi(t,x)$; if one demands that the system be unchanged after a local change of phase, the electron field needs to be *coupled* to another field: the coupling is determined by the requirement that at every point this field can be transformed in such a manner so as to compensate for the change of phase of the electron field.

This field is called a **gauge field** and is denoted by $A_\mu(t,x)$. The gauge field is gauge transformed to compensate for the change of phase of the electron field; taken together the system of $\Psi(t,x), A_\mu(t,x)$ has what is called *local* gauge invariance; the term local is used to indicate that the gauge symmetry transformation depends on the spacetime point, varying from point to point.

Gauge fields can come in many types and varieties; the simplest exemplar is the Maxwell gauge field, which consists of the electric and magnetic fields. All the interactions of the Standard Model, discussed in Chapters 12 and 13, are mediated by different gauge fields that are coupled to quarks and leptons and respect the principle of **local gauge symmetry**.

Global gauge invariance leads to the law of charge conservation, and hence the *electric charge* Q of the electron is a conserved quantity. The electron's state function can be labeled as $|Q; p, \ell, E, m, s\rangle$. Note that the quantities Q, ℓ, s have discrete values and are quantum numbers, which differentiates them from other conserved quantities like p, E, m that take values in a continuous range.

11.7 Baryon and Lepton Quantum Numbers

As discussed in Sec. 11.4, an elementary particle is described by its quantum numbers; we need to determine what are the relevant quantum numbers. In Sec. 11.6, a number of symmetries were discussed that are related to the transformation of a particle in space and time — resulting in the conservation of energy, angular momentum and so on. Conserved quantum numbers that describe elementary particles are global symmetries that consist of transformations that do not refer to spacetime — called **internal symmetries** — and yield discrete conserved quantum numbers.

In summary, quantum numbers of elementary particles are called **internal quantum numbers** since these are not the result of the invariance under transformations in spacetime, which was the case for the conserved quantities discussed in Sec. 11.6.

Let us start our analysis with the hydrogen atom that is composed of an electron orbiting a proton and held together by their electromagnetic interaction.[3] The electron has a mass $m_e \simeq 0.5\,\text{MeV/c}^2$ which is about 2000 times lighter than the proton. So, to start with, we have two distinct families of particles: those like the electron which are 'light-weighted', called **leptons**, and heavy particles like the proton and neutron called **baryons**. All leptons and baryons are spin 1/2 fermions.

Amongst the most important quantum numbers that describe elementary particles are baryon and lepton number, and which take the value of either 0 or ± 1. Lepton number is discussed in this section. The baryons are more complex than the leptons, and we will need to analyze the behavior of quarks, as in Sec. 11.15, to be able to assign baryons their appropriate quantum numbers.

The assignment of lepton and baryon quantum numbers to elementary particles would be meaningless if not for the fact that in all processes in Nature, both lepton

[3]As mentioned earlier, the electromagnetic interaction itself is mediated by another elementary particle, namely the photon, which belongs to a class of elementary particles known as bosons, and discussed in Chapter 12.

Table 11.1 All known leptons and two baryons.

Leptons	Neutrinos	Baryons
electron e^-	ν_e: electron neutrino	p
muon μ^-	ν_μ: muon neutrino	n
tau τ^-	ν_τ: tau neutrino	

Table 11.2 Three generations of the lepton, with each generation occurring in pairs.

l	Q	L_e	L_μ	L_τ	Rest energy: mc^2
e^-	-1	1	0	0	511 KeV
ν_e	0	1	0	0	<30 eV
μ^-	-1	0	1	0	107 MeV
ν_μ	0	0	1	0	<0.5 MeV
τ^-	-1	0	0	1	1784 MeV
ν_τ	0	0	0	1	<250 MeV

number and baryon number are exactly conserved! Hence these quantum numbers are permanent qualities.

The electron is the only lepton that appears in the periodic table of atoms. The electron has a companion particle called the **neutrino** — a charge neutral massless fermion.[4] The electron together with its neutrino is called a **lepton pair**. In the Standard Model, discussed in Chapter 13, all of Nature's lepton are organized in three generations of lepton pairs, with each generation being more massive than the earlier one, and identical in all other respects.

Table 11.1 summarizes our knowledge of leptons, listed together with two of the most important baryons, namely the proton and the neutron.

To differentiate leptons from baryons it is postulated that leptons carry zero baryon number and carry a unit of a quantum number called **lepton number** L; lepton numbers come in three types L_e, L_μ and L_τ, as shown in Table 11.2. Similarly baryons carry zero lepton number and a unit of a quantum number B called **baryon number** and which is zero for leptons.

The assignment of the lepton quantum numbers is a typical example of how one identifies elementary particles. For example having the quality of 'lepton-ness' or having the quality of 'baryon-ness', is expressed in terms of the respective quantum number, and which indicates the presence or absence of these qualities. Whenever a truly new process is found which cannot be accounted for by the existing quantum numbers, a new quantum number is postulated.

[4]The neutrino may have a small mass, but this will not change any of the results that we will obtain.

The conservation of all the quantum numbers like lepton and baryon numbers is an experimental fact and is theoretically realized by a certain 'global phase symmetry' of the elementary particle's state function. The following quantum numbers are *absolutely conserved* in Nature.

- B: Baryon number
- Q: Electric charge
- L_e, L_μ, L_τ: Lepton numbers are each separately conserved.

In summary, an elementary particle is described by a state function ψ. The quantum numbers unique to a particle are carried by the relevant state function that is written as $|Q; \mathbf{p}, \ell, E, m, s, L, B\rangle$. For example, a proton with momentum $\mathbf{p}$ has a state function given by $|Q = 1; \mathbf{p}, \ell, E, m, s = 1/2, L = 0, B = 1\rangle$.

11.8 Antiparticles

Antiparticles are forms of matter that exist in spacetime and carry energy and momentum like all forms of matter. All elementary particles come in pairs of a particle and its antiparticle; there are a few neutral elementary particles like the photon that are their own antiparticle.

To start with, what is an antiparticle? Intuitively, an antiparticle should have all the internal quantum numbers of a particle but with the *opposite sign* so that when we combine or annihilate a particle with its antiparticle, *all internal quantum numbers* of the resultant quantum state should be zero — and which indeed is the case. Of course the final quantum state conserves energy and momentum and is *pure energy* since it has zero for all the quantum numbers.

For example, when an electron and its antiparticle (a positron) collide, they annihilate and release their energy as a photon, indicated by the symbol γ; a virtual process is illustrated in Fig. 11.13 and was earlier represented in Fig. 2.11. To conserve momentum and energy, this virtual process leads to a physically observable process only in the vicinity of a heavy nucleus, as discussed earlier and shown in Fig. 2.12.

Note that the photon is an example of a particle for which all the internal quantum numbers are zero and hence is a form of pure energy. The vacuum state,

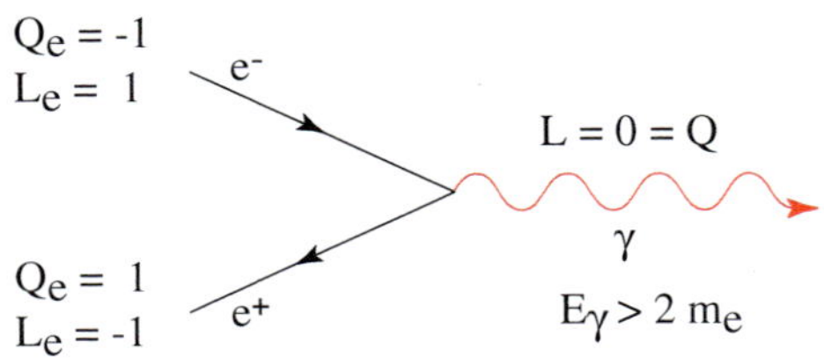

Fig. 11.13 A virtual process for the annihilation of electron–positron.

also called the ground state, is the quantum state with the lowest possible energy; for the vacuum state of quantum electrodynamics, discussed in Sec. 12.4, all of its quantum numbers are equal to zero.

Particle and antiparticle must have exactly the same (positive) mass. Hence, for an elementary particle described by state function $|Q; \mathbf{p}, \ell, m, s, L, B\rangle$, its antiparticle has a state function given by $|-Q; -\mathbf{p}, \ell, m, s, -L, -B\rangle$, with both particle and antiparticle having *positive energy* given by $E = \sqrt{\mathbf{p}^2c^2 + m^2c^4}$.

11.9 Antiparticles and Causality

The combination of quantum mechanics and special relativity necessarily leads to the existence of antiparticles, and is briefly discussed in this section.

For a particle of mass m, how precisely can we specify its position? Recall that a particle moving with velocity $v \ll c$ has a **de Broglie wavelength** given by

$$\lambda = \frac{h}{p} = \frac{h}{mv} \propto \Delta x$$

where Δx indicates the extent to which its state function is spread over space. We need to generalize the de Broglie wavelength, which holds for non-relativistic particles, to the relativistic case, where it is called the **Compton wavelength**. For very high momentum, the relativistic limit of momentum is given by $|\mathbf{p}| \simeq E/c$. Hence

$$\Delta x \sim \frac{h}{E/c} = \frac{hc}{\sqrt{\mathbf{p}^2c^2 + m^2c^4}}. \tag{11.1}$$

For a particle at rest, $\mathbf{p} = 0$ and yields the following:

$$\lambda = \Delta x \sim \frac{h}{mc} \quad : \text{ Compton wavelength of a particle.}$$

A quantum particle's state function is spread over a distance λ about its average position. For example, the Compton wavelength of an electron is $\lambda = h/(m_e c) \sim 2.4 \times 10^{-12}$ m. The Compton wavelength implies that the lighter is the particle, the more 'spread out' is its likelihood of being found around its average position, as illustrated in Fig. 11.14.

If we combine this feature of quantum mechanics with relativity we have a non-classical possibility, with a possible violation of causality; we discuss how this possible violation is, in fact, avoided in a remarkable manner by quantum mechanics.

Consider a quantum particle at a spacetime point (t, x) that is at a distance less than its Compton wavelength from an observer located at the origin (0, 0), as shown in Fig. 11.15. Since the quantum particle's indeterminate states are spread over a radius of the Compton wavelength, there is a likelihood of the quantum particle being absorbed at the origin by the observer. This possibility is ruled out

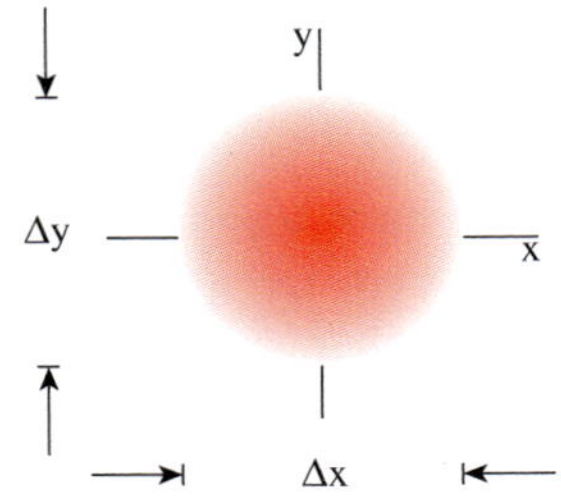

Fig. 11.14 The location of a particle is not definite.

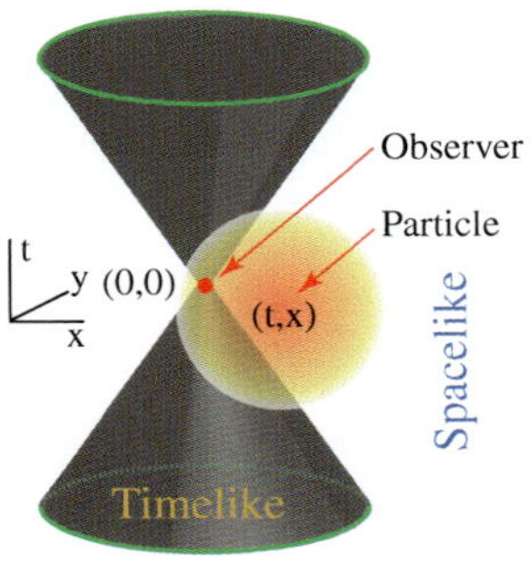

Fig. 11.15 Particle's state function overlaps with origin of light cone.

classically since the particle is at a spacelike distance away from the origin. Recall that spacelike distances have been discussed in Sec. 3.8 and it was shown that a particle can never classically travel across a spacelike distance as this would require the particle to travel at a speed greater than the speed of light.

In classical language, to reach the origin the quantum particle has to have a virtual velocity *higher* than the velocity of light and instantaneously cover distances less than or equal to its Compton wavelength. The quantum particle is not really traveling in a classical sense, but its indeterminate states are spread over space, and hence the emergence of a non-classical possibility. Any physical signal traveling faster than the speed of light can run into problems with causality.

In other words a quantum particle can make a transition from (t, x) to $(0, 0)$ even if the spacetime points are separated by a spacelike distance, as long as the distance covered is less than the Compton wavelength of the particle. In symbols, it can make a transition from (t, x) to $(0, 0)$ as long as

$$x^2 - c^2t^2 \le \left(\frac{h}{mc}\right)^2 . \tag{11.2}$$

This has the makings of a possible inconsistency, and we now examine how quantum mechanics resolves it.

Consider the two spacetime points $(0,0)$ and (t,x) as illustrated in Fig. 11.16(a) that have a *spacelike* separation. Consider a frame in which, at $t > 0$, a proton at position $(0,0)$ decays into a neutron and a pion. The pion is an elementary particle with mass m_π, discussed in Sec. 11.14, and appears in subnuclear processes; the pions form a family of three particles, namely positively and negatively charged pions, denoted by π^+ and π^-, respectively, and a neutral pion denoted by π^0. The decay process is given by

$$\text{p} \to \text{n} + \pi^+ \ : \ \text{First event at } (0,0).$$

The event (t,x) is at a distance *less* than the Compton wavelength of the pion. The virtual states of the pion 'propagate' in a virtual sense to (t,x), where there is a neutron already present, as shown in Fig. 11.16(a). The neutron absorbs the π^+ and is transformed into a proton by the following process

$$\text{n} + \pi^+ \to \text{p} \ : \ \text{Second event at } (t,x).$$

Hence the initial state at time $t = 0$ is a proton at $(0,0)$ and a neutron at $(0,x)$, followed by the emission of a pion and its subsequent absorption by the neutron, yielding the final state at time t with a neutron at $(t,0)$ and a proton at (t,x). Figure 11.16(a) is a Feynman diagram that represents the probability amplitude for this process.

As discussed in Secs. 3.7 and 3.8, since the separation of the neutron and proton is spacelike, we can always find an inertial frame such that the time ordering of the emission and absorption of the pion is *reversed*. In particular, since the events are spacelike we can do a Lorentz transformation $(0,0) \to (0,0)$ and $(t,x) \to (t',x')$ such that $t' < 0$. In the new frame, the Feynman diagram for the process is shown in Fig. 11.16(b): since $t' < 0$, the neutron at (t',x') absorbs a π^+ that *seems to be*

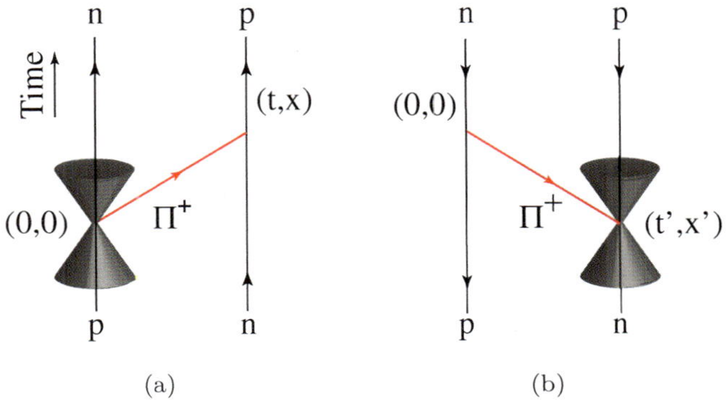

Fig. 11.16 (a) A π^+ traveling forward in time, from the proton to the neutron. (b) Pion π^+ traveling backwards in time in another frame of reference.

propagating backwards in time and is apparently absorbed *before* it is emitted by the proton at $(0, 0)$.

How can the neutron absorb a pion *before* it is emitted? How could the pion travel backward in time? This seems to violate causality (and common sense as well!).

The only way out of this apparent inconsistency between two frames of references is to postulate the existence of *another* particle π^-, a close partner of the π^+ pion. The absorption of a pion π^+ propagating backwards in time is re-interpreted as the *emission* (by the neutron) of particle π^- at (t', x') propagating *forward* in time, as shown in Fig. 11.17.

$$\mathrm{n} \rightarrow \mathrm{p} + \pi^- \quad : \ \text{First event at } (t', x')$$

followed by the absorption of π^- by proton at the later time.

$$\mathrm{p} + \pi^- \rightarrow \mathrm{n} \ : \ \text{Second event at origin } (0, 0).$$

In other words, π^+ propagating backwards in time is identical to π^- propagating forward in time. Experiments have shown that π^-, in fact, exists and has all the properties required for it to be the antiparticle of π^+.

The reason π^- is the antiparticle of π^+ arises from the requirement of consistency. Consider the process given in Fig. 11.16(a), namely

$$\mathrm{p} \rightarrow \mathrm{n} + \pi^+ \tag{11.3}$$

whereas from Fig. 11.17

$$\mathrm{p} + \pi^- \rightarrow \mathrm{n}. \tag{11.4}$$

'Adding' π^- to both sides of Eq. (11.3) gives

$$\mathrm{p} + \pi^- \rightarrow \mathrm{n} + (\pi^+ + \pi^-) \tag{11.5}$$

which implies that $\mathrm{p} + \pi^- \rightarrow \mathrm{n}$ given in Eq. (11.4) can hold only if

$$\pi^+ + \pi^- \ : \ \text{pure energy}. \tag{11.6}$$

The vacuum state of the Universe is postulated to have zero for all the quantum numbers; the antiparticle has to have the internal quantum numbers that are the *negative* of the particle, since only then can we have $\pi^+ + \pi^- = 0 + \text{pure energy}$. Hence, consistency demands that π^- is the *antiparticle* of the pion π^+.

As mentioned earlier, in the new frame the neutron decays at time $t' < 0$ into a proton *before* the proton at $(0, 0)$ changes into a neutron. We can think of all antiparticles, in particular π^-, propagating *forward* in time as being equivalent to the particle, π^+, propagating *backwards* in time, as shown in Figs. 11.16(b) and 11.17 *In the first frame* π^+ *is observed at the origin, and in the second frame it is* π^-. The result is consistent with quantum field theory, which is the synthesis of special relativity and quantum mechanics.

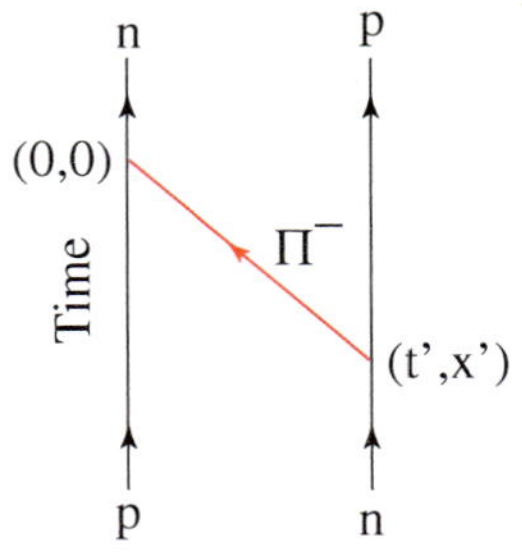

Fig. 11.17 π^- exchange.

11.10 The Yukawa Interaction

We investigate whether the existence of the pion and its antiparticle can affect the subnuclear forces. This is determined by the Compton wavelength of the pion λ_π. The mass of the pion $m_\pi \simeq 140\,\text{MeV}/c^2$; hence the Compton wavelength of the pion is given by

$$\lambda_\pi = \frac{h}{m_\pi c} \simeq 6.0 \times 10^{-15}\,\text{m}. \tag{11.7}$$

The length λ_π is also the size of the typical nucleus. If the proton is at (x_1, t_1) and the neutron at (x_2, t_2), then inside the nucleus

$$(x_1 - x_2)^2 - c^2(t_1 - t_2)^2 \le (10^{-15})^2\,\text{m}^2. \tag{11.8}$$

Hence, inside the nucleus, the neutron and proton are at a distance *smaller* than the Compton wavelength of the pion: the protons and neutrons are separated by a distance for which the existence of the pion and its antiparticle is important.

The Feynman diagram given in Fig. 11.18 shows the processes going on inside the nucleus. Namely, the proton emits a virtual pion π^+ and changes into a neutron; the neutron absorbs the π^+ and changes into a proton; the neutron in turn now emits the antipion π^- and converts into a proton with the proton absorbing the antipion and changing into a neutron; one now has completed one cycle in the multiple exchange of pions, which goes on indefinitely inside the nucleus.

The exchange of virtual pions and antipions gives rise to the nuclear **Yukawa force**, and is similar to the exchange of photons that gives rise to the Coulomb attraction of the proton to the electron in the hydrogen atom, discussed later in Sec. 12.5.1 and illustrated in Fig. 12.14. The *nuclear force* between protons and neutrons is approximately explained by the attractive Yukawa force that holds the neutrons and protons together to form the tightly bound nucleus.

The **Yukawa interaction** for the nucleons is an effective force described by an effective field theory, similar to the case of Debye screening discussed in Sec. 12.6. The pion is represented by a scalar field and has a coupling, denoted by g, to the

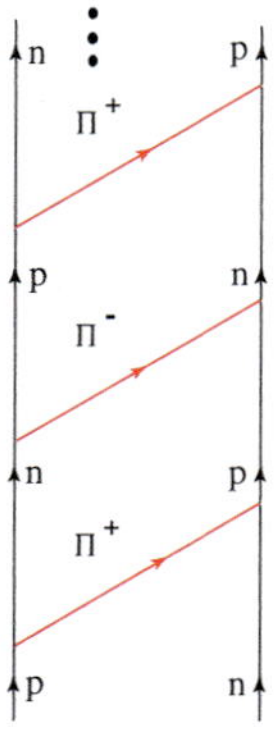

Fig. 11.18 The Yukawa interaction leads to the multiple exchange of π^+, π^- pions.

proton and neutron — considered as fundamental fermions. The attractive Yukawa potential is spherically symmetric which, for the neutron and proton separated by a distance r, is given by

$$U(r) = -\frac{g^2}{r}\exp(-r/\lambda_\pi). \tag{11.9}$$

Note that the **Yukawa potential** has a *shorter* range of interaction than the Coulomb potential due to the exponential damping of the potential as in Eq. (11.9).

As discussed in Secs. 11.14 and 11.16, from the point of view of the quark model, the pions are a special case of mesons, being the bound state of a quark and an antiquark, and neutrons and protons are the bound states of three quarks. The Yukawa scalar field is an effective low energy representation of the pion and the Yukawa force between neutrons and protons arises from the underlying interaction of quarks and gluons, discussed in Sec. 12.11.

The salient property of the Yukawa interaction is that it couples a scalar field to the fermions. In the case we have discussed, the scalar field is the pion and the fermions are the proton and neutron.

The Yukawa interaction in the Standard Model — unlike the case discussed above — is a *fundamental interaction*: the scalar field is the fundamental Higgs field, which is directly coupled to the fundamental fermions (quarks and leptons) via the Yukawa coupling. The Yukawa coupling of the Higgs field to the fermions is the lynchpin in generating the mass for all quarks and leptons, and is discussed in Chapter 13.

11.11 Antiparticles and Quantum Field Theory

Antiparticles have many other effects. The total number of particles is no longer fixed due to the existence of the antiparticle since virtual processes allow the creation of virtual states containing antiparticles. The creation of a particle and

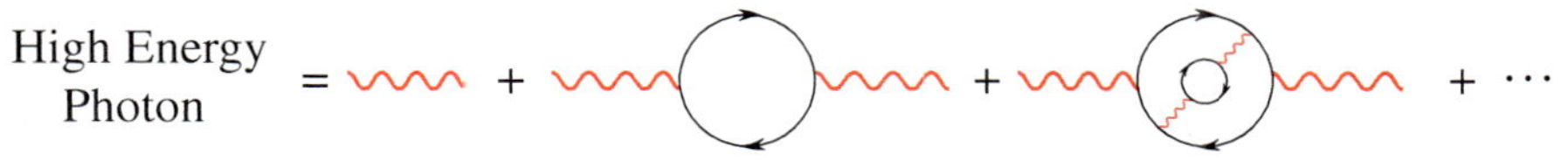

Fig. 11.19 Photon expressed in terms of virtual processes.

its antiparticle was discussed in Sec. 2.10 and illustrated Fig. 2.12; in the discussion of Hawking radiation in Sec. 5.17, the metaphor of the virtual creation of a particle and an antiparticle near black hole's horizon was used to explain Hawking radiation.

In non-relativistic quantum mechanics, the system always consists of a determinate and *fixed number* of particles; in contrast, a quantum system that is relativistic has an indeterminate (fluctuating) number of particles, and is described by **quantum field theory**.

More generally, once the concept of the antiparticle is introduced as an elementary particle in its own right, the concept of a single particle existing in isolation and by itself — say a single photon — is no longer applicable. If the photon has high enough energy, then it has a finite probability to make the following virtual transitions: $\gamma \to \mathrm{e}^+ + \mathrm{e}^-$, $\pi^+ + \pi^- \to \gamma$, and so on. The Feynman diagram for this representation of a high energy photon is shown in Fig. 11.19.

The state function of a photon represents this behavior; consider a *high energy* photon; it can be shown that this high energy photon is 'equivalent' to photons, electrons and antielectrons and other multiparticle states at *lower energies*. Symbolically, the quantum states of the high energy photon can be represented as follows

$$|\text{high energy } \gamma\rangle = |\gamma\rangle + |\gamma; \mathrm{e}^+ + \mathrm{e}^-\rangle + |\gamma; \pi^+ + \pi^-; \cdots\rangle + \cdots . \tag{11.10}$$

One can think of the state function of the photon as containing pairs of particles and antiparticles. If m is the mass of the particle, then to create a particle and antiparticle pair we need energy of $E \sim 2mc^2$; the virtual pair will have a half-life of Δt, which is given by the uncertainty principle as

$$\Delta t \simeq \frac{\hbar}{2mc^2}. \tag{11.11}$$

To physically materialize the pair, we need to carry out an experiment similar to the pair creation discussed for the electric field in Sec. 2.10. In the language of state vectors, the pair creation due to an intense electric field is that the photon state makes a quantum transition, called tunneling, from a state vector given by Eq. (11.10) to a state with a lower intensity electric field and a physical pair of particles. The electric field is a measure of the number of photons (photon density) at a particular position in space.

11.12 Energy Conservation and Quantum Numbers

Consider a process, shown in Fig. 11.20, with elementary particles A, B, C and D

$$\mathrm{A} + \mathrm{B} \to \mathrm{C} + \mathrm{D} \tag{11.12}$$

that conserves all internal quantum numbers such as B, L, s and Q, and hence is allowed.

The existence of antiparticles vastly increases the number of allowed physical processes at the subnuclear level. The reason being that all derived processes, obtained by 'moving' the particles across the arrow in Eq. (11.12) — which entails replacing a particle by its antiparticle — are allowed as well; in particular, denoting the antiparticle by a bar, the following processes are also allowed.

$$\begin{aligned} \mathrm{A} &\to \bar{\mathrm{B}} + \mathrm{C} + \mathrm{D} \\ \mathrm{A} + \mathrm{B} + \bar{\mathrm{C}} &\to \mathrm{D} \\ \bar{\mathrm{C}} + \bar{\mathrm{D}} &\to \bar{\mathrm{A}} + \bar{\mathrm{B}} \\ &\cdots \end{aligned}$$

It may be the case that although all the quantum numbers are conserved, nevertheless the process can be ruled out kinematically by energy conservation. Energy conservation requires

$$E_A + E_B = E_C + E_D.$$

We can always bring A and B together with zero momentum and obtain particles C and D moving with momentum $\mathbf{p}$ in opposite directions. Recall that energy is given by $E = \sqrt{\mathbf{p}^2c^2 + m^2c^4}$; since the incoming particles A and B have zero momentum, their energy is given by their mass; hence, we have

$$m_Ac^2 + m_Bc^2 = \sqrt{\mathbf{p}^2c^2 + m_C^2c^4} + \sqrt{\mathbf{p}^2c^2 + m_D^2c^4}.$$

$$\text{or} \qquad m_A + m_B \geq m_C + m_D. \tag{11.13}$$

Hence a process (11.12) that conserves all the requisite quantum numbers is allowed only if, *in addition*, it also satisfies the above condition on the masses as in Eq. (11.13).

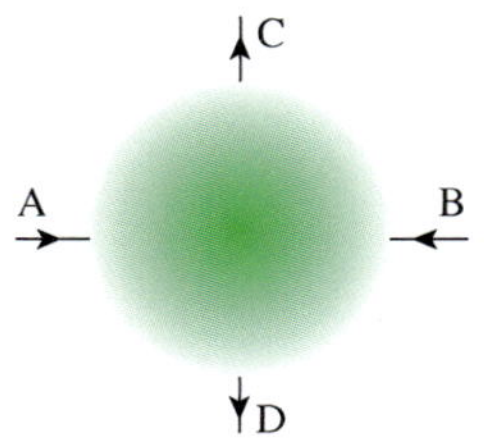

Fig. 11.20 Quantum numbers are conserved.

The proton is the lightest baryon and baryon number conservation is considered to be absolute. Since there is no lighter mass baryon that the proton can decay into, Eq. (11.13) states that the proton *cannot* decay — and hence it is stable. In contrast, the neutron, in principle, can decay since it has a mass greater than the proton; this does not necessarily mean that the neutron must be unstable but it turns out, if fact, that a free neutron is unstable — and decays into a proton in about 882 seconds.

11.13 Antiparticles: Baryons and Leptons

Antiparticles for charged particles are denoted by reversing the sign of the charge; for example the electron, namely e^-, has the positron, denoted by e^+, as its antiparticle. Similarly, the muon is given by μ^- and its antiparticle by μ^+.

The baryons p and n, with baryon number $B = 1$, have $\bar{p}$ and $\bar{n}$ as their antiparticles, which have baryon number $B = -1$.[5]

All the leptons have antileptons and are given below.

$$
\begin{aligned}
e^- &\leftrightarrow e^+ \ ; \quad \nu_e \leftrightarrow \bar{\nu}_e \\
\mu^- &\leftrightarrow \mu^+ \ ; \quad \nu_\mu \leftrightarrow \bar{\nu}_\mu \\
\tau^- &\leftrightarrow \tau^+ \ ; \quad \nu_\tau \leftrightarrow \bar{\nu}_\tau .
\end{aligned}
$$

In all processes one needs to take into account the quantum numbers of both the particles as well as of the antiparticles, which recall have the same internal quantum numbers as particles, but with the opposite sign.

To illustrate how the internal quantum numbers constrain the possible processes in Nature, we examine a few processes in which an elementary particle decays into a set of particles and antiparticles.

- Example 1: Allowed: beta decay of a neutron

$$
\begin{aligned}
& n \rightarrow p + e^- + \bar{\nu}_e \quad \text{beta decay} \\
B &: 1 = 1 + 0 + 0 \quad \text{OK} \\
L &: 0 = 0 + 1 - 1 \quad \text{OK.}
\end{aligned}
$$

- Example 2: Allowed: the decay of a muon

$$
\begin{aligned}
& \mu^- \rightarrow e^- + \bar{\nu}_e + \nu_\mu \\
L_\mu &: 1 = 0 + 0 + 1 \quad \text{OK} \\
L_e &: 0 = 1 - 1 + 0 \quad \text{OK.}
\end{aligned}
$$

[5]Note that although the neutron is an electrically neutral particle, its antiparticle is not necessarily the particle itself; and in fact, the antineutron is a distinct particle, namely $\bar{n} \neq n$.

- Example 3: Forbidden: the decay of a pion into the following leptons

$$\begin{aligned}\pi^+ &\rightarrow \mu^+ + \nu_\mu + \nu_e \\ B &: 0 = 0 + 0 + 0 \quad \text{OK} \\ L_\mu &: 0 = -1 + 1 + 0 \ \text{OK} \\ L_e &: 0 \neq 0 + 0 + 1 \quad \text{Not OK.}\end{aligned}$$

11.14 Hadrons: Strangeness Quantum Number

Protons and neutrons are the particles that make up the nucleus and carry more than 99.9% of the mass of an atom. In studying how a proton and neutron behave inside the nucleus, we introduced, in Sec. 11.10, yet another elementary particle, namely the pion π^+ and its antiparticle, the antipion π^-.

The proton, neutron and pions are members of a large family of elementary particles, and are collectively called **hadrons**. Particles such as the pion are produced in high energy collisions. The defining property of *all hadrons* is that they have *strong interactions* with each other.[6] We will see later, in Sec. 11.16, that all hadrons are made out of various quarks and antiquarks.

Hadrons are further classified into two groups, namely mesons and baryons, as shown in Fig. 11.21. The pion is a typical meson and the proton and the neutron are the prime examples of baryons. All mesons, such as $\pi^\pm, \pi^0$, have integer spin (0, 1, ...) and hence are bosons; mesons have zero baryon number, namely $B = 0$. Baryons, such as p and n, have half-integer spin (1/2, 3/2, ...) and hence are fermions; they have unit baryon number, namely $B = 1$. Mesons generally are light, with masses of about 200–400 MeV/c^2, whereas baryons in general have larger masses of about 1000–1300 MeV/c^2.

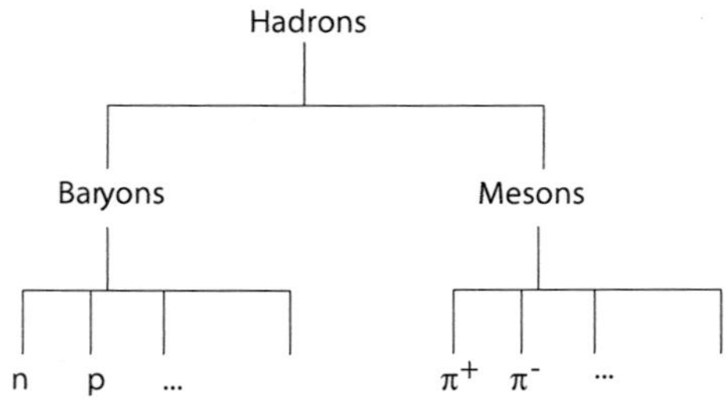

Fig. 11.21 Hadrons consist of baryons and mesons.

[6] Three fundamental interactions, namely the strong, weak and electromagnetic interactions, are discussed in Chapter 12.

By the beginning of the 1960's, more than 400 hadrons had been observed in high energy collisions; it was becoming quite clear that hadrons could not be fundamental particles — there were simply too many of them. The attention of high energy theorists naturally turned towards examining this zoo of elementary particles with a view of classifying and understanding these particles.

All hadrons except the proton are unstable, and typically have incredibly short lifetimes of about 10^{-23} s or less — the characteristic time scale of processes mediated by strong interactions, and discussed in the Sec. 12.3. A notable exception to this general statement about hadrons is the free neutron, which has a lifetime of about 882 s — more than 25 orders of magnitude greater than the lifetime of other unstable hadrons (the *bound* neutron inside a nucleus, however, is stable).

There is another set of hadrons with a lifetime of around 10^{-10} s, which are called 'strange' hadrons due to their long lifetimes compared to the other hadrons. Although 'strange' mesons and baryons are *produced* on a time scale of 10^{-23} s, which is the time scale of strong interactions, strange hadrons, for example Λ^0 shown in Fig. 11.22, always *decay* via a process such as

$$\Lambda^0 \rightarrow \mathrm{p} + \pi^- \tag{11.14}$$

that take place *slowly*, on a time scale of $\sim 10^{-10}$ s, which is typical of processes mediated by the *weak interactions*.

In the high energy collision of hadrons, strange hadrons always appear in *pairs*. For example the neutrally charged strange hadrons K^0, Λ^0 are produced in the annihilation of a pion with a proton, as shown in Fig. 11.23, namely

$$\begin{aligned} \pi^- + \mathrm{p} &\rightarrow \mathrm{K}^0 + \Lambda^0 \\ B \ : \ 0 + 1 &= 0 + 1 \quad \text{OK} \\ Q \ : \ -1 + 1 &= 0 + 0 \quad \text{OK.} \end{aligned}$$

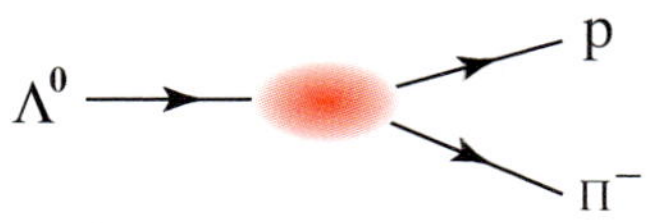

Fig. 11.22 Decay of a strange baryon: $\Lambda^0 \rightarrow p + \pi^-$.

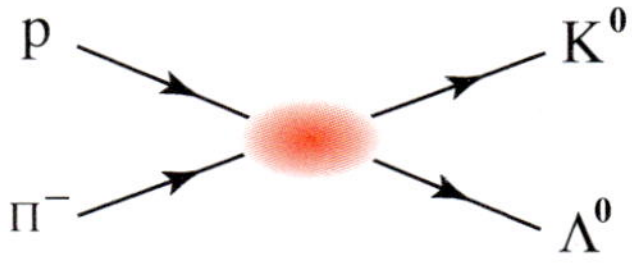

Fig. 11.23 Proton–pion interaction.

The annihilation never occurs in a process such as

$$\pi^- + \mathrm{p} \not\rightarrow \mathrm{K}^0 + \mathrm{n} \ : \ \text{Forbidden}$$

even though B and Q are conserved.

To understand the behavior of strange hadrons Gell-Mann and Nishijima proposed, in 1964, that there exists *another quality* of hadrons and assigned an internal quantum number to this quality, which they named as **strangeness** and denoted by S. Unlike Q, B and L that are absolutely conserved, strangeness is *conserved* by strong and electromagnetic interactions but it *is not conserved* by weak interactions.

All particles are assigned the quantum numbers of B, L, S and Q. The strangeness quantum number of all leptons is zero, and strangeness is assigned consistently to all hadrons, with only strange hadrons having non-zero strangeness number.

For example, the kaon K^0 is assigned strangeness number $S = +1$ and the lambda particle Λ^0 is assigned strangeness number $S = -1$. Hence we now can decode the earlier process in which strangeness is conserved as being the production of strange hadrons mediated by strong interactions. More precisely, since the following strong interaction conserves strangeness, it is allowed.

$$\pi^- + \mathrm{p} \rightarrow \mathrm{K}^0 + \Lambda^0 \ : \ \text{Strong interactions allowed}$$
$$S \ : \ 0 + 0 = 1 - 1 \ : \ \text{OK, strangeness conserved.}$$

However, the following strong interaction cannot take place since it does not conserve strangeness, namely

$$\pi^- + \mathrm{p} \not\rightarrow \mathrm{K}^0 + \mathrm{n} \ : \ \text{Strong interactions forbidden}$$
$$S \ : \ 0 + 0 \neq 1 + 0 \ : \ \text{Not OK, strangeness not conserved.}$$

In contrast to strong interactions, the decay of a strange hadron can violate strangeness number conservation only if it decays via weak interactions; we can consequently deduce that the following decay must be mediated by weak interactions:

$$\Lambda^0 \rightarrow \mathrm{p} + \pi^- \ : \ \text{Weak interactions allowed}$$
$$S \ : \ -1 \neq 0 + 0 \ : \ \text{Strangeness not conserved.}$$

The conservation of strangeness quantum number is *violated* by weak interactions; all strange hadrons decay via weak processes into the stable hadrons that have zero strangeness quantum number, and hence strangeness is completely erased from ordinary matter. It is only in high energy collisions and in a special category of stars and deep in the core of heavy nuclei that one can directly observe the strange baryons.

In summary, in addition to the quantum numbers that are absolutely conserved, namely Q: charge, B: baryon number, and L_e, L_μ, L_τ: lepton numbers, to adequately describe and classify hadrons one needs to add the strangeness quantum number S.

Unlike the other quantum numbers, strangeness is conserved in processes mediated by strong and electromagnetic interactions but is not conserved in weak interactions.

11.15 Quark Model

Faced with the proliferation of hadrons, both mesons and baryons, it became clear by the 1960's that neither the mesons nor the baryons are elementary particles. In 1964 Murray Gell-Mann and George Zweig postulated that all hadrons are the bound states of three quarks, named the up quark (u), the down quark (d) and the strange quark (s). Figure 11.24 shows the assignment of the charge Q and strangeness quantum numbers S for these three quarks and their antiquarks.

Quarks are the point-like spin 1/2 particles that are the excitations of the underlying quark quantum field that permeates all of spacetime. When sufficient energy is supplied to the quark field, a quark–antiquark pair is brought into existence and carries the energy supplied to the field.

In 1974, experiments indicated the existence of a fourth quark with a new quantum number called charm. The fifth quark called the bottom quark was produced in 1977, and has quantum number bottom; in 1995 the quark model was completed with the experimental production of the top quark with quantum number top. All quarks are spin 1/2 fermions. Each quark has its own distinct quantum number, namely $U, D, S, C, \mathcal{B}, T$, and are given in Table 11.3, together with their masses.[7] Reversing the sign on all the internal quantum numbers yields a similar table for the quantum numbers of the antiquarks.

In summary, all hadrons are bound states of quarks. All mesons are the bound states of a quark and an antiquark whereas all baryons are the bound states of three quarks, and antibaryons the bound states of three antiquarks.

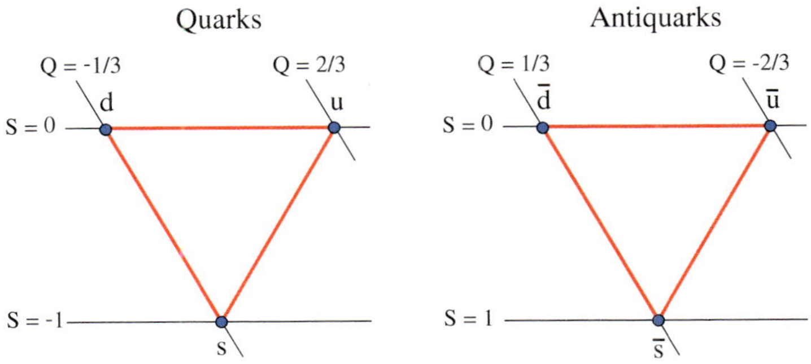

Fig. 11.24 Quarks and antiquarks, classified by their charge and strangeness quantum numbers.

[7] Bottom quark quantum number $\mathcal{B}$ is not to be confused with baryon quantum number B.

Table 11.3 Three generations of the quark that occur in pairs. All quarks have baryon number = 1/3 and have spin = 1/2. The quantum numbers $U, D, S, C, \mathcal{B}, T$ refer to the quantum number of specific quarks.

Quark	Q	D	U	S	C	$\mathcal{B}$	T	Rest energy: mc^2 (MeV)
d	$-\frac{1}{3}$	−1	0	0	0	0	0	360
u	$\frac{2}{3}$	0	1	0	0	0	0	360
s	$-\frac{1}{3}$	0	0	−1	0	0	0	540
c	$\frac{2}{3}$	0	0	0	1	0	0	1500
b	$-\frac{1}{3}$	0	0	0	0	−1	0	5×10^3
t	$\frac{2}{3}$	0	0	0	0	0	1	173×10^3

One of the most significant properties of the quarks is that they carry fractional charge, namely $\pm 1/3e$, $\pm 2/3e$; an isolated fractional charge has never been observed in any experiment, and this fact was one of the biggest mental obstacles for high energy theorists in coming up with the idea of quarks. It is the current wisdom that the strong interactions of the quarks result in the permanent confinement of quarks inside the hadrons, and is discussed in Sec. 12.14. The permanent confinement of quarks results in quarks occurring only in bound states (such as hadrons) that always have integer charge.

Below are typical examples illustrating how some well known mesons and baryons are made from quarks. The quantum numbers of the hadrons are obtained by simply adding the quantum numbers of the constituent quarks. For now, consider only the up, down and strange quarks and their electric charges and strange quantum numbers.

Mesons

The quark content of the meson π^+ is shown in Fig. 11.25(a).

$$\text{Pion} : \pi^+ = \mathrm{u\bar{d}}$$
$$Q : 1 = \frac{2}{3} + \frac{1}{3}; \quad B : 0 = \frac{1}{3} - \frac{1}{3}; \quad S : 0 = 0 + 0$$

$$\text{Antipion} : \pi^- = \mathrm{\bar{u}d}$$
$$Q : -1 = -\frac{2}{3} - \frac{1}{3}; \quad B : 0 = -\frac{1}{3} + \frac{1}{3}; \quad S : 0 = 0 + 0.$$

A few more mesons are the $\bar{\mathrm{K}}^0 = \mathrm{\bar{d}s}$; $\mathrm{K}^0 = \mathrm{\bar{s}d}$.

Baryons

The quark content of the proton p is shown in Fig. 11.25(b).

$$\text{Proton} : \mathrm{p} = \mathrm{uud}$$
$$Q : 1 = \frac{2}{3} + \frac{2}{3} - \frac{1}{3}; \quad B : 1 = \frac{1}{3} + \frac{1}{3} + \frac{1}{3}; \quad S : 0 = 0 + 0 + 0.$$

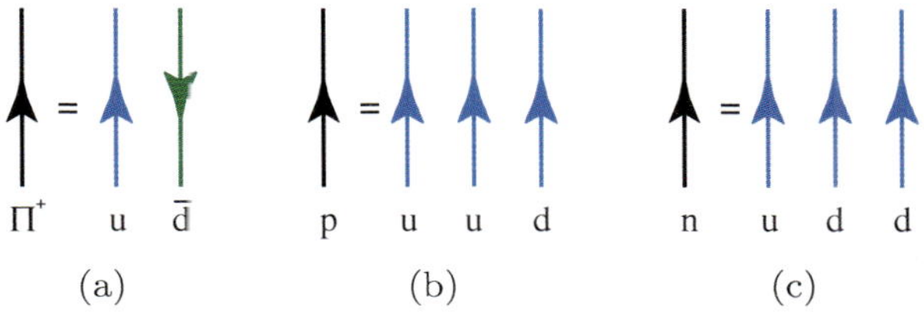

Fig. 11.25 (a) Quark constituents of the pion. (b) Quark constituents of the proton. (c) Quark constituents of the neutron.

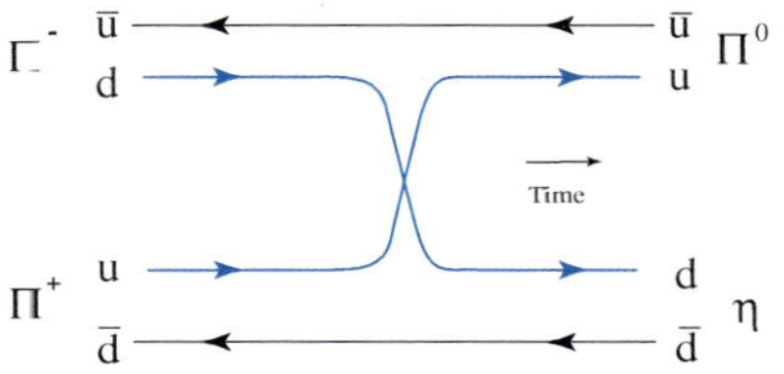

Fig. 11.26 Scattering of π^+ off a π^- producing π^0 and η.

The familiar neutron, shown in Fig. 11.25(c), has the following quark content:

$$\text{Neutron} : \mathrm{n} = \mathrm{udd}$$
$$Q\ :\ 0 = \frac{2}{3} - \frac{1}{3} - \frac{1}{3}; \quad B\ :\ 1 = \frac{1}{3} + \frac{1}{3} + \frac{1}{3}; \quad S\ :\ 0 = 0 + 0 + 0$$

and a strange baryon has the following quark content:

$$\text{Lambda} : \Lambda^0 = \mathrm{uds}$$
$$Q\ :\ 0 = \frac{2}{3} - \frac{1}{3} - \frac{1}{3}; \quad B\ :\ 1 = \frac{1}{3} + \frac{1}{3} + \frac{1}{3}; \quad S\ :\ -1 = 0 + 0 - 1.$$

The probability amplitudes for high energy hadron–hadron scattering are represented using Feynman diagrams in terms of the quarks and antiquarks that compose the hadrons. For all hadronic scattering processes the total number of quarks and antiquarks must be the same before and after collision.

Suppose we scatter a π^+ off a π^- — what do we expect? Since $\pi^+ = \mathrm{u\bar{d}}$ and $\pi^- = \mathrm{\bar{u}d}$, one possible reaction is that the pion and antipion simply exchange a u and a d quark, transforming themselves into a $\pi^0 = \mathrm{u\bar{u}}$ and an η meson given by $\eta = \mathrm{d\bar{d}}$; the Feynman diagram for this scattering process is given in Fig. 11.26.

In other words

$$\pi^+ + \pi^- \to \pi^0 + \eta.$$

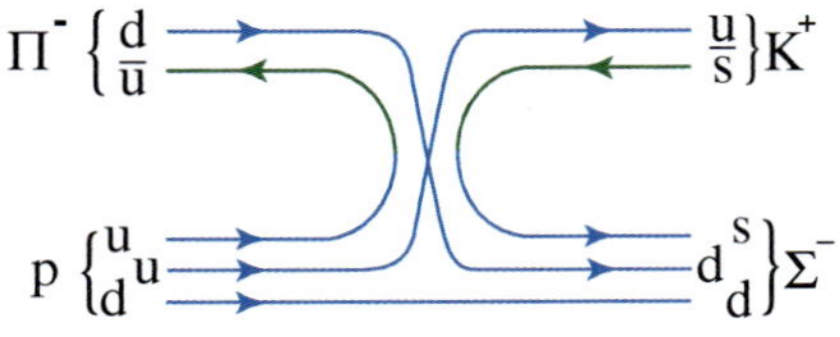

Fig. 11.27 Strong interaction of strange quarks.

For the above process, the quantum numbers work out as shown below.

$$\left.\begin{array}{ll} Q: & -1+1=0+0 \\ B: & 0+0=0+0 \\ S: & 0+0=0+0 \end{array}\right\} \text{Allowed!}$$

Consider the scattering of a pion off a proton in which a strange hadron is produced; this scattering is mediated by strong interactions since strangeness is conserved. For example

$$\begin{array}{rcl} & \pi^- + \mathrm{p} \to \mathrm{K}^+ + \Sigma^- & \\ & \mathrm{d\bar{u}} + \mathrm{uud} \to \mathrm{u\bar{s}} + \mathrm{dds} & \\ B: & 0+1 \to 0+1 & \\ Q: & -1+1 \to 1-1 & \\ \text{number of u}: & -1+1+1 \to 1 & \text{OK} \\ \text{number of d}: & 1+1 \to 0+2 & \text{OK} \\ \text{number of s}: & 0 \to -1+1 & \text{OK.} \end{array}$$

The Feynman diagram, given in Fig. 11.27 for this process, shows that the ū antiquark from the pion and a u quark contained in the proton annihilate on colliding. Furthermore, a strange quark s and its antiquark s̄ are produced from the vacuum; the strange anitquark s̄ combines with a u quark coming from the proton to form a strange meson $\mathrm{K}^+ = \mathrm{u\bar{s}}$; and the strange quark s combines with a d quark coming from the π^+ and another d quark coming from the proton to form a strange baryon $\Sigma^- = \mathrm{dds}$.

11.16 The Eight Fold Way

So far we have simply grouped all the hadrons into mesons (spin 0, 1...) and baryons (spin 1/2, 3/2...). The quark model succeeded in bringing order to the entire zoo of 'elementary' particles that appear in high energy collisions. The arrangement of mesons and baryons into a regular structure is a result of the fact that they are composed out of a quark and an antiquark and three quarks, respectively.

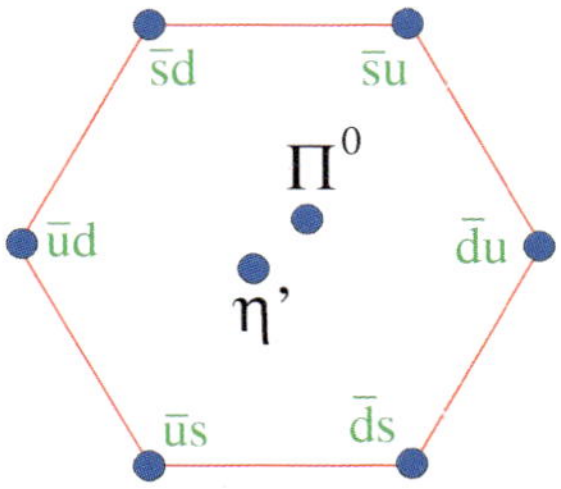

Fig. 11.28 Quark constituents of the meson octet **8**. $\pi^0 = (\bar{u}u - \bar{d}d)/\sqrt{2}$; $\eta' = (\bar{u}u + \bar{d}d - 2\bar{s}s)/\sqrt{6}$.

Since only three quarks, namely u, d, s, compose most of the low mass hadrons, for now we ignore the rest of the (heavy) quarks. The three quarks, taken together, are represented by the symbol **3**, called a *triplet*, and their antiquarks by the antitriplet denoted by $\bar{\mathbf{3}}$. These symbols stand for all the information contained in Fig. 11.24.

The mesons with the lowest masses are formed out of the bound states of a quark and an antiquark; they are formed by taking any one quark from the triplet **3** and combining it with any antiquark from the antitriplet $\bar{\mathbf{3}}$. This gives rise to nine possibilities that are further organized according to their charge and strangeness number, and represented as follows

$$\bar{\mathbf{3}} \otimes \mathbf{3} = \mathbf{8} \oplus \mathbf{1}.$$

The **8** mesons, called an octet of eight mesons, are the ones that are organized as a hexagon in Fig. 11.28, and the single **1** meson is called a singlet.[8] All the nine mesons, namely the octet **8** and the singlet **1**, have been experimentally detected.

To illustrate the composition of the mesons, note that the following *six* combinations are quite obvious, namely $\bar{s}d$, $\bar{s}u$, $\bar{u}d$, $\bar{u}s$, $\bar{d}u$ and $\bar{d}s$; to these six combinations one adds the following *two* combinations $\pi^0 = (\bar{u}u - \bar{d}d)/\sqrt{2}, \eta' = (\bar{u}u + \bar{d}d - 2\bar{s}s)/\sqrt{6}$ yielding the *eight* mesons that form the octet **8** of mesons shown in Fig. 11.28. The ninth combination, namely $\eta = (\bar{u}u + \bar{d}d + \bar{s}s)/\sqrt{3}$, is a meson that is a symmetric combination of the three quarks and antiquarks — which is orthogonal to the other eight mesons — and forms the singlet **1** meson.

Baryons are combinations of three quarks and to form the baryons one takes all possible combinations of one quark from each of the three triplets, and which yields 27 possible baryons; these baryons, based on their behavior under SU(3) transformations, are classified in the following manner

$$\mathbf{3} \otimes \mathbf{3} \otimes \mathbf{3} = \mathbf{10} \oplus \mathbf{8} \oplus \mathbf{8}^* \oplus \mathbf{1}. \tag{11.15}$$

[8]Lie groups are discussed in Noteworthy 14.1 and the quarks yield a representation of the SU(3) Lie group. Under 'rotations' (transformations) of the quarks and antiquarks by SU(3), the octet of mesons **8** transform amongst each other, with the singlet meson **1** remaining unchanged under such transformations.

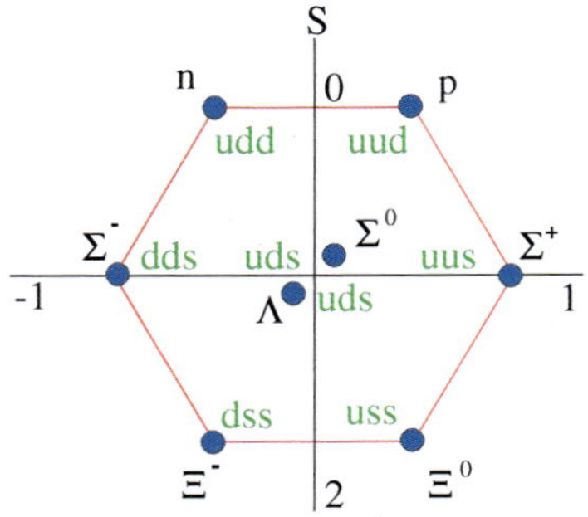

Fig. 11.29 The baryon octet. The horizontal axis is the so-called isospin quantum number.

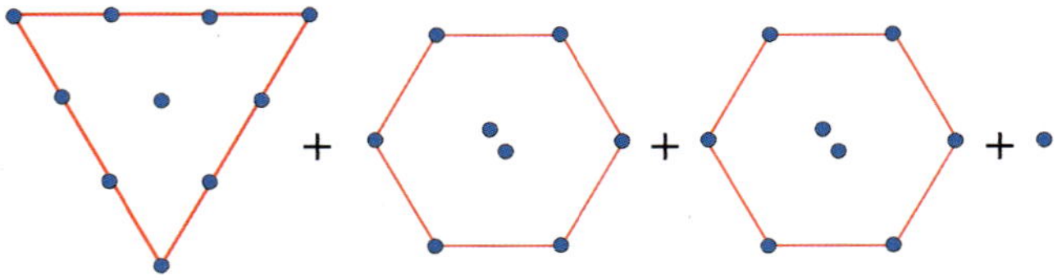

Fig. 11.30 Organization of low mass baryons in the quark model.

The result of composing three quarks, given by the right-hand side of Eq. (11.15), is shown in Figs. 11.29 and 11.30.

There are two distinct baryon spin 1/2 octets (hexagons) given by **8** and **8***, each consisting of eight baryons, with the baryon singlet given by **1**. The familiar proton (p = uud) and neutron (n = udd) appear in the **8** octet, given in Fig. 11.29, together with six other baryons. We discuss the baryon decuplet **10** in Sec. 11.16.1.

Consider only the mesons and baryons given in Figs. 11.28 and 11.29, respectively. The well known hadrons, which include the proton and neutron, are organized into two distinct (six-sided) hexagons — based on their electric charge and strangeness. Both the mesons and baryons occur in a hexagon pattern with eight hadrons in each hexagon. This classification of the mesons and baryons into hexagons is called the *eight fold way* (no connection with Buddhism!) because it is an octet of particles which form the hexagon.

The classification of mesons and baryons as bound states of quarks has tremendous predictive power, and is the repetition of the idea of the periodic table for atoms but at the subnuclear level. Similar to Mendeleev's periodic table, if any of the positions in the decuplet or hexagon are empty, then it is predicted that a hadron with those particular quantum numbers must exist. This was the case of the $\Omega^- =$ sss, given in Fig. 11.31, and which is made out of three strange quarks.

11.16.1 *The Omega Minus* Ω^-

There are 10 baryons with spin 3/2 which are organized into a decuplet **10**, as given in Fig. 11.31. Nine of these baryons were already known; however the last entry, namely the **Omega Minus** Ω^-, had not yet been discovered. Gell-Mann predicted that Ω^-, which is in the baryon decuplet, must exist and have the following quantum numbers: spin 3/2, $Q = -1$, $S = -3$ and with a mass which is about three times the mass of a strange quark, namely $3 \times 540\,\text{MeV}/c^2 = 1620\,\text{MeV}/c^2$; Ω^- also needed to have a long lifetime of 10^{-10} s since the decay of a strange quark is mediated by the slow acting weak interactions.

Experiments found the Ω^- particle in 1964, which was produced in the following process:

$$\begin{array}{ccccc} K^- + & p & \rightarrow \Omega^- & + K^+ & + K^0 \\ \bar{u}s & uud & sss & u\bar{s} & d\bar{s} \end{array}$$

with the particles resulting from the decay shown in Fig. 11.32. Note that strangeness number is conserved in the production of Ω^- as is required by

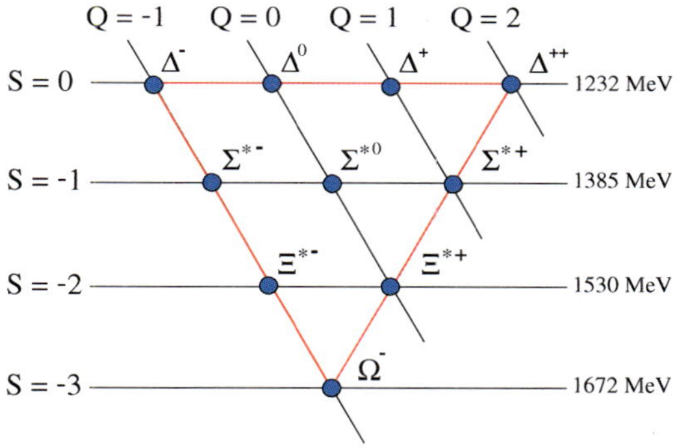

Fig. 11.31 The baryon decuplet.

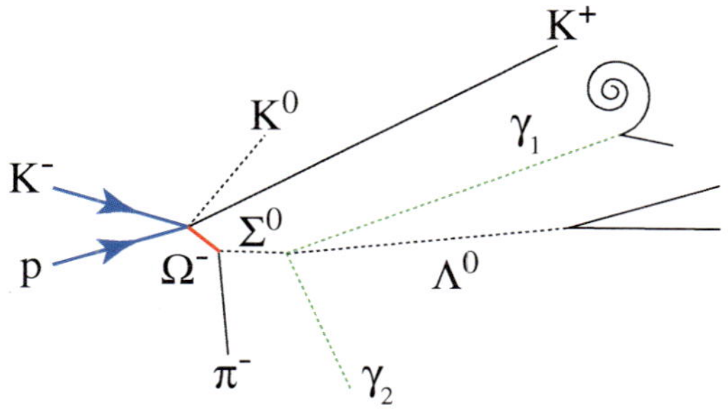

Fig. 11.32 The production and decay of the Ω^- baryon (shown by the red line).

strong interaction; in contrast, the weak process by which the Ω^- decays violates strangeness conservation. The Ω^- thus produced then decays, as shown in Fig. 11.32, via the following process

$$\Omega^- \rightarrow \Sigma^0 + \pi^-.$$

In terms of their quark content, the above weak interaction process is written as

$$\mathrm{sss} \rightarrow \mathrm{uss} + \bar{\mathrm{u}}\mathrm{d}.$$

The quantum numbers, lifetime and observed mass of $1680\,\mathrm{MeV}/c^2$ of Ω^- agreed with the predicted values![9] This discovery was a great triumph of the quark model since the Ω^- was found based on the prediction of Gell-Mann, and for this he was the single winner of the Physics Nobel Prize in 1969 — a rare honor even amongst Nobel laureates.

11.17 Experimental Evidence for Quarks

An isolated quark has never been observed. What evidence do we have that quarks actually exist inside the nucleus? That a proton for example is the bound state of quarks?

The proton is considered to be a bound state of uud quarks. One way of detecting the presence of quarks inside the proton is to follow the footsteps of Rutherford that led to the discovery of the nucleus inside the atom. He accelerated alpha particles (two protons and two neutrons) to high energies and a beam of such alpha particles was made to hit a thin film of gold, resulting in the incident alpha particles being scattered in all directions. Rutherford's atomic experiment is shown in Fig. 11.33, where some of the alpha particles are seen to scatter through *wide angles*, with a few alpha particles even back-scattering and completely reversing their direction of motion.

The wide angle scattering cannot be caused by a smooth distribution of charge inside the gold nucleus since such a charge distribution would only cause a small deviation in the paths of the alpha particles. The wide angle scattering can be

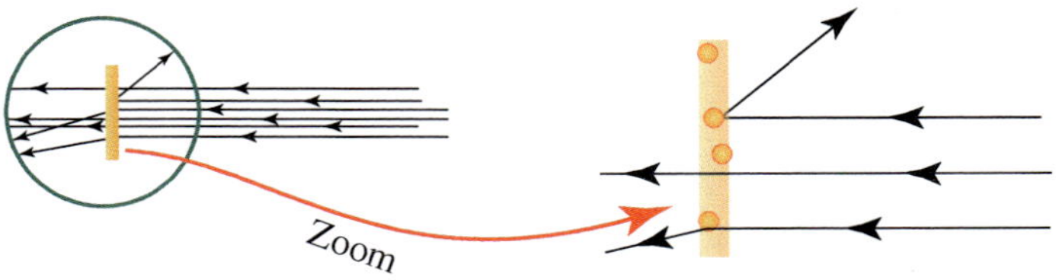

Fig. 11.33 Wide angle scattering of alpha particles.

[9]The observed mass is within 3% of the expected value.

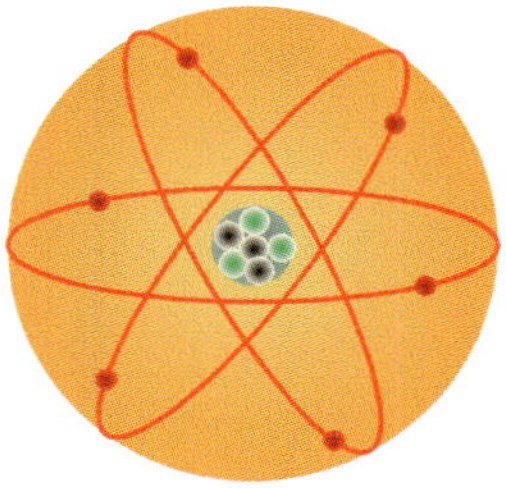

Fig. 11.34 Structure of atom.

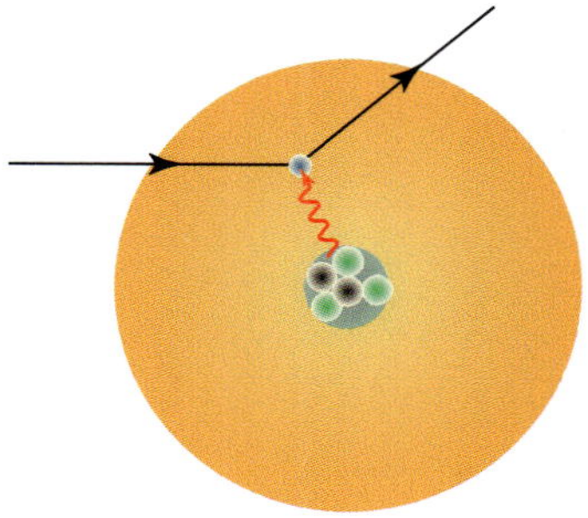

Fig. 11.35 The scattering of electrons off a nucleus.

accounted for by a concentration of positive charge occupying only a small volume, which was then identified as the nucleus of the atom. This led to the discovery of the structure of the atom, as shown in Fig. 11.34, as consisting of a tiny nucleus having a positive charge and being surrounded by negatively charged electrons bound to the nucleus through the Coulomb interaction.

The existence of the quarks in the early 1970's was established using reasoning identical to that followed by Rutherford, but carried out at much higher energies and at the nuclear and not the atomic scale. Electrons at very high energies were scattered off a proton target, called **deep inelastic scattering**, as shown in Fig. 11.35. The distribution of wide angle scattered electrons led to the conclusion that the proton is constituted of point-like quarks. The charges of the proton's constituents were shown to be 2/3 and $-1/3$ the charge of the electron, as predicted by the quark model and that all quarks are spin 1/2 fermions. The high energy collisions of hadronic matter led to the production of many other hadrons, and all the quantum numbers of the quarks were determined by the decay modes of the hadrons produced.

11.17.1 *Quark jets*

Even though quarks cannot be seen as isolated particles in any experiment, there are many experiments similar to wide angle scattering, in which indirect evidence of the existence of quarks can be obtained. A particularly clean signal for the existence of quarks is obtained from the so-called two jet experiments. Consider very high energy head on collisions of electrons e^- and antielectrons e^+ that annihilate into a quark and an antiquark, as shown in Fig. 11.36.

The reaction has the form

$$e^- + e^+ \to \bar{q} + q \to \text{two jets of hadrons.}$$

Due to the nature of strong interactions, the quark and antiquark produced in the e^-, e^+ collision, namely $\bar{q}$ and q, are converted to ordinary hadrons, called hadronization, by secondary processes; the time scale for the formation of jets is about 10^{-23} s, which is the typical time scale for strong interactions and discussed in

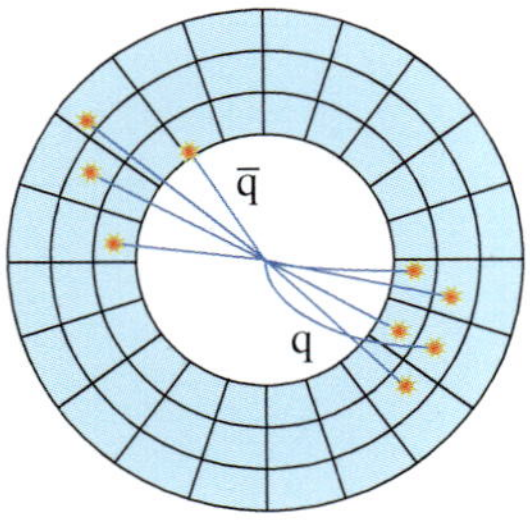

Fig. 11.36 High energy collisions of electrons and positrons giving rise to two jets of quarks and antiquarks.

Sec. 12.3. What appears in the particle detector, as shown in Fig. 11.36, are two separate 'jets' of various particles, formed as the final state of the two quarks produced in the high energy collision. Quark jets have been experimentally detected in CERN.

11.18 The Answer: Three Generations of Particles

The answer from particle physics to the ancient question as to what constitutes all of matter around us is the following: all tangible matter is composed out of quarks and leptons. As of 2014, experiments indicate that the leptons and quarks have no structure down to a size of 10^{-18} m. These are fundamental and elementary point-like particles that are described by their quantum numbers and come in six different varieties, called flavor.

There is a remarkable fact that there are an equal number of leptons and quarks — organized in three generations, with each generation (family) being a pairing of two quarks and two leptons; the three generations of fermions are shown in Fig. 11.37. An explanation for the pairing of quarks and leptons,

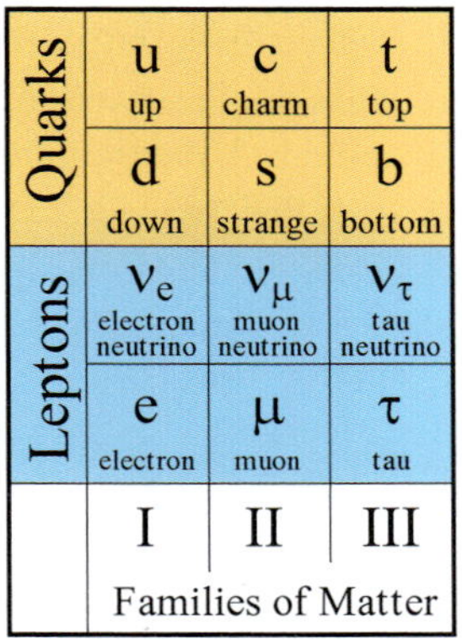

Fig. 11.37 Three families of quarks and leptons.

based on the requirement of consistency of weak interactions, is discussed in Chapter 13.

When the quark model was proposed in the 1960's, Gell-Mann stated that the quarks need not be physical objects, but rather can be thought of as a mathematical scheme for classifying elementary particles. However, it was discovered by the 1970's that the quarks are indeed physical objects — point particles carrying energy and momentum, and having all the quantum numbers attributed to them by the quark model. Quarks are enigmatic since, unlike the leptons, they are never observed in isolation and are permanently confined inside the nucleus. We discuss this enigma in Chapter 12.

Chapter 12

Fundamental Interactions

What holds matter together?

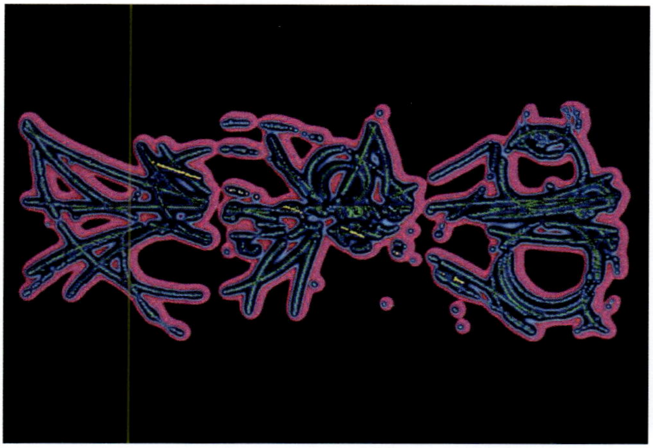

12.1 The Question

If we take matter to be constituted by quarks and leptons, the question naturally arises as what holds them together to make nuclei, atoms, molecules and large conglomerations of matter. In other words, what holds matter together?

12.2 Interactions in Nature

The constituents of matter are held together in a variety of bound states. The bound states of the quarks give rise to the proton and neutron — which, in turn, give rise to the nuclei of various elements. The electrons are in bound states with nuclei (composed of protons and neutrons) forming the electrically neutral atoms of the periodic table. Furthermore, the attractive force of gravity causes large masses of matter to exist in bound states, such as the solar system and the galaxy composed of stars.

All the bound states of matter result from the forces between the constituents of matter; the forces that hold the constituents in bound states are themselves special forms of energy that exist in their own right, and with unique and remarkable properties. We now examine more closely these forces that we have encountered in previous chapters.

Forces are also called **interactions**. Although the term 'force' may evoke Newtonian concepts, it is a term that denotes interactions, which are both relativistic and quantum in nature, between the different elementary particles that have been discussed in Chapter 11.

The *four fundamental interactions* between the constituents of matter are the following:

- **The electromagnetic interactions**
- **The weak interactions**
- **The strong interactions**
- **The gravitational interactions**

All other forces that appear in Nature are ultimately reducible to these four fundamental forces. In the broadest classification of all forms of physical entities, the constituents of matter, namely quarks and leptons, are *fermions* and it will be shown that all the fundamental forces of Nature are *bosons*.

The electromagnetic and gravitational forces are familiar from classical physics and from our direct perception of the world around us. We have vision due to our perception of electromagnetic waves and we feel our weight due to gravity.

The strong force is required to confine the quarks into bound states of protons and neutrons that form the nuclei of all atoms. Recall that the nucleus being composed of protons and neutrons has a net positive charge with a very short distance separating the protons; the positive charges do not fly apart since another force *stronger* than Coulomb repulsion binds the protons in the nucleus, namely the strong force.

The existence of weak interactions is a bit more difficult to observe. Neutrons are stable inside the nucleus but are unstable when isolated outside a nucleus, with a lifetime of 882 s. The decay of the neutron, along with strangeness changing decays of hadrons, as discussed in Sec. 11.14, are so slow that only a weak force (and not the electromagnetic or strong force) can account for it. The weak force also describes the (non-electromagnetic) interactions of leptons with each other and with the quarks.

Gravitation — as discussed in Chapter 4 — is not a force in the Newtonian sense, but instead is a manifestation of the curvature of spacetime; in contrast, the other three interactions are completely described by quantum mechanical processes. We will return to analyzing the force of gravity in Chapter 15 and for now, the focus is on the remaining three forces, namely the strong, electromagnetic and weak forces — all of which are explained by relativistic quantum field theory.

12.3 Strength and Duration of Interactions

The forces in Nature have their own **characteristic strength** and duration of the interaction — called the time scale of the interaction. A summary of these characteristics is given in Table 12.1.

The strength of an interaction is an intrinsic property and is given by its coupling constant that yields the *dimensionless numbers* $\alpha_s, \alpha_e, \alpha_w$ for the strong, electromagnetic and weak interactions, respectively. The duration of the interaction, namely the time taken for the interaction to be completed, is related to both the strength of the interaction as well as the (microscopic) distances over which these fundamental interactions operate.

As shown in Table 12.1, the interactions vary greatly in strength, from $\alpha_s \approx 1$ for strong interactions to $\alpha_w \approx 10^{-6}$ for weak interactions. One consequently expects that the explanation for the three interactions, namely strong, electromagnetic and weak, should be quite different from each other.

To quantify the duration of the strong interactions, consider the decay of a Δ^{++} particle, composed of three up quarks, namely uuu, into a proton and a pion, by a process that is mediated by the strong interaction — and is shown in Fig. 12.1. The volume over which the strong interaction takes place is denoted by the red blob; due to the interaction, a u quark is ejected from the Δ^{++} and a $\mathrm{d}, \bar{\mathrm{d}}$ pair is created from the energy of the interaction, leading to a final state of a proton uud and a pion π^+ given by $\mathrm{u}\bar{\mathrm{d}}$.

Table 12.1 Summary of the four fundamental interactions of Nature.

Force (interaction)	Relative strength	Range of force	Time scale for interaction	Mediating particle
Strong	$\alpha_s = 1$	short ($\approx 10^{-15}$ m)	10^{-23} s	Gluon
Electromagnetic	$\alpha_e = 10^{-2}$	∞	10^{-18} s	Photon
Weak	$\alpha_w = 10^{-6}$	short ($\approx 10^{-18}$ m)	10^{-10}–10^{3} s	$\mathrm{W}^{\pm}, \mathrm{Z}^0$ bosons
Gravitational	10^{-39}	∞	—	(Graviton)

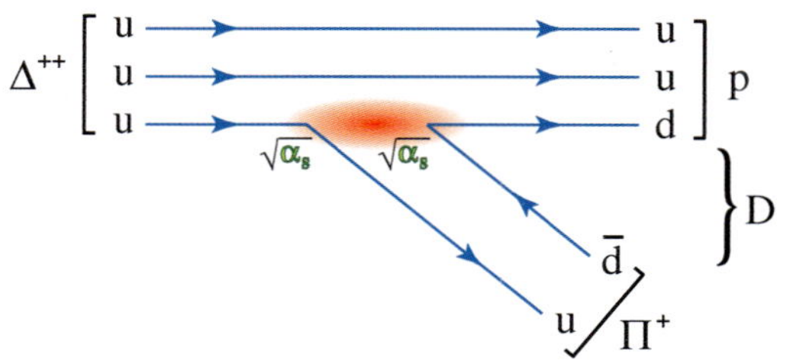

Fig. 12.1 Δ^{++} decays into a proton and a pion. The time scale of strong interactions is determined by distance $D \simeq 10^{-15}$ m.

The Feynman diagram (introduced in Sec. 2.9.1 and shown in Fig. 2.11) that describes the **probability amplitude** for the decay process of Δ^{++} has a coupling constant $\sqrt{\alpha_s}$ at each point (vertex) where the strong interaction occurs. Strong interactions occur at each of the vertices, namely the point where the up quark leaves Δ^{++} and the point where pair creation of $\mathrm{d}, \bar{\mathrm{d}}$ occurs.

The strong interactions act over a distance that is the size of the nucleus, which is about 10^{-15} m; the duration for which the strong interactions are operational is the time it takes the quarks to be separated by a distance greater than the size of a nucleus, and hence is the time it takes for the proton and pion to physically separate to a distance greater than 10^{-15} m. The quarks are moving close to the velocity of light; hence the time τ for the separation of the proton and pion to about the distance of $D \simeq 10 \times 10^{-15}$ m, as shown in Fig. 12.1, is about the time it takes for light to cross the distance D and yields

$$\tau = \frac{D}{c} = \frac{10^{-14}}{3 \times 10^{9}}\ \mathrm{s} \approx 10^{-23}\ \mathrm{s}.$$

The *probability* for the process is the modulus square of the Feynman diagram and hence, for the diagram given in Fig. 12.1, is proportional to α_s^2. The time taken for the separation, namely τ, is given by the probability amplitude for the process, namely α_s^2, multiplied by τ_s — the duration of time for which the strong interaction acts on the quarks. Hence, the time scale of strong interactions τ_s is given by

$$\tau = \alpha_s^2 \tau_s \approx 10^{-23}\ \mathrm{s}. \tag{12.1}$$

From Table 12.1, $\alpha_s \approx 1$. Hence, unstable particles whose decays are mediated by the strong interactions have a lifetime τ_s, of about 10^{-23} s.

Noteworthy (optional content) 12.1: Stability of the Proton

One may wonder why the proton does not decay, and that too in 10^{-23} s? The reason is baryon number conservation; the proton is the lightest baryon and hence, due to energy conservation, it can only decay into another baryon having a lower mass; in the absence of such a baryon, the proton is stable. There were attempts in the 1980's to model the decay of the proton by violating baryon number conservation, discussed in Sec. 14.2.2, but experiments have ruled out all such models.

Recall that the neutron decays at a time scale much greater than that of strong interactions because its decay is mediated by the weak interactions, which has a much longer duration of interaction as compared to the strong interactions.

For electromagnetic interactions involving an electric charge e the strength of the interaction is given by

$$\alpha_e = \frac{e^2}{\hbar c} \approx \frac{1}{137} \tag{12.2}$$

where c is the velocity of light and $\hbar$ is Planck's constant divided by 2π.[1] The time scale for the electromagnetic interaction is hence given, using a reasoning similar to Eq. (12.1), by the following.

$$\begin{aligned} \frac{\tau_e}{\tau_s} &= \frac{\alpha_s^2}{\alpha_e^2} \Rightarrow \tau_e = \frac{\alpha_s^2}{\alpha_e^2}\tau_s \\ \Rightarrow \tau_e &= (137)^2 \times 10^{-23}\,\text{s} \approx 10^{-16}\text{–}10^{-18}\,\text{s}. \end{aligned} \tag{12.3}$$

The range for the electromagnetic interaction reflects different time scales depending on the process. For example, a neutral pion, namely π^0, is a bound state of the $\mathrm{u\bar{u}}$ and of the $\mathrm{d\bar{d}}$ quarks (more precisely $\pi^0 = (\mathrm{u\bar{u}} - d\bar{\mathrm{d}})/\sqrt{2}$) and decays into two photons, namely

$$\pi^0 \to 2\gamma$$

via the electromagnetic interaction in about 10^{-16} s. Chemical reactions mediated by the electromagnetic interaction, for instance the absorption of light in photosynthesis, take about 10^{-14} s to be completed.

We can use the time scale of the decay of strange hadrons to estimate the strength of weak interactions. From Sec. 11.14, the lifetime of a strange hadron yields the time scale of weak interactions given by $\tau_w \approx 10^{-10}\,\text{s}$; hence, the weak interaction strength α_w is given by

$$\frac{\alpha_w^2}{\alpha_s^2} = \frac{\tau_s}{\tau_w} = 10^{-13} \Rightarrow \alpha_w \approx 10^{-6}. \tag{12.4}$$

The time scale of weak interactions, in fact, has a wide range, given by $10^{-10}\text{–}10^3\,\text{s}$; this is because the strength of coupling α_w varies with energy and will be discussed in the derivation of Eq. (12.14).

12.4 Quantum Electrodynamics

The quantum theory for one of the most important fundamental interactions, namely the electromagnetic force mediating the interactions of electrons and antielectrons, is quantum electrodynamics; this theory describes the interaction of photons with charged elementary particles in general, and which includes the quarks and leptons. Most of chemistry and biology, and most of atomic and many body physics, is explained by the interaction of photons, electrons and protons.

[1] For brevity, in all the Feynman diagrams that follow, the units are taken such that $\hbar = 1 = c$ and hence $\alpha_e = e^2$ in appropriate units.

We study **quantum electrodynamics** in some detail since the other two subnuclear interactions, namely weak and strong interactions, have a very similar structure: all three interactions are described by vector bosons that are gauge fields obeying gauge symmetry, discussed in Sec. 11.6. There are many significant differences as well, and these will be addressed subsequently.

The electromagnetic force, together with gravitation, is the most well known of the forces, and can be experienced directly by our five senses and has been briefly discussed in Chapter 2 on fields. The lighting that one sees is a manifestation of the electric field, and magnets are the manifestation of the magnetic field; taken together the electric and magnetic fields constitute the electromagnetic field, which was discovered by Maxwell in the 1860's. As mentioned in Sec. 11.6, the electromagnetic force of interaction is carried by the Maxwell gauge field.

Maxwell's field is a classical field that obeys the laws of special relativity, although it was discovered almost forty years before the special theory of relativity was formulated in 1905. When Maxwell's field is combined with quantum mechanics and is coupled to the relativistic electron field, one obtains (relativistic) quantum electrodynamics. As is the case for all quantum field theories, the fundamental excitations of the field are quantized, and the particle that carries one quantum of energy of the Maxwell field, for a given momentum, is the photon; the lowest energy excitation of the *electron field* is the single electron or positron (antielectron).

The photon is a fundamental elementary particle having zero mass, is its own antiparticle and travels at the speed of light; it is a spin 1 particle and hence is a boson. The photon, like other particles, has a momentum $\mathbf{p}$, which in turn is related to the wavelength of the photon. For the photon, the energy and momentum are given by the following

$$E = pc = h\nu; \quad p = |\mathbf{p}|$$

$$\lambda = \frac{c}{\nu} = \frac{hc}{E}.$$

The photon couples to any charged particle — the strength of the coupling is given by the electric charge. The dimensionless coupling of photons to electrons is given by α_e, as in Eq. (12.2). The photon does not couple to itself since it is electrically neutral, having zero electric charge.

12.4.1 *Photons and electrons*

The interaction of photons with electrons is best represented using **Feynman diagrams** — for which purpose these diagrams were invented in the first place — and which are discussed briefly in Sec. 2.9.1.

Let us indicate the propagation of a photon by a wavy line, and the propagation of an electron by a straight line, as shown in Fig. 12.2. The interaction of a photon with an electron is given by a vertex of the wavy and the straight line; shown in

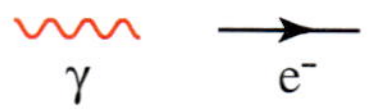

Fig. 12.2 Feynman diagrams for the photon and electron.

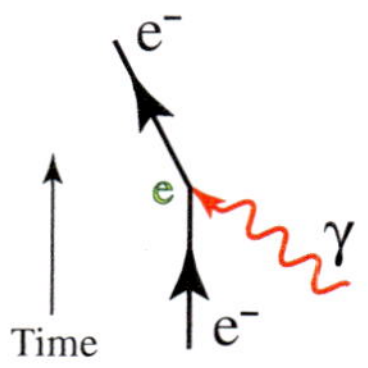

Fig. 12.3 Feynman diagram for photon absorption.

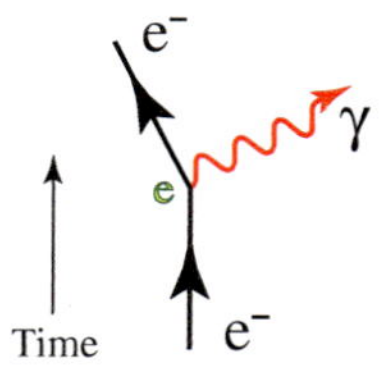

Fig. 12.4 Feynman diagram for photon emission.

Fig. 12.3 is a Feynman diagram for the photon that meets an electron at a spacetime point. A factor of electric charge $e = \sqrt{\alpha_e}$ is placed at every vertex where the photon line meets the electron line, measuring the strength of the electromagnetic interaction. An arrow pointing upwards on a straight line is a fermion propagating forward in time; antifermions propagate backwards in time, with the arrow pointing downwards. Fermion lines do not end, showing the conservation of the number of fermions.

The photon has the remarkable property that the number of photons at any moment is not constant; namely, the number of photons is indeterminate. For this reason photons can be absorbed and emitted by electrons, as illustrated in Figs. 12.3 and 12.4, respectively.

The interpretation of a Feynman diagram depends on how the diagram is laid out in time. In Fig. 12.3 an electron absorbs a photon and the photon imparts its energy and momentum to the electron: thus the kink in the electron line; Fig. 12.4 is the converse of this process, in which the electron emits a photon imparting energy and momentum to the photon; in Fig. 12.5 an electron and an antielectron meet and annihilate into a photon that carries the energy and momentum of the annihilating pair. Figure 12.6 shows a photon disintegrating into a pair of electron and antielectron. The basic Feynman diagrams represent virtual processes, which are mathematical processes that appear in the underlying theory of quantum electrodynamics. For the emission and absorption of a physical (observable) photon,

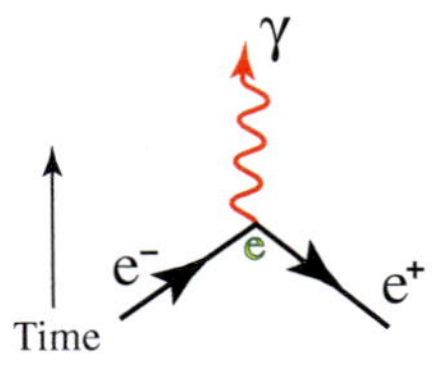

Fig. 12.5 Annihilation of electron and positron.

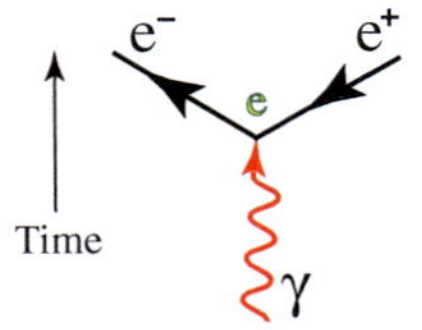

Fig. 12.6 Creation of electron and positron from photon.

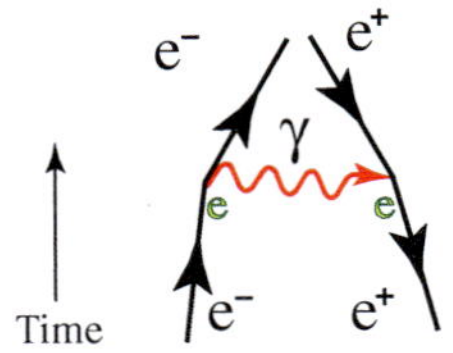

Fig. 12.7 Electron–positron interaction.

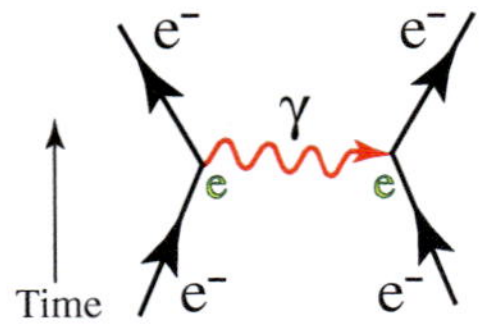

Fig. 12.8 Electron–electron interaction.

energy and momentum need to be conserved and this puts additional constraints on how to apply and interpret the Feynman diagrams, as in Fig. 2.12.

One can see that the photons in fact mediate the *interaction* between charged particles. Consider the Feynman diagram given in Fig. 12.7 for the probability amplitude that an electron and antielectron approach each other, with the electron emitting a photon that in turn is absorbed by the antielectron. On computing this probability amplitude, one can show that an attractive force is exerted between the oppositely charged particles, and the Feynman diagram shows this attraction. Similarly, Fig. 12.8 is the Feynman diagram for the amplitude that an electron will emit a photon that is absorbed by another electron; a calculation shows that there is repulsive force between the two similarly charged particles, and the Feynman diagram shows that the two electrons scatter away from each other.

Feynman diagrams are the building blocks for describing the scattering of quarks and leptons via the fundamental interactions. As one can imagine, the basic vertices given in Figs. 12.3 and 12.4 can be combined to form Feynman diagrams of arbitrary complexity.

The Feynman diagrams for quantum electrodynamics will be seen to be repeated in weak and strong interactions, but with many new features and structures added to the basic Feynman diagram.

12.5 Renormalization

One can have a process, with Feynman diagram given in Fig. 12.9, in which a photon first disintegrates into a virtual electron–antielectron pair, and the electron–antielectron pair then recombines back into the photon! Similarly, a free electron can emit a virtual photon and then re-absorb it, as in Fig. 12.10.

These virtual processes can be of higher and higher order of complexity, as shown in the diagrams in Figs. 12.11 and 12.12, and can be seen to be composed of many loops within loops. Each loop contributes a factor of α_e to the Feynman diagram, and hence one can see that the larger the number of loops the higher the power of α_e and hence the smaller the contribution, since $\alpha_e \simeq 1/137$.

What do these virtual processes mean? The loop diagrams, as they are called, show that once two quantum fields are coupled, it is meaningless to talk of either of them in isolation and what is observed is the total quantum state of the coupled

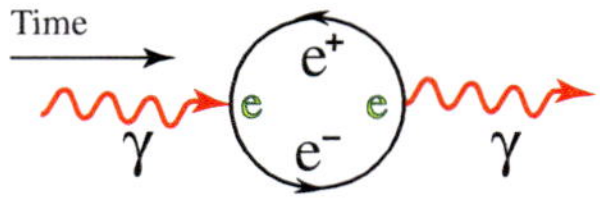

Fig. 12.9 Photon, transforming into an electron and antielectron pair, and back to a photon.

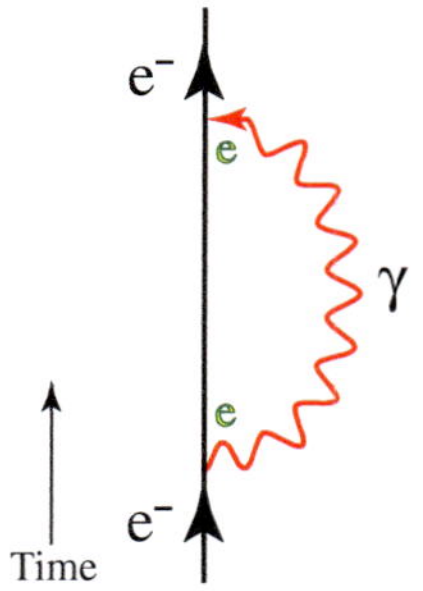

Fig. 12.10 An electron propagating in space, with virtual interactions with the photon.

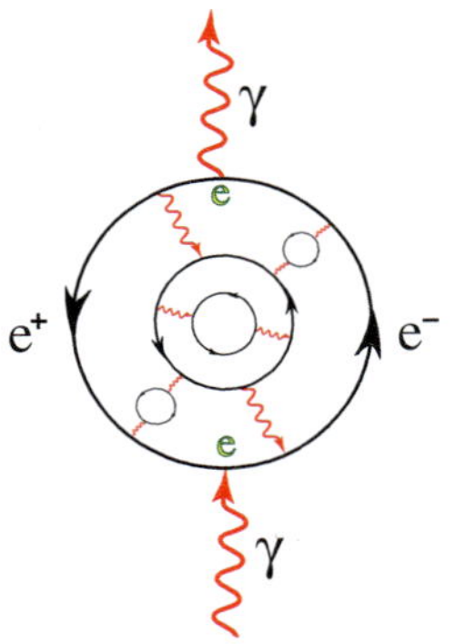

Fig. 12.11 Higher order virtual process of an electron interacting with a photon.

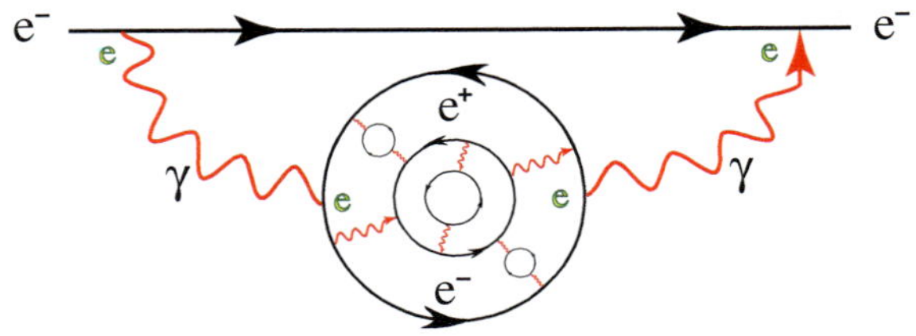

Fig. 12.12 Further expansion of virtual processes of a propagating electron.

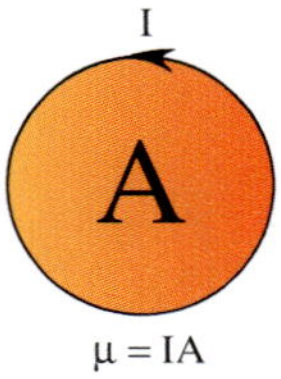

Fig. 12.13 Magnetic moment of a loop.

quantum fields. In particular, photons and electrons redefine, technically called renormalize, each other. A physically observed electron or photon is what results from this process of **renormalization**. The deep and intuitive explanation of quantum field theory — including the concept of renormalization — was given by K. G. Wilson, and earned him the Nobel Prize in Physics in 1982.

To understand the significance of interaction and renormalization, consider the concept of the *magnetic moment* of the electron. For a classical current I flowing in a loop and enclosing area A, as shown in Fig. 12.13, the magnetic moment μ of the circuit is given by

$$\mu = IA.$$

In Dirac's relativistic theory of free electrons (that is, *not interacting* with photons), the electron has an *intrinsic* spin and which results in a magnetic moment for the non-interacting case being given by

$$\mu_e^0 = 2\mu_B \quad \text{and} \quad g_0 = \frac{\mu_e^0}{\mu_B} = 2.$$

where μ_B is the unit of the **Bohr magneton** given by $\mu_B = e\hbar/2m_e c = 9.274 \times 10^{-24}\,\text{Am}^2$, and m_e is the mass of the electron.

When the electrons are coupled to photons, the magnetic moment of an electron moving freely (not in a bound state) is renormalized by its interactions with the photon field. The experimentally observed value of the electron's magnetic moment is *changed*, due to its interactions with the photon given in Figs. 12.10, 12.11 and 12.12, from its non-interacting value of μ_e^0 to its experimentally observed value of μ_e, known as the Lamb shift, and is given by

$$g = \frac{\mu_e}{\mu_B} = 2.002319304377(4). \tag{12.5}$$

The value of g has been computed to eight loops and hence to eighth order in the parameter α_e; its theoretical value is given by[2]

$$(g-2)/2 = 1159652175.86(0.10)(0.26)(8.48) \times 10^{-12}. \tag{12.6}$$

The agreement of theory with experiment for quantum electrodynamics is up to the eleventh decimal point. Quantum electrodynamics is currently the most accurately verified theory in science; although every theory in science is ultimately expected to have a limit to its accuracy, this limit has not yet been reached for quantum electrodynamics.

12.5.1 *The hydrogen atom revisited*

Now that we have a theory of electrons interacting via photons, we re-consider how the hydrogen atom is formed. In the derivation using the concept of a potential, the (non-relativistic) Schrödinger equation is solved with the electron moving in the Coulomb potential of the proton (nucleus), and results in the electron being in a bound state with the proton. This derivation is only an approximation to the full story.

From the point of view of quantum electrodynamics, the bound state of the electron and proton is due to the continual exchange of photons between the proton and the electron. Figure 12.14(a) shows the Feynman diagrams for the hydrogen

[2] Reference T. Kinoshita, arXiv:hep-ph/0507249.

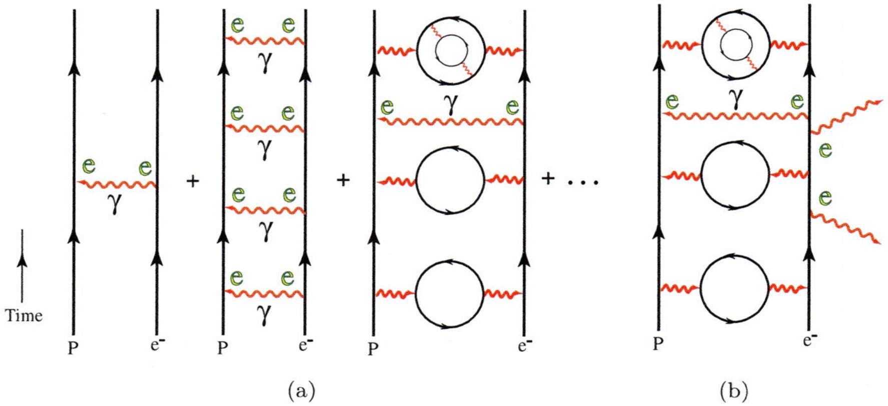

Fig. 12.14 (a) Hydrogen atom. (b) Typical absorption and emission of a photon by an electron of the hydrogen atom.

atom as an expansion in the coupling constant $e = \sqrt{\alpha_e}$. The absorption and subsequent emission of a photon by the electron in the hydrogen atom is shown in Fig. 12.14(b).

The full calculation of the hydrogen atom using quantum electrodynamics shows that the average velocity of the electron in the hydrogen atom $v \approx \alpha_e c$; since $\alpha_e \approx 1/137$, we see that the electron moves fairly slowly inside the hydrogen atom, in effect making the velocity of the electron non-relativistic. This is the reason that the non-relativistic Schrödinger equation can be used to study the quantum mechanics of most of the atoms.

For atoms with high atomic number, the inner shell electrons are moving at relativistic speed and quantum electrodynamics has to be used for performing accurate calculations. The yellowish color of gold is due to the effect of the relativistic motion of the gold atom's inner shell electrons.

12.6 Photons in a Plasma

We have so far discussed the properties of photons and electrons in vacuum, that is, in the complete absence of any kind of matter. One can also study how photons behave inside bulk matter. The special case of the behavior of photons in a plasma is an important and interesting example of how interactions are modified by the medium in which it takes place.

A **plasma** is a collection of an equal number of positive and negative charges in which the charges are free to move. One can for example consider the special case of a plasma consisting of positively charged atoms occupying equilibrium positions on a regular lattice, and with the electrons being free to move around.[3]

One measure of the plasma is the number of electrons per unit volume, denoted by n; an equivalent measure is r_s, which is the volume occupied by each electron, which is proportional to $1/n^{1/3}$. Many metals such as copper, aluminum and magnesium, which conduct electric currents, are modeled quite well by considering their positively charged nuclei to be occupying lattice sites and with their 'valence' electrons being free to move and form a plasma.

Consider a positive charge located in a vacuum; the (electrostatic) Coulomb potential ϕ at a distance r from the positive charge is given by $\phi = e/r$; the long range of the electrostatic potential is due to the photon being massless. Now consider a positive charge in a plasma; the electrons will be attracted to the positive charge and will move around until the positive charge is *screened.* This is called Debye screening: the **screened Coulomb potential** due to the positive charge is *short ranged.* The Coulomb potential and its screened case are both shown in Fig. 12.15. For metals the screened Coulomb potential is given by

$$\phi_s = \frac{e}{r} e^{-k_s r}; \quad k_s = \left(\frac{2.95}{r_s/a_0} \right) 10^{-10}\,\mathrm{m}^{-1} \tag{12.7}$$

where the Bohr radius $a_0 = \hbar^2/m_e e^2 = 5.292 \times 10^{-10}\,\mathrm{m}$. The screened Coulomb potential is shown in Fig. 12.15. For copper, the screening wave vector is given by $k_s = 0.55 \times 10^{-10}\,\mathrm{m}^{-1}$.

The Debye potential has a finite mass m_γ (or equivalently a Debye length which is proportional to the inverse of the Debye mass) which is the **effective mass** of

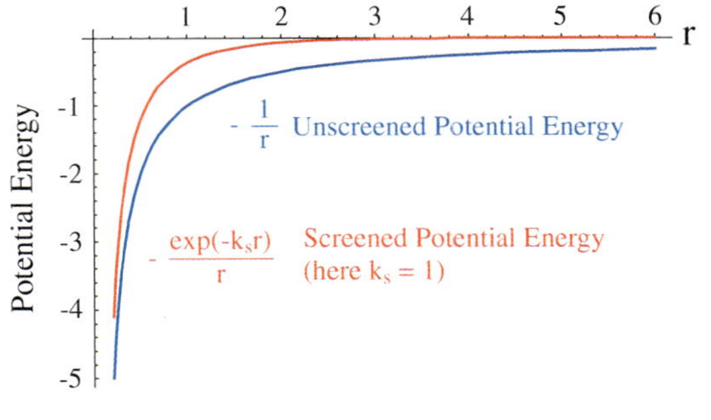

Fig. 12.15 Screened Coulomb potential.

[3] A plasma of greater generality, in which both the positive and negative charges are free to move as is the case with all stars including our Sun, is more difficult to analyze and will not be discussed any further.

the photon in the plasma. From Eq. (12.7)

$$\phi_s = \frac{e}{r} e^{-(m_\gamma c/\hbar)r} \Rightarrow m_\gamma = \frac{\hbar k_s}{c}.$$

For copper, the effective mass of the photon is given by $m_\gamma = 1.998 \times 10^{-35}$ kg $\simeq 2.19 \times 10^{-3} m_e$.

The discussion on the screened Coulomb potential illustrates the fact that — although the fundamental particle, namely the photon, is massless — an effective interaction can arise such that the photon has an effective mass m_γ due to the medium that it is in. The screened Coulomb potential is similar to the Yukawa potential given in Eq. (11.9); both of them are effective potentials, the screened Coulomb potential arising from the photon being in the medium of a plasma and the nuclear Yukawa potential being the manifestation of the strong (gluonic) interactions taking place inside the medium of the nucleus.

The photon inside the plasma continues to be described by a (transverse) vector potential that is gauge invariant and yields a zero mass for the photon. Similar to the screened Coulomb potential, the **Yukawa potential**, discussed in Sec. 11.10, is the effective interaction of protons and neutrons inside the nucleus that can be described by the scalar pion quantum field, which approximately represents strong interactions (gluonic) inside the nucleus.

12.7 Electroweak Interactions

The two fundamental processes that one addresses in weak interactions are the decay of a neutron and the decay of strange hadrons. Since the leptons do not form the nucleus of any stable 'leptonic atoms', the most interesting interaction of the leptons is with the quarks.

Recall, from Table 11.2, lepton number $L_e = 1$ for e^- and $L_e = -1$ for $\bar{\nu}_e$. Neutron decay is given by

$$\mathrm{n} \rightarrow \mathrm{p} + \mathrm{e}^- + \bar{\nu}_e. \tag{12.8}$$

Note that charge, lepton and baryon numbers are conserved in the above process.

In the Fermi theory of weak interactions, the process given in Eq. (12.8) takes place at a single spacetime point, as shown in Fig. 12.16, with strength of coupling given by Fermi's constant G_F. Fermi's theory, however, turns out to be mathematically inconsistent.[4]

In terms of quarks that compose all the hadrons, a neutron is the bound state of udd quarks, and the decay of a neutron into a proton, namely uud, is the result

[4]It can be shown that $G_F = \frac{\alpha_w}{4\sqrt{2}m_w^2}$, where m_w is the mass of the W boson, discussed later.

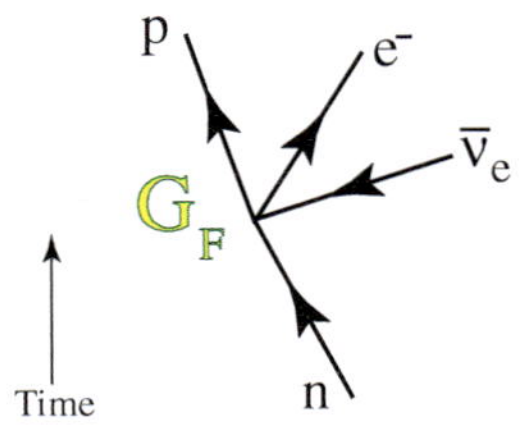

Fig. 12.16 Fermi's theory of weak interaction. G_F is Fermi's coupling constant. This theory is not correct, see Sec. 12.9.

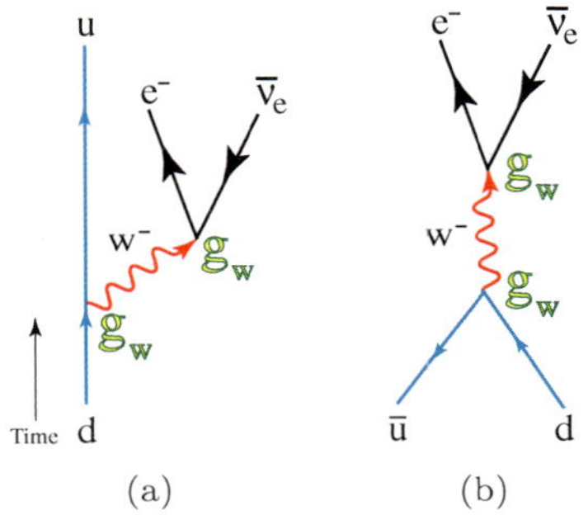

Fig. 12.17 (a) Change of quark from d to u. (b) Quark–antiquark annihilation into leptons.

of a d quark changing to a u quark, shown in Fig. 12.17(a). In terms of underlying quarks

$$\mathrm{d} \rightarrow \mathrm{u} + \mathrm{e}^- + \bar{\nu}_e \Rightarrow \mathrm{d} + \bar{\mathrm{u}} \rightarrow \mathrm{e}^- + \bar{\nu}_e. \tag{12.9}$$

In 1973, S. Glashow, A. Salam and S. Weinberg, all working independently, proposed a theory that looked similar to quantum electrodynamics; they postulated that similar to the photon, quarks and leptons interact with each other via a photon-like particle, called the $\mathrm{W}^{\pm}, \mathrm{Z}^0$. All the bosons are the generalization of the Maxwell gauge field, discussed in Sec. 11.6. Similar to the electric charge that determines the strength of the photon–electron coupling, weak interactions have a weak coupling constant $g_w = \sqrt{\alpha_w}$ that determines all the interaction vertices of the weak force.

Hence, for instance, one would have to modify Eq. (12.9) to the following which, together with the antiparticle case, yields

$$\mathrm{d} + \bar{\mathrm{u}} \ \rightarrow \mathrm{W}^- \rightarrow \mathrm{e}^- + \bar{\nu}_e \tag{12.10}$$

$$\bar{\mathrm{d}} + \mathrm{u} \ \rightarrow \mathrm{W}^+ \rightarrow \mathrm{e}^+ + \nu_e. \tag{12.11}$$

The Feynman diagrams for Eq. (12.10) is given in Fig. 12.17(b).

12.7.1 *Electroweak bosons:* $\mathbf{W}^{\pm}$ *and* $\mathbf{Z}^0$

Note that in Eq. (12.10), the electric charge of d is $-e/3$ and that of the $\bar{\mathrm{u}}$ antiquark is $-2e/3$; the quarks decay via the W^- boson into an electron and an electrically neutral antineutrino. Hence, the W^- bosons carry one unit of negative electric charge, namely $-e$, to ensure charge conservation in this process. Due to relativity, the W^- boson must be accompanied by its antiparticle, namely the W^+ boson that is positively charged. The process given in Eq. (12.11) shows the coupling of the W^+ boson to the quarks and leptons.

Furthermore, the Feynman diagram for neutron decay, given in Fig. 12.18, shows that the d quark changes to the u quark and also points to the existence of the W^- boson. Figure 12.19 shows the coupling of the W^- boson to the second generation leptons, namely the muon and the muon antineutrino.

Are there any more weak bosons? If one collides a W^+ boson with its antiparticle W^-, the resulting particle must have zero quantum numbers. The photon couples to the charged bosons $W^{\pm}$ with strength of coupling given by e. Figure 12.20(a) shows that the W^- and W^+ weak bosons can annihilate to yield a photon.

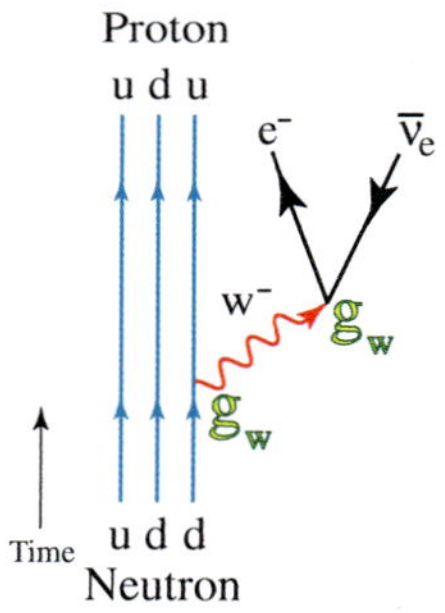

Fig. 12.18 Neutron decay. The d quark changes to a u quark.

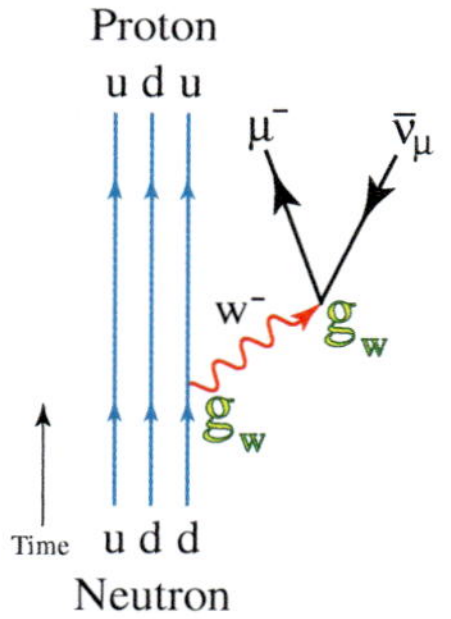

Fig. 12.19 Neutron decays to a muon μ^- and the muon antineutrino $\bar{\nu}_\mu$.

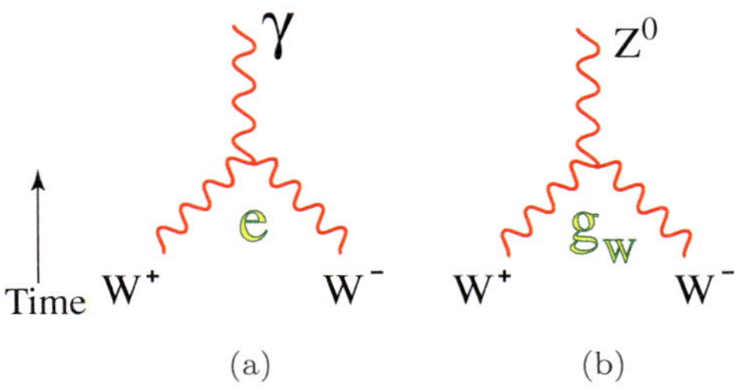

Fig. 12.20 (a) Annihilation of charged weak bosons $W^{\pm}$ into γ. (b) Annihilation of charged weak bosons $W^{\pm}$ into Z^0.

The second possibility, shown in Fig. 12.20(b), is that the annihilation can yield a *new particle*, namely the Z^0 boson which has zero for all its quantum numbers, similar to the photon, but has mass similar to the $W^{\pm}$ bosons. The annihilation into a Z^0 is given by the reaction

$$W^- + W^+ \to Z^0$$

with the Feynman diagram given in Fig. 12.20(b) and the strength of the vertex given by the weak coupling constant g_w. The annihilation of the $W^{\pm}$ bosons into a photon is due to the electromagnetic interaction, and into the Z^0 is due to the weak interaction.

The existence of the $W^{\pm}$ and Z^0 bosons have all been confirmed by experiments. In the Glashow, Salam and Weinberg theory, the photon γ is combined with the $W^{\pm}$ and Z^0 bosons and yields a unified theory of electromagnetic and weak interactions called the **electroweak theory** — and which is one of the great achievements of particle physics.

All the electroweak carriers of interaction, namely the photon γ, Z^0 and $W^{\pm}$, carry spin one and hence are bosons, collectively called the electroweak bosons. The $W^{\pm}$ and Z^0 bosons are all photon-like bosons with two remarkable distinctions, namely

- Both $W^{\pm}$ carry a single unit of electric charge $\pm e$ respectively.
- The $W^{\pm}$ and Z^0 bosons are all massive.

The weak force carriers $W^{\pm}$, Z^0 have masses given by

$$\begin{aligned} m_w &= m_{w^\pm} \simeq 91.1887 \pm 0.0022\,\text{GeV} \\ m_z &\simeq 80.410 \pm 0.180\,\text{GeV}. \end{aligned} \tag{12.12}$$

Note that the masses of the $m_{w^\pm} = m_w$ must be exactly equal since one is the antiparticle of the other; this constraint that the masses of a particle and antiparticle pair must be exactly equal has been verified by experiments.

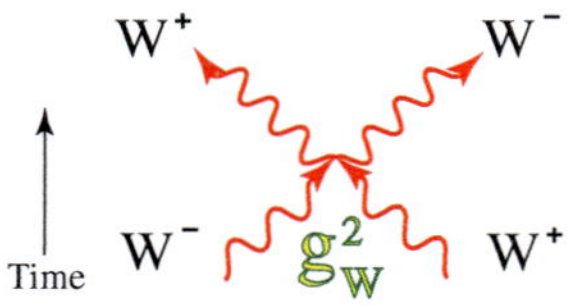

Fig. 12.21 Self-interaction of the $W^{\pm}$ bosons.

The range of weak interactions, namely the distance over which the weak bosons interact, determines how close the quarks and leptons must be to weakly interact; the range is given by the inverse of the weak boson mass, and is approximately 10^{-18} m.

A major difference between the weak bosons and photons is that the weak bosons are self-interacting; the Feynman diagrams for the interaction vertices are given in Figs. 12.20(b) and 12.21. The $W^{\pm}$ bosons couple to all particles that have electric charge, and are themselves carrying electric charge. For this reason, the weak bosons are unlike the photon, which does not carry any charge and hence does not have any direct self-interactions.

Note that, from Fig. 12.21, the strength of interaction of the four weak boson vertex for $W^{\pm}$ is g_w^2. Since $W^{\pm}$ are charged, all the vertices must conserve charge; hence for example in Fig. 12.21, a pair of $W^{\pm}$ annihilate and create another pair of $W^{\pm}$ at the same point.

The interaction of the charged weak bosons $W^{\pm}$ with the neutral boson Z^0 — with the vertex having a strength of g_w given in Fig. 12.20(b) — is similar to that of a photon with an electron, as in Eq. (12.2). However, unlike the case of the photon-electron interaction, the interaction of the $W^{\pm}$ with the Z^0 is a form of self-interaction due to the nonlinearity of the weak boson fields.

12.8 Electroweak Coupling Constants

The electroweak interaction is the unification of the electromagnetic interaction and the weak interactions. The electric charge (coupling constant) e that couples the photon to electric charge carriers and the two masses of the weak bosons, namely m_w and m_z, are the *three independent parameters* of the electroweak theory. The interaction vertex of the $W^{\pm}$ and Z^0 bosons is given by coupling constant g_w; furthermore

$$g_w = \frac{e}{\sin\theta_w}$$

$$\sin\theta_w = \sqrt{1 - \left(\frac{m_z}{m_w}\right)^2} \approx 0.22 \tag{12.13}$$

where θ_w is the **Weinberg weak mixing angle**.

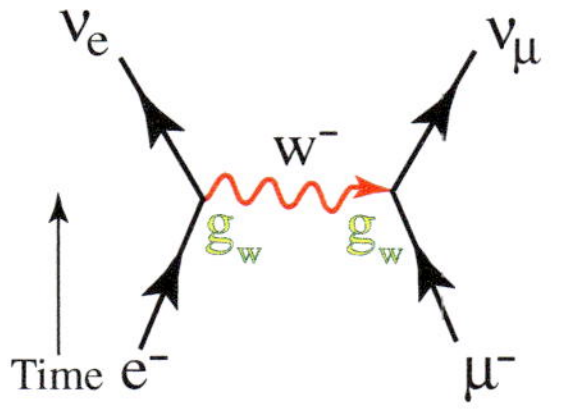

Fig. 12.22 Weak interaction at energy E.

We see that the photon has been unified with the bosons of the weak interactions since the coupling constant of the weak bosons is fixed by electric charge that originates in the photon's coupling to charged particles. Furthermore, the coupling constants e and g_w are of the same order, implying that the weak bosons interact with quarks and leptons with about the same strength as the photons interact with electrons. The reason that the weak interactions are so weak, compared to electromagnetic and strong interactions, is because — similar to α_e being the effective coupling constant for quantum electrodynamics — the effective weak coupling constant is α_w given by

$$\alpha_w = \left(\frac{E}{m_w c^2}\right)^2 \alpha_e \tag{12.14}$$

where E is the energy involved in the weak process. The energy E, for example, can be the centre of mass energy with which leptons approach each other in a high energy scattering experiment, as in Fig. 12.22.

We see that α_w, the effective coupling constant for weak interactions, is small for hadronic decays since the *large mass* of the W boson m_w in the denominator strongly damps the process, yielding $\alpha_w \approx 10^{-6}$ in the energy range of $E \simeq 10\,\text{Gev}$ which is typical for hadronic processes. One expects the electroweak theory to break down for energies for which $\alpha_w \approx 1$, and which yields $E \simeq 1.7$ trillion eV.

12.9 Coupling of Weak Bosons to Fermions

The coupling of photons to charged quarks and leptons has been discussed in Sec. 12.4 and we now focus on the coupling of weak gauge bosons to the fermions, namely the quarks and leptons. The coupling of the weak bosons to only three quarks, namely $\mathrm{u}, \mathrm{d}, \mathrm{s}$ and three leptons (and their neutrinos) $\mathrm{e}^-, \mu^-, \tau^-$ (as well as all the antiquarks and antileptons) is sufficient to illustrate all the important features of the weak bosons' couplings.

The couplings of the weak bosons to the quarks and leptons are given in Figs. 12.23(a) and (b). The couplings are determined by the gauge symmetry of the

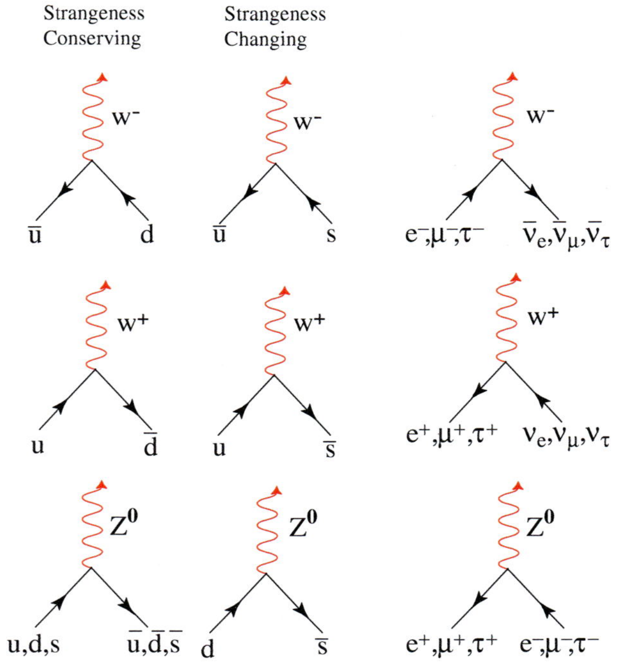

(a) Quark Couplings (b) Lepton Couplings

Fig. 12.23 Couplings of weak bosons to quarks and leptons.

theory — and has been confirmed by experiments to correctly describe the weak processes. The couplings of quarks and leptons to the weak bosons are organized based on the generation to which they belong, with the three generations given in Fig. 11.37.

12.9.1 *Lepton–lepton couplings*

The Feynman diagrams for the couplings are given in Fig. 12.23(b) and show that all lepton–lepton couplings are only within a *single generation*, with there being *no coupling* between the different generations of leptons. The neutral boson Z^0 couples to the particle and antiparticle pairs of all the quarks and leptons whereas the charged bosons $W^{\pm}$ couple the leptons $e^{\pm}, \mu^{\pm}, \tau^{\pm}$ to their respective neutrinos. Figure 12.24 shows the decay of a muon μ^- mediated by the coupling of the weak boson to the lepton–lepton sector.

All leptons interact via the weak interaction if they approach each other to distances less than 10^{-18} m — the distance over which the weak interaction operates. Since electrons and other leptons carry charge, they interact via the electromagnetic interaction over large distances since the photon's range of interaction goes out to infinite distances.

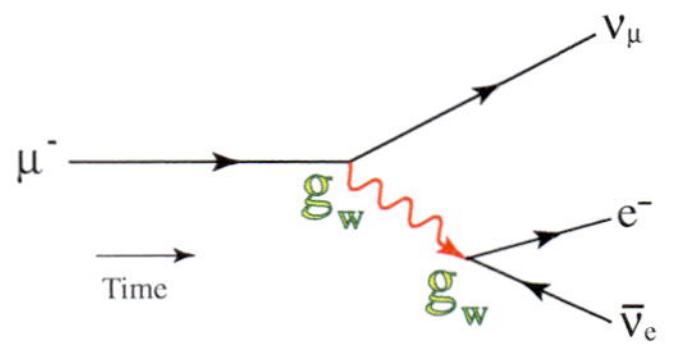

Fig. 12.24 Decay of a muon to other leptons via the weak interactions.

All couplings involving neutrinos (for all three generations) involve only 'one hand' of the neutrinos since they break parity (reflection) symmetry. This is discussed later in Sec. 13.4. All quarks and leptons respect parity except for the neutrinos. Parity violation in Nature was first postulated by C. N. Yang and T. D. Lee in 1956.

12.9.2 *Quark–quark couplings*

The intricacy of weak interactions takes place in the coupling of the quarks to the weak bosons, which couple quarks within a generation as well as quarks of different generations.

The left most column in Fig. 12.23(a) shows the coupling of quarks to the weak bosons *within* a single generation; these couplings conserve the strangeness quantum number. For example, the coupling of $\bar{\text{u}}\text{d}$ to $\text{e}^-\bar{\nu}_e$ via the W^- boson is a coupling within a *single* generation of quarks and leptons. There is a similar coupling for the remaining two generations of quarks and leptons, all mediated by the W bosons.

There is another set of couplings of quarks to the weak bosons that couple quarks of *different* generations, with couplings of the u and d quarks to the s quark shown in the second column of Fig. 12.23(a). This coupling **violates strangeness conservation** and is discussed in Sec. 12.10 below.

In general, the weak interaction vertices can be combined in any consistent manner to get a possible allowed process. Whether this process will be significant depends on the energies available. For example, consider the neutron's decay given by $\text{n} \to \text{p} + \text{e}^- + \bar{\nu}_e$; one can also have another decay mode shown in Fig. 12.19, in which

$$\text{n} \to \text{p} + \mu^- + \bar{\nu}_\mu$$

and which only becomes relevant as one goes to higher energy collisions. Similarly, a few diagrams are given in Fig. 12.25 showing the interactions of the weak bosons with the fermions as well as with itself.

Note that there is no four fermion coupling of quarks to leptons at a *single point* — as shown in the interaction given in Fig. 12.16 — that was erroneously

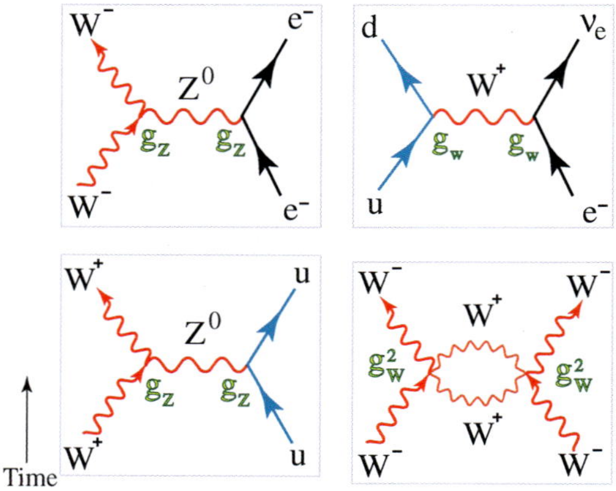

Fig. 12.25 Electroweak interactions.

proposed by Fermi.[5] In other words, there is no direct coupling of the quarks with the leptons; the coupling of quarks to leptons is always via one of the weak bosons or via the photon.

All experimentally observed processes involving quarks and leptons are accounted for by the couplings shown in Fig. 12.23 and the other similar couplings for the other generations that we have omitted for simplicity.

12.10 Strangeness Changing Processes

As discussed in Sec. 12.9 above, unlike the weak interactions of leptons that do not couple leptons from different generations, the weak interactions of quarks couple quarks from different generations. This was the reason that in hadron physics, as discussed in Sec. 11.14, one encountered strange hadrons that decayed much slower than the usual fast time scale of 10^{-23} s characteristic of strong interactions.

Strangeness changing interactions couple quarks across two different generations in a very special way. For the first and second generation quarks, consisting of the u, d and c, s quarks respectively, only the u and d quarks couples to the s quark, with no coupling of the c quark with the u, d quarks. A similar structure is repeated for the second c, s and third t, b generation quarks; the charm and strange quarks c, s couple with the b quark, and there is no other inter-generational coupling. The Feynman diagrams for the strangeness changing couplings of u, d and s connecting the first to the second generation of quarks is shown in Fig. 12.23(a).

[5]The four fermion inteaction violates the requirement of renormalizability.

There is another coupling connecting the second to third generation quarks that is not shown for simplicity. Note that there is no direct coupling between the first generation quarks u, d and the third generation quarks t, b.

The very specific interaction for the quark generation changing interaction has been tested by experiments and found to be correct. In particular, it is able to explain the decay of all the strange hadrons. For example, recall the decay of strange hadron given in Eq. (11.14)

$$\Lambda^0 \to \mathrm{p} + \pi^- .$$

Writing the decay of the $\Lambda^0 = \mathrm{uds}$ in terms of the quark content, we have

$$\begin{aligned} \mathrm{uds} &\to \mathrm{udu} + \bar{\mathrm{u}}\mathrm{d} \\ \Rightarrow \mathrm{s} &\to \mathrm{u} + \bar{\mathrm{u}}\mathrm{d} \\ \Rightarrow \mathrm{s} + \bar{\mathrm{u}} &\to \mathrm{W}^- \to \bar{\mathrm{u}}\mathrm{d}. \end{aligned} \tag{12.15}$$

The quark couplings of both the columns in Fig. 12.23(a) yield the two different Feynman diagrams (interaction vertices) that yield the strangeness changing decay of Λ^0 — as given in Eq. (12.15) above.

12.11 Quantum Chromodynamics

So far we have studied the interaction of quarks and leptons. The electromagnetic force creates the binding of electrons to the nucleus (composed of the bound states of quarks), whereas the weak interactions are responsible for the disintegration of an isolated neutron.

The nucleus itself is composed solely of quarks and the forces of the strong interaction that hold them together; there are no leptons inside a stable nucleus, and electroweak interactions cannot explain the binding of quarks to form mesons and hadrons. In fact the electromagnetic force is a force that acts to break up a nucleus since the nucleus has net positive charges carried by the protons, and the Coulomb force causes the positively charged protons to want to fly away from each other.

So what interaction is strong enough to overcome the repulsive Coulomb force as well as bind the quarks to form the protons and neutrons? Also, recall that quarks appear only as mesons, which is a bound state of a quark and its antiquark, or in groups of three quarks bound together to form baryons. How do we explain the fact that although the nucleus is thought to be made out of quarks, an isolated quark has never been detected in any experiment? These are the puzzles any theory of strong interactions has to address.

In 1974 it was postulated by several physicists that the strong force holding the nucleus together consists of quarks interacting with eight spin one bosons collectively called *gluons*. The range of the strong force is the size of the nucleus, which is

10^{-15} m; light takes about 10^{-23} s to cross a nucleus and is the typical duration for which the gluons act on the quarks.

The quarks have another quantum number called the **color charge**; color comes in *three* distinct varieties: green, blue and red. So for example the up quark u comes in three varieties, namely $\mathrm{u}_g, \mathrm{u}_b, \mathrm{u}_r$, and similarly for the other quarks. If one takes all three generations of quarks into account, one now has $3 \times 6 = 18$ quarks and their antiquarks. The leptons and all other carriers of forces carry zero color charge.

The gluons couple to color charge with coupling constant g_s, and is similar to how the photon couples to electric charge. The gluons also carry color themselves; the eight gluons can be thought of as carrying all possible combinations of anticolor and color, namely

$$g : \begin{pmatrix} \bar{b}b & \bar{b}g & \bar{b}r \\ \bar{g}b & \bar{g}g & \bar{g}r \\ \bar{r}b & \bar{r}g & \bar{r}r \end{pmatrix}. \tag{12.16}$$

Gluons interact with themselves since they themselves carry color charge. Figure 12.26 shows the coupling of gluons to themselves. Each leg of the Feynman diagram shown in Fig. 12.26 is a gluon propagator, with arrows pointing in opposite directions to differentiate between color and anticolor. The scattering of gluons off gluons is given by the interaction vertices similar to the self-interactions of the weak bosons given in Figs. 12.20 and 12.21.

Two colors, namely a color and another anticolor, taken together are carried by a single gluon, as can be seen from the legs of the Feynman diagram shown in Fig. 12.26. If one takes all possible combinations of a color and an anticolor charge, as shown in Eq. (12.16), one would have nine combinations.

However, there are in fact only eight gluons; one of the nine possible combinations separates out, similar to separation out of one singlet in the representation of the mesons, discussed in Sec. 11.16, using $\bar{\mathbf{3}} \times \mathbf{3} = \mathbf{8} + \mathbf{1}$; in particular, the symmetric combination of $\bar{b}b + \bar{g}g + \bar{r}r$ is removed from the allowed gluonic particles, leaving only eight gluons.

One example of Feynman diagrams that describe the scattering of quarks and gluons is given in Fig. 12.27, with the four lines labeled $\mathrm{u}_r, \mathrm{d}_g$ being a colored red

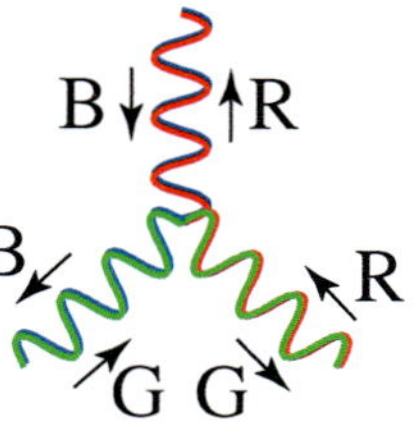

Fig. 12.26 Three gluon vertex.

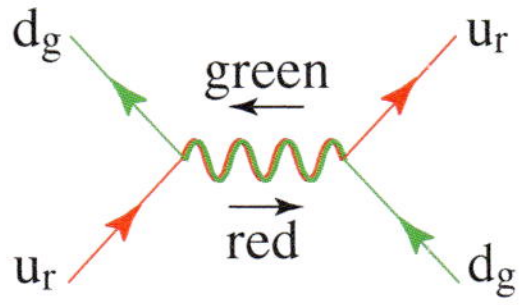

Fig. 12.27 Colored quarks interacting via colored gluons.

up and green down quark, and the line running between them being a gluon made from red color and green anticolor. The vertex shows that the coupling of the colored quarks to the colored gluons exactly conserves color charge.

Quantum chromodynamics is the theory describing strong interactions. Quantum chromodynamics is a quantum field theory of colored quarks interacting with (nonlinear) colored gluons. Note that in the Feynman diagrams shown in Figs. 12.26 and 12.27, none of the color lines end, showing the *absolute conservation of color* in quantum chromodynamics.

When quarks are very close, the interactions of the gluons with the quarks can be neglected and quarks behave as if they are free; however, when the quarks are pulled apart, the gluons interact more and more strongly with the quarks. In other words, the further is the separation of the quarks, the *stronger* is the coupling of the gluons to the quarks: attracting the quarks and binding them together. This behaviour of the quark–gluon coupling constant, going to zero for short distances and becoming very large for large distances is called **asymptotic freedom** and **infrared instability**.

Recall in the case of weak interactions the effective coupling constant depends on the energy of the process being studied, and was given by Eq. (12.14) as $\alpha_w = \left(E/m_w c^2\right)^2 \alpha_e$, where E is the energy scale of the process being mediated by the weak interactions. In a similar manner, the effective coupling constant of the strong interactions also depends on the scale of energy E that is being probed by the gluons. The effective strong coupling constant is also called the **running coupling constant** since it changes ('runs') with energy.[6]

The strong coupling α_s has the following dependence on E, the energy of interaction:

$$\alpha_s(E) = \frac{12\pi}{21\log(E^2/\Lambda^2)}; \quad \Lambda \approx 0.2\,\text{GeV}.$$

[6]For the more advanced reader, note that for a gauge theory with SU(n_c) gauge group and for n_f number of quarks, the running coupling constant is given by

$$\alpha_s(E) = \frac{12\pi}{(11n_c - 2n_f)\log(E^2/\Lambda^2)}.$$

For the case of quantum chromodynamics, $n_c = 3$ and for six quark flavors, which is the case of the Standard Model, $n_f = 6$.

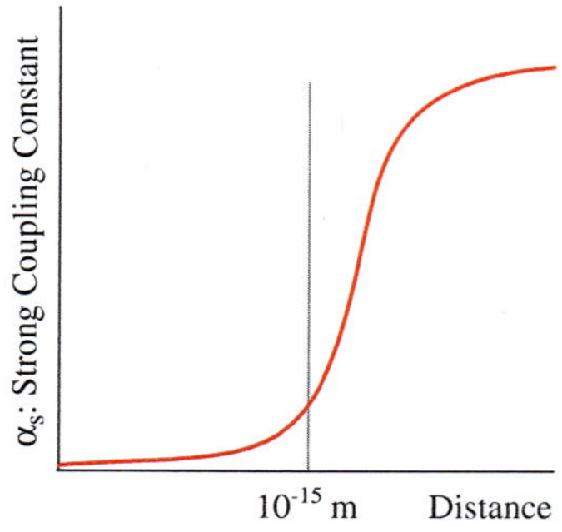

Fig. 12.28 The dependence of the strong coupling constant α_s on distance between quarks.

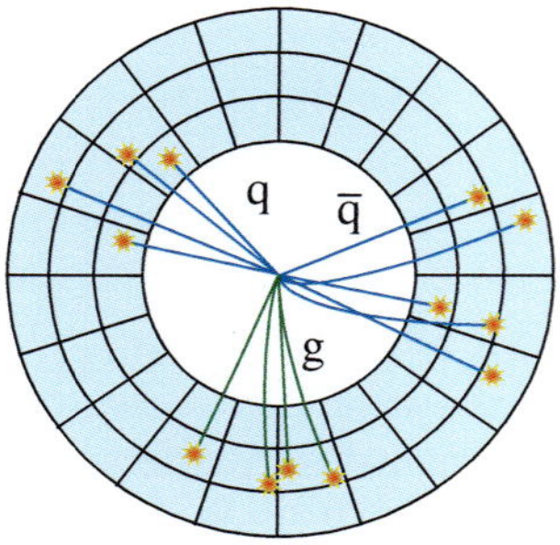

Fig. 12.29 High energy collisions of electrons and positrons giving rise to two jets of quarks and antiquarks and a third jet due to the glouns.

Figure 12.28 shows the value of α_s as a function of distance; the effective strong coupling constant increases with increasing distance (decreasing energy); in other words the further the quarks are separated the stronger is the quark–gluon interaction and the larger is the strong coupling constant. The energy parameter Λ is fixed by the cross-over distance, given by 10^{-15} m, from high energies for which quarks have free-like behavior to lower energies for which they are strongly interacting.

Gluons are quantum particles in their own right, and can be detected in high energy experiments. A bound state of the gluon is possible due to the self-interaction of the gluons and is discussed further in Sec. 12.13.

Indirect evidence of gluons is found in experiments, similar to those pointing to the existence of quarks and given in Fig. 11.36. Consider very high energy collisions of electrons e^- and positrons (antielectrons) e^+ that annihilate into quarks and antiquarks; since the collision has such high energy, quarks and gluons are both produced copiously, shown in Fig. 12.29. The reaction has the form

$$e^- + e^+ \to \bar{q} + q + g \to \text{three jets of hadrons.}$$

Due to the nature of strong interactions, the three particles produced in the e^-, e^+ collision, namely $\bar{q}, q$ and g, are converted to ordinary hadrons by secondary processes. The signature for the production of gluons appears in the particle detector. As shown in Fig. 12.29 there are *three* separate *quark–gluon jets* of various particles bunched around the three particles $\bar{q}, q$ and g produced in the e^-, e^+ collision. **Gluon jets** have been experimentally detected in CERN.

12.12 Charmonium: Linear Potential

The dynamics of quarks and gluons greatly simplifies in the case of the **charm quark**. The charm quark's mass is $1.5\,\text{GeV}/c^2$; in high energy collisions charmed mesons (bound states of charm and anticharm quarks) are produced. There are two major simplifications for charm quark bound states.

- Due to the high mass of the charm quark, the creation of virtual pairs of quarks in a charmed meson is negligible. This in turn implies that one can consider the number of quarks in a charmed meson to be approximately *fixed.*
- The average velocity of the charm quark in a charmed meson is about $0.5c$ and hence effects due to relativistic speeds can be treated as perturbations.

Due to the two features above, the charmed meson can be treated using *non-relativistic* quantum mechanics. In particular, the bound state of the charm and anticharm, called charmonium, can be studied using an appropriate potential $V(r)$, where r is the distance between the charm quarks.

Note that, in contrast to charmed mesons, a meson state made of low mass quarks and antiquarks, such as u and d and their antiparticles, cannot be described by non-relativistic quantum mechanics. For example, consider the neutral pi meson $\pi^0 = (\bar{\text{u}}\text{u} - \bar{\text{d}}\text{d})/\sqrt{2}$; since both the u, d are light, they move almost at the speed of light c inside π^0; moreover, the number of quarks in the meson is indeterminate, and constantly fluctuates due to pair creation. Hence for such a state the full quantum field theory needs to be employed to derive its state function.

At short distances we know that the gluons' self-interaction can be ignored and they behave as a collection of eight photon-like particles; hence at short distances we expect the potential $V(r)$ to be an attractive Coulomb-like $1/r$ potential. For large distances the increasing strength of the coupling constant can be modeled by a linear term r in the potential.[7]

[7]The linearly growing potential is the reason that quarks are permanently bound inside the nucleus, and this point will be discussed later.

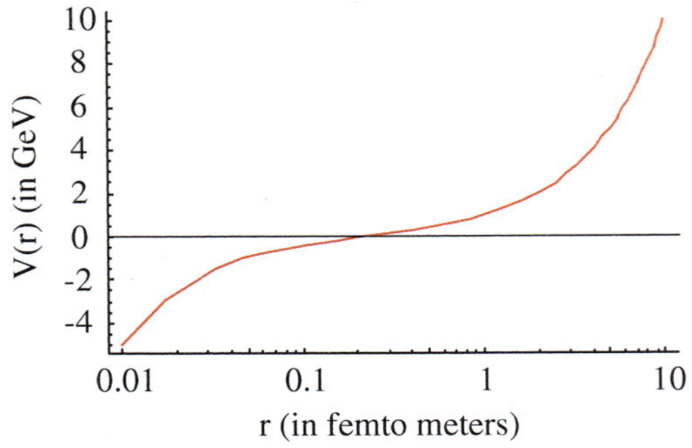

Fig. 12.30 The strong interaction inter-quark potential due to gluons. [1 femtometer = 10^{-15} m.]

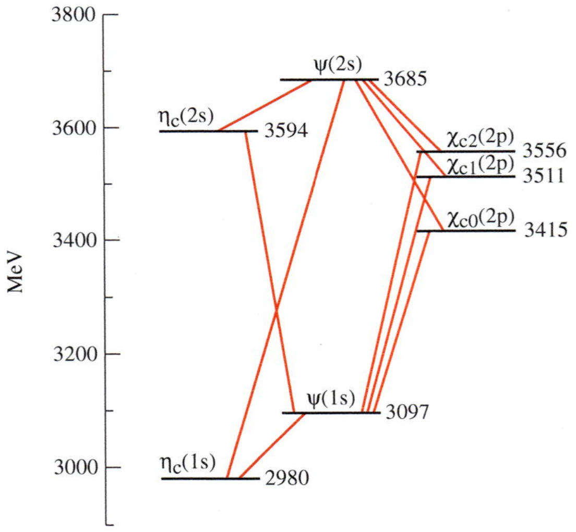

Fig. 12.31 The energy levels of charmonium.

The strong potential — due to the gluons — between charm and anticharm quarks is shown in Fig. 12.30, and is given by

$$V(r) = -\frac{a}{r} + br$$

$$a \approx 0.05\,\text{GeV} \times 10^{-15}\,\text{m}; \quad b \approx 1\,\text{GeV} \times 10^{15}\,\text{m}^{-1}.$$

The discovery of charmonium has revolutionized the study of mesons. Every excited state of the quantum mechanical system governed by the linear potential is a possible meson, and all the techniques of spectroscopy used in atomic physics at the scale of 10^{-10} m can now be lifted in its entirety to study of meson physics at the scale of 10^{-15} m! Figure 12.31 shows some of the rich details of the predictions of the charmonium model, and which find remarkable confirmation in experiment.

12.13 Gluonic Strings: Mesons and Baryons

The description of quarks and gluons using Feynman diagrams is appropriate for describing the short distance properties of strong interactions. However, as one separates the quarks the strength of the interaction increases indefinitely, leading to a *qualitative* change in the behavior of the gluons.

Hadrons, and the nuclei in general, are the bound states of quarks and gluons. These bound states emerge when the quarks are separated to large distances and the gluon field interacts strongly with itself and with the quark field. To describe the quark bound states, a **strong coupling representation** of quarks and gluons is required.

Consider separating a quark and antiquark which compose a meson; in response to the quarks being separated, the gluon field becomes strongly coupled to the quarks and vastly increases the energy in the gluon field. The energy that was initially spread uniformly around the quarks, as shown in Fig. 12.32(a), tends to get collimated into a **gluonic string** — a string leading from the quark to the antiquark. The formation of a gluonic string, as shown in Fig. 12.32(b), is energetically favored by the gluon field since, in doing so, the gluon field energy is lowered. Figures 12.32(a) and (b) show how the gluonic field in a meson gets collimated into a string-like configuration as the quarks are separated.

All the eight gluons need to be taken *together* to form the gluonic string. The representation of the gluonic string in Fig. 12.33 is not a Feynman diagram; instead, the figure is the strong coupling representation of the gluonic string that is equivalent

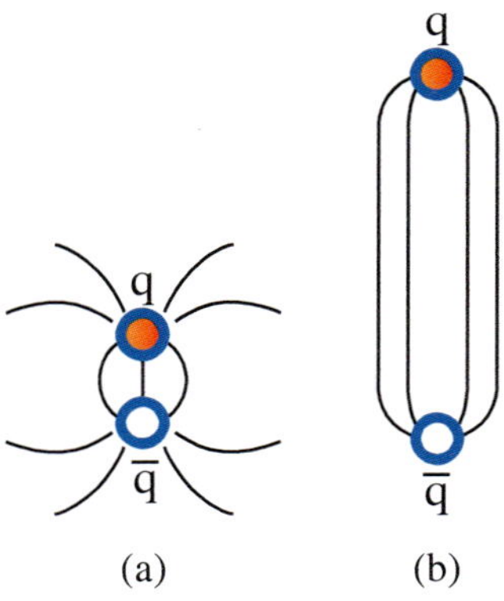

Fig. 12.32 (a) The gluon field is spread out in space when the quarks are close to each other. (b) The gluon field is collimated into a gluonic string as the quarks are separated.

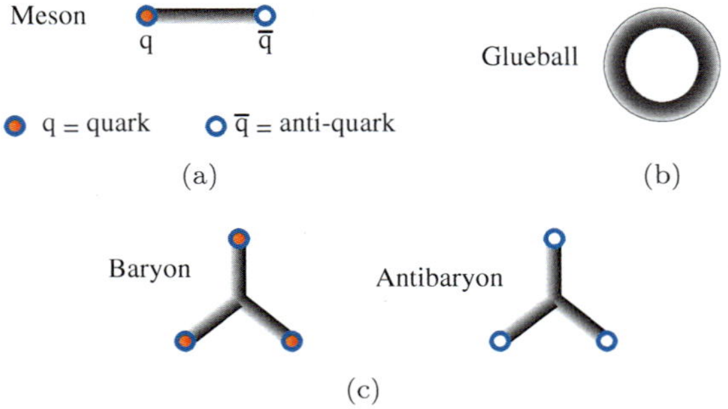

Fig. 12.33 (a) Quark confined in mesons by gluonic strings. (b) The loop is a pure gluon state, called a glueball. (c) A baryon and antibaryon confined by gluonic strings.

to an infinite collection of Feynman diagrams, which are appropriate for representing the eight glounic fields only when they are weakly coupled.

In the strong coupling representation of the gluon, the open gluonic string carries color and anticolor charges only at the endpoints; the colors of the gluonic string at the two ends are exactly canceled by the colors of the colored quarks. Hence, in the strong coupling representation all mesons can be thought of as consisting of a quark and an anitquark connected by a gluonic string.

Figure 12.33(a) shows a meson, in the strong coupling representation of the gluon field, consisting of a quark and antiquark bound together by a gluonic string. Figure 12.33(b) shows a bound state of pure gluons, which forms a loop to have zero color charge; the bound state is called a **glueball**. It is the strong coupling state of the glueball that one expects to observe, and the lightest glueball is estimated to have a mass in the range of $1-1.7\,\mathrm{GeV}/c^2$.

The gluonic string representation clearly displays why the strength of interaction increases with distance: as one separates the quarks, the string grows linearly; this requires a consequent increase in energy that depends linearly on the inter-quark distance and hence the emergence of the linear potential discussed in Sec. 12.12.

Three gluonic strings, with end points being all color or anticolor, can meet at a single point, as shown in Fig. 12.33(c), to yield a colorless junction.[8] The end points of the three string junction *all* carry either color or anticolor. So one can attach three quarks to the endpoints of the three junctioned gluonic strings to form baryons, as shown in Fig. 12.33(c) — which also shows how one can form antibaryons by starting with a three string junction with color at the the three endpoints and attaching antiquarks to these.

In both the short distance weak coupling and large distance strong coupling quark–gluon theory, color charge is *exactly* conserved at *every point* of spacetime.

[8]The three gluonic string junction is possible due to the fact that the gluons are derived from gauge fields with SU(3) symmetry. Lie group SU(3) is discussed in Noteworthy 14.1.

What this means is that every point has its own **conservation of color**, and that color is conserved separately and independently at every point. All physical states of quarks and gluons have zero net color charge at every point.

The exact local conservation of color implies that no object can have any color charge. This is the reason that a colored quark *cannot* propagate, either in time or move around in space, as this would entail the violation of local color conservation. In particular, an isolated colored quark *cannot* exist as an observable state.[9]

The gluonic string, sometimes called the QCD string, is not a fundamental entity but, instead, is an *effective* description of the behavior of quantum chromodynamics when the quarks and gluons become strongly coupled. The string-like description of the gluons is in contrast to fundamental strings, which are entities in their own right and are discussed in Chapters 14 and 15.

12.14 Permanent Confinement of Quarks

The requirement of exact local color conservation requires that no isolated quark carrying color can exist in Nature. We can now address the *dynamical mechanism* of how quarks are *permanently confined* inside hadrons and the nuclei.

In order to understand the **confinement of quarks**, we need to first analyze why the electron is *not* confined inside an atom. The electron is bound to the nucleus due to the Coulomb interaction between the two. If one hits an electron, bound to the nucleus, with a photon, the photon imparts momentum (and energy) to the electron which then tends to move away from the nucleus; as it moves away from the nucleus, the electron experiences a slowly *decreasing* Coulomb force. Hence, for large enough energy the electron can completely escape from the nucleus and appear as a free particle.

In contrast, suppose one imparts a large amount of momentum (and energy) to a quark inside a proton, as shown in Fig. 12.34, by hitting it with a high energy photon; as the quark moves away from the other quarks the increasing distance makes the gluonic force of attraction *stronger* and *stronger* — and thus *increasing* the energy in the gluon field, as indicated by the green color in Fig. 12.34.

When the quark has receded from the other quark to about 10^{-15} m, there is so much energy in the gluon field that it can lower its energy by creating a quark and antiquark pair from the vacuum. The receding quark gets bound to the antiquark of the produced quark pair and becomes a colorless meson; the other created quark of the pair combines with the other quarks and creates a colorless proton.

[9]If Nature chooses to *break* the local conservation of color charge, the quarks could be observed as free particles carrying color, but this is not the case.

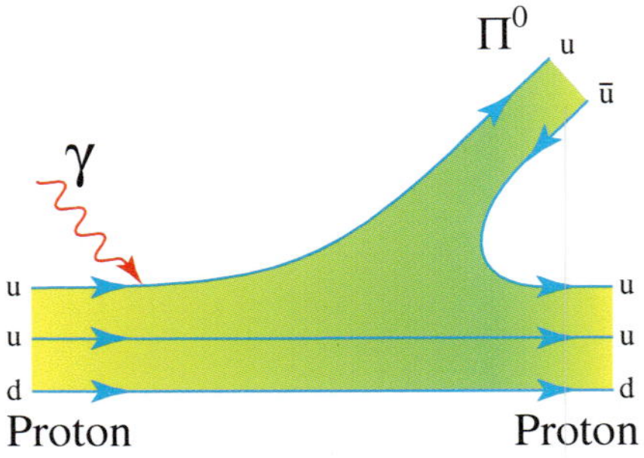

Fig. 12.34 When a photon with sufficient energy is absorbed by a quark in a proton then, instead of obtaining a separated quark, the energy is used to create a new quark–antiquark pair in addition to the original proton.

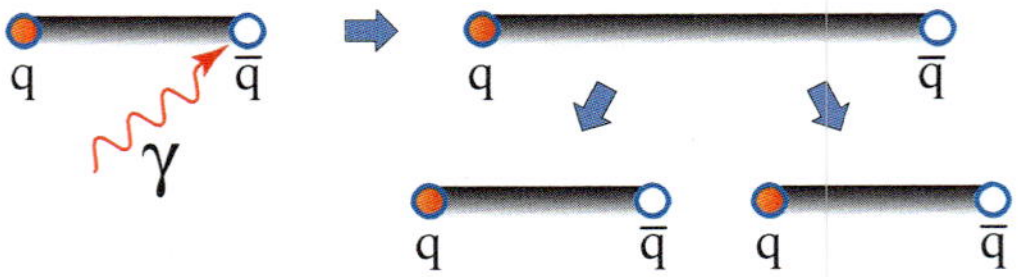

Fig. 12.35 High energy photon being absorbed by a quark leads to the elongation of the of gluonic string — and then to the breaking of the string and the creation of a new quark–antiquark pair.

Since both the meson and proton are colorless, the interaction between them is approximated by the Yukawa potential which, recall from Eq. (11.9), has a range of interaction that is shorter than the Coulomb potential. In this manner, the meson separates from the proton, carrying away energy and momentum supplied by the photon to a quark inside the proton — and yielding *two* colorless states (particles), namely the proton and the pion. This process is shown in Fig. 12.34. A similar process occurs if one tries to free a quark from any meson or baryon.

The confinement of quarks can also be understood in the strong coupling representation of the gluon field becoming collimated into a gluonic string, as discussed in Sec. 12.13. As one separates the quarks in a meson, it is energetically favorable for the gluon field to *break* the string by creating a quark–antiquark pair from the vacuum and thus creating two strings — as shown in Fig. 12.35 — and hence giving rise to two mesons; the energy imparted to the quarks is partly taken up by the creation of the quark and antiquark pair, and any left over energy can take the form of the kinetic energy of the mesons.

In summary the energy supplied to the quark in an attempt to free it from its bound state ends up creating pairs of quark and antiquark from the vacuum, and results in the creation of more hadrons. In essence, quarks are permanently confined inside hadrons due the nature of the gluon field. The colored quarks are coupled to the colored gluon field, which in turn is a self-interacting field for which the strength of interactions *increases*, without limit, with increasing distance of the color charge carriers.

It is a testament to the power and depth of quantum field theory that the counter-intuitive, enigmatic and mysterious phenomenon of quark confinement — a phenomenon that has no analogy or exemplar in the rest of science — finds a precise and quantitative explanation in quantum chromodynamics. The success of quantum chromodynamics has once again shown the tremendous power of science to describe and predict new and unexpected behavior of Nature.

In particular, it is worth noting that the power of prediction in high energy physics is entirely predicated on the sophisticated and powerful mathematical structures of quantum field theory — which is the means for, both, describing Nature as well as for making predictions that can subsequently be tested, namely verified or falsified, by experiments. Of course, Physics is an empirical science at its very core, and it is experiment that is the final and sole criterion of scientific truth; quantum field theory is one of the major theoretical tools for exploring and finding scientific truth.

12.15 The Answer

To answer the question "What holds matter together?" we had to explore the subnuclear structure of matter and found the four fundamental forces of Nature that mediate all interactions and, in particular, that also hold matter together. Of the four fundamental forces, we focused only on three of these, leaving aside gravity.

The first step in our journey was to determine what are the carriers of matter, which turned out to be fermions that come in two varieties, namely quarks and leptons.

The force holding the nucleus together is the strong interactions that are mediated by gluons; the electron is held to the nucleus by the electromagnetic force mediated by photons. The third force turned out to be the weak interaction, which has quite an intricate structure and, in particular, is responsible for the decay of the neutron. Quarks and leptons, together with the three fundamental forces, succinctly summarize the constituents and forces of Nature, except for gravity.

The strong force is simple to express, consisting of the coupling of colored quarks with colored gluons. However, due to the strong interactions that result in the permanent confinement of quarks, it is very difficult to apply the theory for predicting experimental results, specially when the strong coupling effects become significant.

The electroweak interaction, as the name indicates, is a force that is weak since all the dimensionless weak coupling constants are less than unity; the weak coupling makes the study of electroweak interactions amenable to perturbation expansions using Feynman diagrams. However, the complexity of electroweak interactions lies in the intricate manner in which the three generations of fermions are coupled to the weak gauge bosons. Parity violation and the breaking of gauge symmetry, to be discussed in Chapter 13 on the Standard Model, introduce numerous new conceptual and computational challenges; in particular, the masses of weak gauge bosons as well as of the quarks and leptons are all generated by the electroweak interactions. Thus, the theories of strong and electroweak interactions are both equally interesting and equally formidable theoretical constructs.

The only piece of the jigsaw puzzle of fundamental particles and forces not addressed so far is the necessity of the Higgs boson's existence, which is discussed in Chapter 13 on the Standard Model.

Chapter 13

The Standard Model

Why is the Higgs field necessary?

13.1 The Question

A stir was caused in the world of Physics, and amongst the public at large, by the announcement by **CERN** in July 2012 that the Higgs boson, which had been postulated to exist in 1964 by Peter Higgs and other physicists, had been experimentally detected. The detection of the Higgs boson is truly a historic event for Physics, and for society at large, since it completed the Standard Model of particle physics that identifies all the fundamental constituents of matter, together with their interactions. The validation of the Standard Model brings to a closure the "atomic dream" of finding out what matter is made out of.[1]

---

[1]Does the realization of the atomic dream mean that matter is not infinitely divisible? We will revisit this question in Noteworthy 14.4 and find a rather surprising answer, namely, that matter *is*, in fact, infinitely divisible!

In Chapters 11 and 12 we discussed elementary particles and fundamental forces, respectively, but there was no discussion on the Higgs field. In this chapter, it is shown how the particles and forces are combined in an essential manner in the Standard Model — and the central role played by the Higgs field in making the Standard Model consistent with all the experiments designed to calibrate and verify it.

13.2 The Standard Model of Particle Physics

The characteristic strengths and time scales of the four fundamental forces of Nature are summarized in Table 12.1. Three of the fundamental forces of Nature, together with all the elementary particles, are shown in Fig. 13.1, which is a grand summary of the knowledge that has been gained about the microscopic composition of Nature. The so-called **Standard Model** of particle physics consists of the constituents of matter, namely quarks and leptons, together with the fundamental interactions holding them together, namely the electroweak and strong interactions. The only fundamental interaction that is not included in the Standard Model (and does not appear in Fig. 13.1) is gravity.[2]

The Standard Model consists of quantum chromodynamics that models the *strong interactions* of quarks and gluons which comprise the nucleus, and electroweak theory that unifies the weak and electromagnetic forces into the *elecroweak interactions* and includes electroweakly interacting leptons.

The focus of this chapter is on the electroweak sector of the Standard Model. In particular, we will discuss the crucial role played by the neutrino in breaking the symmetry of parity in Nature. Furthermore, the coupling of quarks and leptons

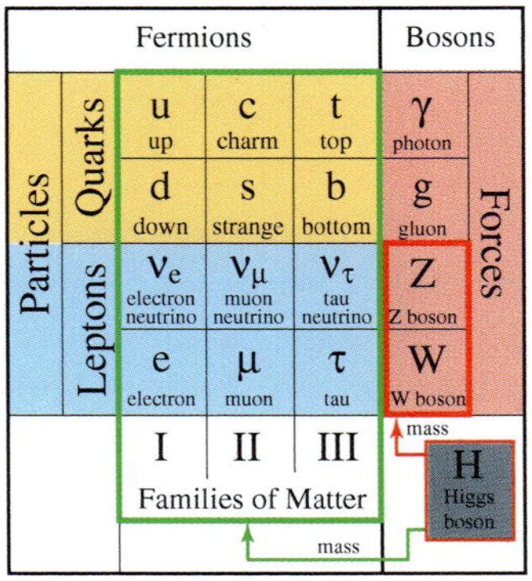

Fig. 13.1 The Standard Model of particle physics.

[2]Superstring theory, discussed in Chapters 14 and 15, has been developed precisely to address the relation of gravity to the Standard Model and to further unify our understanding of Nature. In particular the aim of superstring theory is to derive gravity as well as the Standard Model from even more fundamental entities, namely superstrings.

to the electroweak interactions does not respect parity symmetry and hence needs a fairly complicated construction. We analyze the rather unexpected role played by the Higgs field in endowing *all* the fundamental particles (except the gluon) with mass.

Quantum chromodynamics has been discussed in Chapter 12. Recall that each quark in Fig. 13.1 comes in *three* colors, and it is the color charge that is the foundation of quantum chromodynamics and, in particular, couples quarks to gluons. The color charge of the quarks does not play a major role in the structure of the electroweak interactions and hence will be discussed only if necessary. Furthermore, quantum chromodynamics is symmetric under the parity transformation and hence there is no change in the theory when quarks are coupled to leptons and does not need any further elaboration.

As mentioned, the properties that are unique to the theory of electroweak interactions revolve around two particles of Fig. 13.1, namely the *neutrinos* and the *Higgs boson*. The neutrinos and the Higgs field hold the key to Standard Model. The neutrinos, which are treated as massless fermions in the Standard Model,[3] explicitly break the symmetry of parity, and are discussed in the next section. The Higgs particle is a massive scalar boson and belongs to a category by itself since, as shown in Fig. 13.1, it is not a spin 1/2 fermion like the other elementary particles and neither is it a force like the other vector bosons.

Noteworthy (optional content) 13.1: Mirror Reflections, Parity Symmetry

---

We discuss the symmetry of parity as it is broken in Nature and leads to many new features of the Standard Model. In our discussion on symmetries in Sec. 11.5, we saw that there is a class of symmetries in which only discrete changes are allowed. One of the most important discrete symmetries is parity. In Physics, **parity symmetry** refers to the invariance of a system under a parity transformation, which is defined below.

One can stand in front of a mirror and view one's right hand; it will be seen that the mirror **reflection** has the shape of the left hand. One can easily convince oneself that the left hand — whose mirror reflection is the right hand — *cannot* be transformed into the right hand by a rotation. They are related by a mirror reflection that is *not* equivalent to any rotation. Denoting mirror reflection by $\mathcal{M}$ yields

$$\begin{aligned} \mathcal{M}(\text{left hand}) &= \text{right hand} \\ \text{Also}\ \ \mathcal{M}(\text{right hand}) &= \text{left hand} \\ \Rightarrow \quad \mathcal{M}^2 = 1, &\quad \mathcal{M} = \pm 1. \end{aligned} \tag{13.1}$$

---

[3] The neutrino is thought to have a small mass of a few eV's — but this does not affect the discussion of the Standard Model in any way.

Fig. 13.2 In a mirror, left and right hands are reversed.

A remarkable fact about Nature is that it violates the symmetry of parity; namely, there are process in Nature which happen only in say the left-hand direction and not in the right-hand direction. Denote the operation of parity by $\mathcal{P}$. If one applies the operation of $\mathcal{P}$ to all processes in Nature, certain processes will not be left unchanged (invariant); more precisely, if two frames of reference are related by parity, then some of the phenomena in Nature will not appear to be the same in both the frames. The processes that violate parity are the ones mediated by the weak interaction.

A parity transformation is a mirror reflection of the coordinates followed by a rotation R, namely $\mathcal{P} = R\mathcal{M}$. Mathematically, parity transforms a coordinate vector $\vec{r}$ as follows

$$\mathcal{P}(\vec{r}) = -\vec{r}; \quad \mathcal{P}^2(\vec{r}) = \vec{r}. \tag{13.2}$$

Hence, on any eigenstate of parity, the action of $\mathcal{P}$ is to multiply the state by either 1 or -1.

In quantum mechanics, a particle is represented by its state function, and

$$\mathcal{P}\psi(\vec{r}, t) = \psi(-\vec{r}, t). \tag{13.3}$$

For a scalar field (even), we have

$$\mathcal{P}\psi_s(\vec{r}, t) = \psi_s(-\vec{r}, t) = \psi_s(\vec{r}, t) \tag{13.4}$$

and for a pseudo-scalar field (odd)

$$\mathcal{P}\psi_{ps}(\vec{r}, t) = \psi_{ps}(-\vec{r}, t) = -\psi_{ps}(\vec{r}, t). \tag{13.5}$$

For a system of two or more particles, parity is multiplicative; for example

$$\begin{aligned} \psi &= \psi_s\psi_s\psi_{ps} \Rightarrow \mathcal{P}(\psi) = -\psi \\ \mathcal{P} &: -1 = 1 \times 1 \times (-1) \\ \psi' &= \psi_{ps}\psi_{ps}\psi_s \Rightarrow \mathcal{P}(\psi') = \psi' \\ \mathcal{P} &: 1 = (-1) \times (-1) \times 1. \end{aligned} \tag{13.6}$$

13.3 β-Decay: Parity Violation in Nature

Consider the β-**decay** of cobalt-60, denoted by ^{60}Co; the decay mode is

$$^{60}\text{Co} \rightarrow {}^{60}\text{Ni} + e^- + \bar{\nu}_e. \tag{13.7}$$

^{60}Co decays into ^{60}Ni by emitting an electron (ignore for now the anti-neutrino $\bar{\nu}_e$). As in Eq. (12.8), the neutron decays as $n \rightarrow p + e^- + \bar{\nu}_e$; the proton changes ^{60}Co to ^{60}Ni and the leptons are emitted. An experiment is done by cooling a cobalt sample in the presence of a magnetic field so that all the ^{60}Co spin angular momenta are *aligned* along the magnetic field **B**, as shown in Fig. 13.3(a). The electrons are emitted mostly in the direction opposite of the magnetic field **B**. In other words, electrons emitted due to the β-decay have a remarkable *preferential direction*, being anti-aligned with the direction of the magnetic field.

Due to angular momentum conservation, in the β-decay, shown in Fig. 13.3(a), the intrinsic spins of the emitted electrons are aligned *parallel* to the spin of ^{60}Co.

The parity transformed version of the experiment is shown in the Fig. 13.3(b). Under a parity transformation, the direction of the magnetic field and the direction of the spin are unchanged. Since $z \rightarrow -z$, under a parity transformation, if parity

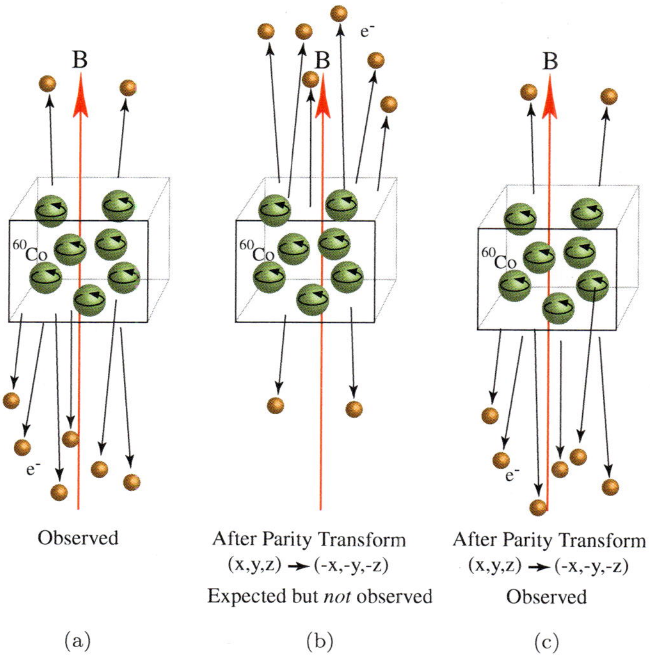

Fig. 13.3 Parity is not conserved in ^{60}Co beta decay.

were to be preserved, the electrons for the parity transformed case should *flip* their direction and be emitted along the $+z$-axis, as shown in Fig. 13.3(b).

However, since neither the direction of the magnetic field nor the direction of the spin of ^{60}Co changes under the parity transformation, there is no change in the direction of the emitted electrons.[4] The result of the experiment after a mirror reflection (parity transformation) is shown in Fig. 13.3(c). It is experimentally observed that the direction of the electron emitted by ^{60}Co is different from what is expected in the mirror reflected case, showing that parity is violated, namely that parity is not a symmetry of the Nature.

Note that the β-decay of ^{60}Co is also accompanied by the emission of photons. In contrast to the electrons and neutrinos that come out mostly in one direction, the photons are emitted in *all* directions with equal likelihood; and, in particular, an equal number of photons are emitted in the $+z$- and $-z$-direction. The emission of photons in all directions shows that the process is invariant under parity — as is to be expected since the electromagnetic interaction does not violate parity.

In summary, Fig. 13.3 shows the original experiment and the parity reversed experiment. Instead of showing the expected result that would conserve parity, the experiment gives a result that violates parity.

The violation of parity is due to a subnuclear process going on inside the nucleus of ^{60}Co; furthermore, since an antineutrino is emitted, the process is mediated by the weak interactions. Hence, we need to analyze the structure of weak interactions to see how a parity violating process can take place in Nature.

13.4 Fermions and Parity

The fermions of the Standard Model are analyzed to understand how they transform under the action of parity.

As can be seen from Fig. 13.1, fermions in the Standard Model come in three families. The other key feature of the Standard Model is that fermions do not respect the symmetry of parity, discussed in Sec. 13.3. We analyze these two key features of the fermions.

Consider a typical fermion, say an electron; the electron quantum field describes both the electron and the positron (antielectron); in addition, both the electron and positron have spin 1/2 that can point 'up' or 'down'; hence, the electron field has *four* components.

---

[4]If formulated like this, parity violation appears to be rather obvious. However, in light of the many fundamental phenomena that do preserve parity, this was nevertheless a fairly shocking discovery. Indeed Lee and Yang won the 1957 Nobel Prize for their work on parity violation in weak interactions.

The characterization of the components of the electron field in terms of particle and antiparticle, each pointing up or down, is a result of the way the field is represented. The four components of the electron field can be re-arranged in many ways depending on which feature of the field we are interested in.

The manner in which particles behave under the parity transformation is an intrinsic property of the particles; a representation is chosen for fermions for obtaining a transparent realization of the behavior of the fermions under the parity transformation. Instead of writing the components of the electron field in the electron and positron representation, we can use another representation in which the electron is viewed as being composed of a **left-handed** component e_L and a **right-handed** component e_R.

For a massless fermion traveling with momentum $\vec{p}$, the component e_R is spinning clockwise along the direction of motion and the component e_L is spinning counter-clockwise. The term right- and left-handed for e_R and e_L alludes to this property of their intrinsic spin; for massive fermions this is not the case.

Keeping in mind the action of parity on fermions, we write the electron field in the following manner

$$\mathrm{e} = \begin{pmatrix} e_R \\ e_L \end{pmatrix}. \tag{13.8}$$

Note that both e_R and e_L have *two* components and, as expected, the electron field has a total of four components.

The parity transformation, as discussed in Sec. 13.3, yields the following action on the fermions

$$(t, \vec{x}) \to (t, -\vec{x}) \tag{13.9}$$

$$\begin{aligned} &\Rightarrow \begin{pmatrix} e_R \\ e_L \end{pmatrix} = e \to P(e) = \begin{pmatrix} e_L \\ e_R \end{pmatrix} \\ &\Rightarrow P(e_R) = e_L; \quad P(e_L) = e_R. \end{aligned} \tag{13.10}$$

One can see from Eq. (13.11) that, under the parity operation, the left- and right-handed components e_L and e_R are *interchanged.* This simple action of the parity transformation on fermions, namely the interchange of the e_L and e_R components of the fermion, is the reason that the representation given in Eq. (13.8) was chosen for the fermions.

To illustrate the result of Eq. (13.11), note that under a parity transformation, the coordinate vector undergoes the transformation $\vec{x} \to -\vec{x}$ and the momentum of the fermion also flips a sign, that is, $\vec{p} \to -\vec{p}$; however, angular momentum remains unchanged under a parity transformation and hence the spin of the fermion does not change.

13.4.1 *Parity: Electron and neutrino*

The electron and neutrino are a canonical pair of leptons; their structure is repeated for the remaining two generations; to simplify the discussion we focus on only these two leptons.

Figure 13.4(a) shows that the electron has both a left- and a right-handed component; Fig. 13.4(b) shows that under parity the left- and right-handed electron components are interchanged; note that the spin assignment after the parity transformation is consistent with the definition of handedness, since a clockwise spin along the direction of motion goes into a counter-clockwise spin after reflection and *vice versa*. In contrast to the electron, the neutrino is a *single-handed fermion* and is shown in Fig. 13.4(c); under a parity transformation, the left-handed neutrino needs to go into a right-handed neutrino that, in fact, does *not* exist — as shown in Fig. 13.4(d).

The electron field has the possibility of being invariant under parity since it has *both* left- and right-handed components e_L and e_R; if all the interactions of the electron also respect the parity symmetry, then after applying the parity transformation, one can simply re-name the left- and right-handed components e_L and e_R and hence recover the original theory. In contrast, by its very construction, the neutrino can never be left unchanged under a parity transformation and has to break the symmetry of parity.

An analogy of parity violation is a person with one hand as contrasted with a person having two hands. Unlike the person with two hands, the appearance of the person with one hand can never be seen to be unchanged if a mirror reflection of that person is observed.

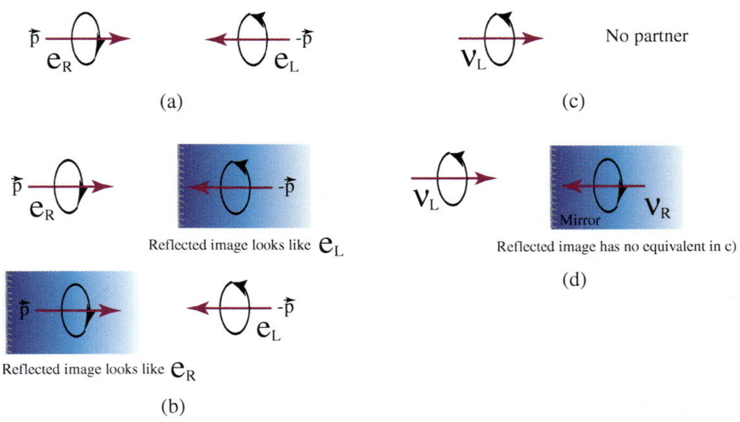

Fig. 13.4 Electrons have both left- and right-handed components whereas the neutrino has only the left-handed component.

All the quarks and leptons, with the exception of the neutrino, have both the left- and right-handed components and hence have the possibility of being parity symmetric.

The parity violation in β-decay is realized in electroweak theory by postulating that neutrinos violate the parity symmetry. **Parity violation** is also present in the coupling of fermions to the weak bosons, with the charged weak bosons $W^{\pm}$ *coupling only to the left-handed fermions*, and having no coupling to the right-handed fermions. This is called as having the **maximal violation of parity**.

We now examine, in some detail, how parity violation is realized in the electroweak theory by the mechanism of the neutrinos.

For notational simplicity, let ν denote the *electron neutrino*, that written out fully is denoted by ν_e in Fig. 13.1. All the subsequent discussions in this chapter will focus on the Family I of the fermions and hence the simplified notation should not conflict with any of the general notation.

Parity violation takes place in Nature due to the fact that the neutrino has only *one hand*, which by convention is taken to be the left hand. The neutrino is said to be a **chiral fermion** to remind ourselves that it has only one of the 'hands' of a fermion.[5] The neutrino field, namely $\nu_e = \nu$, has the following structure

$$\nu = \begin{pmatrix} 0 \\ \nu_L \end{pmatrix}.$$

In summary, the neutrino, in principle, must break the symmetry of parity since a parity transformation yields $\nu_L \rightarrow \nu_R$ and hence requires the existence of ν_R, which does not exist. A representation of parity violation is illustrated in Fig. 13.4(d).

13.5 Fermions and Weak Bosons: Parity Violating Couplings

As mentioned earlier, the electroweak bosons have a coupling to the quarks and leptons that violates parity symmetry. We now work out the parity violating coupling of the fermions with the electroweak bosons.

The three first generation fermions u, d and e have both the left- and right-handed components, with only the neutrino ν having a single hand. The u, d quarks, the electron e and the neutrino ν are written as follows[6]

$$\mathrm{u} = \begin{pmatrix} u_R \\ u_L \end{pmatrix}; \quad \mathrm{d} = \begin{pmatrix} d_R \\ d_L \end{pmatrix}; \quad e = \begin{pmatrix} e_R \\ e_L \end{pmatrix}; \quad \nu = \begin{pmatrix} 0 \\ \nu_L \end{pmatrix}. \tag{13.11}$$

---

[5]Chiral means handedness in Greek.

[6]Recall that each hand, say u_R, e_L and so on, stands for two components.

To couple the first generation fermions to the electroweak bosons, the left- and right-handed fermions are separated out and the first generation (Family I) fermions are organized as follows

$$\begin{pmatrix} \nu_L \\ e_L \end{pmatrix}, e_R; \quad \begin{pmatrix} u_L \\ d_L \end{pmatrix}; \quad u_R, d_R.$$

Note that the fundamental *asymmetry* between the leptons and the quarks is due to the absence of ν_R. Furthermore, the left- and right-handed quarks and leptons have asymmetric (parity violating) couplings to the weak bosons. Fermions of Families II and III have a similar parity violating structure.

The electroweak bosons $\gamma, Z^0, W^\pm$ have the following **parity violating couplings** to the fermions. The strangeness changing couplings given in Fig. 12.23(a) are ignored for simplicity.

- **The $W^\pm$ weak bosons couple to *only* the left-handed fermions**, namely e_L, ν_L, u_L, d_L; in particular, they have *no coupling* to the right-handed fermions, namely e_R, u_R, d_R. The coupling of $W^\pm$ to the fermions breaks the symmetry of parity. The coupling constant is denoted by g_w.
- **The Z^0 couples to *all* the fermions**, namely with e_L, ν_L, u_L, d_L as well as e_R, u_R, d_R. The coupling constant is denoted by g_z.
- **The photon γ couples to e_L, u_L, d_L and e_R, u_R, d_R**; the coupling constant is denoted by e. Note that γ does not couple to the neutrino ν_L since it has zero electric charge.

The detailed structure of the couplings are given in Secs. 12.7, 12.10 and 12.8.

Note that the couplings of the fermions to the electroweak bosons are clearly asymmetric, with the left-handed sector containing the neutrino ν_L having different couplings compared to the right-handed sector, for which there is no right-handed neutrino. What is remarkable is that the asymmetry observed in Nature due to parity violation can be represented by this choice of couplings.

13.6 Pairing of Fermions: Chiral Anomaly Cancellation

The fermions of the Standard Model, namely the quarks and leptons are coupled to the electroweak bosons $\gamma, Z^0, W^\pm$ in a manner so as to reproduce the parity violating processes that are occurring in Nature. In particular, recall from Sec. 13.5 that the $W^\pm$ weak bosons couple to only the left-handed **chiral fermions**, with no coupling at all with the right-handed fermions — thus violating parity symmetry. This asymmetric coupling results in the parity violating decay of the neutron. Furthermore, the fact that one has only the left-handed neutrino ν_L explicitly breaks parity invariance.

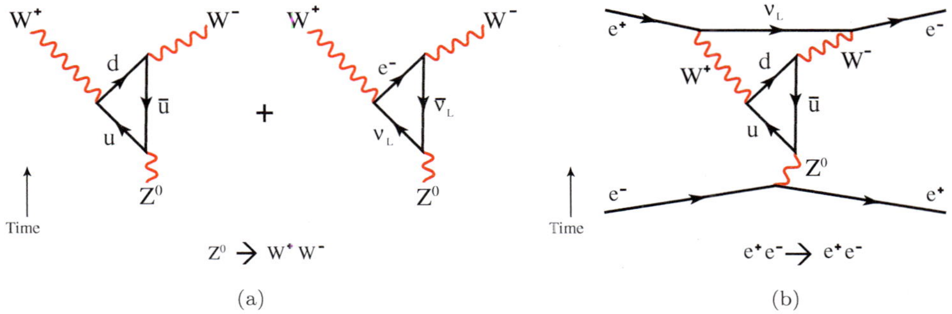

Fig. 13.5 Some Feynman diagrams containing the chiral anomaly.

The theory of electroweak interactions has to be renormalizable for mathematical consistency.[7] Certain infinities, called **chiral anomalies**, arise due to the parity violating couplings of the electroweak bosons $Z^0, W^\pm$ with the leptons and quarks and could spoil the renormalizability of electroweak interactions. The chiral anomalies must cancel for consistency.

To understand how the cancellation of the chiral anomaly takes place, note that whenever a chiral fermion appears inside a Feynman diagram that contains a fermion loop with *three* external gauge bosons — called a triangle diagram, of which two examples of this are shown in Fig. 13.5(a) — an anomaly can occur in the Feynman diagram.[8] Due to the parity violating nature of chiral fermions, the anomaly occurring in the triangle diagram cannot be removed by the procedure of renormalization and has to be absent in the theory. The only way that the anomaly due to the chiral fermions can be removed from the theory is if the anomalies occurring in *all* the allowed triangle diagrams somehow sum up and exactly cancel. Two Feynman diagrams containing the chiral anomaly are shown in Fig. 13.5(a).

There are four fermions in Family I — given in Fig. 13.1 — namely the u, d quarks and the e, ν leptons.[9] Each of the family (generation) of fermions generates a piece of the chiral anomaly. The chiral anomalies arise *separately* for each generation, so one can examine the problem generation by generation. One way that the total anomaly — recall that it arises due to the coupling of the chiral fermions to the electroweak bosons — can be canceled is that the anomaly arising from the u, d quarks be equal, but have the *opposite* sign, to the anomaly coming from the leptons e, ν.

As can be seen from the examples of triangle diagrams shown in Fig. 13.5(a), the Feynman diagrams show that the cancellation of the chiral anomaly depends

[7]The requirement of renormalizability is what led to the rejection of Fermi's theory of weak interactions as being inconsistent.

[8]It is a non-trivial result that if a chiral fermion occurs in a fermion loop with *more* than three external gauge bosons, then these Feynman diagrams do not have any chiral anomalies.

[9]Recall that we are using the notation $\nu = \nu_e$.

on the charges carried by the fermions, as this determines their coupling to the gauge bosons. Furthermore, as shown in Fig. 13.5(a), all the four fermions, namely, $\nu, \mathrm{e}, \mathrm{u}, \mathrm{d}$, occur in the Feynman diagrams; in other words, *all* the members of the first generation of quarks and leptons are necessary for **anomaly cancellation**. The anomalies that arise in triangle diagrams, by themselves, are finite, but when appearing inside higher order Feynman diagrams, as shown in Fig. 13.5(b), they lead to divergences that cannot be removed by the procedure of renormalization.

The anomaly cancellation requires that a **doublet of quarks** u, d must always be *paired* with a doublet of leptons e, ν_e. Note the non-trivial requirement that the electric charges of the leptons and quarks must be measured in the same unit of charge, namely e, for the anomaly cancellation to take place.

The electric charges of the quarks and leptons are given in Tables 11.2 and 11.3. Let Q_q be the electric charge of the quark q and let Q_ℓ be the electric charge of the lepton ℓ; then the anomaly cancellation requirement is given in the following equation

$$\sum_\ell Q_\ell + 3\sum_q Q_q = 0 \ : \ \text{Chiral anomaly cancellation} \qquad (13.12)$$

where the sum is over all leptons and quarks. The factor 3 multiplying the quark term in Eq. (13.12) arises from the three colors carried by the quarks, and over which the sum has been explicitly performed. For a mysterious and magical reason that is not yet understood, the fermions that appear in the Standard Model obey the condition of anomaly cancellation given in Eq. (13.12).

Recall that the anomaly cancellation takes place generation by generation, with no cross-generation mixing of the chiral anomalies. To illustrate the cancellation given in Eq. (13.12), it suffices to consider the charges (in units of electric charge e) of the first generation fermions, given as follows:

$$\begin{pmatrix} \mathrm{e}; Q_e = -1 \\ \nu; Q_\nu = 0 \end{pmatrix}; \quad \begin{pmatrix} \mathrm{u}; Q_u = 2/3 \\ \mathrm{d}; Q_d = -1/3 \end{pmatrix}.$$

In other words, in units of electric charge, the charge of the electron is $Q_e = -1$ and of the neutrino is $Q_\nu = 0$; of the u quark is $Q_u = 2/3$ and of the d quark is $Q_d = -1/3$; hence, from Eq. (13.12)

$$Q_e + Q_\nu + 3(Q_u + Q_d) = -1 + 0 + 3(2/3 - 1/3) = 0 \ \text{ as required.}$$

The anomaly cancellation is not just a matter of mathematical consistency devoid of physical content. Up until 1974, only three quarks, namely the u, d and s quark, had been discovered. A calculation of the chiral anomaly showed that electroweak theory is inconsistent unless a fourth quark — the partner of the s quark — also exists; an experimental search led to the discovery of the charm quark c completing Family II of the Standard Model, as shown in Fig. 13.1. The fermions of Family III also exactly cancel the chiral anomaly.

In summary, the three generations of fermions of the Standard Model, generation by generation, are arranged in such a manner so as to guarantee the cancellation of the total chiral anomaly.

13.7 Unification of the Weak and Electromagnetic Interactions

The **unification** of the photon γ, accounting for electromagnetism, with the weak forces $Z^0, W^\pm$ is expressed in there being a *single* coupling constant e determining both these forces.

As given in Eq. (12.13), the coupling constants are unified by the following relation

$$e = g_w \sin(\theta_w) = g_z \cos(\theta_w)$$

where θ_w is the so-called weak mixing angle; this angle arises due to the physical photon being a linear combination of two of the boson fields that go into building the electroweak theory.

The unification of the electromagnetic and the weak interactions is related to the necessity of having a finite value for the Feynman diagrams that occur in high order processes — in which the photon γ and the $Z^0, W^\pm$ appear inside loops of Feynman diagrams. For example, consider the annihilation of an electron and positron to a muon and antimuon given by

$$e^+ + e^- \rightarrow \mu^+ + \mu^-.$$

The one loop Feynman diagram for this process is given in Fig. 13.6; the photon γ appears in one of the loops together with a Z^0, whereas in the other two diagrams the $Z^0, W^\pm$ are in the loop; the sum of the divergences of these three Feynman diagrams must cancel, and for the cancellation to occur the charges that couple γ and $Z^0, W^\pm$ to the fermions have to be unified. And in general, in order for all the infinities that appear in the high order Feynman diagrams to cancel, the coupling constants of the electromagnetic and weak interaction have to be unified as given by Eq. (12.13).

The unification of the electromagnetic and the weak interactions also arises from the necessity of canceling all the chiral anomalies, as discussed in preceding Sec. 13.6. The anomaly cancellation condition given in Eq. (13.12) relates the charge of the

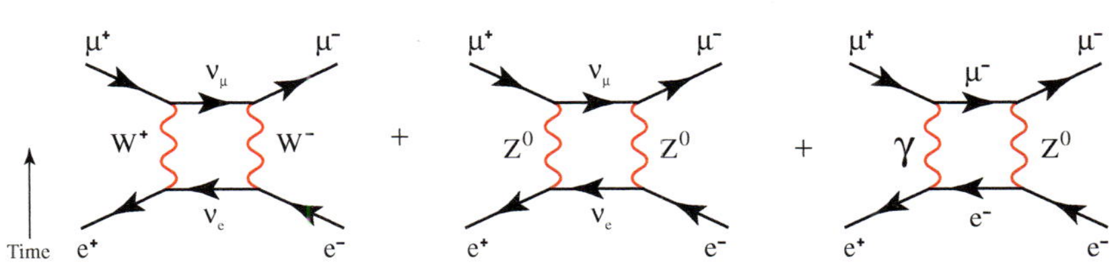

Fig. 13.6 Some Feynman diagrams for the reaction $e^+ + e^- \rightarrow \mu^+ + \mu^-$.

leptons and the quarks, and the cancellation is only possible due to a fine balance of all the fermion fields that couple to the weak and electromagnetic interactions.[10]

It is worth noting that both the unification of the charges of weak and electromagnetic interactions as well as the cancellation of the chiral anomalies are necessary for removing infinities from physically measurable quantities. This requirement of obtaining finite results from quantum field theory may seem mysterious since why should the *mathematical* consistency of a physical theory impose conditions of what can occur in Nature? A possible reason is that there may be some higher physical principle that appears in a more comprehensive theory, such as superstring theory discussed in Chapters 14 and 15, and that these principles manifest themselves in terms of mathematical requirements in the Standard Model.

13.8 The Higgs Field and Phase Transition

There is one last ingredient that goes into the theory of electroweak interactions, namely the **Higgs field** and its associated quantum particle, called the **Higgs boson**. The Higgs field is a scalar quantum field, and is denoted by Φ (not to be confused with the Coulomb potential ϕ). We briefly discuss the reason why this field is required.

Recall that the electromagnetic and weak interactions are unified by requiring that all the bosons of electroweak interactions have a strength of coupling to quarks and leptons that is determined by the coupling constant e. The weak coupling constants g_w, g_z are derived from e, as discussed in Sec. 12.8 and given in Eq. (12.13).

In unifying the weak and electromagnetic interactions, a question that one is confronted with is the following: if the electroweak interaction is truly a unified theory, why do the weak interactions have a range of interaction that is 10^{-18} m whereas the electromagnetic interaction has an infinite range? Or what is the same thing, why do the weak bosons $\mathrm{Z}^0, \mathrm{W}^\pm$ have very large masses compared to the photon that is massless? In other words, how does the unified theory produce such a dramatically different behavior for the weak and electromagnetic sectors of the theory?

More precisely, how do the weak bosons $\mathrm{W}^\pm, \mathrm{Z}^0$ — which of necessity must 'start off' as being massless due to the need for renormalizability (similar to the photon) — become massive, as is required by the weak interactions? The Higgs field was introduced into the electroweak theory precisely for the sake of giving mass to the weak bosons and remarkably enough, it rather unexpectedly gives masses to the all the fermions (except the neutrino) as well.

The mechanism by which the Higgs field gives masses to the weak bosons and fermions lies in the phenomenon of *phase transition*. The electroweak theory at

[10] A possible reason that the lepton and quark electric charges are unified is discussed in Sec. 14.2.2.

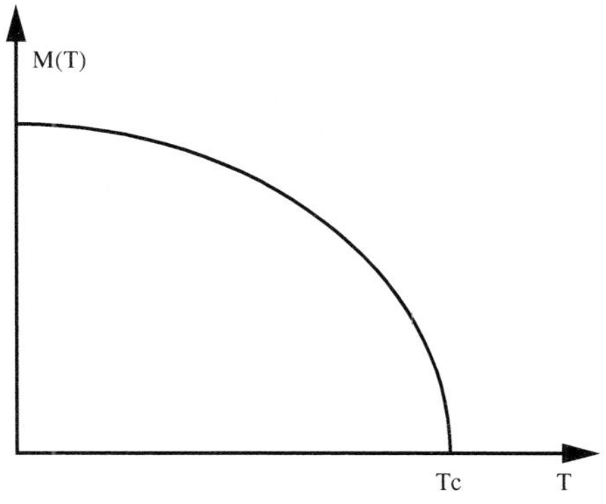

Fig. 13.7 A non-magnetic to magnetic phase transition at critical temperature T_c, giving rise to spontaneous magnetization $M(T)$.

the Big Bang starts off with massless photons, massless weak bosons and massless fermions; everything is massless. The Higgs field undergoes a condensation — a phase transition — as the Universe cools. It is the condensed phase of the Higgs field that gives rise to the masses of the particles.

13.8.1 *Phase transition*

We briefly discuss phase transitions and the associated phenomena of symmetry breaking and restoration. A leading exemplar of a phase transition is that of a ferromagnet; at high temperatures a ferromagnet is non-magnetic. As shown in Fig. 13.7, when it is cooled below a certain critical temperature T_c, the ferrogmagnet undergoes a phase transition and develops a spontaneous and permanent magnetic field M. Iron, cobalt, nickel and certain rare earths are ferromagnetic.

The inter-atomic interactions of a ferromagnet are spherically symmetric; when iron is cooled below 1043 K, called the Curie temperature, iron develops an average value for its macroscopic magnetic field M. Since the magnetic field must point in a *specific* direction, the ferromagnet *spontaneously* chooses a direction for its magnetization and, in doing so, breaks spherical symmetry. If one heats a ferromagnet above its Curie temperature, it loses its magnetism and spherical symmetry is restored.

13.8.2 *Higgs condensation*

The behavior of the Higgs field is similar to the ferromagnet in that it also undergoes a phase transiton. The condensed phase of the Higgs field is analogous to the spontaneously magnetized phase of the ferromagnet.

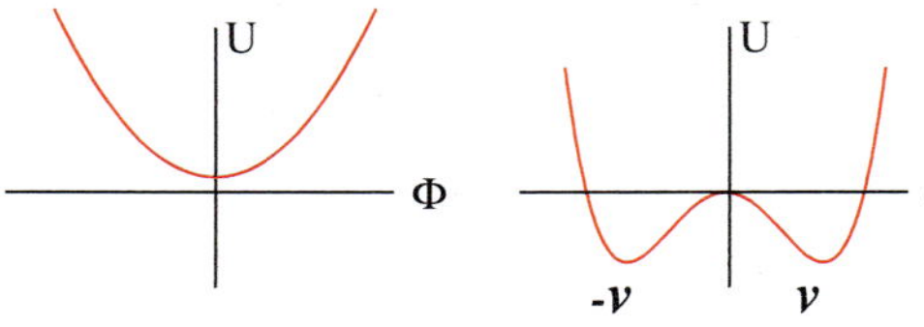

(a) $T > T_c$ (b) $T < T_c$

Fig. 13.8 A double well potential for the Higgs boson.

The reason for the phase transition is related to the vacuum state, namely, the lowest energy state, as has been discussed in Sec. 6.11 on inflation; we repeat this discussion in the context of the Higgs field. The potential for the Higgs field, for both $T > T_c$ and $T < T_c$, is shown in Figs. 13.8(a) and (b), respectively. For $T > T_c$, the potential has a minimum at $\Phi = 0$, and the Higgs field is in its non-condensed phase with $E[\Phi] = 0$. For $T < T_c$, the potential has a minimum at $\Phi = \pm v$, and the Higgs field is in its condensed phase with $E[\Phi] = \pm v$.

At the **Big Bang**, since $T > T_c$ the Higgs field is in its **non-condensed phase** — and all the fermions and weak bosons are massless; this fact has observable implications for the Physics of the early Universe, as discussed in Sec. 6.9. It is estimated that 10^{-12} s after the Big Bang, the Universe cools to the critical temperature $T_c = 10^{16}$ K, mentioned in Eq. (6.41), with the Higgs field undergoing a phase transition and spontaneously condensing throughout the Universe. The reason being that for $T < T_c$, the potential in Eq. (13.13), shown in Fig. 13.8(b), develops two minima for the field — with the constant value of either v or $-v$.

The potential energy $\mathcal{U}[\Phi]$ of the Higgs field, for temperatures less than $T < T_c$, is given by[11]

$$\mathcal{U}[\Phi] = -\lambda \left(\Phi^2 - v^2\right)^2; \quad \lambda > 0. \tag{13.13}$$

Hence, when the Universe cools to $T_c = 10^{16}$ K, the vacuum state of the Higgs field shifts, similar to what is shown in Fig. 6.10, from the Higgs field having an average value of zero to having an average value of either v or $-v$. The Higgs field *spontaneously* chooses the value of v and yields the average vacuum value of the field as given by

$$E[\Phi] = v. \tag{13.14}$$

The empirical value of v is equal to $246\,\text{GeV}/c^2$.

The phase transition that the Higgs field undergoes is quite different from the phase transition in a ferromagnet or any similar phase transitions that can take

[11]For $T > T_c$, the parameter $\lambda < 0$ and hence there is no minima and no phase transition.

place in a laboratory. The main difference being that the Higgs vacuum permeates the entire Universe — with the Higgs field having a constant value v at every point of spacetime.

It can seen from Eq. (13.14) that the Higgs field could have chosen to have its equilibrium value at $-v$ or at $+v$ and it spontaneously chooses one of them, which is taken to be $+v$. The Higgs field is said to spontaneously break 'gauge symmetry', since having a vacuum value of v breaks the symmetry of weak interactions under gauge transformations. If the Higgs field is heated to its critical temperature of 10^{16} K, its vacuum expectation v will go to zero and the broken symmetry will be restored.

We summarize the difference of weak interactions in the non-condensed and condensed phase of the Higgs field.

13.8.2.1 *Non-condensed phase: $T > T_c$*

- The temperature of the Universe is higher than 10^{16} K and the average vacuum value of the Higgs field is zero, namely $E[\Phi] = 0$.
- The Higgs field consists of two *complex* charged scalar fields, with one field being coupled to the electromagnetic field by electric charge e and the other field coupled by charge $-e$. Each complex scalar field consists of two real scalar fields and hence, in the non-condensed phase, the Higgs field consists of a total of *four* real scalar fields.
- The weak bosons W^+, W^-, Z^0 are all massless. Because they are massless, the charged bosons W^+, W^- each consists of two real fields and carries a charge of $+e$ and $-e$ respectively; the neutral weak boson Z^0 consists of two neutral real fields.
- All the quarks and leptons are massless.

13.8.2.2 *Condensed phase: $T < T_c$*

- The temperature of the Universe is less than 10^{16} K and the average vacuum value of the Higgs field is non-zero, namely $E[\Phi] = v$. The present temperature of the Universe, in fact, is 2.7 K, and so we are well within the condensed phase.
- The charged boson W^+ can be massive only if it has three independent fields. The boson W^+ absorbs one real scalar component of the Higgs field as well as its coupling via $+e$ charge to the electromagnetic field; in doing so, the boson W^+ becomes *massive*. The electric charge carried by the additional component of W^+ is crucial since all three components of W^+ must carry $+e$ charge.
- The W^- boson similarly absorbs one of the components of the Higgs field and inherits its negative electric charge and thus becomes a massive charged boson, with three real and charged components. A similar point — relating to the electric charge of the additional component discussed for W^+ — holds for W^- as well.
- The weak boson Z^0 absorbs one neutral real scalar field and also becomes massive.

- Moving from the non-condensed phase to the condensed phase, the Higgs field loses *three* of its real scalar components as well as its coupling to the electromagnetic field and becomes an electrically neutral scalar field, not carrying any electric charge.
- Due to their Yukawa coupling to the non-condensed Higgs field, in the condensed phase all the quarks and leptons, except for the neutrino, acquire a mass proportional to the average vacuum value of the Higgs field, namely proportional to v.

13.8.3 *The Higgs mechanism*

In high energy physics, the acquisition of masses by bosons and fermions through their interaction with the Higgs field is known as the **Higgs mechanism**.

In the absence of their interaction with the Higgs field, the weak bosons as well as all the fermions are massless. The interaction of the weak bosons and fermions with the Higgs field does not necessarily result in the bosons and fermions acquiring a mass; instead, it is the spontaneous breaking of gauge symmetry by the Higgs field due to the Higgs field acquiring a vacuum expectation value v that results in the weak bosons and fermions (except for the neutrino) becoming massive. The Higgs mechanism is the result of the subtle interplay of gauge invariance and spontaneous symmetry breaking.

As shown in Fig. 13.9, due to the vacuum value of the Higgs field, the bosons are in an environment that 'impedes' their unhindered propagation and results in giving them a mass. One can think of the fermions and bosons moving in a viscous fluid, and viscosity 'slowing' down their motion thus resulting in an effective mass.

13.8.4 *Higgs interactions*

The coupling of the Higgs particle to the other particles of electroweak interactions in the condensed phase is given in Fig. 13.10.

The strength of the Higgs coupling to the electroweak gauge bosons goes as m_b^2/v^2, where m_b is the mass of a boson and v is the Higgs field's vacuum expectation value; the Higgs coupling to the fermions goes as m_f/v, with m_f being the fermion mass.

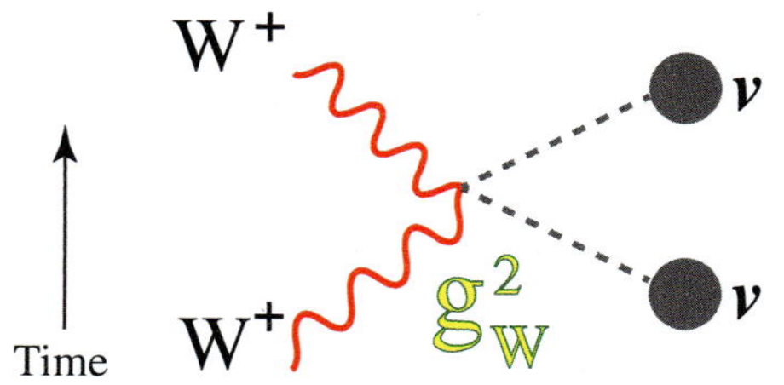

Fig. 13.9 Weak boson acquiring a mass by its interaction with the Higgs vacuum.

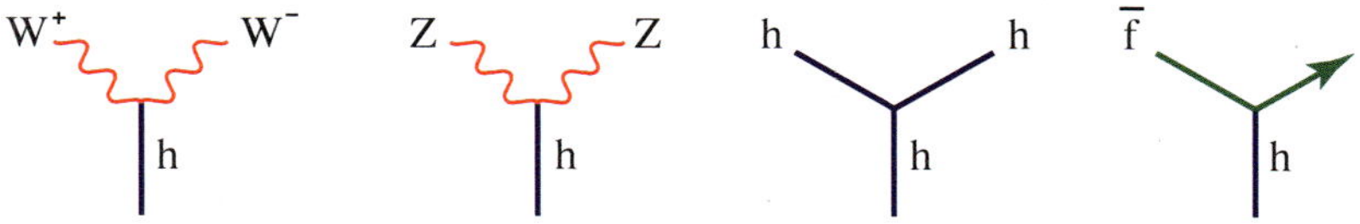

Fig. 13.10 The Higgs particle, with its line denoted by h, coupling to other particles and to itself. Fermions and antifermions are denoted by $f, \bar{f}$.

The heaviest boson is the Higgs particle with a mass m_H of about 125 GeV/c^2; and its coupling constant λ is given by

$$\lambda_H = \frac{m_H^2}{v^2} \simeq 0.25.$$

The heaviest fermion is the top quark with a mass m_t of about 175 GeV/c^2 and yields

$$\lambda_t = \frac{m_t}{v} \simeq 0.71.$$

The significance of the fact that λ is less than unity is far reaching: since the Higgs coupling is small, all the observable effects of weak interactions can be computed and analyzed using Feynman diagrams. In particular, the physical particles that can be observed in experiments, together with their quantum numbers, can be read off from the Feynman diagrams. Otherwise, if the couplings λ were larger than unity, weak interactions would be like strong interactions, for which its particle content is not given by Feynman diagrams; the computations would need other techniques, such as numerical simulations, and would make the analysis and predictions much more difficult to carry out.

13.9 The Masses of Electroweak Particles

As discussed above, in the condensed phase, the Higgs field Φ is completely specified by *one* neutral scalar field.

As is the case with other quantum fields, the excitations of the Higgs scalar field are quantized and the first excitation of the Higgs field above the vacuum is the Higgs particle. In other words, the Higgs particle is a quantum particle that is an excitation of the underlying Higgs field Φ. The Higgs particle is a boson since the Higgs field is a bosonic field and the particle has zero spin since the Higgs field is a scalar.

The **Higgs boson** acquires a mass due to the non-zero vacuum expectation value v of the Higgs field, and is given by

$$m_{_H} = kv = (125.3 \pm 0.6)\ \text{GeV}/c^2 \tag{13.15}$$

where k is a constant.

The great difficulty encountered in the detection of the Higgs boson is because its mass is not directly related to any of the weak interaction processes from which one could estimate its mass. So the design of an experiment and detector to find the Higgs boson had to consider a wide range of possibilities. This is one of the reasons that it took so long to find the Higgs boson.

13.9.1 *Masses for the weak bosons*

The vacuum value of the Higgs field, namely $E[\Phi] = v$, gives rise to the masses of the weak bosons $\mathrm{W}^{\pm}, \mathrm{Z}^0$ while keeping the mass of γ, the photon, equal to zero. The precise expression is given by

$$m_w = m_{w^-} = m_{w^+} = \frac{1}{2} v g_w; \quad m_\gamma = 0 \tag{13.16}$$

$$m_z = \frac{1}{2} v g_w \sqrt{1 + \tan^2 \theta_w} \Rightarrow m_w = m_z \cos \theta_w \tag{13.17}$$

$$v \equiv E[\Phi] \simeq 246 \text{ GeV}/c^2.$$

Note that the weak boson masses are all proportional to v and their experimental values are given in Eq. (12.12); for completeness, the masses are given below.

$$m_w \simeq 91.1887 \pm 0.0022 \,\text{GeV}$$

$$m_z \simeq 80.410 \pm 0.180 \,\text{GeV}.$$

The photon remains massless even though the other weak bosons pick up a mass via the Higgs mechanism. The reason for this is that the Higgs field acquiring a vacuum expectation value breaks the gauge symmetry only for the weak bosons $\mathrm{Z}^0, \mathrm{W}^{\pm}$ but preserves the gauge symmetry of γ, the photon, which results in the photon remaining massless.

For the theory to be consistent with renormalization requires that the weak bosons be exactly gauge invariant. Introducing a mass term directly for the weak bosons breaks the gauge symmetry of the theory — a symmetry that is indispensable for obtaining a finite renormalized theory. Renormalization removes infinities that occur for quantum field theories that are solely due to the interactions at very short distances. In particular, the properties of the vacuum are large distance effects that do not affect the short distance behavior of the weak bosons. The key feature of the Higgs mechanism that preserves the renormalizability of the weak bosons is that it endows mass to the weak bosons by changing the properties of the vacuum, and hence does not spoil the renormalizability of the weak interactions.

13.9.2 *Masses for the fermions*

A mass term of the fermions, for example the electron, couples the left- and right-handed components of the electron field and is given by the following

$$m_e(\bar{e}_L e_R + \bar{e}_R e_L) \ : \ \text{Electron mass term} \tag{13.18}$$

where $\bar{e}_R, \bar{e}_L$ are the conjugate fields of e_R, e_L.

A mass term like the one in Eq. (13.18) is perfectly allowed by the renormalizability of the fermion field.

Since the mass term given in Eq. (13.18) couples the left- and right-handed components of the fermion, it comes into conflict with the coupling of the fermions to the weak bosons. As discussed in Sec. 13.5, the left- and right-handed components, namely e_L, e_R, couple very differently to the weak bosons. The fermion mass term given in Eq. (13.18) would violate the gauge symmetry of the electroweak theory — and is forbidden by the requirement of its renormalizability.

To have exact gauge invariance, the Standard Model starts off with all the fermions being massless. However, it is a well known experimental fact that quarks and leptons have masses. So how can one make the Standard Model yield the observed fermionic masses? The Higgs field comes to the rescue once again. The Higgs field has a coupling to the fermions, namely the Yukawa coupling, discussed in Sec. 11.10. The Higgs vacuum value v interacts with the fermions via the Yukawa coupling — by flipping a left-handed fermion e_L into a right-handed e_R one, as shown in Fig. 13.11; this interaction provides the cross-term of e_L with $\bar{e}_R$ and hence yields the required fermion mass term. The vacuum expectation v induces a mass term for the fermions by yielding a term of the form given in Eq. (13.18).

All the fermion masses are *proportional* to vacuum expectation v; in other words, for a typical fermion mass m_f, we have

$$m_f = \Gamma v$$

where Γ is a constant.

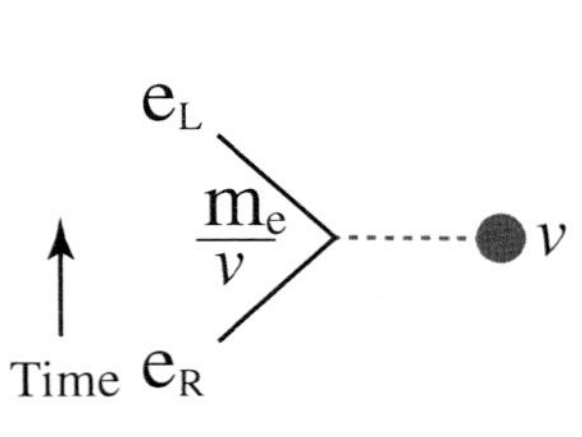

Fig. 13.11 Fermions interacting with the Higgs vacuum results in the left-handed fermion being flipped into a right-handed one and hence picking up a mass.

Noteworthy (optional content) 13.2: Determining the Electroweak Parameters

- The value of the mixing angle θ_w can be determined by calculating the likelihood of the annihilation of an electron and a positron — of sufficiently high energy — leading to the creation of a pair of muons, namely $e^+ + e^- \to \mu^+ + \mu^-$.
- Once θ_w is determined, both g_w, g_z can be obtained from Eq. (12.13).
- The vacuum expectation value of the Higgs field $E[\Phi] = v$ can be determined from the lifetime of a antimuon μ^+.
- From g_w and v, one can calculate m_w, the mass of the $W^\pm$ bosons, given by Eq. (13.16).
- Once m_w is known we can obtain mass m_z from Eq. (13.17).
- The maximum number of fermion families is experimentally fixed by the production of the Z^0 weak boson as an intermediate state in the process $e^+ + e^- \to \ell^+ + \ell^-$, where ℓ is any charged lepton. Experimental result shows that there are only *three* fermion families, and in particular, excludes four or more families.

13.10 Superconductivity and Higgs Mechanism

Superconductivity was discovered in 1911. At temperatures below a critical temperature T_c, which ranges from 1 K to 20 K, an ordinary conductor like copper or mercury becomes a perfect conductor and has zero electrical resistance.

The explanation of superconductivity based on quantum field theory is known as the BCS-theory. To make the analogy with the Higgs mechanism and spontaneously broken gauge symmetry, it is more convenient to use the phenomenological Landau–Ginzburg description of superconductivity.

In the Landau–Ginzburg approach, superconductivity is described by the interaction of two quantum fields, namely the Maxwell gauge field A_μ and a *complex* valued scalar field Φ (which can be thought of as two real valued scalar fields). Above the critical temperature T_c the average value of all fields is zero, namely

$$E[\Phi] = 0 = E[A_\mu]; \quad T > T_c.$$

When the conductor is cooled below T_c, the system undergoes a *phase transition*. The average value of the scalar field below the critical temperature T_c is no longer zero and the superconducting phase is given by the following

$$E[\Phi] = a; \quad E[A_\mu] = 0; \quad T < T_c.$$

The Landau–Ginzburg approach is the simplest way of obtaining the superconducting phase: one *postulates* that the scalar field Φ has a potential similar to the

Higgs potential given in Eq. (13.13), namely[12]

$$U(\Phi) = -g(|\Phi|^2 - a^2)^2$$

where $|\Phi|$ is the absolute value.

In summary, the following are the two phases for the superconductor.

- In the normal phase, the average value of the scalar field is zero, namely

$$E[\Phi] = 0; \quad E[A_\mu] = 0; \quad T > T_c.$$

- In the superconducting phase

$$E[\Phi] = a; \quad E[A_\mu] = 0; \quad T < T_c.$$

Under a gauge transformation, the complex scalar field undergoes the following transformation

$$\Phi \rightarrow e^{i\chi}\Phi.$$

This transformation changes a to $e^{-i\chi}a$ and hence no longer leaves the superconducting phase unchanged; since a is an experimentally measured quantity, its value cannot change and hence the superconducting phase is not invariant under a gauge transformation and consequently *breaks gauge invariance.*

To understand the significance of the superconducting phase, consider for simplicity a semi-infinite superconductor occupying the positive x-axis, as shown in Fig. 13.12. We examine the Meissner effect, which states that the magnetic field inside a superconductor is zero.

On imposing an external magnetic field B, the magnetic field does not suddenly drop to zero inside the superconductor; instead the magnetic field is non-zero in a *boundary layer* of the superconductor that is specified by a **penetration depth** of λ. The scalar field is also affected by the magnetic field to a depth into the

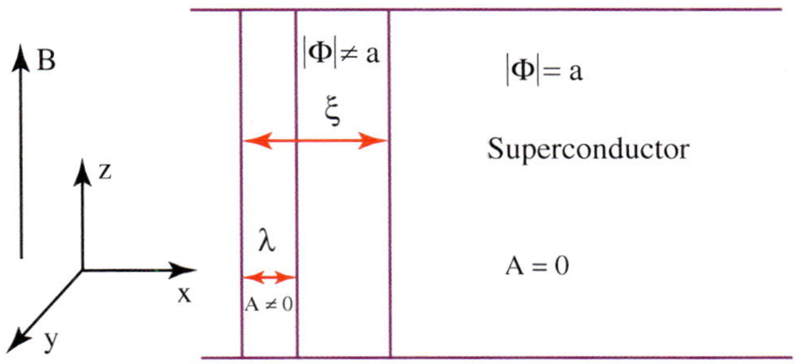

Fig. 13.12 Boundary layer for a superconductor. The correlation length ξ and penetration depth λ are distances determined by the behavior of the scalar and the Maxwell gauge fields.

[12]Note the *interpretation* of the potential is very different from the Higgs' case.

superconductor given by ξ, called the **correlation length**. Hence the boundary layer is defined by *two* length scales, namely λ and ξ. The behavior of the fields in the boundary layer is the following

$$x > 0 \; : \; A \simeq e^{-x/\lambda}; \quad |\Phi| = a(1 - e^{-x/\xi}) \tag{13.19}$$

where A is the magnitude of the gauge field.

In the boundary layer, the material is perturbed away from its condensed phase for which $|\Phi| = a$; the two length scales determine the distance (scale) over which the external magnetic field 'disturbs' the superconducting phase and are given by

$$\lambda = \frac{1}{2ea}; \quad \xi = \frac{1}{2a\sqrt{g}}$$

where e is the electric charge. The typical values of ξ and λ are 10^{-8} or 10^{-9} meters.

A detailed study based on the BCS-theory shows that the scalar field Φ is a (simplified) representation of the bound state of a *pair of electrons*, called the **Cooper pair**. The correlation length ξ is seen to be the average separation of the electrons in the Cooper pair.

13.10.1 *Analogy with the Higgs mechanism*

The Higgs mechanism is similar to the mechanism responsible for superconductivity. The complex scalar field Φ plays the role of the Higgs field. Above the critical temperature, the conductor is ordinary — having a finite electrical resistance — and is similar to the non-condensed phase of the Higgs field. The conductor makes a phase transition to its superconducting phase when it is cooled below its critical temperature. The superconducting phase is analogous to the condensed phase of the Higgs field.

The superconducting phase is determined by a non-zero value, given by a, of the average value of the Φ field. The mass of the Higgs particle is analogous to the inverse of the correlation of the scalar field, namely

$$m_H \; : \; \frac{1}{\xi} = 2a\sqrt{g}.$$

Due to the superconducting phase transition, a photon inside the superconductor becomes massive by the breaking of gauge invariance and acquires a longitudinal component by absorbing one of the components of the (complex) scalar field Φ. The mass of the photon inside the superconductor is the analog of the mass of the weak bosons, namely

$$m_W \; : \; \frac{1}{\lambda} = 2ae.$$

Note that both the lengths λ, ξ are proportional to the value of $a = E[\Phi]$ and this is the reason that both the masses m_H, m_W are proportional to v.

The analogy of superconductivity with the Higgs mechanism shows that the phase transition of the Higgs field leads to a mass for the Higgs particle. Furthermore, due to the breaking of the gauge symmetry by the superconducting phase transition, the photon gains a mass inside the superconducting material. In a similar manner, the condensation of the Higgs field leads to the weak gauge bosons — which in the case of the Standard Model are the bosons $W^{\pm}, Z^0$ — all acquiring a mass.

The Landau–Ginzburg description of the bound state of two fermions (electrons) by a scalar field Φ is analogous to the Yukawa scalar pion field that yields π^+, π^-, π^0, discussed in Sec. 11.10, and which is a low energy representation of the bound state of a quark and an antiquark. There have been attempts to follow further the analogy with superconductivity and to model the Higgs scalar as the bound state of some — as yet undiscovered — underlying fermions. All these attempts have yielded no results and the Higgs scalar field is now considered to be a fundamental field, on par with the other fields that yield the elementary particles and fundamental forces of Nature.

13.11 Masses of Quarks and Leptons

The Higgs vacuum expectation value v yields finite masses for quarks and leptons, and which have been experimentally measured; Tables 11.2 and 11.3 give a summary of the results. The antiparticles for quarks and leptons have another similar table, with their masses being exactly equal to the particles but with all the quantum numbers having the opposite signs. Note that the Standard Model yields zero for the masses of all three neutrinos, and the non-zero upper bounds for their masses, given in Table 11.2, are based on models that are variations of the Standard Model.

If we ignore the color charge of the quarks, there are equal number of leptons and quarks — six at present — as required by the mathematical consistency of the Standard Model; and this is borne out by experiments. If one includes color charge, there are $3 \times 6 = 18$ quarks and 6 leptons, yielding a total of 24 fundamental fermions.

A shown in Fig. 13.13, quarks and leptons are grouped into 'generations' which only differ from generation to generation by having progressively higher masses. The fundamental units of the leptons are the electron and its neutrino that is matched by the fundamental units for quarks made from the up and down quarks, namely

$$\begin{pmatrix} \mathrm{e} \\ \nu_e \end{pmatrix}; \quad \begin{pmatrix} \mathrm{u} \\ \mathrm{d} \end{pmatrix}.$$

The other two generations of leptons and quarks replicate this structure of matching the lepton and quark doublet, generation by generation. So far all experiments show that there are only three generations of fermions and no more.

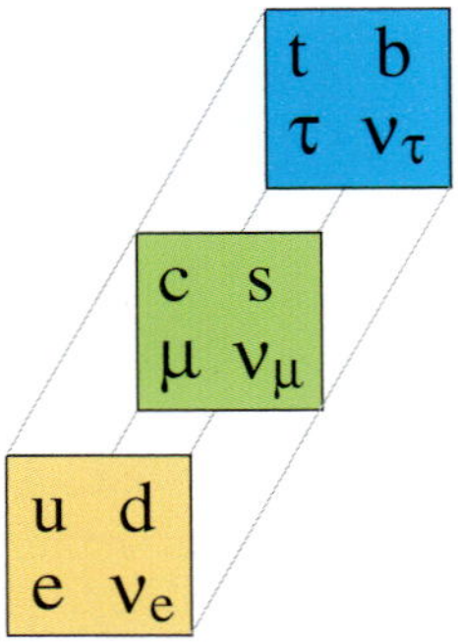

Fig. 13.13 Three generations of particles.

So far, no theoretical explanation has been made as to why Nature chooses to have only three generations of quark–lepton doublets — and this is one of the major unanswered questions of the Standard Model. In Chapter 14 we will see that, in one of the allowed models of superstring theory, the number of fermion generations is fixed by the topology of the extra hidden dimensions of space that are postulated to exist.

13.12 Large Hadron Collider

All the particles in Fig. 13.1, except the Higgs boson, were experimentally detected by 1994, with the last one being the top quark. So the detection of the Higgs boson in 2012 brings the Standard Model of the fundamental particles and forces of Nature to a closure.

There are many aspects to the detection of the Higgs boson. For starters, the sheer experimental complexity required for detecting the Higgs boson is staggering and we now review the accelerators and the detectors.

The **Large Hadron Collider** (LHC) is a particle accelerator that is located in Geneva; an aerial view of its surface is shown in Fig. 13.14. It took 10 years to complete in 2008, and lies in a tunnel 175 m below ground and with a circumference of 27 km. It has been built with the participation of more than 10,000 scientists and engineers from over 100 countries. At a cost of 7.5 billion euro, the LHC is the most costly scientific equipment ever built.

As illustrated in Fig. 13.15, the LHC accelerates protons in two beams, one going clockwise (shown in red) and the other going counter-clockwise (shown in green), to an energy of almost 7 TeV (trillion electron volts) and colliding at four intersection points. In the center of mass frame, the collision energy can reach up to 14 TeV — equal to energy that prevailed 10^{-13} seconds after the Big Bang, as discussed in Chapter 6.

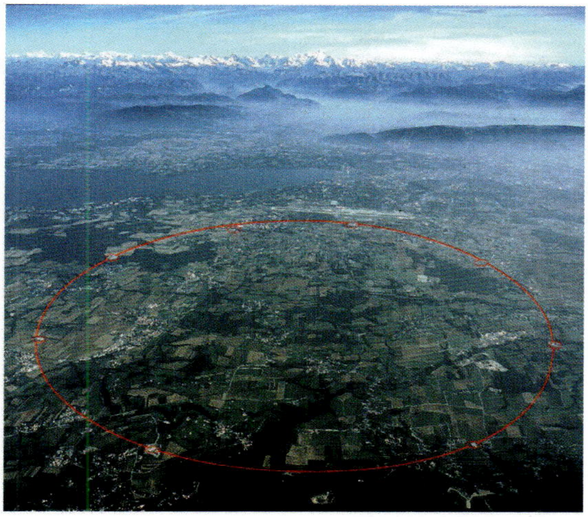

Fig. 13.14 An aerial view of the Large Hadron Collider.

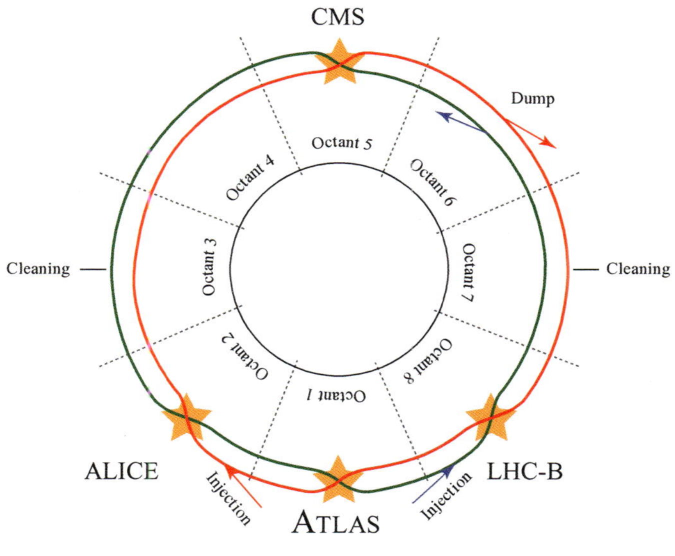

Fig. 13.15 The LHC collider, showing the four collision sites.

Straddling the LHC accelerator are four giant detectors — positioned at the points where the two proton beams moving in opposite directions intersect — and designed to study different properties of the particles created by the collisions.

Two independent experiments to detect the Higgs boson were conducted by two separate groups, namely the **Compact Muon Solenoid** (CMS) experiment and the **Atlas** experiment. The two experiments are independent of each other for two reasons: firstly, they are located at different locations and hence the readings are independent of each other; secondly, and more importantly, the designs and detectors of the two experiments are completely different and hence probe the physics of the Standard Model using very different features of the theory.

The **Higgs boson** was detected by both the Atlas and CMS experiments. The independent confirmation of the experimental result is an important criterion for the validity of the detection of the Higgs boson. It is remarkable that both the experiments reported the detection of the Higgs boson with the same mass, within the experimental errors.

13.13 The Atlas Experiment

For the purpose of illustrating the complexity of the experiment, a discussion in some detail is given of the Atlas experiment; the CMS experiment is equally complex and important and uses a different set of detectors.

Atlas is the largest collider detector ever built, having dimensions of 46 m × 25 m × 25 m and weighing 7000 tons. It is almost half the size of the Notre Dame Cathedral in Paris and weighs as much as the Eiffel Tower. The experimental team running Atlas consists of 1900 scientists and engineers from 35 countries.

State-of-the-art technology, including superconducting magnets and advanced software, is essential for the functioning of Atlas. The customized software used for detecting the Higgs boson was running on a worldwide distributed computing

Fig. 13.16 The Atlas experiment.

grid operating at 285 sites in 48 countries and having 10^{15} bytes of disk space and working at the limits of what was possible.

Every second, 1 billion collisions take place in the LHC; of these 40 million collisions take place inside Atlas and most of which can be detected; recording all this data would amount to about 100,000 GB per second. The enormous amount of collision data is scanned in real time and the software then makes a decision, in a few millionths of a second, whether to reject or record the result of a particular collision.

For the experiment that detected the Higgs boson, all the events that could be explained by the existing theories of particle physics were rejected, and the detector recorded only a small set of data, equal to about 20 GB per minute, directly relevant to the detection of the Higgs boson.

The recorded data was then analyzed for the tell-tale signals of the Higgs boson. The production of the Higgs particle is a very rare event; for the decay modes that were detected by the Atlas and CMS detectors, namely the Higgs boson decaying to either two photons or four leptons, only one Higgs boson is produced for every billion (10^9) proton–proton collisions; ultimately over 300 trillion proton–proton collisions at the LHC were recorded and analyzed in confirming the particle's discovery in July 2012.

13.14 Detection of the Higgs Boson

So what is Atlas experiment looking for? Atlas was primarily designed to detect the Higgs boson. Finding a new phenomenon or a new particle is analogous to looking for a needle in a haystack, and in the case of the Higgs boson it is like looking for an infinitesimal size needle in the metaphorical haystack! So how can any experiment find this 'needle'? The role of theoretical physics is crucial and central in all experiments in particle physics, and it is theory that guided experiments on how to detect the Higgs boson, denoted by H^0.

All the indirect limits to the **Higgs mass** that had been obtained by other experiments, together with the possible likely decay modes of the Higgs boson, were factored into the design of Atlas.

In principle, the Atlas detector can observe any event that cannot be explained by the current theories of particle physics. After the discovery of the Higgs boson, Atlas is now engaged in studying other phenomena, such as signals for supersymmetry as well as studying the formation of a quark–gluon plasma in the collision of heavy nuclei and so on.

The collision of high energy protons inside the LHC converts the kinetic energy of the colliding protons into a ball of incredibly high energy density, as shown in Fig. 13.17; this ball of high energy can materialize in all the different possible ways allowed by the laws of Nature, and which the theories of high energy physics try to model.

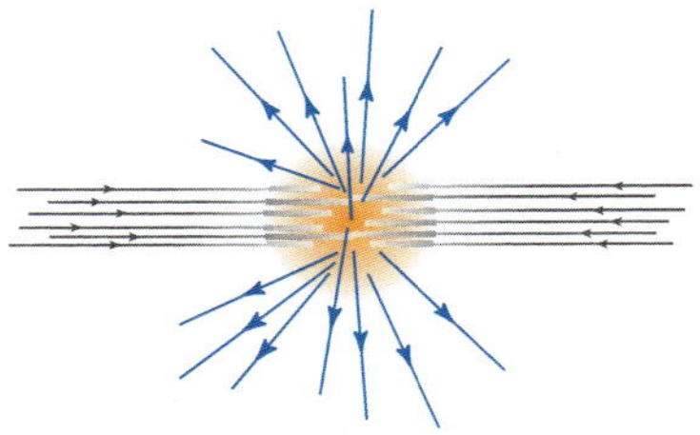

Fig. 13.17 Collision of particles and the resultant spray of particles.

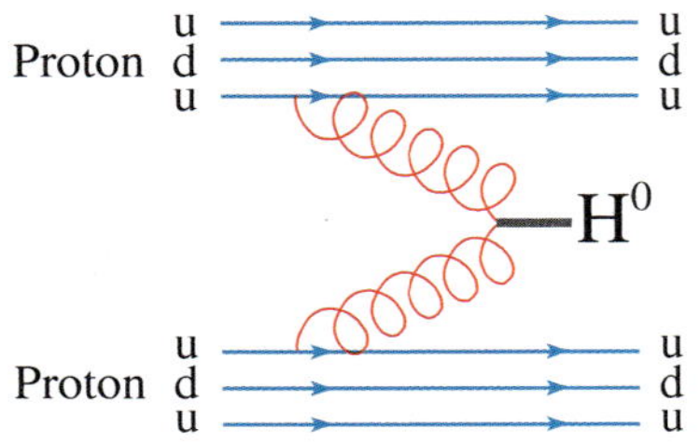

Fig. 13.18 Gluon fusion in a proton–proton collision leading to the production of the Higgs boson.

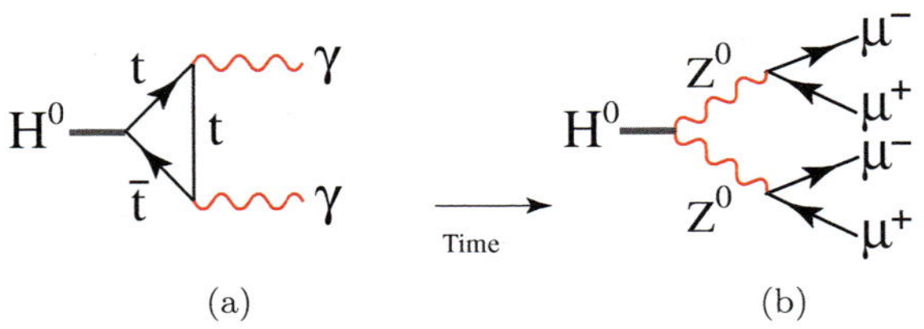

Fig. 13.19 (a) The decay of the Higgs boson via two photons; the Atlas experiment was designed to detect the two photon final state. (b) The decay of the Higgs boson via four muons; the CMS experiment was designed to detect this decay mode.

As shown in Fig. 13.18, the energy released by colliding protons in the form of gluons can lead to the fusion of gluons that creates a Higgs boson. In the LHC, gluon fusion was the main process used for the production of the Higgs boson. There are two dominant modes for the decay of the Higgs boson, one of them results in a final state of two photons, as shown in Fig. 13.19(a) and the other results in the production of a four muon final state, as shown in Fig. 13.19(b).

In Fig. 13.20, the number of all photon pairs detected in Atlas are plotted against the energy of the photon pairs, denoted by $m_{\gamma\gamma}$; there is a small increase in the number of photon pairs at the energy of 125 GeV/c^2, and which is attributed to the production and subsequent decay of the Higgs boson. Hence, the mass of the

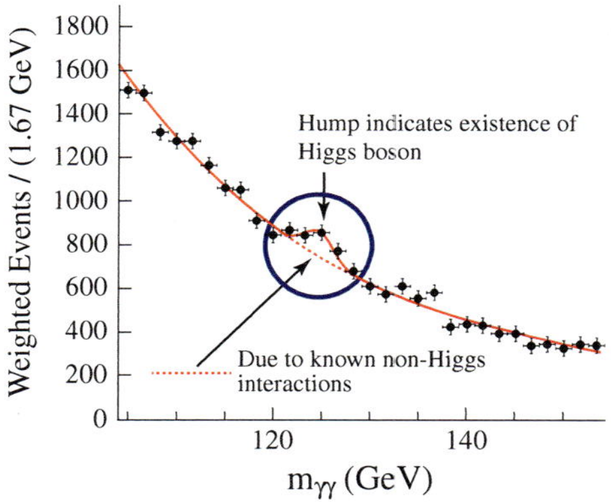

Fig. 13.20 The experimental evidence.

Higgs boson m_H, detected almost 50 years after being postulated, is given by

$$m_H = 125\,\text{GeV}/c^2 \; : \; \text{Mass of the Higgs boson.}$$

13.15 The Answer

It is indeed a great vindication of the Standard Model, and of the electroweak theory of Glashow, Salam and Weinberg in particular, that the Higgs boson has been detected. The gauge fields are nonlinear and arise from the principle of gauge symmetry; the fermions acquire their properties due to the synthesis of the quantum principle with the special theory of relativity. In constrast, the Higgs field looks very 'ordinary' — the only fundamental scalar field in Nature, having a very simple Yukawa coupling to the fermions and coupled to the gauge fields in a manner that preserves gauge invariance. But as the old saying goes, "looks can be deceptive": the Higgs field plays a crucial role in the electroweak theory, and in particular, it is indispensable in endowing masses to both the weak bosons and the fermions. The Higgs field is necessary for the empirical success of the theory since without the particles having masses the Standard Model would have no physical content.

The success of the Standard Model is also a landmark event for quantum field theory. The entire edifice of the Standard Model stands on the properties and behavior of quantum fields. In particular, all the cases of Feynman diagrams having loops, an example of a one-loop diagram being given in Fig. 13.19(a), are the result of the quantum nature of the fields that constitute the Standard Model. The rather arcane and opaque procedure of renormalization, another essential feature of relativistic quantum fields, is necessary to obtain finite results for high energy processes. Furthermore, the unification of the weak interactions with the electromagnetic interactions is a necessary requirement for removing unwanted infinities from the quantum field theory that describes the Standard Model. So, in sum, the mathematical structure of quantum fields, together with the theoretical postulates of the quantum principle and special relativity, have achieved a great and singular victory in the empirical validation of the Standard Model.

The Standard Model has successfully withstood hundreds of precise quantitative and qualitative experimental tests, both in the sector of quarks and gluons as well as for electroweak interactions. It would be fair to say that the discovery of the Higgs boson answered the final question of the Standard Model — namely how the leptons and weak bosons acquire mass — and has brought the theory to a closure.

A number of outstanding problems still need to be addressed. Why are there so many fermions and why are there *three* generations of quarks and leptons? Why is the electric charge of the quark a fraction of the same electric charge carried by the leptons? There are many masses and coupling constants in the Standard Model; can all these be combined into a single constant? Can all the fermions and bosons of the theory be further unified into a single entity? And lastly, how does one reconcile the classical gravitational field with the quantum principle? All these questions are addressed in the final two Chapters 14 and 15, namely in superstring unification and superstring gravity.

Chapter 14

Superstring Unification

How does all of Nature fit together?

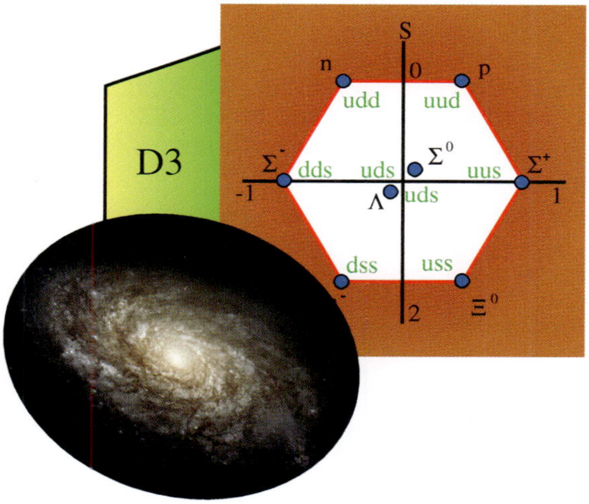

14.1 The Question

One of the key goals of science, and of Physics in particular, is an attempt to find principles that are more and more universal, and that explain greater and greater domains of phenomena in a unified manner. Indeed, the quest for a theory that underlies all natural phenomena has been at the core of theoretical physics since the early days of the scientific revolution. In a wider context, the quest for a unified explanation of Nature is perhaps as old as mankind itself.

This chapter and the next discuss the prospect of achieving unification based on superstring theory. A successful superstring theory should be able to explain the Standard Model with only a few parameters — ideally with no free parameters — and provide testable predictions. This ideal is yet to be achieved, but an impressive beginning has been made by superstring theory, and which will be our focus.

Various types of open and closed superstring theories in various spacetime dimensions are discussed in this chapter, and one may wonder if one theory is better than another. It will only emerge towards the end of this chapter, in Sec. 14.15, that all the apparently distinct superstring theories are, in fact, representations and approximations of a *single* underlying and unique theory.

A number of points regarding superstring theory need to be noted.

- The superstrings we discuss are all relativistic **quantum superstrings** and are based on the quantum principle of indeterminacy. In particular, the necessity of going to 26 spacetime dimensions for the purely bosonic string and 10 spacetime dimensions for superstring is required by the consistency between the quantum priciple and the special theory of relativity.
- There is currently no empirical evidence for the existence of superstrings.
- Even from the theoretical point of view the 'correct' superstring theory has not yet been found.
- As mentioned above, the various open and closed superstring theories in various dimensions that we will discuss are all expected to be related to each other, since we expect all superstrings theories to be different representations of the still sought for complete superstring theory.
- The discussion on superstrings is meant to bring the reader close to one of the frontiers and cutting edge of theoretical physics.

One of the most important achievements in theoretical physics was the unification of light, electricity and magnetism by James Clerk Maxwell in the late 1860's. Later, Einstein unified space and time and went on to formulate his theory of gravitation that unified energy and geometry into a single entity. More recently, in the 1960's, electromagnetism and weak interactions were unified by Glashow, Salam and Weinberg into the electroweak model. In the 1970's strong and electroweak interactions were unified into what is now known as the Standard Model, and discussed in Chapter 13. An overview of the various steps taken on the path to **unification** is shown in Fig. 14.1, together with two possible future steps indicated by question marks.

In the Standard Model, all known matter as well as three of the four fundamental forces (the electromagnetic, weak and strong interactions) are described in a consistent manner and the model has stood the test of an extensive range of experiments. Thus it would seem that the quest for a unified theory is coming to an end, with perhaps some patchwork remaining to be done. However, contrary to what one would have expected, no sooner had the Standard Model been postulated, and even while it was being tested, theorists enumerated another set of problems that seemed as fundamental as those that had been solved — and it became clear that a unified theory was still far out of reach.

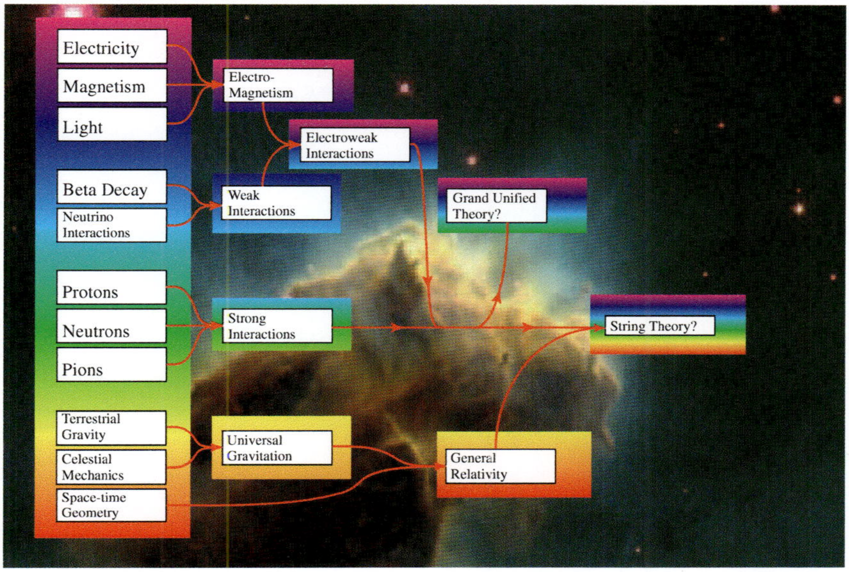

Fig. 14.1 An overview of unifiying theories.

Unification is a key goal of Physics.

In the Standard Model, there are 24 fermionic and 11 bosonic quantum fields that are 'elementary' and over 25 free parameters — such as masses and coupling constants that have to be fixed from experiments. The plethora of fundamental quantum fields has led to an impasse of sorts in high energy theory. There are simply too many 'fundamental' entities and a new challenge for theoretical physics is to synthesize and integrate these entities into a simpler and unifying underlying structure. Furthermore, another major outstanding challenge is to incorporate the fourth fundamental force (gravity) with the other three forces.

The methods and techniques used in the Standard Model are based on the quantum principle and the special theory of relativity — with spacetime being flat and requiring no special geometrical considerations. Gravity, on the other hand, was shown by Einstein's general theory of relativity to be intrinsically geometrical — being the result of curved spacetime. Furthermore, the Standard Model is a quantum theory whereas general relativity is a classical theory. Hence for a theory encompassing all known matter and forces, including gravity, it is necessary to incorporate both — the Standard Model as well as the geometry of spacetime — in a single theory based on the quantum principle. Table 14.1 shows some phenomena of current theories that need to be unified. Ideally, the final theory of everything will unify all the phenomena in the table.

Table 14.1 Some of the phenomena that need to be unified. SM = Standard Model, GR = general relativity, QT = quantum theory.

Currently individual phenomena	Current theories	Future unified phenomena
Bosons and fermions	SM	Single 'higher' entity
Electroweak and strong forces	SM	Grand Unified Theory (GUT)
Gravity and quantum mechanics	GR, QT	Quantum gravity
Electroweak, strong and gravitational forces	GR, SM	Single unified force

In our view, currently the most promising approach to a unified theory is based on so-called superstrings and branes, and which are the main content of this chapter.

14.2 On the Road to Unification

As a prelude to our discussion on superstrings, we review a few important results that we expect to obtain from superstring unification.

- It is expected that **supersymmetry** should play a key role in the unification.
- Grand Unified Theories emerge naturally by combining all the three interactions of the Standard Model in a straightforward manner. The Grand Unified Theory of interactions based on gauge fields is expected to form one of the cornerstones of any unified theory.
- One of the main objectives of unification is to obtain a quantum theory of gravity that combines the quantum principle with gravity. We postpone discussing this aspect of unification until Chapter 15.

14.2.1 *Supersymmetry*

Symmetries play a key role in our understanding of Nature due to their close relationship with conservation laws. The symmetries that we have encountered in previous chapters always act either on bosons or on fermions. For example, under rotation a boson remains a boson, with rotational symmetry leading to angular momentum conservation for the boson. One may wonder whether it is possible to have a symmetry that transforms bosons into fermions and *vice versa*. Such a symmetry is indeed possible and is called **supersymmetry**, often abbreviated as **SUSY**. The reason for adding the term "super" is to indicate that the symmetry that combines bosons and fermions forms a superset that includes all the symmetries of the Standard Model.

A particle is a boson or a fermion if its intrinsic spin is either integer or half-integer, respectively. For example, all bosons in the Standard Model have either spin zero as in the case of the Higgs particle, or spin one as in the case of the electroweak vector bosons and strongly interacting gluons. Similarly, all the fermions in the Standard Model, in particular quarks and leptons, are spin 1/2 particles.

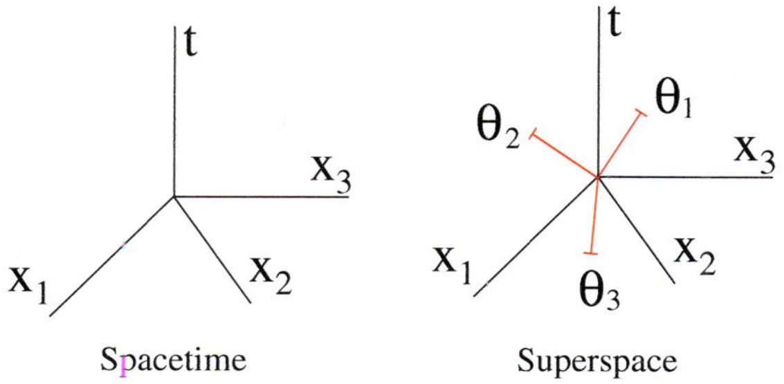

Fig. 14.2 The superspace is constructed by adding fermionic spacetime dimensions labeled with θ.

The property of having an intrinsic spin, be it integer or half-integer, is the result of the particle existing in Lorentz spacetime. One way of introducing supersymmetry is by considering an extension of Lorentz spacetime by adding new 'fermionic' dimensions to spacetime and obtain what is called **superspace** (see Fig. 14.2). The coordinates of superspace are given by $(t, x_1, x_2, x_3, \theta_1, \theta_2, \ldots) = (x, \theta)$, where we have added new fermionic coordinates given by $\theta = (\theta_1, \theta_2, \ldots)$ to the regular four coordinates used for Lorentz spacetime $x = (t, x_1, x_2, x_3)$. The number of fermionic coordinates depends on the type of superspace we would like to construct. Fermionic coordinates θ_i have the defining property that $\theta_i^2 = 0$ and $\theta_i\theta_j = -\theta_j\theta_i$.

A quantum field defined on superspace, called a **superfield**, is denoted by $\Phi(x, \theta)$ and contains both fermionic and bosonic fields. Fermions and bosons are combined into a single entity using the concept of a superfield. Superspace allows for a larger class of transformations that includes Lorentz transformations as a special case and also has new supersymmetric transformations that connect bosons with fermions.

Fermions and bosons are transformed into each other under supersymmetry transformations implemented by the supersymmetry charge operator Q; depending on the dimension of superspace there can be more than one supersymmetry generator, labeled by $Q_a, Q_b, Q_c, \ldots$. Denoting fermionic states by $|F\rangle$ and bosonic states by $|B\rangle$, we have $Q|F\rangle = |B\rangle$; hence we see that the supersymmetry charge Q is a *fermionic* **charge operator**. Supersymmetry transformations on the superfield are implemented by the fermionic charge operator Q.

For Nature to have supersymmetry all fermions and bosons must come in pairs with equal masses; this is analogous to particle and antiparticle having exactly equal masses. Hence, one should expect to find **superpartners** for all observed particles. For example, a quark and an electron should have bosonic superpartners called the squark and the selectron, with the photino and gluino being the fermionic superpartners of the photon and the gluon, and so on. The pairing of bosons and fermions is a consequence of the supercharge operator Q being a symmetry of Nature and hence, similar to electric charge, being conserved in all interactions.

However, the observed particles do not show such a pairing of masses; for example, the supersymmetric fermionic partner of the photon, namely the photino,

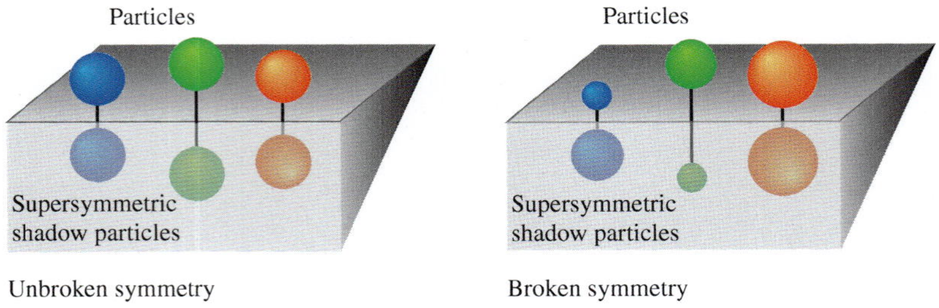

Fig. 14.3 A schematic rendition of supersymmetry, with unequal masses being a reflection of the breaking of supersymmetry in Nature.

has not been observed. The absence of the superpartners of the observed particles implies that, in Nature, supersymmetry is broken, analogous to the breaking of gauge symmetry by the Higgs boson; the breaking of supersymmetry is further thought to be the origin of the unequal masses of all the observed particles. Figure 14.3 shows an artistic rendition of broken and unbroken supersymmetries.

14.2.2 *Grand Unified Theories (GUTs)*

Supersymmetry is expected to be a symmetry in a fully unified theory. Besides the relationship between fermions and bosons, the task of a unified theory is to also bring the fundamental forces of Nature together. A theory which attempts to unify the three gauge inteactions of the Standard Model (*i.e.* the electromagnetic, weak and strong forces) is generally referred to as a **Grand Unified Theory (GUT)**, proposed mostly in the 1980's.

As discussed in Chapter 12, both the electroweak and strong interactions are mediated by gauge fields; hence, one is tempted to consider all the interactions as arising from a single force, with the electroweak and strong interactions being components of a single gauge field. A hint of how this may be possible can be obtained by considering the strengths of the forces' coupling constants at various interaction distances.

Due to the vacuum quantum fluctuations of the gluon field, the value of the effective color charge varies with distance; for the case of quantum chromodynamics, the effective charge becomes stronger as one probes the gluon field at larger and larger distances, as shown in Fig. 12.27. In contrast, the effective coupling strength of the photon to the electrons becomes stronger as the electrons get closer and closer.

In other words, at shorter and shorter distances, the electroweak coupling constant becomes stronger while the strong coupling constant becomes weaker. We therefore expect that strong interactions should become equal in strength to the electroweak interactions at sufficiently small distances. Although the changing behavior of the coupling constants has only been experimentally probed to about

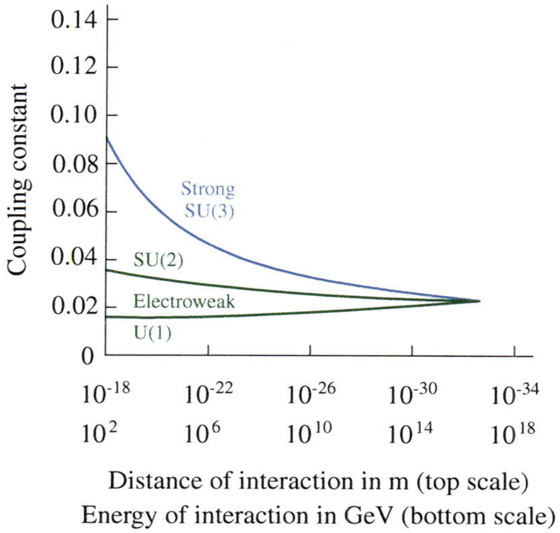

Fig. 14.4 At very high energies (that is, at very short distances), the strengths of the strong and electroweak forces become equal.

10^{-20} m, we can extrapolate the observed behavior of the various running coupling constants to shorter distances.

The extrapolation of the running coupling constants, as shown in Fig. 14.4, shows that the strengths of the electroweak and strong interactions *converge* to approximately the same value at distances of about $\simeq 10^{-32}$ m. This convergence can be made exact if we consider the supersymmetric extension of the Standard Model. We interpret Fig. 14.4 to mean that, at a distance of about 10^{-32} m, there is a *single* coupling constant that fixes the strength of all the electroweak bosons and gluon couplings.

It is convenient to express distances in terms of energy with the help of the speed of light c and Planck's constant $\hbar$. More precisely, we have the relation

$$\frac{\hbar c}{\mathrm{MeV}} = 197 \times 10^{-15}\,\mathrm{m}$$

$$\Rightarrow 10^{14}\,\mathrm{GeV} \to \frac{197 \times 10^{-15}}{10^{17}}\,\mathrm{m} \simeq 10^{-30}\,\mathrm{m}.$$

The corresponding distances and energies are shown in Fig. 14.4.

Recall from our discussion in Sec. 11.6 that to have theories with local gauge invariance, a gauge field has to be coupled to the fermions. All the interactions of the Standard Model are mediated by gauge fields, with the weak and strong forces being nonlinear generalizations of the electromagnetic gauge field. The nonlinear generalization of Maxwell's theory is given by the Yang–Mills gauge field — which forms the backbone of the Standard Model as well as of superstring theory.

The result shown in Fig. 14.4 indicates that there is a *single* gauge field determining all the interactions of the Standard Model at the distance of $\simeq 10^{-32}$ m;

on being probed at larger and larger distances, this single interaction appears as three different interactions, namely the electromagnetic, weak and strong interactions.

The analysis of the coupling constants of the Standard Model leads us to conclude that unification would consist in unifying the interactions at the very short distance of $\simeq 10^{-32}$ m. This is a tremendous extrapolation since data on the behavior of the various coupling constants is known only up to an energy of about 10^4 GeV, whereas we are extending the known behavior out to energies of 10^{15} GeV — which is often referred to as the GUT energy scale.

Noteworthy (optional content) 14.1: Lie Groups

Group theory is a vast subject of mathematics and forms one of the theoretical cornerstones of the Standard Model. A group $\mathcal{G}$ is a collection of elements $g \in \mathcal{G}$ that obey the following four axioms.

- A multiplication $*$ is defined such that, for $g_1, g_2 \in \mathcal{G}$, we have $g_1 * g_2 \in \mathcal{G}$.
- The multiplication $*$ is associative in that, for $g_1, g_2, g_3 \in \mathcal{G}$, $(g_1 * g_2) * g_3 = g_1 * (g_2 * g_3)$.
- The identity element I exists such that for $g, I \in \mathcal{G}$, $I * g = g * I = g$.
- The inverse exists, namely for $g \in \mathcal{G}$, there exists a $g^{-1} \in \mathcal{G}$ such that $g * g^{-1} = g^{-1} * g = I$.

A **Lie group** is a group with the additional property that it is constituted by elements that form a continuous differentiable space, with group multiplication having continuity for elements that are continuously varied.

The most familiar Lie groups in Physics are the unitary (matrix) groups, denoted by U(N), which are the set of $N \times N$ matrices with complex elements that satisfy the condition that $U^\dagger U = I$, where $U^\dagger$ is the complex conjugation and transposition of the matrix. Group multiplication $*$ is the ordinary matrix multiplication. The group SU(N) is the special unitary group that has the additional constraint that $\det U = 1$. The Lie groups SU(5), SU(3), SU(2), U(1), SO(10), E_6 and $E_8 \times E_8$ appear in the discussions in this and other chapters.

To combine the different gauge fields that appear in the Standard Model, we need to describe them in a more mathematical language, so that we can then discuss the manner in which they can be unified. Gauge transformations discussed in Sec. 11.6 form what is called a Lie group. Each gauge field is completely described by its Lie group of gauge transformations. The following are the Lie groups of the gauge transformations — called the gauge groups — of the Standard Model.

- The electromagnetic gauge field : U(1)
- The weak gauge bosons $W^\pm, Z^0$: SU(2)
- Gluons : SU(3)

The procedure for combining the strong, weak and electromagnetic interactions is to combine the specific gauge transformations of the different interactions into a larger gauge transformation that includes all the specific ones. This is mathematically implemented by enlarging the Lie group of gauge transformations of the Standard Model into the following: SU(3) × SU(2) × U(1).

The Lie group of gauge transformation of the Standard Model is: SU(3) × SU(2) × U(1).

The gauge group of the unified theory needs to generate a gauge field that contains all the gauge fields of the Standard Model, something like a large box containing smaller boxes. More precisely, the gauge field that can account for the observed SU(3) gauge field of strong interactions and the SU(2) × U(1) gauge field of the electroweak interactions must contain at least the gauge group SU(3) × SU(2) × U(1). The smallest gauge group that fulfills this condition is the Lie group SU(5), and is called the **Grand Unified Group**. The observed gauge fields are derived from the gauge field having the SU(5) gauge group by a series of gauge symmetry breaking, and is shown schematically in Fig. 14.5.

Calculations based on a Yang–Mills quantum field theory with a SU(5) gauge group shows that the various coupling constants converge in the energy range of 10^{14} GeV to 10^{16} GeV, but do not actually meet at a single value as desired. Hence improvements are necessary. One can further refine a GUT by incorporating supersymmetry. What this means is that one first fixes the gauge fields using the SU(5) gauge group, and then chooses a specific collection of fermion fields so that the theory has exact supersymmetry (SUSY). In other words, a supersymmetric GUT is invariant under the interchange of the boson fields (which includes the gauge fields among other boson fields) with the fermion fields.

SUSY GUTs have the positive feature that all three gauge couplings of the Standard Model converge to the *same* coupling constant — at an energy of

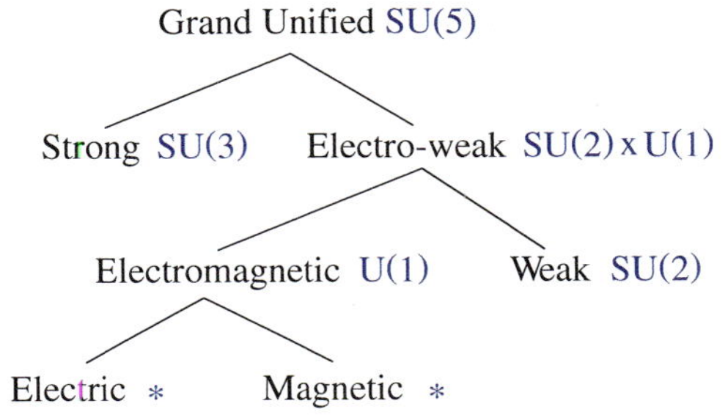

Fig. 14.5 The gauge symmetry groups underlying the main theories. The "*" indicates that there is no gauge group.

10^{16} GeV. The unification of all the coupling constants explains why the electric charge carried by the leptons, which is measured in units of e, is the same unit of the electric charge measured for the quarks — a fact that appeared in the anomaly cancellation of the Standard Model, discussed in Sec. 13.6, but which had no explanation.

The important result that all the gauge interactions and coupling constants can emerge from a single coupling constant is an indication of the important role that supersymmetry plays in the search for a unified theory of all forms of matter and energy. Later in this chapter when we discuss strings, it is argued that supersymmetry is essential for obtaining theories that are finite and mathematically self-consistent.

SUSY GUTs carry the unification scheme of the Standard Model to its logical conclusion, but make predictions that are ruled out by experiments. All models that unify electroweak and strong interactions at the GUT scale are based on local quantum fields that combine quarks and leptons into a single entity and, in doing so, allow for processes that violate baryon number conservation. This in turn implies that the proton would decay via the Higgs boson or other new boson fields introduced by GUTs, with an estimated lifetime of about 10^{30} years. All experiments for testing whether the proton decays have shown that the proton in fact is stable and has a lifetime, as of 2014, greater than 1.29×10^{34} year. The fact that the proton does not decay rules out all GUTs.

Even though GUTs failed, it moved the attention of theorists to the incredibly small distances of $\simeq 10^{-30}$ m in looking for the new physics of unification. On a more technical level, it was also found that one of the best choices for a group that contains the gauge group SU(3) × SU(2) × U(1) of the Standard Model is E_6 (an exceptional Lie group). Furthermore, the group SO(10) also contains SU(5) and hence is also a possible starting point for unification. These Lie groups, namely E_6 and SO(10), will reappear in superstring unification.

14.2.3 *Gravity and unification*

The gravitational force is very weak, indeed it is the weakest of all forces, being 10^{39} times weaker than the strong force at distances of the nucleus, which is about 10^{-15} m; however, at distances of about 10^{-35} m, gravity becomes equal in strength to the strong and electroweak forces. This length scale is of the order of the Planck length $l_p = 1.61615 \times 10^{-35}$ m, which we already encountered in our discussions of black hole entropy in Sec. 5.14.

The Planck length therefore seems to be the length scale at which we need a quantum theory to describe a unified electroweak-strong-gravitational force. The theory which attempts to combine the notions of quantum mechanics with general relativity is usually referred to as **quantum gravity** and is discussed in Chapter 15 (see also Fig. 15.1).

14.3 Superstrings

Now that we have had a brief look at some of the approaches to unification and their problems, let us proceed to discuss what has become the dominant, and in our view currently the most promising, attempt at unification: superstring theory. We give a brief overview of what is a superstring and the main outcomes of superstring theory.

A one dimensional string 'smears' out matter (energy) into a line — as opposed to local quantum fields that have point-like constituents. A quantum string resolves the problems of mathematical consistency for a quantum theory of gravity, discussed in Sec. 14.7. We end our exploration of the string world with the culminating unification of gravity and quantum mechanics in Chapter 15.

Superstring theory was formulated in 1984 by M. B. Green and J. Schwarz. It is based on two key notions. First that a unified theory needs to be supersymmetric, and second, more radically, that the point particles of the Standard Model need to be generalized to one dimensional relativistic quantum objects called **strings** as illustrated in Fig. 14.6. The strings in superstring theory can be of two types: open (ends do not connect) or closed (the string forms a loop), as illustrated in Fig. 14.7. Superstrings are massless and the endpoints of the open superstring are traveling at the speed of light. Similar to a point particle tracing out a world line as it evolves in time, shown in Fig. 14.8, a string spans out a world sheet as shown in Fig. 14.9.

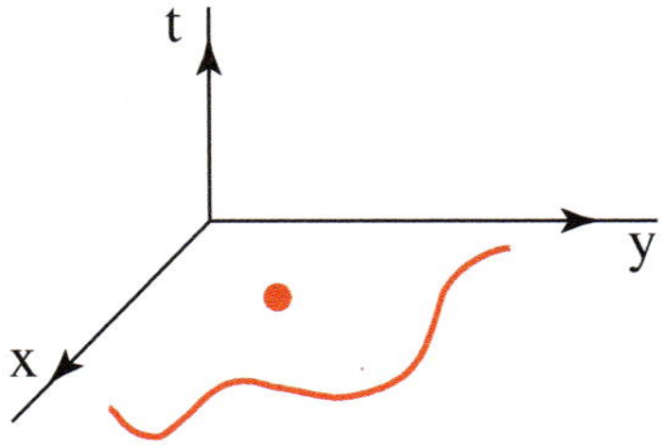

Fig. 14.6 From point to a string.

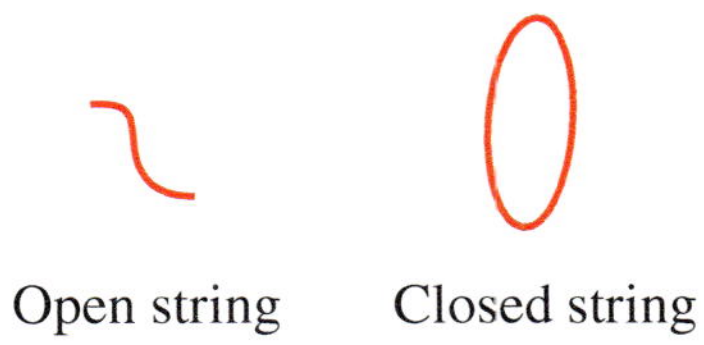

Fig. 14.7 Open and closed strings.

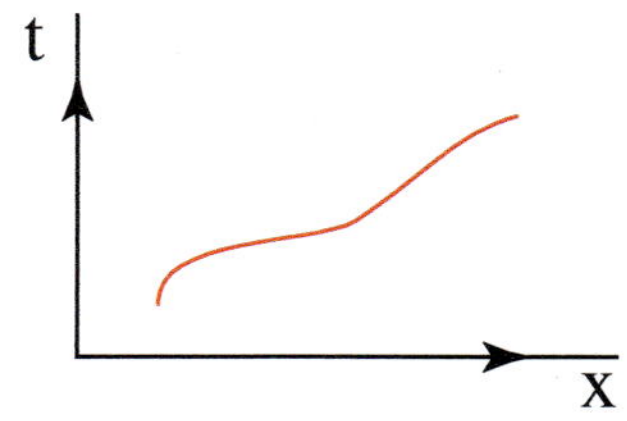

Fig. 14.8 World line of particle.

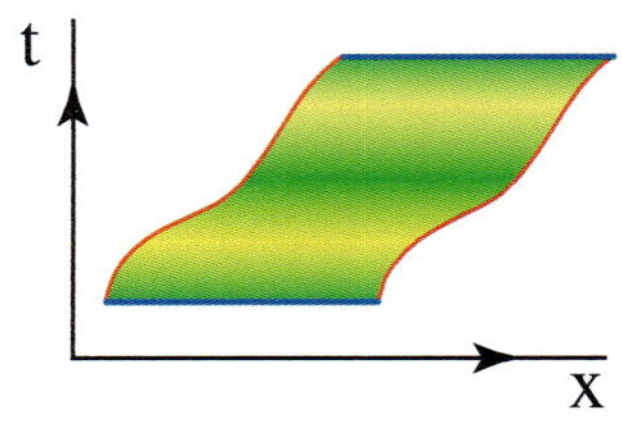

Fig. 14.9 World sheet of string.

All the fermions and gauge fields (forces) of the Standard Model arise as excitations of the underlying superstring, and spacetime itself is the manifestation of one of the vibrational modes of the superstring. In summary:

Every fundamental particle — be it a fermion or a boson — is one of the possible states (vibrations) of the superstring.

Given below is a brief overview of some of the remarkable features that emerge naturally from superstring theory.

- **Grand Unified Theory:** Superstring theory is sufficiently complex to produce — at the GUT scale — all the particles and forces of the Standard Model.
- **Higher Dimensional Spacetime:** Superstring theories generalize the concept of spacetime. All the generalizations of spacetime and of its topology lead to a spacetime that has 10 (or 11 or 12) dimensions: there is in general only one time direction, with space having dimensions greater than the three dimensions that we observe.
- **Spacetime Compactification:** Superstring theory has solutions that yield spacetime manifolds with four large dimensions, with the remaining (six) dimensions being compactified into a space that is the size of the Planck length.
- **Brane Worlds:** In the so-called brane world point of view, our Universe is considered to be a three dimensional subspace of a higher dimensional space, with all directions being large. This may explain the phenomena usually ascribed to dark matter.
- **Supersymmetry:** Consistent superstring theories require spacetime supersymmetry.
- **Parity Violation:** Superstring theories can generate parity violating couplings of gauge fields (forces) with fermions, as is required for the parity violating couplings of leptons and quarks to the weakly interacting gauge fields.
- **Gravity:** Superstring theory contains a massless spin 2 particle, whose low energy interactions reduce to the gravitational field of general relativity.

- **Quantum Gravity:** Superstring theory provides a consistent theory of quantum gravity, and in fact is the only known theory which can do so. This is in contrast to all local quantum field theories that provide mathematically inconsistent theories of quantum gravity.

Noteworthy (optional content) 14.2: The Landscape Problem

The initial hope for superstring theory was that there is going to be one unique and single superstring theory in higher dimensional spacetime, with the theory having many different 'reductions' to the observable four dimensional spacetime. It was hoped that superstrings would be a theory that is free from any tunable coupling constants or masses and that all the 25 parameters of the Standard Model would be generated internally. This hope has turned out to be too optimistic.

Superstring theory has a true ground (vacuum) state, similar to the vacuum state in quantum field theory, and based on this ground state all the physical particles and parameters are fixed. The landscape problem refers to the fact that there are somewhere between 10^{10} to 10^{100} minima to the potential function of superstring theory that determine many 'false' ground states, with only one of them being the absolute minimum.

Superstring theory can be defined in any of the false ground states, which corresponds to one of the many minima of the effective potential. The superstring Universe may reside in a 'false vacuum' and then decay over a very long time to the true ground state, which is at the absolute minimum. There is no principle which dictates which state that superstring initially chooses, and hence the Universe can be in any of the possible false ground states. This inability to fix the true ground state of the superstring makes it impossible to predict all the parameters that are observed in our Universe.

The anthropic principle can be invoked to choose a ground state that supports the emergence of intelligent life, but this teleological solution is unacceptable to many physicists.

All the conclusions that we have stated above result from the single postulate of superstring theory: that a relativistic quantum object spread out in one dimension is at the foundation of physical reality. In summary, as illustrated in Fig. 14.10, superstrings yield a theory that contains supersymmetry, quantum gravity and constitutes a Grand Unified Theory as well.

At present, there is no experimental evidence that superstring theory is the correct theory of Nature. The reason being the energy scale of 10^{19} GeV, or equivalently distances of the order of the Planck length 10^{-35} m, is far beyond the energy range of existing particle accelerators that can only probe distances of about 10^{-20} m. A string is so tiny that if one were to magnify an atomic nucleus

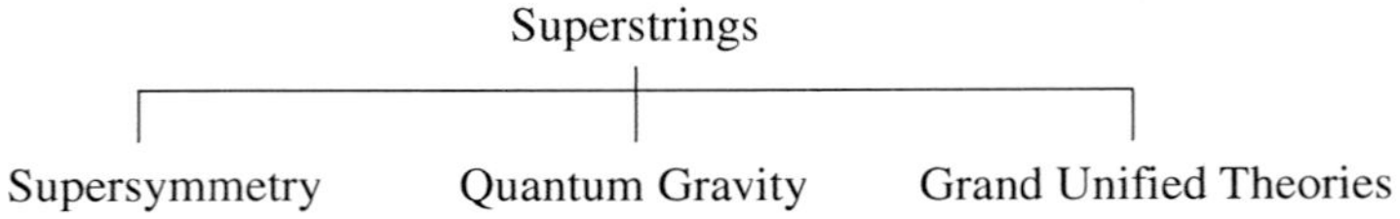

Fig. 14.10 Superstrings unify all other theories.

to the size of our solar system, then a string would still be only about 10^{-7} m in length, the size of a few hundred atoms.

Suppose we want to make an accelerator that can probe distances small enough to resolve a string. The shorter the distance being probed the larger the energy needed. With the currently achievable magnetic field strength, a ring accelerator would have to be about 10 billion light years in diameter — almost the size of the Universe — to accelerate, say an electron, to the Planck energy required to detect a string. A hopeless task indeed!

Rather, what one hopes for is that, instead of direct verification by conducting experiments at the Planck length, theorists can make some definite predictions for the behavior of superstrings at observable energies. This, in turn, could then be tested by experiments within the energy ranges that are available.

The introduction of a string greatly increases the number of independent degrees of freedom (variables) — and creates new problems of mathematical consistency. What is however already non-trivial is that superstring theory is, in fact, a mathematically self-consistent theory — not a small achievement given the numerous anomalies of string theory that must cancel out in order to maintain consistency. A theory with the complexity of superstring theory is completely new in the annals of science and has already produced many new results for hitherto intractable problems in mathematics and has even created new branches of pure mathematics.

14.4 Higher Spacetime Dimensions

In contrast to a particle that exists at a single point, a string is an extended object that exists on a line. A string, be it open or closed, is an extended object that exists simultaneously at many points of space — after all that is what a string means, that it exists along a direction in space. In our discussion on antiparticles in Chapter 11, we saw that once a quantum object 'occupies' a spacelike interval, there is a possibility of a conflict with causality. For the case of particles, since they are point-like when observed, causality is recovered by introducing antiparticles that are also point-like.

However, for the case of a string, it is in general no longer possible to recover causality by say introducing an 'antistring'. It turns out, for reasons that are not entirely clear, that a purely bosonic string is consistent with both relativity and

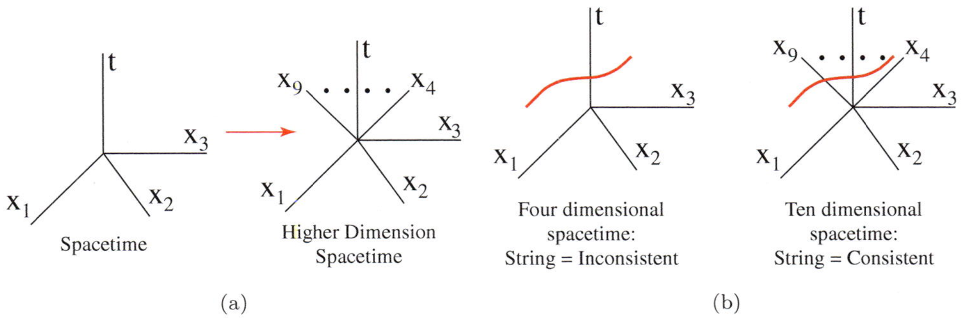

Fig. 14.11 The superstring is inconsistent with quantum mechanics in four dimensional spacetime but is consistent in ten dimensional spacetime.

quantum mechanics in only one unique and critical spacetime dimension d_B, namely $d_B = 26$ (note that this dimension pertains to a purely bosonic string and not to a superstring).

The problem with a purely bosonic string is that its vacuum state turns out to be unstable; furthermore it does not have any fermions and hence cannot generate the quarks and leptons of the Standard Model.

Fortunately, one can extend the bosonic string to a superstring by including fermionic string degrees of freedom that are the supersymmetric partners of the bosonic string degrees of freedom. The supersymmetric string has a critical spacetime dimension given by $d_S = 10$ (see Figs. 14.11(a) and (b)). In 10 dimensions the superstring obeys both the laws of quantum mechanics and special relativity, and has a stable vacuum state as well.

The critical spacetime dimension for open and closed purely bosonic strings is $d_B = 26$, and for open and closed superstrings it is $d_S = 10$. Supersymmetry and, in particular, the existence of fermions in our world make us choose a spacetime having a dimensionality of $d_S = 10$.

> A Minkowski spacetime with one time and $d-1$ space dimensions is denoted by $\mathcal{M}_d$. In particular, $\mathcal{M}_{10}$ and $\mathcal{M}_4$ denote 10 and 4 dimensional flat spacetime, respectively.

14.4.1 *Dimensional reduction and compactification*

The fundamental question of superstring theory that we need to address is: how does one arrive at the spacetime that we perceive in daily life which is only four dimensional? Superstrings in 10 dimensional spacetime $\mathcal{M}_{10}$ have to be 'reduced' to the observed four dimensional spacetime $\mathcal{M}_4$. Two ways that one can do this are (i) dimensional reduction and (ii) compactification.

14.4.1.1 *Dimensional reduction*

Consider some physical quantity f of the theory that depends on all the coordinates of $\mathcal{M}_{10}$, that is $f = f(t, x_1, \ldots, x_9)$. In **dimensional reduction** the spacetime dependence of f is reduced to $\mathcal{M}_4$ by simply ignoring all the six other coordinates, which are set to zero. More precisely

$$f(t, x_1, x_2, x_3, x_4, \ldots, x_9) \rightarrow \text{dimensional reduction} \rightarrow f(t, x_1, x_2, x_3).$$

In superstring theory, dimensional reduction is sometimes used.

14.4.1.2 *Compactification*

A more important procedure for lowering the dimension of the theory — from a physical point of view — is to follow the procedure of **compactification**. This is a strategy of making the extra dimensions very small, so that only four of the dimensions appear at larger scales. This approach of rendering some of the space dimensions small is called compactification since one 'compactifies' a large dimension by curling it up as illustrated in Fig. 14.12.

The reason compactification is preferred over dimensional reduction is because it yields a parameter, the radius of compactification, which one can make very small before taking the limit of zero. This process of taking the compactification scale to zero allows one to retain significant information about the physics of the extra dimensions and is similar to the procedure of taking epsilon to zero in calculus. In contrast, in dimensional reduction one completely loses all the information about the extra dimensions.

One of the simplest examples of compactification is to start with a large sheet of paper; take one side of the sheet and roll it up so that the sheet becomes a cylinder. If one chooses to tightly roll the sheet into a cylinder with a very small radius, the sheet will appear to be approximately one dimensional, as shown in Fig. 14.13. A hose pipe is a good example of compactification: when viewed from a great distance, say from a high floor, the hose appears to be only a line — with the extra circle of the hose being too small to be seen. Mathematically, one can think of the plane as a product of two lines, namely $\Re \times \Re$, and the space of the cylinder is the product of a

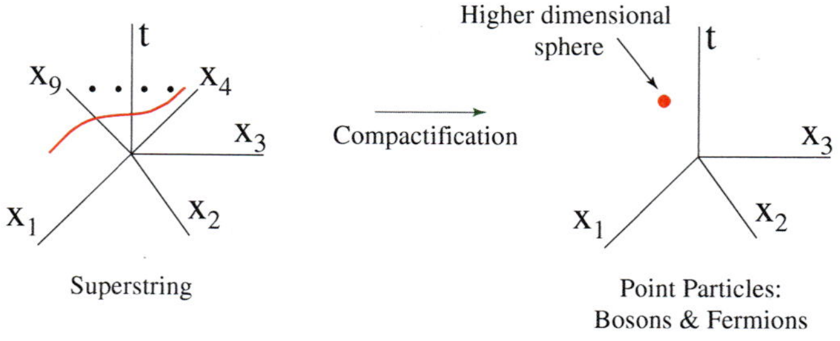

Fig. 14.12 Compactification yields point particles from superstrings.

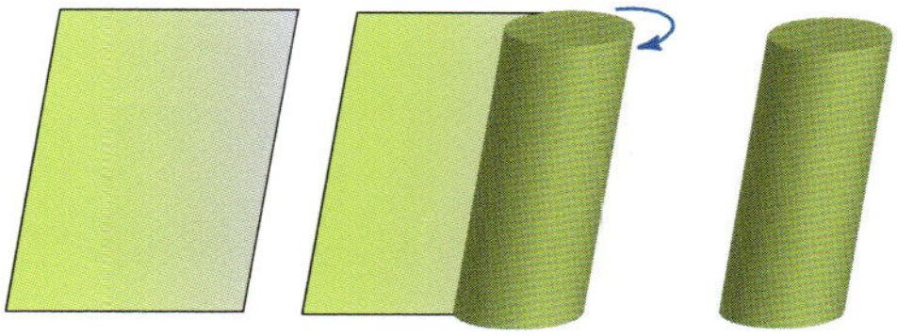

Fig. 14.13 Compatifying a plane into a cylinder: rolling up a sheet into a cylinder.

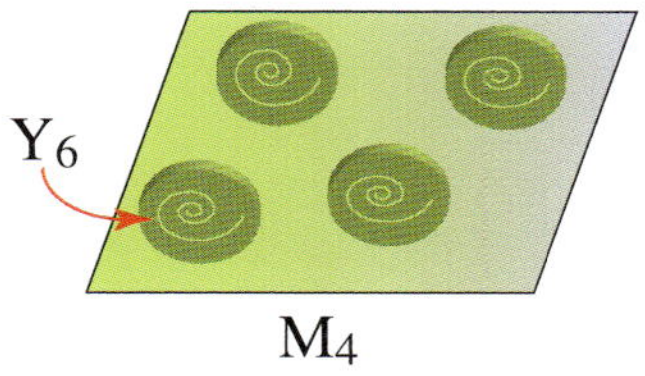

Fig. 14.14 Four large and six small dimensions; the compact space can be a Calabi–Yau space.

line with a circle, namely $\Re \times S^1$. Compactification is then given by $\Re \times \Re \rightarrow \Re \times S^1$, with S^1 having a small radius.

For 10 dimensional superstring theory, at every point of ordinary four dimensional spacetime there are, in orthogonal directions, an extra six dimensions of space. The analogy with the case of the cylinder is that the line corresponds to our observed four dimensional spacetime and the circle corresponds to the extra six small dimensions; the extra dimensions are very small and hence from a very large distance, which is the case for any ordinary experiment, they are invisible.

The line and hose pipe analogy implies that, since one needs to compactify 6 space dimensions, we need a 6 dimensional analog of the circle, which is a 6 dimensional compact space — with its size being of the Planck length. We then obtain the 10 dimensions of spacetime as being composed of four large dimensions and six small space dimensions, written as $\mathcal{M}_4 \times Y_6$ and illustrated in Fig. 14.14. The interpretation of the small compactified space is that at *every* point of ordinary space and time, there is an additional internal space Y_6 — the analog of the small circle S^1 at every point along the length of the hose. An example of Y_6 is the **Calabi–Yau manifold**, which has specific mathematical properties required for a consistent compactification of superstring theory.

14.4.2 *Topology and geometry*

A space is fully described by its geometry that is determined — as discussed in Chapter 3 — by the distance between any two points in that space. For example, a two dimensional space may be the surface of a perfect sphere, or of an ellipsoid, and so on. These two shapes have different geometries since distances between points in the two spaces are quite different as can be seen in Fig. 14.15.

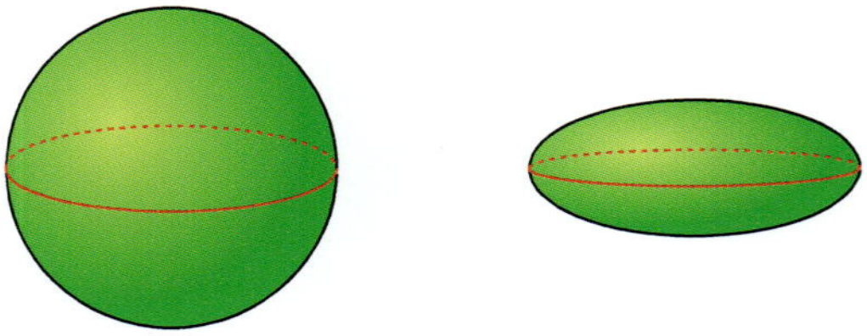

Fig. 14.15 Deformation of a sphere.

Fig. 14.16 A two dimensional Riemann surface.

The **topology** of a space, in contrast, depends only on the global properties of the entire space. In particular, it does not directly depend on the detailed geometry of the space, and spaces having different geometries can have the same topology. For example in Fig. 14.15, both the surfaces of a perfect sphere and an ellipsoid have the same topology.[1]

Riemann surfaces play an essential role in describing the self-interactions of a closed superstring, discussed in Sec. 14.6. The topology of a Riemann surface is entirely specified by how many holes the surface has, and does not depend on the detailed nature of the shape of the surface. An example of a Riemann surface is shown in Fig. 14.16. It looks like a pretzel and the genus g is a measure of how many holes the pretzel has; the sphere and ellipsoid of Fig. 14.15 are both Riemann surfaces with genus 0.

One of the most interesting and intriguing results of superstring theory is that the number of generations of quarks and leptons is determined by only the topology of the extra space Y_6 and, in particular, does not depend on the geometry of Y_6. More precisely, the number of generations depends only on the **Euler characteristic** χ of the manifold Y_6. The Euler characteristic is an integer that describes the topology of a space of any dimension, unlike the genus of a Riemann surface that is applicable only to a two dimensional space.

[1]In topology, objects are distinguished by whether they can be continuously deformed into each other or not. For example, both the sphere and the ellipsoid have no holes and it is intuitively clear that they can be continuously transformed into each other. Specifically, objects with different numbers of holes cannot be continuously transformed into each other and hence are topologically different, but there are other criteria as well.

For example, the circle S^1 has Euler characteristic $\chi(S^1) = 0$ and a three dimensional solid ball has Euler characteristic 1. The Euler characteristic for a genus g Riemann surface is given by $\chi = 2 - 2g$. The two dimensional sphere S^2 has genus $g = 0$ and the two dimensional torus $\mathcal{T}$ has $g = 1$, yielding $\chi(S^2) = 2$ and $\chi(\mathcal{T}) = 0$, respectively.

Superstring theory relates χ to the number of generations of quarks and leptons as follows:

$$N_g = \text{Number of quark and lepton generations} = \frac{1}{2}\chi(Y_6). \qquad (14.1)$$

For $Y_6 = S^6$, the six dimensional sphere, we have $\chi(S^6) = 2$ and hence we have only one generation of quarks and leptons; for $Y_6 = S^2 \times S^4, \chi(S^2 \times S^4) = \chi(S^2) \times \chi(S^4) = 4$ yielding $N_g = 2$, that is, two generations. One can construct a space Y_6 to be a sum and product space of spheres, and produce the $N_g = 3$ ($N_g = 3$ is the number of generations in the Standard Model) — but this is not a unique result: it turns out that there are thousands upon thousands of manifolds M, if not millions, all with $\chi(M) = 6$.

We briefly return to a discussion on Y_6 in Sec. 14.9.1, where we consider Y_6 to be a Calabi–Yau space. However, the question as to what is the nature of the hidden manifold Y_6 — required for matching superstring theory's results with Nature — has not yet been settled.

14.5 Superstrings: Observed Forces and Particles

The fundamental constituents of Nature, namely superstrings, exist in 10 dimensional spacetime. Of course what we observe are particles, such as the electron, in four dimensional spacetime. The ultimate goal of superstring theory is to predict, without any arbitrary parameters, the Physics that is observed in the laboratories. What is observed in experiments are the particles of the Standard Model.

Superstring theory therefore aims to show that the fermions and forces of interaction in the Standard Model — as well as gravity — are all the *vibrations* (excitations) of superstrings. It is thought that superstrings give rise to all the particles and forces of Nature. In other words:

All known particles and forces are thought to be the manifestations of superstrings.

All vibrations of the superstring, both open and closed, are transverse. The superstring vibrates in all the space dimensions, with the vibrational frequencies being determined by the string tension. String tension is the energy per unit length. The only energy scale at the Planck length is the Planck energy of 10^{19} GeV, yielding a string tension T of the superstring to be an astronomical $T = 10^{44}$ N.

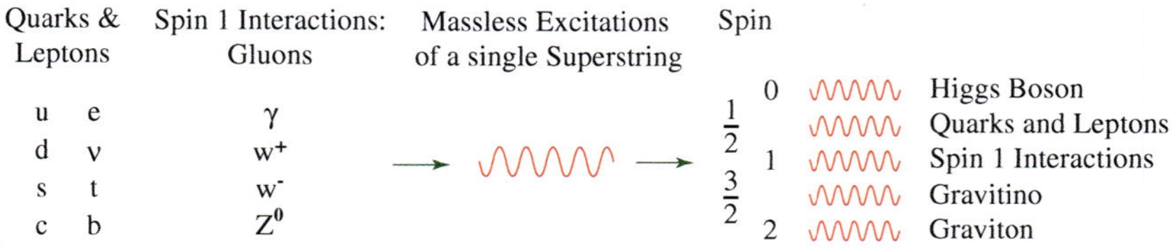

Fig. 14.17 Point particles and interactions are all excitations of superstrings.

All superstrings have two classes of vibrations, namely the lowest vibrational states that are massless and the remaining higher frequency excitations that have mass; due to the enormous string tension, all the higher vibrational modes of the string produce excitations with an effective mass equal to or greater than Planck energy 10^{19} GeV, which is about 10^{-8} kg — the mass of sand grains that are visible to the naked eye. Consequently, excitations with mass have a vanishingly short lifetime and it is only the *massless excitations* that can give rise to all the particles that are observed.

If particles such as the electron are the result of massless superstring excitations, then how can the electron have mass? As we have discussed in Chapter 13 on the Standard Model, endowing a massless particle with mass can be achieved through the Higgs mechanism of symmetry breaking and phase transition. The mass we observe of what start off as massless superstring excitations manifests itself after the symmetries at the Planck scale are broken and a transition is made to our macroscopic world.

Indeed, the entire programme of superstring theory is to find a mathematical structure such that the *massless vibrations* of the superstring uniquely produce gravity as well as the particles and forces of the Standard Model. The massless superstring states that are bosonic correspond to the bosons — namely the Higgs and other bosons that carry the forces of interaction — whereas the massless superstring states that are fermionic correspond to quarks and leptons; see Fig. 14.17.

14.6 Closed Superstrings

There are two types of superstrings: **closed superstrings** discussed in this section, and open superstrings, discussed in Sec. 14.10. Open and closed superstrings are part of a number of different but related superstring theories and it is important to keep in mind that not all physically relevant features have been found in the different theories. However, clear relationships between the theories have been found and it is hoped that in the end a well-defined and comprehensive theory will emerge such that the current theories are its different limits.

Consider a circle, also called a loop, as shown in Fig. 14.18. The loop has a coordinate σ that specifies the different points of the circle. Since the loop is periodic,

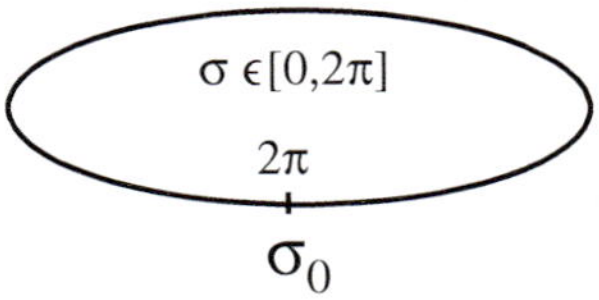

Fig. 14.18 Closed loop with coordinate σ.

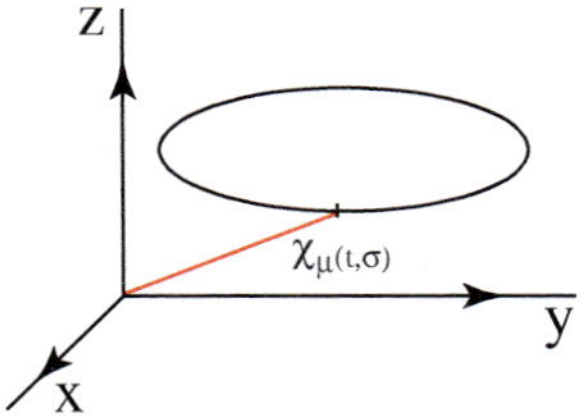

Fig. 14.19 String coordinates $X_\mu(t,\sigma)$ of a loop at a given time t.

the coordinate is also periodic with $\sigma \in [0, 2\pi]$. Although spacetime itself emerges from the dynamics of superstring theory, to simplify our discussion, consider the closed loop to be in a 10 dimensional spacetime (9 space and 1 time dimension); what this means is that the circle is taken to lie in 9 dimensional space and has a time coordinate t, as illustrated in Fig. 14.19, and with coordinates of the world sheet given by (t, σ).

Let X_μ, $\mu = 0, 1, 2, \ldots, 9$, be the position of the loop in 10 dimensional spacetime; each point of the loop then has a coordinate given by $X_\mu(t, \sigma)$. Depending on the excitation, a closed superstring can either be a boson or a fermion. The bosonic component of the superstring can be directly represented by the coordinate $X_\mu(t, \sigma)$, while the fermionic component has spin 1/2 on the string's world sheet and is given by a two component spinor

$$\psi_\mu(t, \sigma) = \begin{pmatrix} \psi_\mu^+(t, \sigma) \\ \psi_\mu^-(t, \sigma) \end{pmatrix},$$

where $\psi_\mu^\pm(t, \sigma)$ are fermionic variables. Taken together $(X_\mu(t, \sigma), \psi_\mu(t, \sigma))$ is the coordinate of the superstring. This type of superstring is called a **closed superstring**.

14.6.1 *Self-interactions and quantum evolution*

How does a superstring evolve in time? Recall that a particle, at any instant, occupies a single point in space; as the particle moves in time t it traces out a curve in space and time — called its path, or more precisely its world line as illustrated in Fig. 14.8. Similarly, as the open and closed superstrings evolve in time, they span out a world sheet as illustrated in Fig. 14.20. As can be seen from Fig. 14.20, the

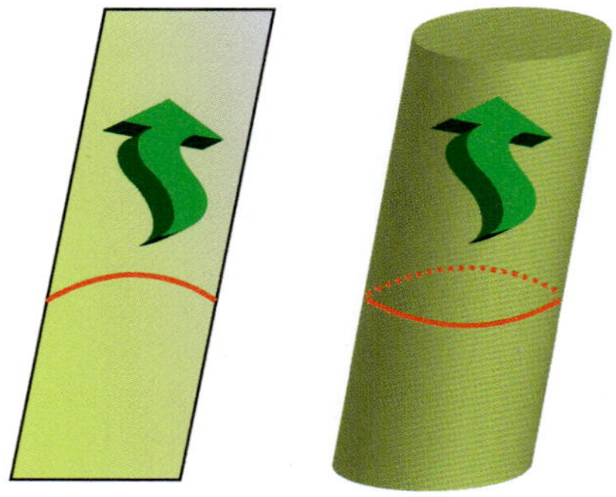

Fig. 14.20 Open and closed superstring world sheet.

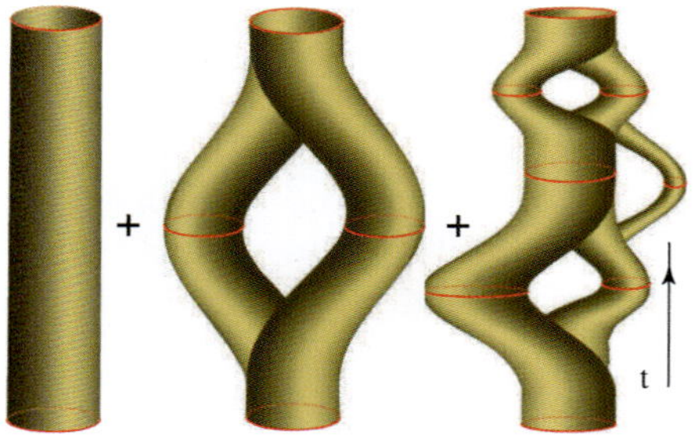

Fig. 14.21 Self-interaction of a single closed superstring. Its initial state is at the bottom of the figure and its final state is at the top, shown by red circles.

open superstring world sheet has a boundary whereas the closed superstring world sheet is a closed cylindrical surface.

In superstring theory there are *no 'non-string' entities* with which superstrings can interact; this is in contrast to, for example, the case of electrons that interact via the photon field. Rather, all the interactions of a single superstring is via **self-interactions**; interactions between two or more superstrings are also mediated by superstrings themselves. Examples of self-interactions are the diagrams of Fig. 14.21: the initial closed superstring spontaneously fissions and splits into two superstrings, with the two closed superstrings subsequently fusing back to form a single superstring; this process leads to the world sheet developing 'handles'. The closed superstring can have more self-interactions with many fissions and fusions taking place between multiple initial and final states. As illustrated in the diagram of Fig. 14.22 closed superstrings with multiple initial and final states can have complicated intermediate states.

The closed superstring is a quantum mechanical object obeying the laws of quantum mechanics. How do we incorporate the quantum aspect into the behavior of the superstring? In quantum mechanics, the probability amplitude for a quantum

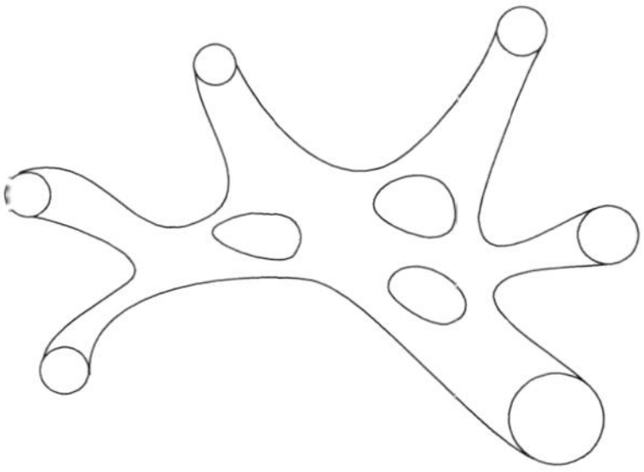

Fig. 14.22 Multiple string interactions with an increasing number of 'handles' for its world sheet.

particle going from its initial to its final position is given by summing over all possible virtual paths that go from the initial to the final position. The two circles at the beginning and end of Fig. 14.21 are the initial and final states of the closed superstring, and belong to the Hilbert space of superstring states; each diagram shown in Fig. 14.21 — with different intermediate configurations — is a possible quantum mechanical transition amplitude from the initial to the final state.

In general, the shape of the world sheet created by the numerous fissions and fusions of closed superstrings yields what is called a *super-Riemann* surface,[2] having two boundaries where the initial and final superstring reside, as illustrated in Fig. 14.21. The probability amplitude for a closed superstring to go from its initial to its final state is given by summing over all possible super-Riemann surfaces that have the initial and final states as their two boundaries as shown in Fig. 14.21 — with the Riemann surfaces having more and more 'handles'.

How are the intermediate world sheets of the closed superstring weighted in the sum over all intermediate super-Riemann surfaces? In analogy with light taking the shortest path between two points, it is postulated that the world sheet that has the *minimum area* is the one that is most preferred and has the highest likelihood of occurrence. The smoother the Riemann surface, the greater its contribution to the probability amplitude.

The superstring vacuum state can also have processes that lead to the **creation and annihilation of superstrings** from the vacuum, similar to the creation of virtual particle–antiparticle pairs in quantum field theory discussed in Sec. 5.17. As illustrated in Fig. 14.23, there can be a spontaneous creation of strings that later annihilate back into the vacuum; the lower part of the figure shows a cross-section of the Riemann surface to illustrate the multiple intermediate superstring states.

[2]The term super-Riemann surface is used to remind ourselves that the world sheet has both bosonic and fermionic variables; the bosonic variables yield a world sheet that is a Riemann surface, and the fermionic coordinates add a fermionic sector. A super-Riemann surface is characterized by the number of 'handles' in the closed sheet.

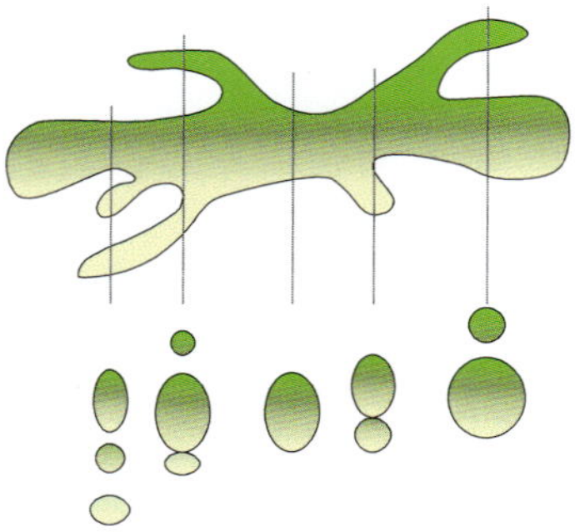

Fig. 14.23 Spontaneous creation and annihilation of superstring states due to a non-trivial superstring vacuum state.

14.7 Superstring Interactions: Geometry versus Topology

The self-interactions of a closed superstring, in particular its fission and fusion, are in essence topological. We illustrate this aspect of the superstring by examining a specific process.

Consider the fusion and fission of two closed superstrings. Figure 14.24 shows two closed superstrings colliding and fusing into a single closed superstring and after a while, fissioning into two closed superstrings. The cross-section of this fusion and fission is shown in the lower part of the figure. We show how the process of the fusion and fission of the closed superstrings is topological.

In a given Lorentz frame the string fusion takes place at a certain point in time, labeled at t and shown in Fig. 14.24(a). However, another observer who is moving at a constant speed and related by a Lorentz transformation to the original frame,

Fig. 14.24 The point of fusion and fission of the closed superstrings is not a physical event but, rather, is frame dependent. What is frame independent is the topology of the entire diagram, reflecting that topology, and not geometry, determines the process.

will see the fusion of the two superstrings happening at a *different* time t', as shown in Fig. 14.24(b).

In other words, the spacetime point at which the two superstrings fuse has no physical significance, being frame (observer) dependent and is not an intrinsic property of string fusion. Similarly, the spacetime point at which the single closed superstring fissions into two closed superstrings is also frame dependent and has no physical significance.

The string interaction is completely fixed by specifying only the *topology* of the world sheet (Riemann surface) that connects the initial and final state of the two closed superstrings; no further details of the geometry of the surface are needed. This topological description is very different from the way in which the point particles of the Standard Model interact.

14.7.1 *Point-like versus topological interactions*

Every fundamental particle is a possible state of the superstring. In the quantum field theory that forms the basis of the Standard Model, all interactions are local, with gauge fields interacting with fermions at a *single point*, as exemplified by the interaction of the photon with the electron shown in Fig. 14.25. In contrast, interactions in superstring theory are purely topological so that the photon–electron interaction becomes extended in spacetime as illustrated in Fig. 14.26, and are described by a Riemann surface.

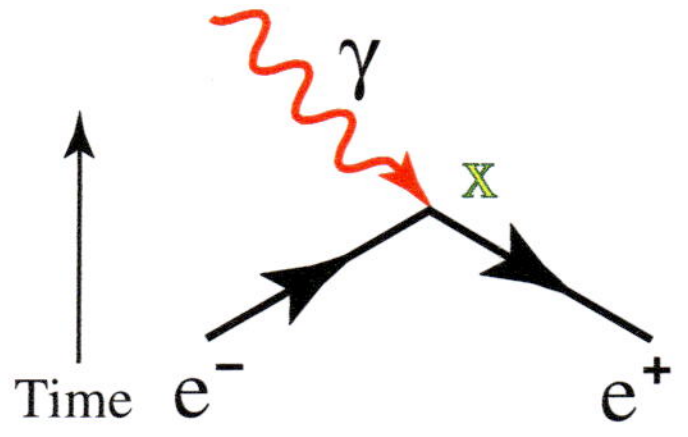

Fig. 14.25 Interaction at a single point as it occurs in the Standard Model.

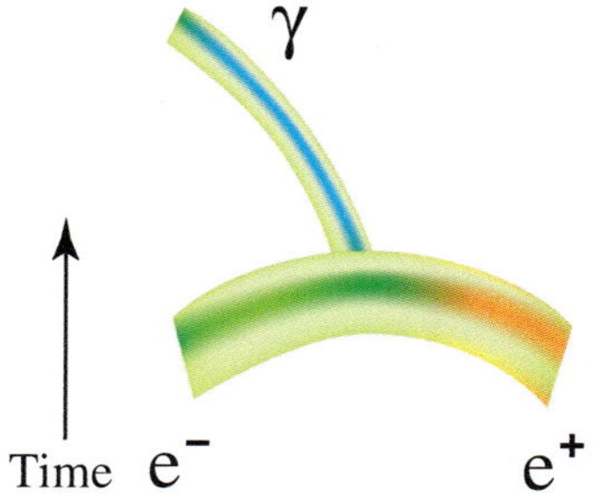

Fig. 14.26 In string theory, the process of interaction is described by a Riemann surface.

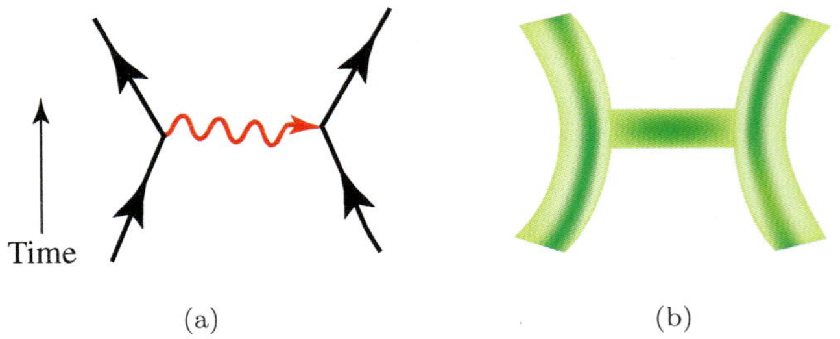

Fig. 14.27 Feynman diagrams for a point-like interaction generalized to a Riemann surface for closed superstrings.

In the Feynman diagram of an electron scattering off another electron by exchanging a virtual photon, the process in Fig. 14.27(a) is replaced by initial and final states being superstrings exchanging another superstring as shown in Fig. 14.27(b). In other words, interactions at a point, as is the case for local quantum field theories, are replaced by interactions that are encoded by the superstring world sheet Riemann surfaces — which are essentially topological. Superstring theory replaces the interactions of the Standard Model, based on distances between interacting points, by interactions that depend only on topology, that is, on the global properties of the underlying processes.

The problems that arise in defining a theory of quantum gravity are mostly due to interactions taking place at a single point. Since a superstring is extended along a line, its energy is smeared out, and this de-localization of energy results in the interactions being topological instead of being geometrical. Moving from a description based on local quantum field theory to one based on superstring theory entails going from the geometrical to a topological framework and is at the root of why we can construct a consistent theory of quantum gravity.

Figure 14.28(a) shows the process of graviton–graviton scattering in a local quantum field theory that leads to incurable divergences; Figure 14.28(b) shows

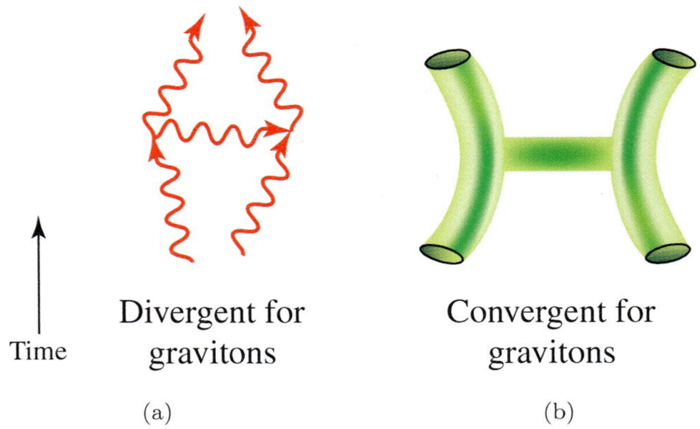

Fig. 14.28 Interaction on a Riemann surface as it occurs in string theory.

the superstring description of the same process, and which yields a finite and well defined result for this process.

14.8 Closed Superstrings: Type IIA and Type IIB

Closed superstrings and their associated theories come in three varieties, namely Type IIA, Type IIB and heterotic. We discuss Types IIA and IIB closed superstring here; the closed heterotic string is discussed in the next Sec. 14.9.

To understand the difference between the three types of closed superstrings, note that all the vibrations of a *closed loop* can be decomposed into the sum of two fundamental and elementary vibrations, namely vibrations propagating (moving) *clockwise* along the closed loop — by convention called *right moving* modes of the superstring — and vibrations that are propagating *counter-clockwise* — called *left moving* modes of the superstring. See Fig. 14.29.

In symbols one has

$$X_\mu(t,\sigma) = X_\mu^L(t-\sigma) + X_\mu^R(t+\sigma)$$
$$\psi_\mu(t,\sigma) = \psi_\mu^L(t-\sigma) + \psi_\mu^R(t+\sigma).$$

The left and right moving modes are *independent* of each other due to the periodicity of a closed loop. The independence of the left and right moving modes leads to a complete factorization of the dynamics of the superstring into left and right sectors. In particular, the state space for the closed string factorizes into a product of two separate state spaces and is denoted by $\mathcal{F}_L \times \mathcal{F}_R$. Each state space $\mathcal{F}_L$ and $\mathcal{F}_R$ provides the smallest representation of the 10 dimensional spacetime supersymmetry algebra given by *eight* supercharges $Q_a, a = 1, 2, \ldots, 8$, called $\mathcal{N} = 1$ supersymmetry. The product of the left and right moving modes' state spaces, namely $\mathcal{F}_L \times \mathcal{F}_R$, contains all the possible states of the closed string and provides *two* representations of 10 dimensional supersymmetry having 16 supersymmetry generators of charge. This is called the $\mathcal{N} = 2$ representation. The notation of IIA and IIB for closed superstrings refers to the $\mathcal{N} = 2$ supersymmetry representation with the A and B suffixes being related to a property of lowest energy states.

Parity and parity violation is one of cornerstones of the Standard Model, discussed in Sec. 13.3, and we analyze how this is realized in superstring theory.

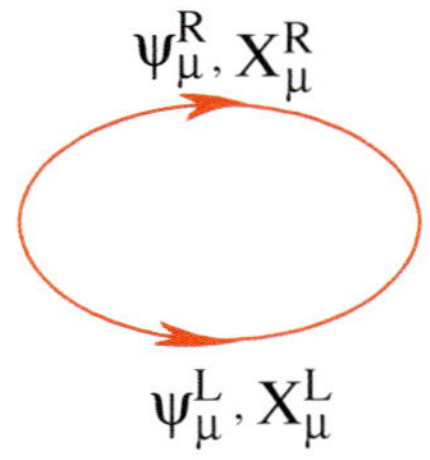

Fig. 14.29 Elementary right and left moving vibrations.

Closed superstrings IIA and IIB are distinguished by the properties of their vacuum states, whose difference is related to the concept of parity. A parity transformation takes a space coordinate to the negative of itself, as in Eq. (13.2); a parity transformation induces a transformation of $\sigma \to 2\pi - \sigma$ for the states of the closed string, leading in effect to the *exchange* of the right and left moving modes. The *vacuum* (lowest energy) states of $\mathcal{F}_L$ and $\mathcal{F}_R$ are called Ω_L and Ω_R respectively. Under a parity transformation on $\mathcal{M}_{10}$ for closed superstrings we have $\Omega_L \to \Omega_R$ and $\Omega_R \to \Omega_L$, that is, one exchanges Ω_L and Ω_R and this leads to the exchange of $\mathcal{F}_L$ and $\mathcal{F}_R$.

The state space of a Type IIA superstring is given by $\mathcal{F}_A = \mathcal{F}_L \times \mathcal{F}_R$. Under a parity transformation, the left and right moving vacuum states are interchanged implying that under a parity transformation the left and right moving state spaces are interchanged, namely $\mathcal{F}_L \to \mathcal{F}_R$ and $\mathcal{F}_R \to \mathcal{F}_L$. Hence, under the parity transformation $\mathcal{F}_A = \mathcal{F}_L \times \mathcal{F}_R \to \mathcal{F}_R \times \mathcal{F}_L = \mathcal{F}_A$. Denoting the parity transformation by $\mathcal{P}$ yields, similar to Eq. (13.10),

$$\mathcal{F}_A = \mathcal{F}_L \times \mathcal{F}_R \Rightarrow \mathcal{P}[\mathcal{F}_A] = \mathcal{F}_A : \text{invariant.} \tag{14.2}$$

Equation (14.2) implies that a Type IIA superstring is invariant under a parity transformation and therefore yields a **non-chiral superstring theory** — in other words, it preserves parity symmetry.

For Type IIB superstrings on the other hand, both the vacua have the same handedness, taken to be Ω_R, yielding the total state space to be $\mathcal{F}_B = \mathcal{F}_R \times \mathcal{F}_R$. Under a parity transformation we have $\mathcal{F}_R \to \mathcal{F}_L$, and this in turn implies that the state space $\mathcal{F}_B$ breaks parity invariance. Hence, in contrast to Eq. (14.2), we have, similar to the parity violation shown in Fig. 13.4, the following:

$$\mathcal{F}_B = \mathcal{F}_R \times \mathcal{F}_R \Rightarrow \mathcal{P}[\mathcal{F}_B] \neq \mathcal{F}_B : \text{non-invariant.} \tag{14.3}$$

In other words, the vacuum state of Type IIB superstrings explicitly breaks parity symmetry and leads to a state space that is not invariant under a parity transformation. Figures 14.30(a) and (b) provide a pictorial representation of Type IIA and IIb superstrings, respectively.

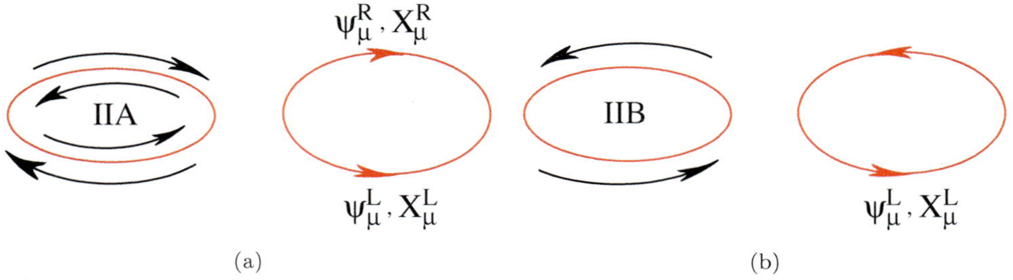

Fig. 14.30 (a) Type IIA closed superstring has bosonic and fermionic degrees of freedom that are moving clockwise and counter-clockwise. (b) In constrast, Type IIB closed superstring has only counter-clockwise moving fermionic and bosonic degrees of freedom.

The vacuum states Ω_L and Ω_R consist of the massless excitations of the superstring. Closed superstrings IIA and IIB differ only in their massless excitations; all the massive excitations for both IIA and IIB are non-chiral and can be shown to be identical.

The fact that Type IIB yields a **chiral superstring theory** is of special importance for constructing realistic four dimensional theories, since we know from the Standard Model that our Universe breaks chiral symmetry. In the D3-brane picture of our Universe discussed in Sec. 14.12, the D3-branes that occur in Type IIB superstring theories are ideally suited for modeling the breaking of chiral symmetry in our four dimensional Universe.

14.9 Closed Heterotic Superstring

We now consider a mixture of a bosonic string and a superstring. Inspired by the Greek term 'heterosis' used for example in breeding where it indicates the improved qualities of a hybrid — that is, a mixed offspring — this type of superstring is called a **heterotic string** and is defined in heterotic string theory.

To understand the theoretical basis of the heterotic superstring, recall that the left and right moving sectors of the closed superstring are completely decoupled. Hence there is no need for the left and right moving modes to originate from the same underlying string. To construct a heterotic string, one takes the left moving modes from the left moving sector of the supersymmetric $d_S = 10$ closed superstring. For the right moving modes of the heterotic string, one uses the right moving modes of a closed purely bosonic (non-super) string that exists in $d_B = 26$.

A superstring theory cannot simultaneously exist in two different spacetimes, that is, it cannot be both in 26 as well as in 10 dimensions. The supersymmetric left moving superstring states are defined in $d_S = 10$ while the right moving bosonic sector originates from a spacetime defined on $d_B = 26$ which therefore needs to be interpreted as also residing in $d_S = 10$. How can this be done? An ingenious construction allows us to re-interpret the coordinates of the 26 right moving bosons as consisting of 10 coordinates $X^R_\mu, \mu = 1, 2, \ldots, 10$, that belong to $d_S = 10$ spacetime, with the remaining 16 coordinates being considered as *bosonic scalar fields* $\phi^R_A, A = 1, 2, \ldots, 16$, that also live in $d_S = 10$. In summary, the 26 coordinates of the right moving bosonic string are interpreted as (X^R_μ, ϕ^R_A) degrees of freedom defined on 10 dimensional spacetime.

The extra 16 coordinates (bosonic fields) are thought to be curled up into small circles, with each coordinate taken to be real periodic variable defined on a circle S^1. These variables are called *compact* variables. Since there are 16 (extra) bosonic coordinates, one needs to find a (compact) space that is a generalization of the 16 circles S^1 in which the bosonic fields take their values.

The 16 dimensional generalization of the 16 circles is chosen to give the maximum possible symmetry for the resulting theory. Only two possible results for the

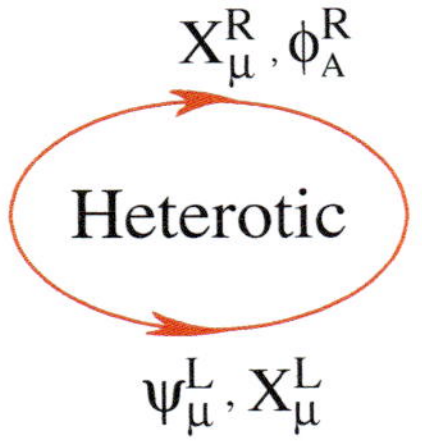

Fig. 14.31 Heterotic superstring. The left moving modes are supersymmetric whereas the right moving modes are purely bosonic, namely (X_μ^R, ϕ_A^R).

heterotic string turn out to be consistent, namely that the 16 bosons have a symmetry defined by either the Lie group SO(32), discussed in Sec. 14.2, or the Lie group $E_8 \times E_8$ — yielding two different heterotic superstring theories.[3]

The heterotic string has only the 10 dimensional $\mathcal{N} = 1$ supersymmetry since only the left moving modes are fully supersymmetric while the right moving modes are purely bosonic without the fermionic degrees of freedom required for supersymmetry. The supersymmetry of the left moving modes is sufficient to remove the vacuum instability that a fully bosonic theory would have.

In summary, the heterotic string exists in $d_S = 10$ dimensional spacetime; the heterotic string has a left moving sector consisting of the left moving modes of a closed superstring that yields a 10 dimensional $\mathcal{N} = 1$ supersymmetry; the right moving modes consist of 10 bosonic string degrees of freedom; in addition, there are 16 bosonic fields curled up into a 16 dimensional generalization of 16 circles. The asymmetric left and right moving sectors are shown in Fig. 14.31.

The right moving sector has no supersymmetry. However, rather unexpectedly, the *massless* states of the right moving purely bosonic sector have the appropriate bosons needed to explain the forces of interaction in the Standard Model and hence there is a possibility that the heterotic superstring may be relevant in understanding the Standard Model.

14.9.1 *Spectrum of the heterotic string*

As mentioned in Sec. 14.4 and illustrated in Fig. 14.17, the key notion of superstring theory is that observed particles of our four dimensional Universe are the massless superstring excitations. The collection of all the possible excitations forms the

[3]Slightly advanced: Note that two real fermion fields on the world sheet are 'equal' to one boson field; hence, the 16 bosonic right moving modes of the heterotic string can be defined to consist of 32 fermion fields. The SO(32) heterotic string can be seen to follow directly from using 32 fermions to represent the 16 bosonic degrees of freedom. However, the $E_8 \times E_8$ heterotic string is easier to understand in the bosonic formulation.

spectrum of the superstring and is therefore an integral part of any string theory. We now have a brief look at the spectrum of the heterotic string.

We are interested in obtaining all the particles and forces of the Standard Model, and also the force of gravity, from superstring theory. We have already discussed that only the massless vibrations of the superstring will be observable at the macroscopic distances at which experiments are carried out. We examine the vibrational spectrum of the heterotic string in some detail as it is one of the more realistic results from superstring theory, since it generates a collection of particles that are close to what we expect from the Standard Model.

The 10 dimensional manifold on which the heterotic string is defined is compactified into $\mathcal{M}_4 \times Y_6$. If one demands that the compactified theory exhibits the smallest four dimensional supersymmetry,[4] then the extra space Y_6 turns out to be the so-called six dimensional **Calabi–Yau space**. The Calabi–Yau space is a compact space (that is having a finite volume) which has a lot of 'holes' in it; the 'size' of the Calabi–Yau (CY) space is of the order of the Planck length; so for all practical purposes this space is invisible.

Taking the six dimensional Calabi–Yau space as being invisible, we obtain a four dimensional theory with the particles and forces given by the massless vibrations of the heterotic string defined on $\mathcal{M}_4 \times CY$.

The massless vibrations of the heterotic string are of two kinds: the left moving supersymmetric closed string variables generate supersymmetric pairs of fermions and bosons and the right moving 26 boson fields generate the Yang–Mills spin 1 gauge bosons that yield the electroweak and strong forces of the Standard Model. The symmetry group $E_8 \times E_8$ can be reduced to E_6 by breaking some of the symmetries of the heterotic string. One way of doing this is to 'trap' some of the boson fields in the 'holes' of the compact dimensions of the Calabi–Yau space as illustrated in Fig. 14.32.

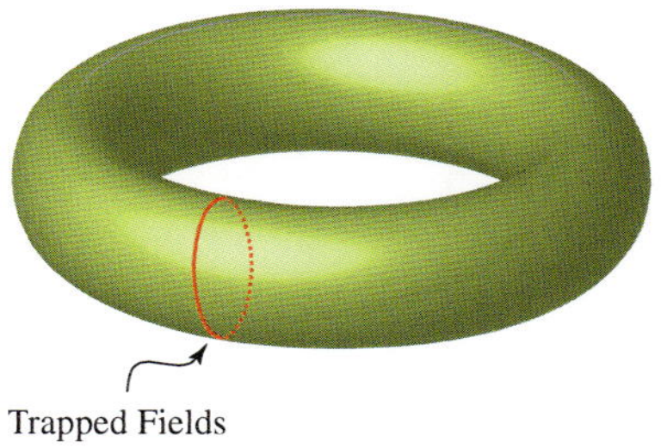

Fig. 14.32 Trapped fields.

[4]For the more advanced reader: having $\mathcal{N} = 1$ supersymmetry in four dimensional spacetime with four anticommuting supercharges Q_a.

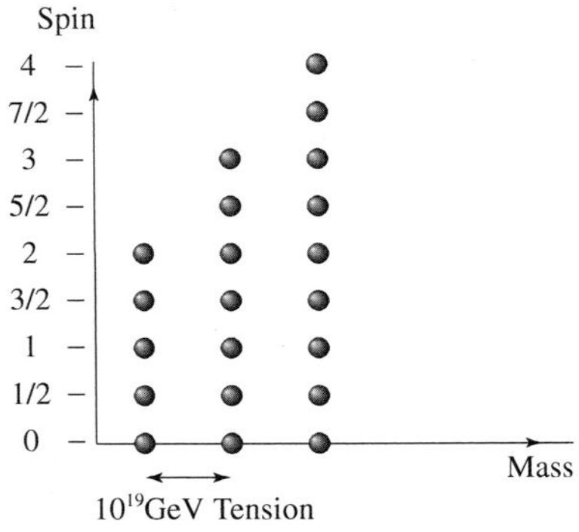

Fig. 14.33 Spectrum of string particles. Only the left most column is massless.

The spectrum (a term used for the complete list of states of a quantum system) showing some of the low lying states of the heterotic string is given in Fig. 14.33. Recall that in the discussion on GUT in Sec. 14.2, it was found that E_6 is a suitable Lie group for unifying all the forces in the Standard Model. The Lie group E_8 contains E_6 as a subgroup and so it is possible that all the particles and forces of the Standard Model can appear as massless vibrations of the heterotic superstring. It turns out that, in fact, the massless sector has all the particles, both fermions and bosons, that one would like to obtain for matching the particles and forces of the Standard Model. All closed superstrings, including the heterotic superstring, contain a massless spin 2 boson that is identified as the graviton, namely the particle that is responsible for generating the force of gravity. The heterotic closed string also has a (massless) gravitino, as expected from supersymmetry.

Table 14.2 below shows the identification of the massless heterotic string states, given in Fig. 14.33, with the particles that occur in the Standard Model. Only the left most column of massless states can be observed.

The massive states of the heterotic string start with masses of the order of the Planck mass, and there is an infinite tower of excited states with higher and higher masses. None of the massive vibrations of the heterotic string can ever be seen since

Table 14.2 Massless particles of the heterotic string.

Particle	Spin
Higgs	0
Quarks and leptons	1/2
Gauge fields	1
Gravitino	3/2
Graviton	2

they exist only at distances comparable to the Planck length — a realm beyond any conceivable experiment.

The **compactification** of a heterotic string on a Calabi–Yau space has a lot of features that are very attractive. But there are many questions that remain unanswered. It remains to be shown how the massless excitations of the heterortic string acquire the masses observed for the Standard Model particles, how is supersymmetry broken, and so on. Furthermore, there are millions of Calabi–Yau spaces that one can choose from; for most of the Calabi–Yau spaces that have been constructed, $\chi(CY)$, the Euler characteristic of the Calabi–Yau space, is not equal to 6 as is required for obtaining the three generations of quarks and leptons of the Standard Model.

14.10 Type I Open Superstrings

Open superstrings are another class of superstrings, and as the name indicates, an **open superstring** has two end points. Open superstrings are consistent with quantum mechanics and special relativity only in 10 dimensional spacetime. Waves traveling along the open superstring are reflected at the end points and hence the left and right moving vibrations are related. This reflection of the modes off the two end point boundaries results in the open superstring having a 10 dimensional $\mathcal{N} = 1$ supersymmetry; the name of Type I for open superstrings is derived from its representation of 10 dimensional supersymmetry. Note that closed strings necessarily exist in any open string theory since an open string can fuse its end points to create a closed string. As mentioned earlier, the end points of the open superstring move at the speed of light.

The open superstring spans out a world sheet with boundaries, as shown in Fig. 14.34(a). An open superstring can be oriented or non-oriented, the two being distinguished by an arrow on the string, as shown in Fig. 14.34(b). Similar to closed superstrings, the open superstring can have self-interactions due to fission and fusion, as illustrated in Fig. 14.35(a). If one examines a slice of equal time in the world sheet, it is clear that a hole appearing in the open superstring's world

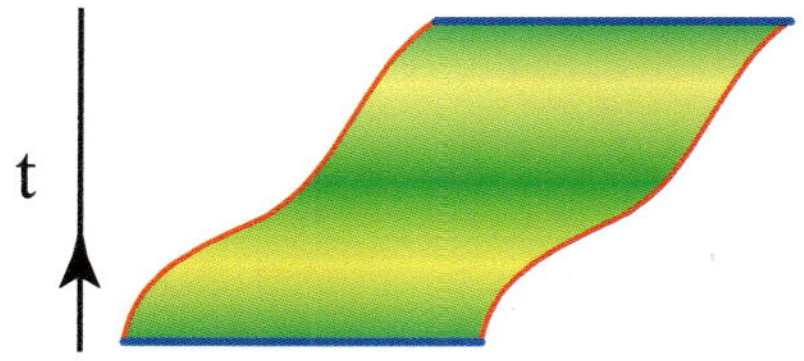

(a)

(b)

Fig. 14.34 (a) World sheet spanned out by an open string, with its boundary indicated in red. (b) A non-oriented and an oriented open superstring.

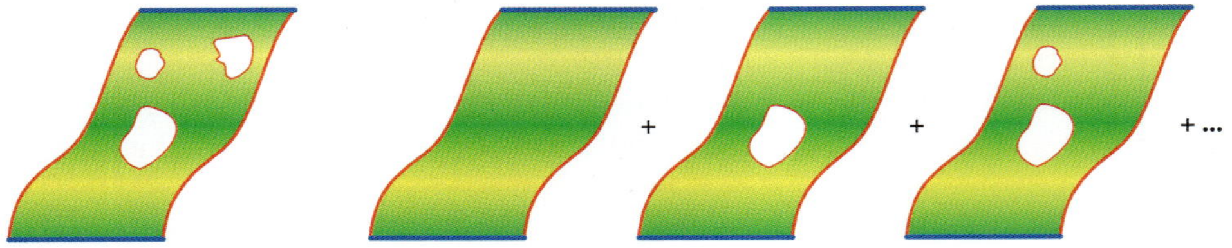

(a) (b)

Fig. 14.35 (a) Self-interaction of an open string. (b) Sum over self-interactions of the open superstring.

sheet is due to the splitting of a single open superstring into two (or more) open superstrings and the subsequent fusion of the endpoints of the open superstrings.

The open superstrings' self-interactions can be classified by the number of holes in the world sheet. The quantum theory of the open superstring is defined similar to that of the closed superstring: the transition amplitude to go from an initial to a final superstring state is obtained by summing over all possible intermediate states, including all possible self-interactions as illustrated in Fig. 14.35(b). The open superstrings with the minimum world sheet area dominate the quantum mechanical transition amplitude.

A key feature of the open superstring is that electric, or in general color, charges can be attached to the end points of the superstring. Mathematical consistency requires that the charges at the end points of an open superstring belong to the SO(32) Lie group. The massless spectrum of the open string does not contain a graviton; it however does contain fermions and analogs of the electroweak and colored gauge vector bosons.

14.11 D-Branes

Superstring theory started in 1984 with the idea of a string-like object being the fundamental constituent of matter, in place of the point-like particles of the Standard Model. From the outset of superstring theory, the question was naturally asked as to why should one stop at a one dimensional object? Why not have fundamental constituents that are higher dimensional extended objects like a membrane, or a three dimensional object like a jello, and so on. Attempts for a direct generalization of the string to higher dimensions have thus far all failed. At first one may therefore surmise that there are no higher dimensional objects in string theory.

Surprisingly, however, it was discovered in 1996 that the very definition of superstring theory does, in fact, contain fundamental extended objects that have higher dimensions than the string. Taking the nomenclature from a membrane, which is now termed a 2-brane, it was discovered that superstring theory contains higher dimensional 'branes' all the way to 9-branes.

Open superstrings can fulfill one of two types of boundary conditions: each end can either be free or can be pinned down. If it is pinned down, it satisfies the so-called Dirichlet boundary condition. Since the open superstring exists in space and propagates in time, the boundary condition on the end points of a string are imposed in 9 dimensional space. The end points of the open superstring maintain their boundary conditions as they evolve in time.

The so-called **D-brane** is defined in the following manner. A D-brane is a physical object that occupies a subspace of spacetime such that one of the end points of the open superstring is restricted to lie in that subspace. The two end points of the open superstring can be restricted to two different branes. When the open strings evolve in time, the end point restricted to a D-brane spans out the world volume of the D-brane.

- All D-branes exist in 9 dimensional space. To distinguish the various D-branes, the *spatial dimension* of the brane is appended; hence a D-brane of two dimensions is denoted as a D2-brane.
- In general, imposing the boundary conditions on open superstrings' end points to be restricted to subspace of different dimensions defines *10* different types of D-branes having different dimensions.[5]
- Given that the maximum dimension of *space* is 9, the D-branes that can exist in a 10 dimensional superstring theory are the following: D0, D1, D2, D3, D4, D5, D6, D7, D8 and D9-brane.
- Most D-branes are strictly flat and are Euclidean spaces of various dimensions; hence, the geometry of a Dk-brane is given by the k-dimensional Euclidean space $\Re_k$, and its world volume is given by $\Re_k \times \Re = \mathcal{M}_{k+1}$, where $\mathcal{M}_{k+1}$ is $(k+1)$ dimensional Minkowski spacetime. More complicated D-branes with non-trivial geometry also exist, but these will not concern us.

Two D2-branes are shown in Fig. 14.36(a); one of the open string starts and ends on the same D-brane and the other open string starts on one D2-brane and ends on another D2-brane. For example, as illustrated in Fig. 14.37, the end points of the open superstring move only in a plane, and are forbidden to leave the plane; the end points of the open superstring span out a two dimensional object called the D2-brane.

A **D-brane** is a dynamical object in its own right; constraining the end points of the open superstring is just one of the ways of describing the D-brane. The end points of an open superstring are restricted to the D-brane due to the physical

[5] An eleventh D-brane can be defined by fixing both the end points to be at given position and at a given time. See Table 14.3.

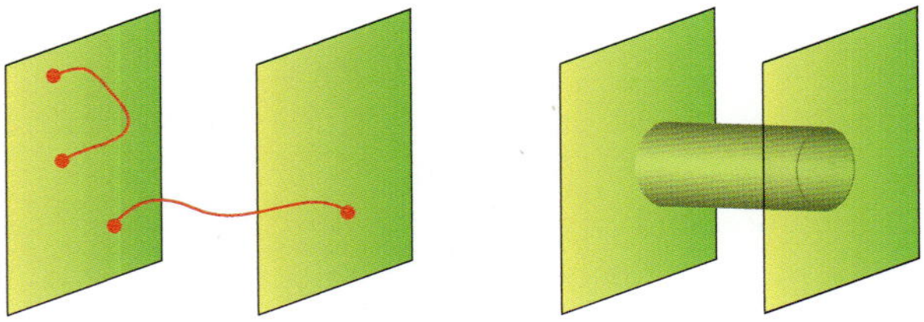

(a) (b)

Fig. 14.36 D-brane interactions.

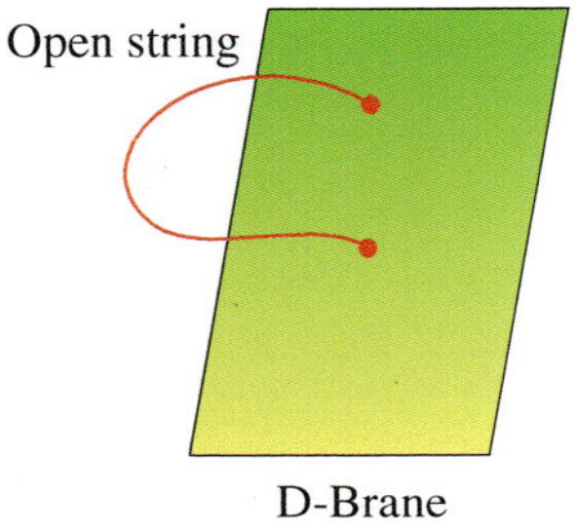

Fig. 14.37 A D2-Brane.

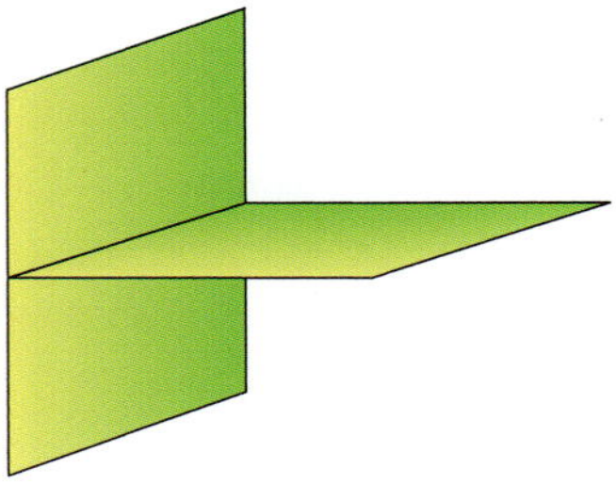

Fig. 14.38 Intersection of two D2-branes.

properties of the D-brane. The presence of a D-brane means that the lowest energy state, namely the vacuum state, of the superstring theory has been changed. The analog of the presence of a D-brane is if one inserts an electric charge into empty space, then the space is no longer 'empty' but instead, an electric charge generates an electric field that permeates all of space.

D-branes can 'interact' with themselves, and with other D-branes, by exchanging virtual open strings as shown in Fig. 14.36(a). D-branes can also interact by exchanging closed superstrings, as shown in Figure 14.36(b). The self-interaction of single D-branes is achieved by the emission and absorption of both open and closed strings. Two or more D-branes can also intersect, as shown in Fig. 14.38.

Table 14.3 Brane-scan: D-branes in superstring theories. The $-1D$ brane is a string configuration that is completely localized in both space and time. 3_D^+ stands for a 'self-dual' D3-brane.

Type IIA		0_D		2_D		4_D		6_D		8_D	
Type IIB	-1_D		1_D		3_D^+		5_D		7_D		9_D
Type I			1_D				5_D		7_D		9_D

Similar to the behavior of electric charges and for preserving one of the fundamental symmetries of superstrings,[6] the presence of a D-brane alters the vacuum state of superstring theory; it generates a fixed non-zero background of a so-called classical **RR (Ramond–Ramond)-field**, which is the analog of the electric field. In other words, the D-branes, are objects carrying charges that generate the RR-fields. The D-branes, together with the RR-fields they generate, define the allowed vacuum state of superstring theory, for both closed and open strings. The string background vacuum state is the analog of the vacuum state with a non-zero value of the Higgs field in the theory of electroweak interactions, discussed in Sec. 13.8.

14.11.1 *D-branes in various superstring theories*

We have so far discussed five types of superstring theories, namely closed superstrings IIA and IIB, closed heteroric string SO(32) and $E_8 \times E_8$, and Type I open superstring. The D-branes that occur in these theories depend on the nature of the RR-fields that are allowed for the vacuum of the theory. In Table 14.3 we have a 'brane-scan' of the various theories and the D-branes that exist in them. The heterotic string theory does not contain any branes and hence there are no entries for it in the table.

14.12 D3-Brane: Our Universe

Let us concentrate on our four dimensional Universe. One can think of the entire three dimensional space of our Universe as a three dimensional *subspace* of the ambient 9 dimensional space of superstring theory; this subspace is *postulated* to be fully occupied by a three dimensional D3-brane. The world volume of a D3-brane is a four dimensional spacetime given by $\Re_3 \times \Re = \mathcal{M}_4$, where $\mathcal{M}_4$ is four dimensional Minkowski spacetime. We identify the word volume $\mathcal{M}_4$ of the D3-brane with our Universe.

This is the brane interpretation of our Universe: all dimensions of the 10 dimensional ambient spacetime are equally large. Our Universe appears to have a

[6]The symmetry is that of super conformal invariance.

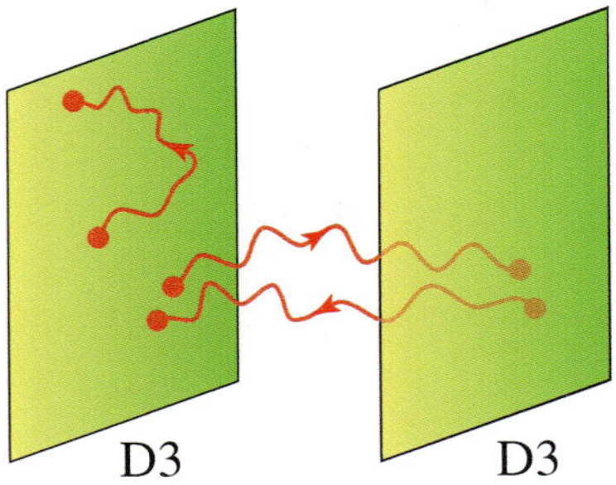

Fig. 14.39 Oriented open superstrings with charges on its end points.

space that is three dimensional since our Universe consists of a D3-brane; the time evolution of the D3-brane gives rise to our four dimensional spacetime, namely $\mathcal{M}_4$.

In this brane picture, there is no need for compactifying the ambient 10 dimensional spacetime into 4 dimensions since our Universe is a D3-brane. Our Universe appears as a four dimensional spacetime because all phenomena in our Universe are *confined* to our D3-brane. The other view that the extra six of nine space dimensions are of Planck size — discussed in the sections on the closed and heterotic superstrings — is also consistent with the formalism of superstring theory. It is ultimately experiments that will decide which one of the two models is correct, or if indeed either of them is correct.

We discuss how quarks, leptons and the forces of the Standard Model are produced in the D3-brane world model of our four dimensional Universe. The only force that is missing is gravity; Chapter 15 discusses how gravity can be incorporated into the D3-brane world.

14.12.1 *Two separated D3-branes*

An open superstring, which carries charges at its end points, starts on one of the D3-branes and ends on another D3-brane, as shown in Fig. 14.39. Recall in Sec. 14.10 it was pointed out that the end points of an open superstring support the gauge group SO(10); the charges of the SO(10) gauge group are attached to the end points of the open superstring. Hence inside the D3-brane the end points of the open superstring appear to be *point charges*, carrying the charge that comes from the SO(10) gauge group, and generating the analog of electric and magnetic fields inside the brane. The carriers of the charge of the SO(10) gauge group are fermions, and the SO(10) charge is the analog of electric charge; the gauge group includes the SU(5) gauge group required for the GUT that unifies all the particles and forces of the Standard Model.

The fermions carrying the SO(10) charges can produce all the quarks and leptons of the Standard Model. In other words, the *end points* of the open superstring carry fermions that appear as the *point-like* quarks and leptons of the Standard Model.

In particular, all the fermions of our four dimensional spacetime are the end points of open superstrings ending on our D3-brane.

The coupling of the fermions at the end points of the open superstring to the gauge fields is a reflection of the self-interactions of the D3-brane mediated by the open superstrings — and this gives rise to the forces in our D3-brane Universe. Consider two D3-branes separated by a small distance d; oriented open superstrings can start on a charge and end on an anticharge on the same D3-brane. Open superstrings can also start on a charge on one of the D3-branes and end on an anticharge on the other D3-brane. See Fig. 14.39.

What will an observer, confined to one of the D3-branes, observe? Since the energy per unit length of an open string is given by the string tension T, an open string of length d has energy of Td; inside the D3-brane, the open superstring will appear as a massive particle of mass Td/c^2: the particle is a spin 1 vector gauge particle that we identify with W_μ^-. Since open strings are oriented, an incoming open superstring will couple to a positive charge; an outgoing open superstring will couple to a negative charge, and will appear as the antiparticle of the incoming open string — and which is identified with W_μ^+. An open superstring coupling to charge and anti-charge that are *coincident* inside a given D3-brane will appear to couple to a neutral charge and hence appear as a neutral particle, identified with Z_μ^0.

Collecting the three particles, $W_\mu^\pm, Z_\mu^0$, we obtain the three spin 1 weak gauge bosons of the weak interactions — showing how gauge bosons in four dimensions can be obtained from open strings and D3-branes.

14.12.2 *Three coincident D3-branes*

Consider *three* coincident D3-branes, which we consider to be our Universe. What happens when D3-branes *coincide*, that is, we set $d \to 0$? The gauge bosons all become massless and we identify these massless gauge bosons with the photon and the gluons. In the string representation, color keeps track of which D3-brane the open string starts from and anticolor determines which D3-brane it ends upon, as shown in Fig. 14.40. Note that this is similar to the scheme we used for labeling the colored gluons in Eq. (12.16). To obtain the eight colored gauge bosons of the strong interactions, we need to *coincide* three D3-branes, yielding nine possible ways that

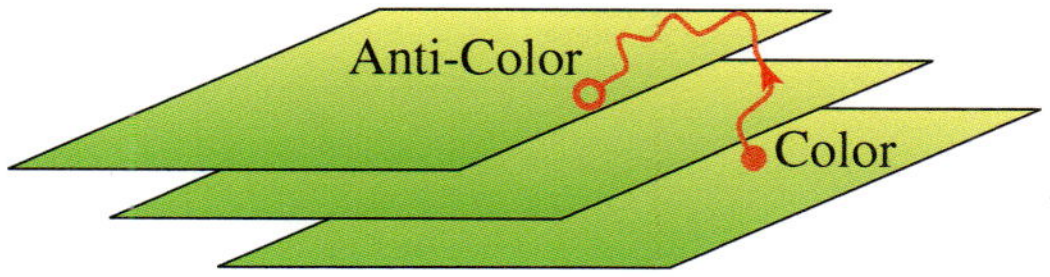

Fig. 14.40 Three *coincident* D3-branes; the open superstring appears as colored fermions and massless gauge bosons.

open strings can connect the three different D3-branes — yielding the eight SU(3) colored gluons of quantum chromodynamics.[7]

In this manner, by considering collections of D3-branes that are close to each other and by making D3-branes coincident, we can produce all the gauge fields and fermions of the Standard Model. The only force not accounted for in this D-brane Universe is gravity, and Chapter 15 addresses this issue.

14.12.3 *Particles and forces in four dimensions*

D-branes have other subtle connections with four dimensional physics. Consider a superstring theory in 10 dimensional spacetime, and let the theory have D2-branes. For simplicity, we compactify the 10 dimensional theory by considering the extra 6 dimensions to be a product of six circles $S^1 \times S^1 \times S^1 \times S^1 \times S^1 \times S^1$, which is a 6 dimensional torus T^6. Wrap a D2-brane around T^2, namely a 2-torus $S^1 \times S^1$, as shown in Fig. 14.41; the mass of the wrapped D2-brane is proportional to the volume of the 2-torus. For comparison, a D2-brane wrapped around a sphere is shown in Fig. 14.42.

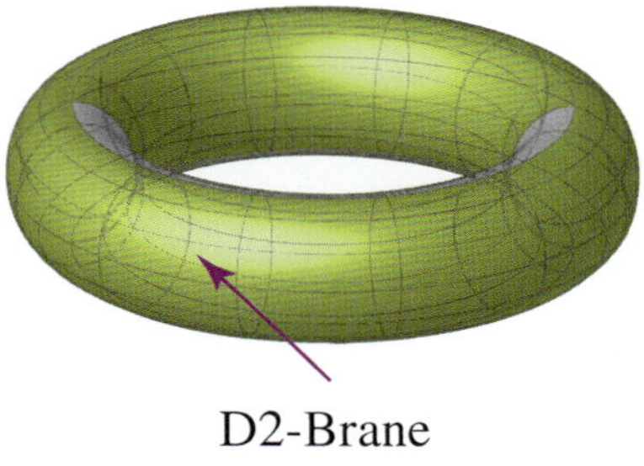

Fig. 14.41 D2-brane wrapping a 2-torus.

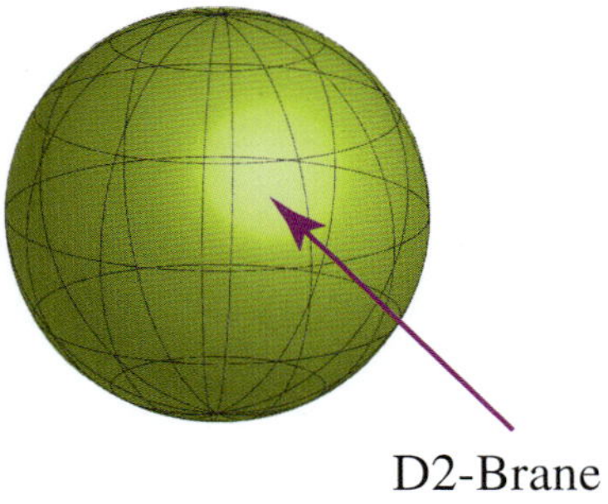

Fig. 14.42 D2-brane wrapping a 2-sphere S^2.

[7]For the case of N coincident D3-branes, the dynamics of open superstrings inside the brane world is described by quarks and gluons described by the SU(N) gauge group.

Suppose the radius of all the circles S^1 — making up the extra 6 dimensions — is r. Let the radius of all the circles goes to zero, that is $r \to 0$: the extra dimensions disappear, and we end up with four dimensional spacetime. If one examines the spectrum of the massless particles of the superstrings in the limit of $r \to 0$, one discovers that the D2 branes wrapped around a 2-torus T^2 become massless, and survive as massless states of the four dimensional theory.

One can make more complicated choices for both the D-branes used for wrapping as well as for the 6 dimensional space, and this gives rise to a rich collection of particles and forces described by local Yang–Mills quantum field theories in four dimensions, and includes all the theories that appear in high energy physics.

Superstrings throw a new and unexpected light on physics in four dimensions, and it seems that to truly understand the deep significance of phenomena in four dimensions — as well as the mathematical structure of quantum field theory — the perspective that superstring theory offers is invaluable.

Noteworthy (optional content) 14.3: D3-Branes and Color Charge: Quantity and Quality

---

As discussed above and shown in Fig. 14.40, inside a Universe consisting of three coincident D3-branes, the properties and interactions of quarks and gluons are determined by the dynamics of the open superstrings. In the D3-brane Universe, color is a pure quantity, a pure number that keeps track of which of the coincident D3-branes the open superstring starts from and on which D3-brane it ends. The end points of the open superstrings appear as quarks that carry the color quantum number, and their interactions via colored gluons is a four dimensional representation of the dynamics of the open superstring.

In the four dimensional world of quarks and gluons, color charge was understood to be a new *quality* of the quarks and the quantum number of color quantified this quality. The remarkable connection between our Universe considered to be a D3-brane world and the higher dimensional Universe of the superstring shows that many *qualities* of our four dimensional Universe may simply be purely *quantitative* features of superstring theory.

The interpretation of color charge that emerges from superstring theory is an example of how many of the key concepts of four dimensional quantum field theory reveal their deeper content when analyzed from the D-brane and superstring point of view.

14.13 M and F Superstring Theories

So far we have discussed closed and open superstrings, namely the heterotic string, Types IIA and IIB strings as well as Type I string, respectively; these superstring theories are described by the behavior of their two dimensional world sheet, which is embedded in the ambient 10 dimensional spacetime. The ambient spacetime can be either 10 dimensional Minkowski flat spacetime, namely $\mathcal{M}_{10}$, or a compactified spacetime $\mathcal{M}_4 \times Y_6$.

A different genre of superstring theories is described by M- and F-theory that are defined, respectively, on 11 and 12 dimensional spacetime. Both M- and F-theory do not directly refer to the superstring's world sheet but instead, are defined directly in the higher dimensional spacetime.

The maximum dimension in which supergravity can be defined is $d = 11$. The accompanying superstring theory, of which the 11 dimensional supergravity theory is considered to be the low energy limit, is known as **M-theory. Supergravity** in $d = 11$ contains two bosonic fields, namely a metric tensor and a three index antisymmetric tensor; the fermionic fields are fixed by the requirement of supersymmetry.

Branes in M-theory are referred to as M-branes. There are no open strings in M-theory and hence, unlike D-branes that are defined by boundary conditions on the end points of open strings, M-branes are instead defined using the concept of the M-brane being an extended charge coupling to the appropriate field.

The three index antisymmetric tensor is the higher dimensional generalization of the Maxwell gauge field, and — similar to the Maxwell field — can couple to charges that reside in an M-brane, which generate the field. The spatial dimension of an M-brane is indicated by the number appended to the letter M. The three index antisymmetric tensor couples to a two dimensional object called the M2-brane. The *dual* of the three index antisymmetric tensor is a six index antisymmetric tensor that couples to a five dimensional object called the M5-brane. The M2- and M5-branes are two and five dimensional flat subspaces inside the 10 dimensional space and yield a three and six dimensional world volume, respectively, as they evolve in 11 dimensional spacetime.

One can consider supergravity on 11 dimensional spacetime that is $\mathcal{M}_{10}$ times a circle S^1, namely $\mathcal{M}_{10} \times S^1$; suppose the radius of S^1 is R_s. It can be shown that the radius R_s is related to the strength of the string coupling constant g_{IIA} of IIA.[8] In the limit $R_s \to \infty$ one obtains the full $d = 11$ supergravity.

A more interesting limit is to take the radius of S^1 to zero, namely $R_s \to 0$. In this limit, M-theory reduces to the 10 dimensional Type IIA non-chiral superstring theory. All the D-branes of Type IIA, namely the D0, D2, D4 and D6-branes are seen to emerge from the dimensional reduction of the M2 and M5-branes from $\mathcal{M}_{10} \times S^1$ to $\mathcal{M}_{10}$ by wrapping the M2 and M5-branes around the S^1 space.

[8] String theory has no coupling constants that are fixed 'from the outside', that is, by experiment; rather the string coupling constant is itself fixed by the dynamics of string theory and is not an external parameter.

M-theory is also related to the other superstring theories in various dimensions less than 10 by dimensional reduction or compactification on various spaces, as shown in Fig. 14.47.

One connection of superstring theory with four dimensional spacetime has been achieved by factoring 10 dimensional space into a product space $\mathcal{M}_4 \times Y_6$, with Y_6 chosen to be a non-trivial space — for example where Y_6 can be taken to be a Calabi–Yau space. A more general construction of superstring theory is not to take spacetime to be a simple product of $\mathcal{M}_4$ with some internal space, but rather, for the spacetime manifold to be some sort of 'twisted' product of underlying spaces.

In the **F-theory** approach, one starts with 12 dimensional spacetime that has a vacuum state consisting of a collection of 24 D7-branes that all are orthogonal to the plane defined by (X_8, X_9). The 12 dimensional spacetime is built out of a technique called **elliptic fibration** leading to a manifold $\mathcal{M}_8 \times K3$, where $K3$ is a four dimensional complex manifold; Type IIB superstring can be consistently defined on this manifold.[9] One obtains IIB from F-theory by compactifying the extra two dimensions to a torus T^2: IIB on $\mathcal{M}_{10}$ is equivalent to F-theory on $\mathcal{M}_{10} \times T^2$, with the size of the torus made vanishingly small, and is shown in Fig. 14.47.

14.14 Superconductor, Vortices and Duality

Duality is an important concept in Physics and has a wide range of applications. A duality transformation is a mapping from one description of a system to another. Duality is a key idea that relates different consistent superstring theories and is discussed in Sec. 14.15. A theory can have two different representations and these are called dual descriptions. One description is more appropriate for describing some particular feature of the theory, and this is illustrated using the example of **magnetic vortices** in a superconductor.

A vortex is the generalization of angular motion of a particle to a fluid, and is defined by the spinning motion of the fluid about an axis that is taken to be the core of the vortex. When one stirs a cup of tea, one is creating a vortex inside the tea cup. A circulating fluid carrying charge creates a magnetic field along its axis of rotation and hence generates a magnetic vortex. The magnetic flux is centered on the core of the vortex.[10]

It was discovered in 1933 that the superconductor, unlike a normal conductor, expels magnetic fields that have a strength less than a critical magnetic field B_c; this is called the **Meissner effect** and is shown in Fig. 14.43. We now analyze the mechanism that underlies the Meissner effect, in which vortices play a key role.

[9] More precisely, one does an elliptic fibration of $\mathcal{M}_{10}$ by a torus T^2 and obtains $\mathcal{M}_8 \times K3$.

[10] Magnetic flux through a surface area is defined, for a constant magnetic field, to be the magnetic field multiplied by the area. In general, magnetic flux $= \int d\mathbf{S} \cdot \mathbf{B}$, where $\mathbf{B}$ is the magnetic field and $\mathbf{S}$ is the (vector for the) surface area.

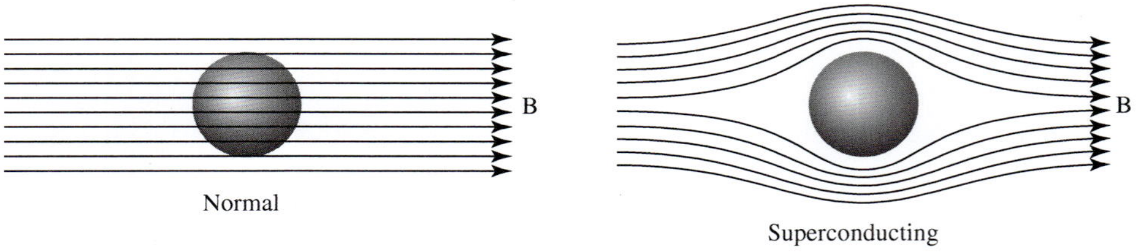

Fig. 14.43 Schematic representation of the Meissner effect for a spherical superconductor. The magnetic field B goes through a normal conductor. The superconductor expels the magnetic field for the values of B that are less than a critical value B_c.

Consider the effect of an external magnetic field B — applied in the z-direction — on a superconductor. As one increases the strength of the magnetic field, the superconductor has two distinct responses depending on the material it is made out of. The correlation length ξ and penetration depth λ, given in Eq. (13.19), play a key role in determining the response of the superconductor. When the magnetic field reaches a critical value, magnetic vortices form inside the superconductor with the magnetic flux lines at their core and the flux lines penetrate the superconductor. The superconductor phase is broken at the core of the vortices.

- **Type I superconductor**, for which $\xi > \lambda$. When the magnetic field reaches a critical value of B_c, vortices form throughout the superconductor and coalesce since, in Type I superconductor, the vortices attract each other. The superconductor abruptly undergoes a discontinuous (called a first order) phase transition and becomes a normal conductor.
- **Type II superconductor**, for which $\lambda > \xi$. When the magnetic field reaches the strength of B_{c1} magnetic vortices that repel each other form deep inside the superconductor. As one increases the strength of the magnetic field, more and more vortices appear, until one reaches a critical value of the magnetic field given by $B_{c2} > B_{c1}$; for magnetic fields equal or larger than B_{c2}, the superconductor undergoes a continuous (second order) phase transition and becomes a normal conductor.

For field strength greater that B_{c1}, the magnetic field penetrates the bulk of the Type II superconductor. The magnetic field is confined to thin filaments in the z-direction and is generated by tiny currents circulating in the xy-plane — and giving rise to a *magnetic vortex*. The magnetic field *breaks* the superconducting phase inside the core of the vortex, where the conductor is in its normal phase. A supercurrent — having no resistance — circulates outside the normal core of the vortex and generates the magnetic field in the filament.

In summary, for Type I superconductors, since the magnetic vortices *attract* each other they combine to form a large vortex — with a normal phase at its core — and this leads to the breakdown of superconductivity throughout the material. In contrast, for Type II superconductors, when the field strength is greater than B_{c1}

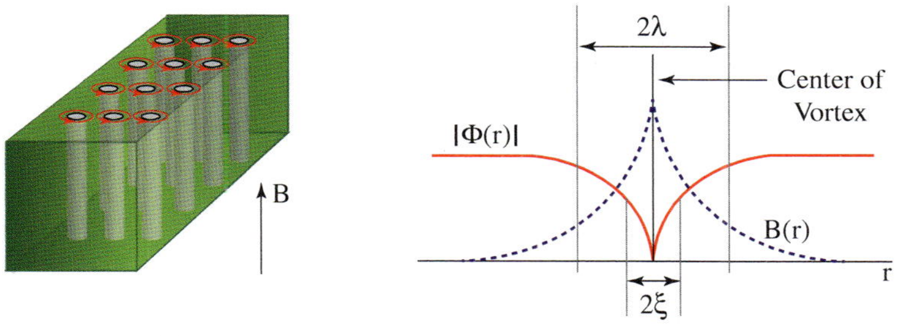

Fig. 14.44 Type II superconductor. (a) Magnetic vortex lines. (b) The order parameter Φ and magnetic field B inside a magnetic vortex line; r is the distance from the center of the vortex.

magnetic vortices that form, called the Abrikosov–Gorkov vortices, *repel* each other and form a regular pattern in the form of a lattice; this allows for the emergence of many filaments in the bulk of the superconductor as one increases the magnetic field beyond B_{c1}.

The magnetic vortices for Type II superconductors are shown in Fig. 14.44(a). Figure 14.44(b) shows that the vortex has a radius given by λ and that the magnetic field rapidly goes to zero on the boundary of the vortex; the Landau–Ginzburg scalar field Φ that encodes superconductivity, and discussed in Sec. 13.10, is zero at the center of the vortex and rises to the value of a when the radius is greater than ξ. Consider a cross-section of the magnetic vortices given in Fig. 14.44(a); the size of the vortex is of the order of λ, given in Fig. 14.44(b), which is about 10^{-9} m; hence for large scale measurements, the vortex can be taken to be a line with a point-like cross-section.

It has been shown that the **Abrikosov–Gorkov vortices** form a two dimensional triangular lattice, as shown in Fig. 14.45. On reaching the strength of B_{c2} the Type II superconductor undergoes a phase transition and becomes a normal conductor.[11]

A remarkable result from quantum mechanics is that the **magnetic flux** of the vortex is **quantized** in units of $\nu = h/(2e)$, where h is Planck's constant. Each vortex is characterized by the flux it carries, which is a positive or negative integer n in units of $\nu = h/(2e)$. In other words, all vortices inside a superconductor can only be $n\nu$, where n is an integer; an antivortex generates a magnetic field in the opposite direction and has negative magnetic flux given by $-n\nu$.

Vortices can attract and repel, as we see for the vortices that appear in Type I and Type II superconductors. Directly modeling the dynamics of the

[11] Type II superconductors are important for applications since they continue to be superconducting for much higher magnetic fields than Type I superconductors. The high speed trains to and from Shanghai airport use Type II superconductors for magnetically levitating the trains using the Meissner effect.

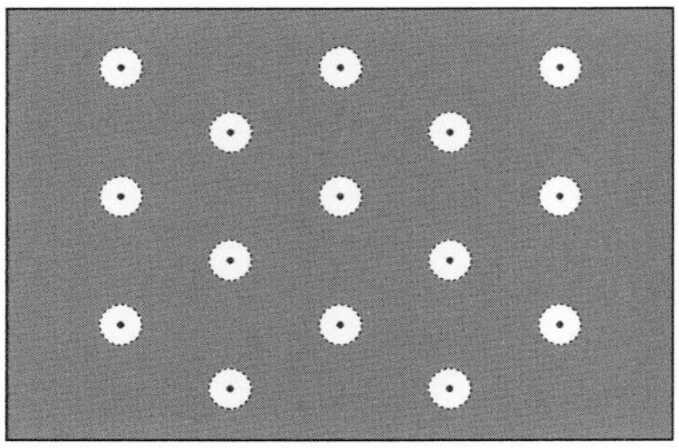

Fig. 14.45 Vortices that can be considered to be point-like.

vortices — instead of deriving them from an underlying theory as we have done — can yield another theoretical description of superconductors. Vortices carry magnetic flux $n\nu$ and the repulsive or attractive interactions of the vortices look similar to the interactions of charges characterized by their electrical charge ne, where n is a positive or negative integer.

Considering a two dimensional cross-section of a superconductor as in Fig. 14.45, one can consider the vortices to be fundamental point-like objects with the analog of charge being the flux they carry. A theory can be written that is entirely based on only the vortices interacting with the Maxwell electromagnetic field — and forms a *dual description* of the theory of magnetic vortices. The **duality mapping** is the following:

$$\text{point charge} \leftrightarrow \text{vortex}$$
$$\text{electric charge e} \leftrightarrow \nu = \frac{h}{2e} \text{ magnetic flux.}$$

If one forms a 'vortex' out of the vortex theory, one will recover the original theory with its point electric charges. In other words, the dual of the dual theory is the original theory, as is required by consistency.

Note that strong and weak coupling behaviors are interchanged by the duality transformation, with e being transformed into $\nu = h/(2e)$. The mapping that *inverts* the electrical charge (coupling constant) from e to $1/e$ is called S-duality in string theory.

14.15 Superstring Theories: Connected by Dualities

The various superstring theories we have discussed are related to each other by a web of interconnections due to what are called S and T-dualities. The concept of duality, which has been discussed for the case of superconductors in Sec. 14.14, is a term for the transformation relating two equivalent descriptions of a given theory.

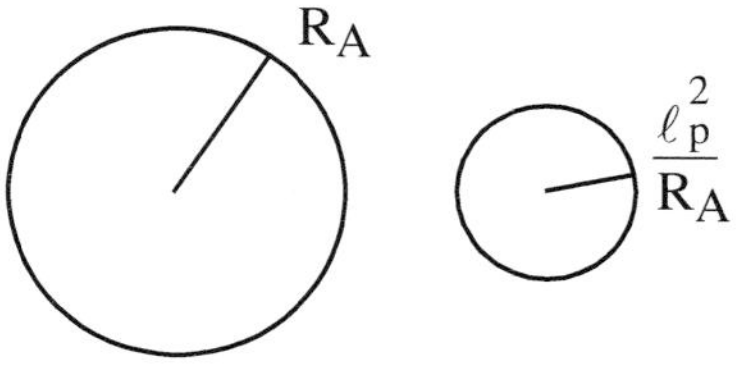

Fig. 14.46 T-duality: the radius $R_A \to 1/R_A$ relates two closed superstring theories.

The most famous duality in superstring theory, and the first to have been studied, is the duality between Type IIA and Type IIB. If we compactify IIA on a manifold $\mathcal{M}_9 \times S^1_A$ having radius R_A and IIB on $\mathcal{M}_9 \times S^1_B$ having radius R_B, then the two theories are identical, including the massless sector, if

$$R_A = \frac{\ell_P^2}{R_B} \Rightarrow R_A R_B = \ell_P^2 \tag{14.4}$$

where ℓ_P is Planck's length — and is illustrated in Fig. 14.46.

The equivalence of two closed superstring theories under the inversion of the radius of compactification is called **T-duality**. T-duality leads to the conclusion that compactifying on a small circle for IIA is equivalent to compactifying IIB on a large circle, and *vice versa*. Hence the *smallest* radius R_m of the S^1 is given by Planck's length since

$$R_m^2 = \ell_P^2 \Rightarrow R_m = \ell_P \ : \ \text{closed superstrings.} \tag{14.5}$$

We see from T-duality that the absolute minimum radius for closed superstrings is the Planck length. The concept of a radius less than ℓ_P is meaningless, since the properties of closed superstrings for any radius d less than ℓ_P is equivalent by a duality transformation to another closed superstring theory with a radius greater than d given by ℓ_P^2/d. Note that the concept of a minimum radius does not apply to the open superstring sector of superstring theory.

There are other interesting dualities. M-theory compactified on a 2-torus T^2, namely on $\mathcal{M}_9 \times T^2$, is equivalent to IIB on $\mathcal{M}_9 \times S^1$, with the radii of the torus T^2 and circle S^1 being chosen appropriately.

The duality that interchanges the weak coupling sector of superstring theory with the strong sector by inverting the strength of the self-interaction is called **S-duality**. In other words, S-duality relates strong coupling superstring of one kind to a weak coupling theory of another kind. For example weak coupling SO(32) heterotic string is S-dual to the $E_8 \times E_8$ heterotic string, and is shown in Fig. 14.47.

The lowering of the dimension of spacetime is done by compactification; in particular, for the examples discussed above, one can reduce $\mathcal{M}_{10}$ to $\mathcal{M}_9$ by first compactifying $\mathcal{M}_{10}$ into $\mathcal{M}_9 \times S^1$ and then taking the radius of the circle S^1 to zero. One can show that compactifying IIA and IIB from $\mathcal{M}_9 \times S^1$ to $\mathcal{M}_9$ yields the same superstring theory, denoted by Type II and shown in Fig. 14.47.

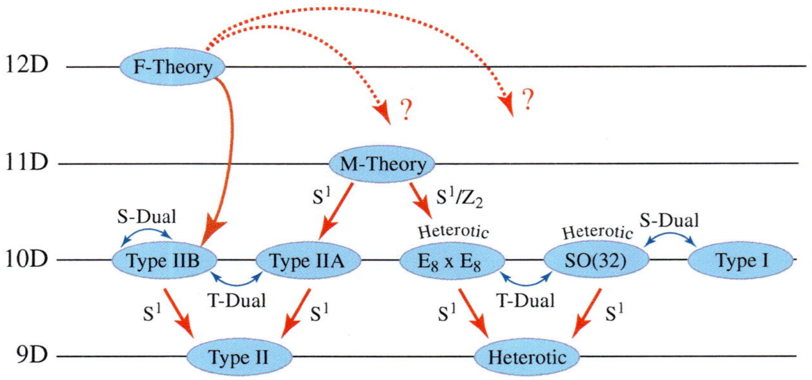

Fig. 14.47 The inter-connections, between the various superstring theories defined in different spacetime dimensions, due to duality.

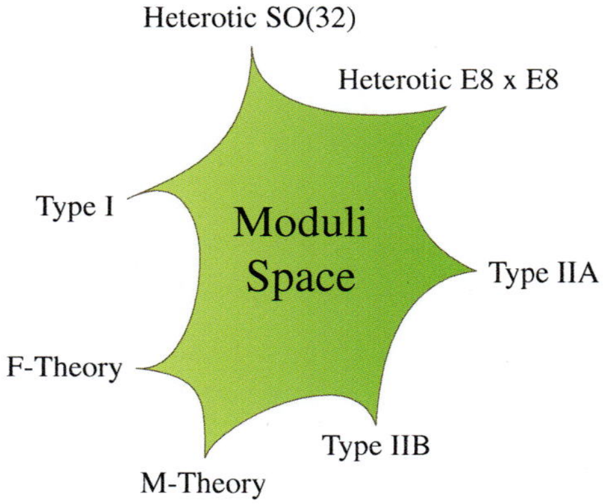

Fig. 14.48 Overview of string theories.

The various dualites and compactifications of superstring theories are summarized in Fig. 14.47.

So how does one understand the various superstring theories that have been obtained? The answer is that the vacuum of 10 dimensional superstring theory has a wide choice of allowed states. Suppose every point of a plane, called the moduli space, corresponds to one of the possible vacua of superstring theory. What we have studied are thought to be the extreme points of moduli space where the vacuum state of superstring theory becomes simple and yields the five types of superstring theory as well as M and F-theory. The moduli space and the various string theories are represented in Fig. 14.48.

As one can see from the discussion in this section, and in particular from Fig. 14.48, the various superstring theories are related by symmetries and transformations. It is our view that the final shape of the underlying superstring theory is not yet clear and all that we know at present is the behavior of superstrings at very special points of its moduli space.

Noteworthy (optional content) 14.4: Is Matter Infinitely Divisible?

Can matter be divided forever, or is there an ultimate constituent that can no longer be further subdivided? This is an ancient debate that finds an expression in the leading ideas of physics.

We mentioned in the beginning of Chapter 13 that the Standard Model of particle physics brings the 'atomic dream' of the ancients to a closure. The quarks, leptons and gauge fields are described by local quantum fields that yield particles that are point-like. Hence, these constituents can no longer be subdivided and hence is a realization of the idea of the 'indivisibility' of the ultimate constituents of matter.

Surprisingly, superstring theory provides another perspective to this ancient question. Open strings can be of any length. D-branes can have any continuous distance between them, and can exchange open strings of any length. The distance between D-branes can go to zero and hence the length of the open string can be infinitesimal. In other words, open strings, in principle, can be subdivided indefinitely — all the way to zero length — providing a realization of the idea that matter is infinitely divisible.[12]

One may conclude that both ideas — the atomic idea that at the fundamental level matter is 'uncuttable' and the opposing view that matter can be infinitely divided — have a place in Physics. Nature seems to have a place for all profound and deep ideas, even if these ideas may appear to be mutually exclusive.

14.16 The Answer

We have had a tour of the various types of superstring theories, and have barely scratched the surface of this vast and intricate theoretical structure; many important results have not been discussed due to the limitations of an introductory discussion. Table 14.4 summarizes the various results.

The present goal in superstring theory is to find the single superstring theory that spans *all* points of moduli space, shown in Fig. 14.48. It is this general theory of superstrings that is unique and that — as yet — has not been found. The underlying and all encompassing superstring theory is what would finally explain how all of Nature fits together and be an outstanding milestone in the effort to achieve the unification of all the forms and manifestations of Nature.

[12]The closed string has a minimum radius of Planck's length due to duality, as given in Eq. (14.5); in contrast, there is no minimum length for open strings.

Table 14.4 Overview of superstring theories and D-branes.

Superstring theory	Type IIB	Type IIA	$E_8 \times E_8$	SO(32)	Type I
String type	Closed	Closed	heterotic Closed, heterotic	heterotic Closed, heterotic	SO(32) Open
Supersymmetry	$\mathcal{N} = 2$ (chiral)	$\mathcal{N} = 2$ (non-chiral)	$\mathcal{N} = 1$	$\mathcal{N} = 1$	$\mathcal{N} = 1$
Gauge groups	none	none	$E_8 \times E_8$	SO(32)	SO(32)
D-branes	−1, 1, 3, 5, 7, 9	0, 2, 4, 6, 8	none	none	1, 5, 9

There is a long way to go before superstring theory can produce any result that can be tested in an experiment. The reasons for this are many. To start with, it cannot yet be decided if the extra dimensions of string theory are taken to be small, as is the case with the approach of compactification, or that all the extra dimensions of string theory are equally large and that the world we live in is a **D3-brane world**. There is also the much more difficult and unsolved problem of deciding which portion of the superstring's moduli space is chosen by Nature, if at all. All we have now are seven windows looking at small parts of the world of string theory. The domain of moduli space that is relevant to the 'real world' may be a domain that none of our present limits of string theory represent.

The initial view was that string theory is a unique theory with no free parameters. However, with a greater understanding of the possible vacuum states of the superstring, a new source of ambiguity, the so-called landscape problem, has emerged and was mentioned in Noteworthy 14.2. In essence, there are many local minima (in an effective potential energy) that would each lead to a quasi-stable 'false' vacuum; the time taken for the 'false' vacuum to decay (tunnel) to the true vacuum state — having the absolute minimum of energy — is extremely long. Hence, one may not have any way of deciding what is the 'true' vacuum of the theory since our Universe might be 'trapped' in one of the local energy minima. This creates many ambiguities for the theory since the 'false' vacuum would be specified by a number of moduli that could not be fixed, *a priori*, by theory.

Superstring theory is still in its infancy; although it is mathematically consistent, which in itself is no mean achievement, superstring theory has as yet no empirical support. It has, nevertheless, yielded many insights into the nature of physical reality. It is also a cornucopia of new results in mathematics. The mathematical edifice of superstring theory is indeed impressive. One hopes that, in addition to many ground-breaking mathematical results, superstring theory will permanently enhance the theoretical structure of Physics and alter the way we perceive our Universe.

Superstring theory postulates a higher dimensional spacetime, with extra dimensions of space that are presently invisible, in which extended objects like superstrings and D-branes emerge naturally. All observed phenomena, including matter and spacetime, are the massless excitations of the underlying superstrings.

It is important to note that the lack of any experimental evidence for superstrings has led some physicists to question the validity of superstring unification and the extra space dimensions that it necessitates. In our opinion, it is too early to discount the approach adopted by superstring theory since superstring theory may yet surprise the naysayers and go on to fulfill all the great promises it holds.

Superstring theory opens completely new vistas for theoretical physics and it is only by understanding its physics more deeply that experimentally testable predictions can be made. What is required is for theorists to have a better understanding of the general formulation of superstring theory, and from this more elevated theoretical vantage point it may become clear how to relate superstring theory to the four dimensional world that we live in.

Superstring theory points to invisible and imperceptible realms of the Universe that exist in higher dimensions. These higher dimensions are causally linked to us and hence are part of the physical reality that constitutes our Universe. Superstrings are almost inconceivably small entities and exist in the higher dimensional spacetime; similar to the case of black holes, superstrings are also invisible objects. Superstrings are inferred to exist based on our theoretical and mathematical conceptions of these objects.

Superstrings and black holes are two leading archetypes of invisible objects; superstrings go further than black holes and exist in a higher dimensional spacetime, which is a realm of the Universe that can never be directly perceived by our five senses. It is for this reason that these two objects have been examined in detail, to try and grasp what is it that makes them so exceptional.

Chapter 15

Superstring Gravity

How is Einstein's gravity superseded by quantum gravity?

15.1 The Question

A major fault line in the foundations of Physics has been the lack of consistency of Einstein's theory of gravity with the quantum principle. The explanation of gravity as being the manifestation of the geometry of spacetime is a *classical* description — since spacetime geometry is taken to be a determinate and classical field, lacking the intrinsic indeterminacy required by the quantum principle. The question as to how and why one can supersede Einstein's classical theory of gravity is addressed by superstring theory.

The requirement of consistency of gravity with quantum mechanics was one of the primary motivations for the search that led to the discovery of superstring theory. The unification of the Standard Model's forces and particles by superstring theory, as discussed in Chapter 14, provides a stepping stone for the quantization of gravity.

This chapter focuses on providing a theoretical framework to show that Einstein's theory of gravity can, indeed, be superseded by superstring gravity. As with all theories of science, whether the theoretical results of superstring theory are valid for Nature can only be decided by experiments.

15.2 Introduction

As discussed in Chapter 14, superstrings exist in 10 or higher dimensions. Superstring theory provides two distinct ways for viewing gravity — both of which are based on the properties of the closed superstring in the higher dimensional spacetime.

- The first view is that the extra dimensions of spacetime are small — of Planck size — and our four dimensional Universe is large. Gravity is the manifestation of the spin two graviton — a massless excitation of the closed superstring — that appears in the spectrum of massless particles for all closed superstring theories. Superstring theory provides a mathematically consistent theory of the graviton that is not possible using four dimensional quantum field theory.
- The other view is that all the (spatial) dimensions of superstring theory are equally large. Our Universe is a D3-brane, namely an object that occupies a three dimensional subspace of the higher dimensional spacetime, as discussed in Sec. 14.12. Gravity is due to the presence of closed superstrings inside the D3-brane.

The discovery of Hawking radiation being a quantum effect is the first phenomenon that combines gravity with quantum field theory. One of the main successes of superstring theory has been to give a derivation of Hawking radiation using the superstring formulation of gravity, and is discussed in Sec. 15.6.

Superstring theory also provides a fresh perspective on cosmology — a description and explanation that is not possible in four dimensional spacetime, and is discussed in Sec. 15.7. In superstring cosmology, our Universe is considered as one of two colliding end of the world 3-branes, oscillating in an ambient 11 dimensional Universe described by M-theory. Such oscillating branes remove the necessity of our Universe originating at the Big Bang. Our Universe is, instead, mathematically represented as a cyclical and endless Universe with no beginning and no end — and with each cycle being different and lasting for about a trillion years.

15.3 Quantum Gravity

In Einstein's formulation, gravity is the manifestation of the curvature of spacetime. Following the approach of the Standard Model where forces are mediated by gauge bosons, such as the photon, one can try and construct a quantum theory of gravity as an interaction that is mediated by a spin 2 massless boson called the **graviton**. In this theory, the graviton is the excitation of an underlying spin 2 quantum field — in the same manner that the photon is an excitation of the quantized electromagnetic field. However, all attempts to quantize gravity along the lines of the electroweak and strong forces have failed: all local quantum field theories containing a graviton have turned out to be mathematically inconsistent and therefore unsuitable.

Nevertheless, there does seem to be a natural connection between gravity and quantum mechanics as can be seen from the following qualitative argument.

Consider the fundamental constants of Nature, namely $\hbar, c$ and Newton's gravitational constant G_N; the only combination having the dimension of length is given by

$$\begin{aligned}\ell_P &= \sqrt{\frac{\hbar G_N}{c^3}} \\ &= 1.6 \times 10^{-35}\,\text{m}\end{aligned} \tag{15.1}$$

where ℓ_P is generally called the **Planck length**. Recall that the Planck length appeared naturally in the study of black hole entropy in Sec. 5.14. The Planck length is roughly 10^{20} times smaller than the size of a proton.

Similarly, a **Planck mass** can be defined as

$$\begin{aligned}m_P &= \sqrt{\frac{\hbar c}{G_N}} \\ &= 1.2209 \times 10^{19}\,\text{GeV}/c^2 = 2.176 \times 10^{-8}\,\text{kg}.\end{aligned} \tag{15.2}$$

The Planck mass has some interesting properties. Let us consider the **Compton wavelength**, given by $\lambda = h/mc$, of an object with Planck mass. Using Eqs. (15.1) and (15.2) we find that the Compton wavelength of a Planck mass particle is given by $\lambda_P = h/m_P c = 2\pi\ell_P$. Or in other words, the Compton wavelength of an object with the Planck mass is close to the Planck length.

The Schwarzschild radius R_S of an object of mass m is given by $R_S = 2Gm/c^2$; if the object has a radius less than or equal to R_S, then it will undergo gravitational collapse and become a black hole with the horizon of the black hole having a

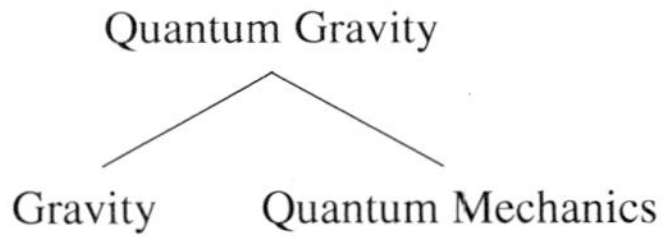

Fig. 15.1 Quantum gravity unifies gravity and quantum mechanics.

radius given by R_S. Interestingly enough, for the case of the Planck mass its **Schwarzschild radius** $R_{S,P}$ is given by $R_{S,P} = 2\ell_P$. Hence, up to a factor of 2, the Schwarzschild radius of a Planck mass is approximately equal to the Planck length!

What does this mean? The Compton wavelength is the approximate 'size' of a quantum particle since a quantum particle has most of its indeterminate (virtual) states within a Compton wavelength of its average position; on the other hand, a black hole is a classical entity, containing a geometric singularity in the spacetime manifold. If the size (Schwarzschild radius) of the black hole becomes equal to its Compton wavelength, this implies that the black hole is a completely quantum mechanical object; and equivalently quantum mechanical particles of Planck mass themselves are like black holes since their mass is sufficiently high and dense to create spacetime singularities.

This argument indicates the possibility of a deep connection between gravity and quantum mechanics at very small scales. Instead of the 'point-like' notion of the elementary particles that emerges from relativistic quantum field theory, a completely new approach to what constitutes a particle is necessary. Thus far, the only feasible approach to go beyond the particle-like description is superstring theory, discussed in Chapter 14.

15.3.1 *Spacetime foam*

The existence of virtual particle–antiparticle pairs is a widely accepted property of the quantum vacuum. Now if such pairs have Planck mass, then this will also lead to the creation of virtual black holes. The vacuum at the Planck scale is therefore full of geometric singularities due to the existence of virtual black holes. Consequently, all quantum field theories for gravity must have quantum fluctuations in the geometry of spacetime at distances of around 10^{-35} m. Due to large quantum fluctuations in energy, virtual pairs of black holes increasingly dominate the geometry of spacetime at the Planck length and the very concepts of a smooth manifold and 'points' of spacetime are no longer valid; spacetime melts into a 'foam' due to quantum randomness of geometry, perhaps similar to the foam in Fig. 15.2.

Fig. 15.2 Perhaps, at the smallest scales, spacetime looks similiar to this foam.

Although attempts to quantize gravity based on local quantum field theory, as discussed in Sec. 14.2 have thus far been unsuccessful, one positive outcome has been the realization that any quantum theory of gravity must be a theory of supergravity — namely a theory containing a spin 2 graviton together with its superpartner, a spin 3/2 gravitino; the existence of a gravitino requires, for consistency, that spacetime must have a dimension greater than four. This requirement of quantum gravity finds a natural expression in superstring theory.

15.4 Superstrings and Gravity

As mentioned earlier, one of the main questions that superstring theory addresses is the relation of gravity to the quantum principle. The graviton, the spin 2 particle that is the quantum of gravitational interactions, appears as a massless particle in the spectrum of a closed superstring. What is the relation of closed strings to the spacetime around us and in particular how does Einstein's theory of general relativity emerge from superstring theory?

The discussion of open and closed strings has assumed an ambient 10 dimensional spacetime. If the only physical entity that exists is the superstring, what is the ambient spacetime? How does it arise? The precise answer is not known and we can only surmise what are our expectations. The quantum mechanical graviton is described in 10 spacetime dimensions by the symmetric metric tensor g_{MN}, which encodes the geometry of spacetime as discussed in Noteworthy 3.4. One scenario is that due to the dynamics of a yet unknown complete theory of superstrings, the gravitons undergo a phase transition and condense to yield a ground state (vacuum) for the geometry of spacetime of superstring theory that is characterized by the product spacetime $\mathcal{M}_4 \times Y_6$. For the product spacetime, the metric tensor also factorizes. In symbols, denoting the expectation value by $E[\ldots]$, one has

$$E[g_{MN}] = \eta_{\mu\nu} Y_{IJ}$$

where $\eta_{\mu\nu}$ is the metric of (flat) Minkowski spacetime given in Eq. (3.16) and Y_{IJ} is the metric of the Calabi–Yau manifold which yields a geometry as given in Eq. (3.16).

How does one obtain a curved geometry for four dimensional spacetime from superstring theory? This is a bit more difficult to obtain and one needs to consider the description of superstrings in a complex setting. Consider, for simplicity, bosonic strings in 10 dimensional curved spacetime; a spacetime point is denoted by X_M, with $M = 1, 2, \ldots, 10$, shown in Fig. 15.3 and its world line is shown in Fig. 15.4; the geometry of the ambient spacetime is given by the classical metric tensor $g_{MN}(X)$. Let the coordinates of the superstring be denoted by $X_M(\sigma)$ and shown in Fig. 15.5. As the string evolves in time, the position of the superstring sweeps out a world sheet in ambient 10 dimensional spacetime and shown in Fig. 15.6, with the coordinates of the superstring's world sheet denoted by $X_M(\tau, \sigma)$.

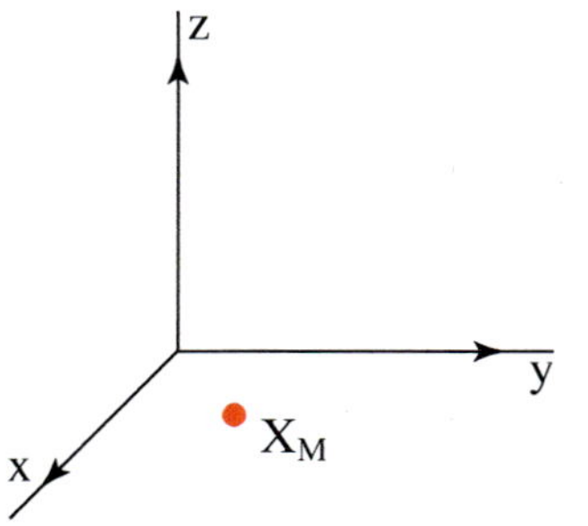

Fig. 15.3 A single point in a multi-dimensional space.

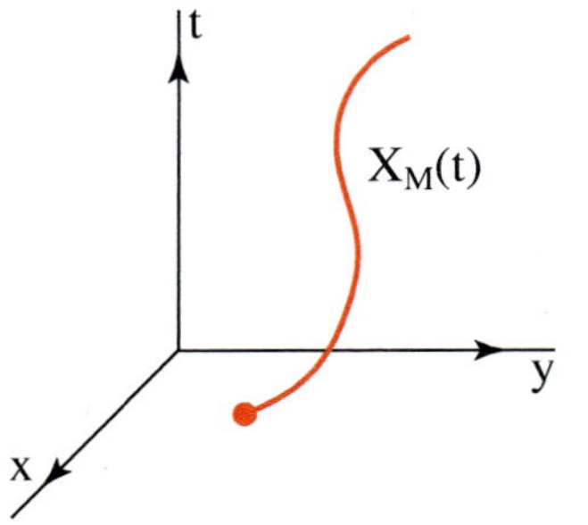

Fig. 15.4 The world line of the point in Fig. 15.3.

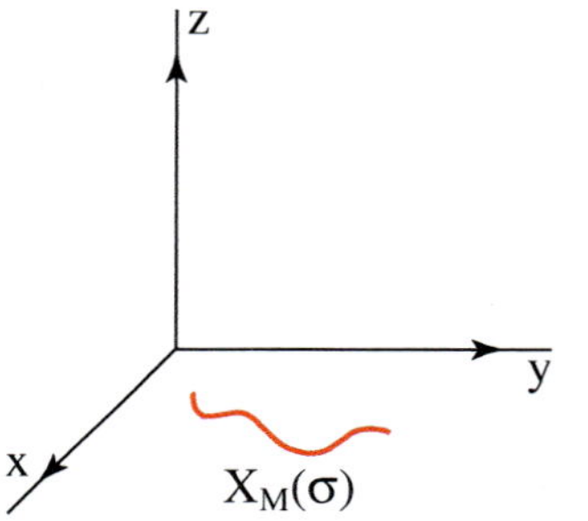

Fig. 15.5 A string in a multi-dimensional space.

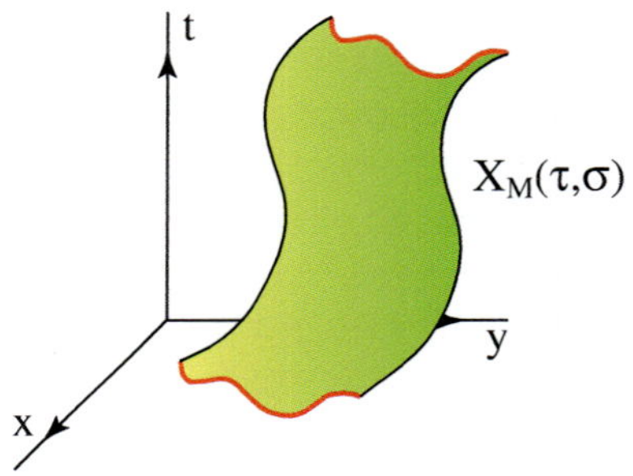

Fig. 15.6 World sheet of the string in Fig. 15.5.

Einstein's field equations for gravity are based on the principle of general coordinate invariance, which states that the choice of coordinates at any point of spacetime is arbitrary and that Nature does not depend on the coordinate system being used. In mathematical terms, Einstein's field equations do not change if one independently changes the coordinates locally — namely, at every point of spacetime. Local coordinate invariance is similar to the invariance under local gauge transformations, discussed in Sec. 11.6.

In superstring theory, a very different principle leads to Einstein's field equations; recall that bosonic strings are consistent with quantum mechanics and relativity only in $d_e = 26$ (and superstrings are consistent only in $d_s = 10$), a result that was derived for flat spacetime. If spacetime is curved we need to show that Physics is independent of the parametrization of the world sheet (τ, σ) that appears in the coordinates of the string, namely in $X_M(\tau, \sigma)$. In other words, Physics needs to be invariant (unchanged) under the re-parametrization of the coordinates of the world sheet; Figs. 15.7(a) and (b) show two ways of placing coordinates on the world sheet and clearly, nothing in Nature should depend on how one defines the world sheet coordinates.

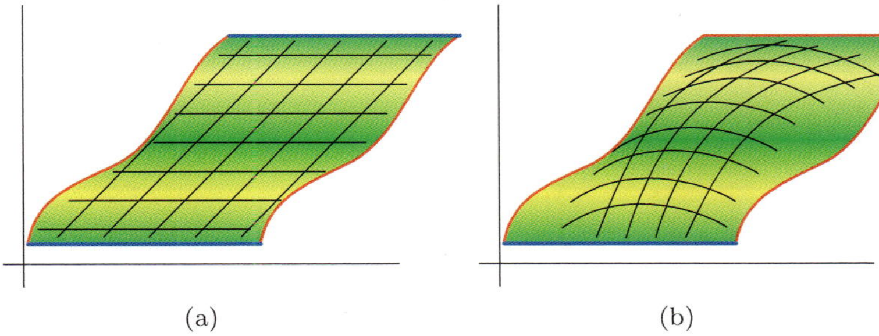

Fig. 15.7 (a) and (b) illustrate how different coordinates can be used for the world sheet.

Re-parametrization invariance needs to be a symmetry of the quantum theory of superstrings and this is realized by making string theory scale invariant, namely the theory is unchanged if all the lengths of the theory are changed. It can be shown that superstring theory is (super)scale invariant only if the metric $g_{MN}(X)$ of the ambient 10 dimensional spacetime satisfies Einstein's field equations. In other words, the propagation of superstrings in a 10 dimensional ambient curved spacetime is consistent only if the metric of ambient spacetime, encoded in $g_{MN}(X)$, satisfies Einstein's field equations given in Eq. (4.10). In summary:

Demanding re-parametrization invariance of the coordinates of the superstring world sheet is the principle that leads to Einstein's theory of gravity.

15.5 Brane Worlds and Gravity

In our earlier discussion on how to relate superstrings defined in higher dimensions to the observed 4 dimensional spacetime, we used the approach of compactification. One possible compactified superstring theory is defined on a product space $\mathcal{M}_4 \times Y_6$, where $\mathcal{M}_4$ is taken to be our observed Universe and there is an additional space Y_6 at every point of $\mathcal{M}_4$. In the approach of compactification it is further assumed that the additional space Y_6 is a tiny compact space that is so small that, for all practical purposes, one can never see it. This view is not too radical, since one can accept, with some effort, that at very short distances spacetime may not be what it seems to be at large distances.

The existence of D-branes points to a radical re-interpretation of our observed Universe that is entirely different from considering the extra dimensions to be of Planck size. In this picture, spacetime has 10 large dimensions and our four dimensional observed Universe is thought of as a D3-brane, whose world volume spans out $\mathcal{M}_4$. This view implies the existence of possibly innumerable brane worlds,

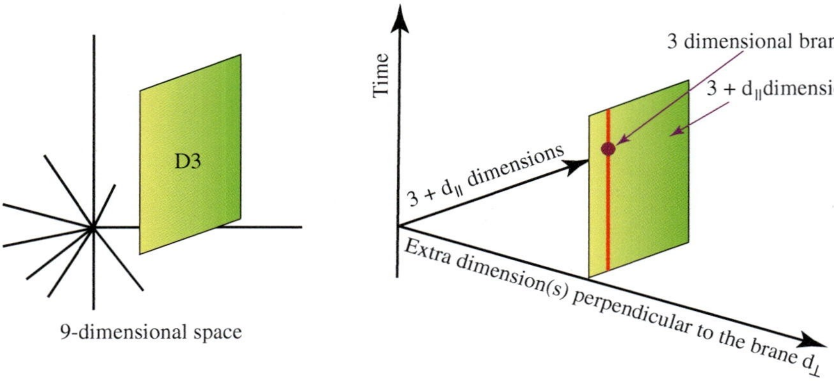

(a) (b)

Fig. 15.8 (a) A D3-brane world. (b) A D3-brane world is a point (red dot) and its world line, indicated in red, spans out a line. The D3-brane world is a slice of a larger D(3+$d_{||}$)-brane world.

similar and dissimilar to ours, with dimensions ranging from 0 to 9 space dimensions; all these brane worlds can, in principle, exist — according to superstring theory.

The most general scenario for a brane world is shown in Fig. 15.8. Our D3-brane world, shown in Fig. 15.8(a), is embedded in a larger D(3+$d_{||}$)-brane world, in which our D3-brane is a subspace. For 10 dimensional superstring theory, there are a remaining $d_\perp = 9 - 3 - d_{||}$ dimensions perpendicular to the D$(3 + \mathrm{d}_{||})$-brane Universe. In Fig. 15.8(b), our D3-brane world appears as a *point*, which is evolving in time and is a part of the larger D(3+$d_{||}$)-brane world that is represented by a line — with $d_\perp$ being the remaining perpendicular dimensions.

Instead of considering the most general brane world, we can consider our Universe to be a single D3-brane, shown in Fig. 15.8(a), that is immersed in a large ambient 9 dimensional *space* — and in which all the dimensions are large and of equal size. The D3-brane interpretation of our Universe has been discussed in Sec. 14.12. All open superstrings start and end on the D3-brane world.

As discussed in some detail in Sec. 14.12, the end points of the open superstring carry both bosonic and fermionic degrees of freedom and appear as point particles on the D3-brane. In the brane picture all the (point) particles and forces of the Standard Model are constructed solely from the end points of open strings and from their interactions. It was pointed out in Sec. 14.12 that open strings do not give rise to the force of gravity inside the D3-brane world. The main problem of the D3-brane world construction is how to incorporate gravity into the brane world picture, which we now address.

15.5.1 *Closed strings and gravity*

All closed strings contain a massless spin 2 boson that is identified as the graviton, the carrier of the force of gravity. As discussed in Sec. 14.9, a closed heterotic

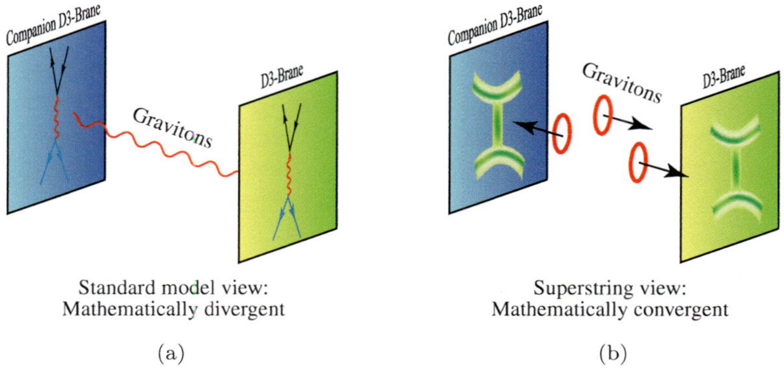

Fig. 15.9 (a) Standard Model with the graviton being point-like, and (b) superstring view of the graviton as closed superstring.

string, defined on compactifed product space $M_4 \times Y_6$, has a massless graviton in its spectrum of states. In compactified superstring theory, this gives rise to the force of gravity.

The force of gravity can also be thought of as arising from the exchange of virtual graviton or of closed strings. In Figs. 15.9(a) and (b), a companion D3-brane has been brought in to provide a source of closed strings, since gravitons or closed strings do not have a significant density inside the D3-brane. (Later on in this section, the role of the companion D3-brane will be seen to have a more physical significance.) A Standard Model description of the particles and forces in the D3-brane uses Feynman diagrams, with the graviton also appearing as another **point particle**, as shown in Fig. 15.9(a); however, the Standard Model description of gravity is mathematically divergent.

For quantum gravity in a D3-brane that is convergent and mathematically consistent, one needs to keep to the superstring view: quantum gravity is irreducibly the result of the exchange of closed superstrings; this exchange has no projection to a point particle representation, as in Fig. 15.9(a). Figure 15.9(b) shows that the point particles of the Standard Model are in fact open superstrings in the D3-brane — that live in the four dimensional spacetime world volume of the D3-brane — and experience the force of gravity due to emission and absorption of closed strings.

In all the superstring theories considered so far, the graviton cannot be confined to a D-brane — and in particular to a D3-brane — since gravitons are massless states that arise from the emission and absorption of closed strings. Closed strings are not confined to any D-brane since they do not have open ends that are restricted to a brane, but rather propagate throughout 10 dimensional spacetime. This leads to the closed strings (gravitational field) 'leaking' out from the D3-brane world and into the 10 dimensional ambient spacetime. This leakage of closed strings leads to a force of gravity that is much weaker and different than what is observed in our Universe considered as a D3-brane.

15.5.2 *Companion D3-brane*

There have been many attempts to enhance gravity inside a D3-brane world by postulating the presence of another 'companion' D3-brane — as illustrated in Fig. 15.10 — that would be the source of gravity, namely closed strings.[1] Although not yet a complete explanation, this idea is still worth discussing since a variant of it may finally explain the mystery of gravity and it also provides an exemplar of the rich theoretical structure of superstring theory.

To obtain gravity inside our D3-brane, one way is to place a *companion* D3-brane close to our D3-brane world to provide for the missing gravitons; the companion D-brane is highly curved — generating a strong gravitational field that results in the large scale emission of closed loops. Inside our D3-brane, a virtual closed string generally breaks into an open string, as shown in Fig. 15.10, since this breaking into open strings leads to an energetically lower energy state. Hence, inside a D3-brane Universe, the force of gravity is very weak, when compared to the strength of the other forces, due to the *decay* of closed strings inside a D-brane.

If the D3-brane moves orthogonal to the companion D3-brane, as shown in Fig. 15.11(a), this will give rise to an exponential change in the flux of closed strings. By adjusting the distance between our D3-brane Universe and the companion D-brane, as well as the curvature of the companion D-brane, we can adjust the flux of closed strings being radiated off by the companion brane and being received by our D3-brane Universe. In doing so we can produce a large enough flux of closed strings coming into our D3-brane Universe, such that it can give rise to the gravitational force which is observed in our Universe.

The companion D-brane can also explain the phenomenon of **dark matter**. Recall that dark matter, which has been discussed in Chapter 7, is the (missing) gravitating matter needed to account for the observed gravitational force associated with every galaxy. The extra gravitational attraction, as shown in Fig. 15.11(b), could be due to the massive flux of closed strings from the companion D-brane

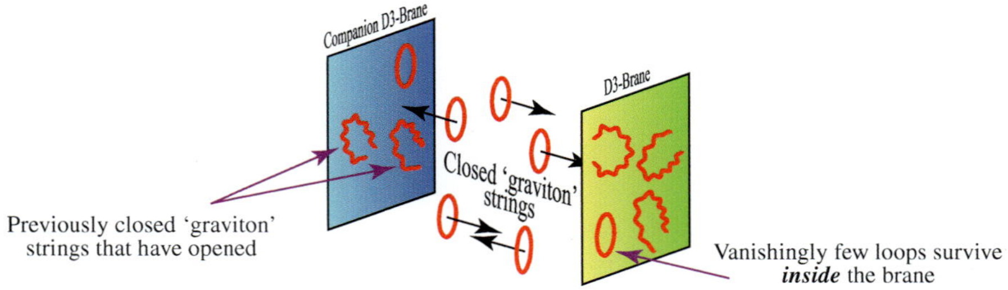

Fig. 15.10 Gravity is weak because only vanishingly few closed superstrings survive inside a D3-brane.

[1]The companion can be any D-brane, and a D3-brane is chosen for definiteness.

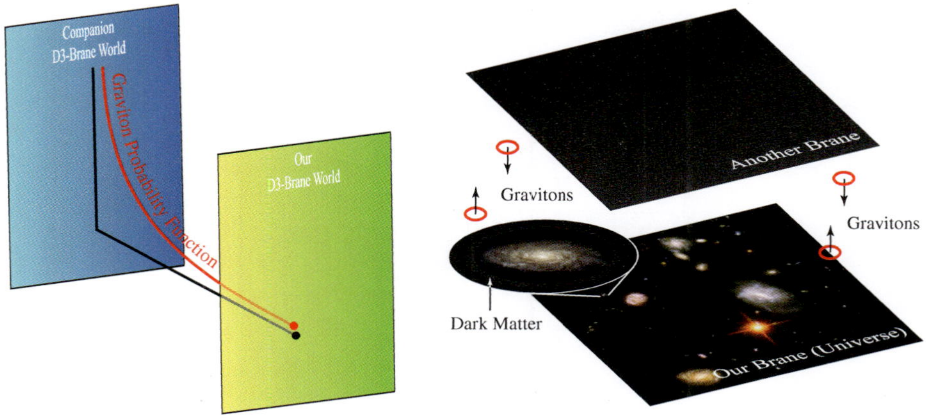

(a) (b)

Fig. 15.11 (a) A strongly curved companion D-brane can be the source of weak gravity in our Universe. (b) We may be part of a greater Universe where our 'Universe' is a D3-brane having another companion D-brane that can account for dark matter.

which has a structure that gives rise to gravitational attraction far in excess of the visible matter found in the galaxies of our Universe.

The notion of our Universe being a D3-brane accompanied by another brane can be extended to a possible explanation of the Big Bang, which is discussed in Sec. 15.7.

15.6 Black Hole Entropy and Superstrings

Recall, from Eq. (5.18), that the entropy of a black hole is given by

$$S_{BH} = k_B \frac{c^3}{4\hbar G} A_H$$

where A_H is the area of the black hole's (outer) horizon.

Black holes have enormous entropy. The entropy of a black hole is proportional to the area of its horizon A_H, and is equal to $A_H/4\ell_P^2$; note that the unit for counting entropy is the Planck area, namely $\ell_P^2 = 10^{-70}\,\mathrm{m}^2$ — an incredibly small area indeed! If one divides the surface of the event horizon into small squares the size of the Planck area, as shown in Fig. 15.12, then each Planck area stands for one unit of entropy. Given the almost vanishing size of Planck area, one can see why the entropy of the black hole must be immense.

The question we need to ask is: how does the black hole generate this enormous amount of entropy? The definition of entropy S, from Eq. (5.15), for a given macrostate is the following

$$S = k_B \ln \Gamma$$

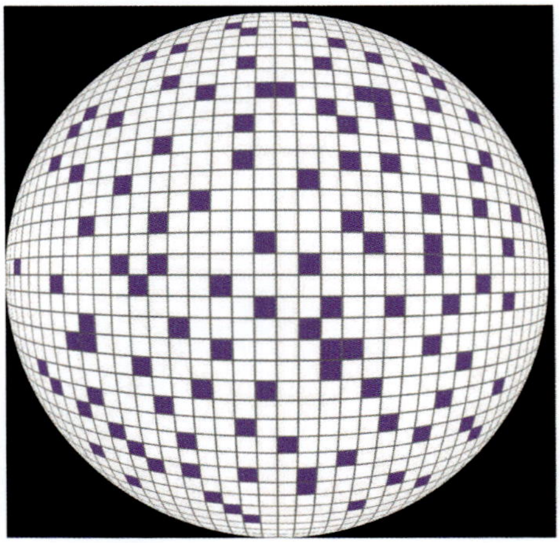

Fig. 15.12 Entropy and information of the black hole.

where Γ is the number of microstates of a thermodynamical system. For a black hole, its macroscopic state is fixed by its parameters such as its dimensionality, its mass M, angular momentum J and charge Q. A microstate, on the other hand, is defined as one possible configuration of the degrees of freedom that constitute the thermodynamical system, namely the black hole.

A black hole with mass equal to our Sun has a Schwarzschild radius of about 3000 m; hence the area of the horizon is about $113 \times 10^6\ \text{m}^2$ and yields a huge entropy of $113 \times 10^{76} k_B$. One can show that the entropy of a black hole formed by gravitational collapse of a star is much greater than the entropy of the star. As discussed in Sec. 6.15, almost the entire entropy of our causally connected Universe, which is about $10^{101} k_B$, is carried by black holes.

The high entropy of a black hole points to an immensely complicated *microstructure* that is reflected in the enormous number of microstates of the black hole, and which yields a very large value for Γ. The gigantic increase in entropy that results from the formation of a black hole can be understood by an analogy with the melting of solid into a liquid.

When a solid melts into a liquid there is a large increase in entropy. The reason being that in a solid the atoms are localized in a regular lattice pattern whereas all the atoms of a liquid can move around freely. Every position and momentum of all the atoms is one possible microstate; since the atoms can move freely for the liquid, there are many more microstates for the liquid as compared to the solid.

Similarly, when ordinary matter — namely the particles and forces of the Standard Model – crosses the black hole's event horizon and enters into the domain enclosed by the horizon, ordinary matter 'melts' (transforms) into a different state: the immense increase in entropy means that there is an 'extraordinary' form of matter inside the black hole's horizon that has to have many more degrees of freedom than ordinary matter — and which can account for the gigantic number of microstates required for generating the black hole's entropy.

One might wonder: what is the form of this 'extraordinary' matter inside the black hole that is so dissimilar to ordinary matter? One expects something quite dramatic since counting black hole microstates requires a quantum theory of gravity. A hint that superstring theory might be required comes from the appearance of the Planck area ℓ_P^2 in measuring the area of the black hole.

The entropy of an extremal **Reissner–Nordstrom** four dimensional black hole is given in Eq. (5.21). An extremal black hole has the minimum mass allowed by the cosmic censorship hypothesis, which states that the singularity of a black hole must always be enveloped by a horizon. For an extremal $d = 4$ dimensional Reissner–Nordstrom black hole, the minimum mass is given by $GM^2 = \hat{Q}^2$, where $\hat{Q} = Q/\sqrt{4\pi\epsilon_0}$ and Q is the electric charge. From Eq. (5.21), the entropy for an extremal black hole simplifies and is given by

$$S_{RS4,\text{Extremal}} = k_B \frac{c^3}{4\hbar G} A_H = \frac{\pi k_B}{c\hbar} \hat{Q}^2. \tag{15.3}$$

Recall that for the extremal $d = 4$ dimensional Reissner–Nordstrom black hole the Hawking temperature is zero, namely $T_H = 0$, as given in Eq. (5.22).

15.6.1 *Reissner–Nordstrom black hole*

In 1995 string theorists C. Vafa and A. Strominger successfully calculated the Bekenstein–Hawking entropy of a five dimensional extremal and supersymmetric Reissner–Nordstrom black hole that is analogous to the extremal Reissner–Nordstrom four dimensional black hole. The reason for considering the supersymmetric extremal $d = 5$ dimensional Reissner–Nordstrom black hole is two fold.

- As discussed in Sec. 5.15, Hawking temperature being zero for the extremal case means that there is no longer any Hawking radiation and the black hole is in equilibrium. Hence one has a much simpler problem with no time dependence.
- The supersymmetric black hole has special properties that allow for an efficient counting of the number of microstates.

The entropy of a supersymmetric five dimensional extremal Reissner–Nordstrom black hole, following Hawking's semi-classical derivation, is given by

$$S_{RS5,\text{Extremal}} = \frac{A_H}{4c\hbar G_5} k_B = 2\pi k_B \sqrt{N Q_1 Q_5}$$

where A_H is the area of the horizon of the extremal Reissner–Nordstrom black hole and G_5 is the five dimensional version of Newton's gravitational constant. Similar to Eq. (15.3), the entropy of the five dimensional extremal Reissner–Nordstrom black hole is written in terms of macroscopic parameters of the black hole that have a transparent interpretation in terms of the parameters of superstring theory.

Note that N is an integer and Q_1, Q_5 are charges of the black hole that couple to Maxwell-like gauge fields that exist in $d = 5$ spacetime.

The superstring derivation showed that a black hole is a highly excited state of the superstrings; the microstates of black holes are made out of a large collection of D-branes and open superstrings. Recall that ordinary matter, comprising the Standard Model, consists of the massless excitations of the superstring. When this ordinary matter crosses the event horizon of a black hole, the other massive superstring degrees of freedom that are usually frozen out are freed up; there are exponentially more superstring microstates than those that exist for ordinary matter, leading to the gigantic entropy of black holes.

To have a qualitative idea of why D-branes are required to explain the microstates of the black hole we briefly outline the reasoning. The derivation requires a supersymmetric black hole, which can be shown to exist only in four and five dimensional spacetime. The case that was studied using superstring theory is a five dimensional charged black hole.

Let us start from the Type IIB closed superstring theory in $d = 10$, which at low energies yields ten dimensional supergravity as an effective field theory. We compactify Type IIB to $d = 5$; the black holes of supergravity are five dimensional supersymmetric Reissner–Nordstrom charged black holes.

A black hole is a gravitational object with strong curvature, whereas most of the discussions on superstrings have been in flat spacetime. So how can superstring theory — formulated in flat spacetime — be used to compute the entropy of a black hole? The key to the answer lies in the concept of extremality and supersymmetry. As discussed earlier in this section, extremal black holes are in equilibrium and do not radiate since their Hawking temperature is zero.

The microstates of the black hole are the allowed quantum states of the superstring. Being in equilibrium means that the entropy of an extremal black hole is a constant and hence the microstates are also time invariant — and one does not need to take into account the dynamics of a black hole to evaluate its entropy. A five dimensional extremal supersymmetric Reissner–Nordstrom black hole is a *special state* since it preserves half of the supersymmetries of the full superstring theory. A remarkable fact about supersymmetric quantum states is that they are independent of the strength of the coupling constants and other parameters of the theory.

Hence, we start with the system having strong coupling with strong gravitational fields and containing an extremal Reissner–Nordstrom black hole; we then reduce the coupling constant of the theory until spacetime becomes flat with no curvature. As the coupling constant is varied, the *number* of special states does not change, although their properties may undergo change. The invariance of the number of the special states is what allows the counting of states to be done in flat spacetime and gives a result that is equally valid for a highly curved object, which in this case is the Reisner–Nordstrom black hole.

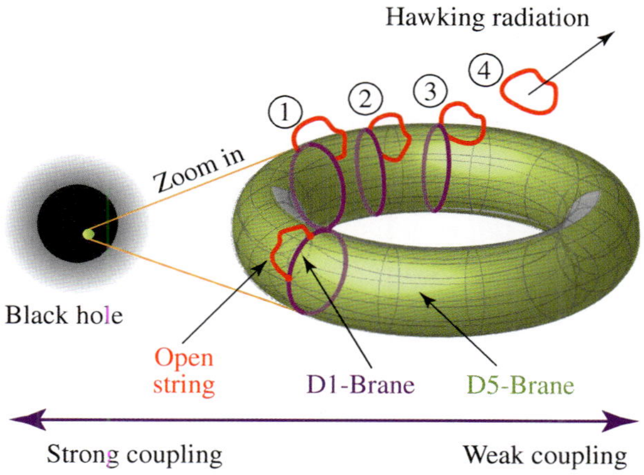

Fig. 15.13 Black holes and superstrings. The circled numbers indicate the sequence in which an open string closes and then detaches from the D1-brane to propagate as Hawking radiation.

In flat $\mathcal{M}_{10}$ spacetime, one uses the formulation of Type IIB superstrings to evaluate Γ: the number of quantum states that corresponds to the extremal black hole. One compactifies Type IIB on $M_5 \times T^5$, where T^5 is the five dimensional torus that is equivalent to 'multiplying' five circles S^1. Every point inside the five dimensional Reisner–Nordstrom black hole is interpreted as having another five dimensional *internal space* T^5, with the radius of the torus being of Planck length. D5-branes wrap the five dimensional torus Q_5 number of times; in addition, there are D1-branes that girdle — Q_1 number of times — one of the circles S^1 inside T^5, and are required to stabilize the D5-brane that would otherwise collapse due to the brane energy per unit volume of five dimensional D5-brane.

A schematic representation of a superstring and brane configuration inside a black hole is given in Fig. 15.13. The length of the D1-brane tends to go to zero and hence the D1-brane is given a momentum equal to N so that the D1-brane does not collapse.

There are many different possible ways of making this D5–D1-brane state, with each possible way being one microstate of the black hole. The collection of all possible microstates is a gigantic number and yields a large value for Γ. The number of microstates of a black hole, as enumerated by superstring theory, is exponentially greater that the number of microstates of any form of ordinary matter occupying the volume enclosed by the horizon.

In the limit of $N, Q_1, Q_5 \to \infty$, that is, having D5 and D1-branes wrapped a large number of times around the internal five dimensional space T^5, it can be shown that black hole entropy is given by

$$S = k_B \ln \Gamma = 2\pi k_B \sqrt{N Q_1 Q_5}.$$

Hawking radiation is now seen as the result of the end points of open strings — that are running inside the brane world — fusing together and forming closed strings that can then leave the brane world. The radiation from the black hole consists of the closed strings that are emitted into spacetime outside the black hole, as shown in Fig. 15.13.

In summary, a black hole is a highly excited state of superstrings and D-branes. *Every* single point inside the horizon of the five dimensional Reissner–Nordstrom black hole has a 'hidden' five dimensional internal space, with a highly excited D5–D1-brane state, as discussed above and shown in Fig. 15.13. This is the reason for the enormous entropy of a black hole. In the superstring derivation of black hole entropy, there is no presence of the geometric singularity. In fact, there is strong evidence that the geometric singularity is completely absent in the string theory description of black holes.

Noteworthy (optional content) 15.1: Black Hole Entropy: A Rosetta Stone of Superstring Theory

The **Rosetta (Rashid) stone**, shown in Fig. 15.14 — found by French soldiers in the town of Rashid, Egypt, in 1799 — was the key for decoding ancient Egyptian hieroglyphs. The decoding was accomplished by comparing the hieroglyphic writing with its translation — all carved in the Rosetta stone — into the known Egyptian Demiotic and Greek scripts.

The Rosetta stone is a metaphor for the role that black holes, and black hole entropy in particular, are playing in decoding the 'hieroglyphs' of superstring theory. One has very little intuition about the world at Planck length; hence one needs to extend the known and relevant concepts of theoretical physics to explain the behavior of Nature at the Planck length.

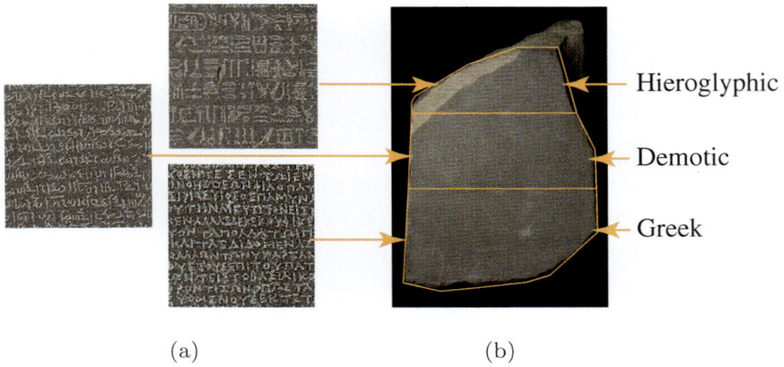

Fig. 15.14 Rosetta Stone. Approx. 196 B.C. The Rosetta stone is a metaphor for the role of black holes in understanding superstring theory. (a) Enlarged areas (non-matching text). (b) The entire stone. Source: The British Museum.

The counting of states to determine the entropy of the black hole is a case where one can connect ideas from superstring theory to known ideas of Hawking radiation, derived from quantum field theory and black holes of general theory of relativity. In particular, the fact that black hole entropy derived from superstring theory agrees exactly, up to numerical factors, with Hawking's semi-classical derivation provides evidence that superstring theory is on the right track.

The existence of a horizon is a macroscopic property of a black hole that emerges from general relativity. One now needs to look and find — until now with no success — the explanation of the horizon in superstring theory. And what does temperature mean for superstring theory and so on. It seems that black holes, especially galactic black holes due to their enormous mass of billions of solar masses and strong gravitational fields, might be the appropriate object to find evidence of superstrings: a true paradox of finding the ultra-small in the ultra-large.

15.7 Colliding Branes, Cyclic Universes and the Big Bang

As discussed in Chapter 6, the Big Bang model of cosmology entails the Universe beginning at a finite time in the past. If one extrapolates Einstein's equations to the beginning of the Universe, one finds that the volume of the Universe shrinks to an infinitesimal size (point) that has infinite curvature and infinite energy density and temperature. Even if one assumes that Einstein's equations do not apply to the very beginning of the Universe, a number of other issues that arise in Big Bang cosmology need to be addressed, namely the flatness, horizon and exotic relics problems. The inflationary model of the Universe, discussed in Sec. 6.11, was developed to address and solve these problems.

Inflationary cosmology is widely considered to be the most successful model of cosmology. It is based on a field that causes inflation and which is coupled to Einstein's classical gravitational field. With the advent of superstring theory, cosmologists have a much larger theoretical framework — a Universe with many additional (hidden) dimensions — to formulate a model of **quantum cosmology** that is completely quantum in its explanation of the Universe and dispenses with the classical gravitational field.

In 2001, a radically new model was proposed for understanding the origin and structure of our Universe, called the cyclic model. The fundamental ingredients of this model are the following:

(1) The starting point of **cyclical cosmology** is the **heterotic M-theory** of an 11 dimensional spacetime, which is the product of a finite interval of length d, namely $[0, d]$, a four dimensional Minkowski spacetime $\mathcal{M}_4$, and a six

dimensional Calabi–Yau manifold Y_6; in other words, 11 dimensional spacetime $= [0, d] \times \mathcal{M}_4 \times Y_6$.[2]

(2) In heterotic M-theory, there are two very special 'end-of-the-world' E3 $\times$ Y_6 branes sitting on the end points of the open interval $[0, d]$, separated by a distance of d. We use the notation of E3 for the three space dimensional end-of-the-world branes of M-theory since these branes are different from the D-branes of open superstring theory. The two E3-branes are not identical, with one E3-brane having a **positive energy density** and its companion E3-brane having a **negative energy density**.

(3) The cyclic Universe model postulates that our Universe is the positive energy E3$\times Y_6$ 'end-of-the-world' brane; the Calabi–Yau manifold Y_6 is small and hence not directly observable and is chosen to reproduce the observed particles of the Standard Model.

(4) The cyclic model assumes that the entire 11 dimensional spacetime has a non-trivial vacuum with non-zero vacuum energy density that is identified with dark energy that is currently thought to be exerting a repulsive gravitational force on all galaxies — making them accelerate away from each other. See discussion on dark energy in Sec. 7.5. Dark energy — the energy of the 'vacuum' (lowest energy) state that is formed out of the condensation of superstrings — is the main actor of the cyclic model.

(5) It is *postulated* that there is an attractive potential energy between the two 'end-of-the-world' branes, as shown in Fig. 15.15. The main weakness of the cyclic model is that, as of now, there is no microscopic derivation of this attractive potential from M-theory.

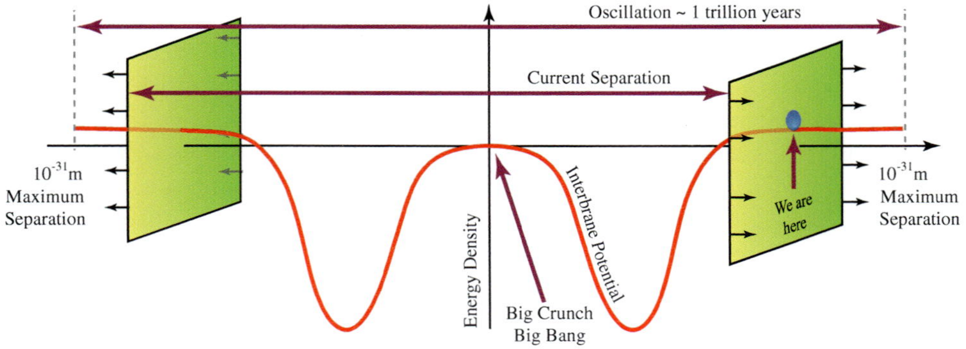

Fig. 15.15 Potential between the two 'end-of-the-world' branes.

[2]Using the identity $[0, d] \equiv S^1/Z_2$, it can be shown that, as $d \to 0$, heterotic M-theory on $[0, d] \times \mathcal{M}_{10}$ is equivalent to the $E_8 \times E_8$ heterotic superstring theory in 10 dimensional Minkowski spacetime $\mathcal{M}_{10}$; see Fig. 14.47.

The cyclic model, as the name indicates, produces a Universe that has endless cycles with no beginning and no end. The problem of resetting the entropy of our four dimensional Universe is discussed later on.[3] To see how this works, we follow one particular cycle and start at a point when the two branes are well separated. By demanding a certain degree of supersymmetry, M-theory implies that, when they are well separated, the E3-branes have infinite extension in all three large dimensions and are exactly flat and parallel. We will ignore the small manifold Y_6 as it plays no role in cosmology.

The following are the steps of one cycle; each step is represented in Fig. 15.16.

(1) Figure 15.16(1). The branes have a maximum separation $d_{\text{max}} \simeq 10^{-31}$ m. The scale is set by making the strength of gravity equal with that of the other forces.
(2) Figure 15.16(2). The branes have a very weak attraction when well separated as shown in Fig. 15.15, and gradually move towards each other, thus reducing the distance between them.
(3) Figure 15.16(3). The branes enter a phase in which the separation of the branes approaches the minimum of the interbrane potential, shown in Fig. 15.15, and causes the branes to rapidly pick up kinetic energy. Ripples form in the two branes, as they approach closer and closer, due to the exchange of M2- and M5-branes of M-theory.
(4) Figure 15.16(4). The branes collide at points of the ripples that have the smallest interbrane distances; the points of collision are shown by red dots. As they approach closer and closer the two E3-branes collide at other points that are closest, creating points of high temperature that later can become the seeding sites for galaxy formation. The distance between the two branes goes to zero. In less than 10^{-23} s after the start of the collision *all points* of the two branes collide. The collision of the two branes is called the Big Crunch, with their recoil being called the Big Bang. It has been shown that the collapse of one of the dimensions of M-theory does not lead to any curvature or other singularities — unlike the singularities that Einstein's equations predict at the Big Bang.
(5) Figure 15.16(5). The branes bounce off each other after colliding, giving rise to the early hot and dense spray of radiation and matter following the Big Bang. The large amounts of kinetic energy picked up by the two branes is released as a plasma of particles and radiation in both branes, giving rise to a Big Bang in each brane. The temperature reached in the cyclic theory's Big Bang is 10^{23} K and not infinite — as is the case with the conventional Big Bang model.

[3]All earlier attempts to model a cyclical Universe failed because if one limits the model to the observed four dimensional spacetime, one can show that due to the ceaseless increase of entropy in each cycle, the cycles slow down and come to a stop. Hence, all cyclical Universes in four spacetime dimension necessarily requires a beginning in time.

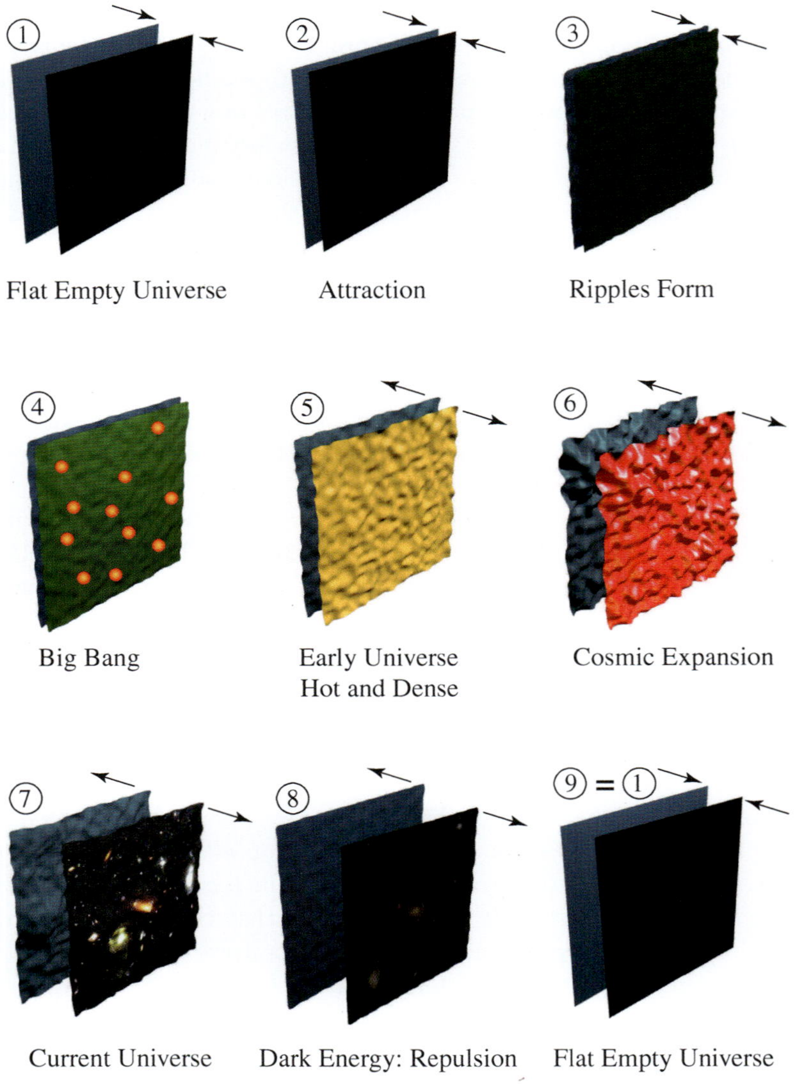

Fig. 15.16 Cyclic Universe and colliding E3-branes.

Of the two branes, the brane that is our Universe has positive tension whereas the companion brane has negative tension. The spray of radiation and matter that results from the Big Bang *reduces* the inertia of the companion brane and hence makes it speed up. This gain in speed is vital for allowing the branes to fully recoil and thus keeping the cycles going on forever. The dimension that had collapsed in the Big Crunch reappears after the branes bounce off each other. The transition from the Big Crunch to the Big Bang happens very fast, in about 10^{-23} s.

(6) Figure 15.16(6). After the Big Bang, the Universe evolves according to Einstein's theory of gravity; in particular the production of particles and radiation creates weak curvature ('creases') that is described by the Friedmann–Robertson–Walker geometry. For next few billions of years, our E3-brane evolves to produce galaxies, stars, planets and so on.

(7) Figure 15.16(7). After about 7 billion years from the occurrence of the Big Bang, dark energy becomes the dominant form of energy and causes all galaxies to repel and recede from each other — leading to the dispersal of matter in the Universe.

(8) Figure 15.16(8). As the branes gradually separate out, they are stretched and all the kinks and creases due to matter are 'ironed out' — becoming more and more flat as all matter is dispersed more and more by the vacuum's dark energy. During each cycle, the branes continually expand due to gravitationally repulsive dark energy. Entropy increases in a given cycle, as required by the Second Law of Thermodynamics. But since the brane is expanding, the density of entropy goes down. Hence, even though the total entropy of the brane Universe goes up, by the end of each cycle the density is driven very nearly to zero.

(9) Figure 15.16(9). After about a trillion years or so, all the nuclear fuel of our known Universe is estimated to run out and the branes are again perfectly flat, parallel and separated to their initial distance of 10^{-31} m — being in exactly the same condition that they started from. The next cycle repeats the steps of the preceding cycle. Whatever residual entropy that is left over from the previous cycle is exponentially diluted and the influx of new matter and radiation in the next collision completely erases the effect of the left over entropy.

There is a loss of kinetic energy in each collision due to dissipative and irreversible processes; so how is it possible to have an exactly cyclical Universe? The answer lies in the attractive nature of gravity; the two branes become more and more tightly gravitationally bound (as each cycle is completed) to compensate for the loss of kinetic energy. Since one can have infinitely negative gravitational binding energy, the E3-branes can recover their initial kinetic energy and hence continue their cyclical collisions indefinitely.

It is important to note that although the branes undergo a perfectly cyclical process, the actual production of matter and galaxies in any particular cycle is not identical to any previous one. The reason is that the conversion of brane potential energy into particles and fields inside an E3-brane can take many different forms due to the quantum nature of the M2- and M5-branes being exchanged between the colliding branes: quantum indeterminateness guarantees that the exchange of virtual M2- and M5-branes is different each time the branes collide. Hence, in each cycle the specific content of our Universe is different from the

Universes that occur in the other cycles, leading to a new and unique Universe for each cycle.

The cyclic theory of cosmology solves a number of problems posed by observational cosmology in a framework completely different from the one employed in the conventional models of cosmology and discussed in Chapter 6; and a few of these are discussed below.

- **The Big Bang.** The two branes collide at all spacetime points of both branes and is the Big Crunch/Big Bang of the cyclic Universe. The collision is more nuanced than simply a collision of say two glass plates. The two branes are flexible and have ripples — created by the exchange of virtual M2- and M5-branes — that make the collision at different points take place at minutely different instants. These minute differences turn out to be crucial in explaining the formation and distribution of the observed galaxies, and is discussed below under the heading of galaxy seeding.
- **The expanding Universe.** The E3-brane that is our Universe as well as its companion brane, at the start of each cycle, have a net *positive energy* stored in the form of the potential, as shown in Fig. 15.15. As the branes approach each other, the potential energy of the E3-branes is converted to energy *within* the branes, and inside our E3-brane this energy behaves like dark energy, causing the spacetime of our E3-brane to expand.

  The branes collide and then separate; as shown in Fig. 15.15, the potential energy of the branes is zero at the Big Crunch and the subsequent Big Bang; hence, all the potential energy of the branes is converted to energy inside the branes. A large fraction of the potential energy is converted into an explosion of matter and radiation and the remaining energy appears as dark energy in the E3-branes.

  As the branes separate, the repulsive dark energy inside our brane causes the expansion of our spacetime to *accelerate* — which is the experimental hallmark of dark energy and discussed in Sec. 7.5 — and thus stretching the brane. The separation of the branes remains nearly constant while both branes stretch exponentially. The stretching continues for the next trillion years, with the wrinkles of the E3-branes being ironed out until they are perfectly flat again.
- **The flatness problem.** The branes have minute deviations from flatness when they collide and hence create a Universe that has only small deviations from flatness, and is consistent with the results from observational cosmology. The inflationary Universe starts from a single point and which leads to a highly curved initial geometry — which then flattens out due to inflation. The cyclic model starts with the two E3-branes being perfectly flat, and which develop minute deviations from flatness due to the exchange of superstrings. A flat Universe is a feature of the colliding E3-branes, which in turn is a result of the properties of the end-of-the-world E3-branes of M-theory. The fact that the observed Universe is almost exactly flat provides a strong motivation for the study of cyclic cosmology.

- **Exotic relics problem**. Monopoles and other exotic particles have not been observed and are hence believed to be very rare. Due to the infinitely high temperature at the beginning of the conventional Big Bang cosmology, plenty of such particles should have been produced. In the inflation model, this is resolved by expanding space exponentially so that the density of the exotics becomes negligible. In the colliding branes model, on the other hand, during the brane collision, the maximum temperature is 10^{23} K and is lower than the temperatures required to produce monopoles and other exotic particles; hence, the absence of these particles.
- **Horizon problem**. The problem is to explain how distant parts of the Universe — that were not causally connected in past — can be in equilibrium and, in particular, have the same 2.73 K cosmic background temperature. In the cyclic model, the Big Crunch is a *non-local event*, with the same conditions prevailing throughout the infinite volume of the entire brane. Hence all parts of the brane Universe have the same properties leading, in particular, to the same background temperature for well separated regions of spacetime.

  One can re-interpret Fig. 6.13(a) in the following manner: although galaxies A and B are far from our galaxy G and are causally disconnected, the reason they have the same temperature, as evidenced by the background radiation, is because they both arise from the same homogeneous E3-brane and hence have the same properties. There is no reason to bring them into causal contact, as is done by inflation and shown in Fig. 6.13(b).
- **Galaxy seeding**. The observed distribution of galaxies requires that at the beginning of the Universe the deviations from exact homogeneity must be 'scale-invariant'. One of the great virtues of the inflation model is its ability to produce a scale-invariant distribution of inhomogeneity.

  The cyclic model, illustrated in Fig. 15.16, produces the required inhomogeneity in the following manner. As the two branes approach each other, quantum fluctuations caused by the exchange of M2- and M5-branes create minute inhomogeneities in the two branes, with irregularities appearing in the surface of the brane in the extra dimension.

  Due to the irregular brane surface, some parts of the branes collide a few instants earlier than other parts leading to the heating and cooling at slightly different times for different parts of space; this leads to the spatial variation in temperature and density required for galaxy formation. A remarkable result of the cyclic model is that the inhomogeneities produced in the brane collision are scale-invariant, hence surviving a crucial test.

The cyclic model provides an explanation for all the main observed features of our Universe. In particular, the cyclic model alters our understanding of the first 10^{-23} s of the life of our Universe and agrees with the existing theory from then on; and of course the cyclic Universe has a very different explanation of the ultimate fate of the Universe, looking a trillion of years into a future for each cycle and the cycles themselves are endless.

Is there any physical phenomenon that can distinguish the cyclic model from the inflationary model? The answer is yes. The two models make different predictions for the polarization of the cosmic microwave background radiation and for the frequency distribution of the gravitational wave spectrum. Since gravitational waves have not yet been detected, the best experiment for deciding which model is correct is the study of the polarization of cosmic photons.

On a more theoretical level, although the cyclic Universe generates a Universe that is almost exactly flat — such as the one we observe — the brane Universe has a cycle of a finite duration. In contrast, an almost exactly flat Universe, according to the four dimensional models of cosmology that result from Einstein's field equations, can expand forever, as discussed in Sec. 6.5 and illustrated in Fig. 6.4.

15.8 The Answer

The need for unifying gravity, or more specifically the general theory of relativity, with the quantum principle is one of the most challenging problems in theoretical physics. Indeed, to date, there is no clear answer as to how such a unified theory can be constructed.

Superstring theory has emerged as the leading candidate for the unification of gravity with the quantum principle — and also unifies it with the strong and electroweak interactions of the Standard Model. Superstring theory has already produced important results, both in physics and mathematics. Superstrings provide a rich theoretical framework for the unification of matter and spacetime; superstring theory produces the unification of all quantum fields from an entirely novel paradigm — namely, that all the observed particles are the massless superstring excitations.

Superstring theory provides a framework for gravity radically different from the geometric view of Einstein in that it explains four dimensional gravity in terms of the topology spanned out by massless excitations of closed strings, which exist in a higher dimensional spacetime. This shift from geometry to topology in the description of gravity, brought about by closed superstrings, yields a well defined quantum theory of gravity. Einstein's general theory of relativity is seen to emerge by the constraint on the geometry of the ambient spacetime required for the consistency of superstring theory.

Superstring theory, using the construction of D-branes and open strings, explains the enormous entropy of black holes as well as the phenomenon of Hawking radiation. Superstring theory removes the spacetime singularity at the center of a black hole and shows that the black hole is a complicated bound state of D-branes and open superstrings, and with Hawking radiation being the emission of closed superstrings.

Superstring provides a mathematical framework for quantum cosmology; in particular, the cyclic Universe that is constructed using M-theory is endless, with no beginning or end. Using the concepts of branes, dark energy and higher dimensions, the cyclic cosmology shows that the Big Bang singularity arises at the instant when two E3-branes collide and bounce off each other — and that this instant is only a moment in a never ending cycle of Universes.

In summary, there is nothing quite comparable to superstring theory for quantizing gravity. It has all the required ingredients for replacing and superseding Einstein's theory of gravity. The physics of black holes might be one of the best places to look for experimental evidence of superstring gravity, since it is only when gravity becomes very strong that the effects of quantum gravity become more pronounced. Superstring cosmology offers another arena where the cyclic cosmology of superstring theory can provide experimental predictions.

The cornerstones of the cyclic Universe are the invisible domains of the ambient 11 dimensional spacetime of M-theory and that of the companion E3-brane, as well as the unseen dark energy. The companion E3-brane is not directly perceived during the normal evolution of our Universe and communicates with our Universe by the exchange of M2- and M5-branes. The companion E3-brane appears in a cataclysmic manner when the cosmic scale Big Crunch/Big Bang occurs, once in a few trillion years.

Invisible realms are at the foundation of the cyclic cosmology and are in keeping with the underlying theme of our book, namely that the visible Universe is immersed in an unimaginably immense and vast unseen and invisible physical reality that comprises the totality of our Universe.

Chapter 16

Epilogue

We have made a long journey in this book, starting from our discussion on fields to our concluding investigation of superstring gravity. It is worthwhile to reflect on the diverse subjects that we have explored, in their entirety, so as to find the underlying threads that weave this tapestry into a single and unified body of knowledge.

Two themes that run through our analysis are the study of the Universe at the very large scale, namely gravity and cosmology, and Nature at the very small scale: the constituents of matter. The study and investigation of these vastly different scales of phenomena are united by the concept of the field — from the gravitational field to the quantum field. The concept of the field, an underlying physical entity spanning the entire Universe, allows us to integrate our exploration of vastly different scales of phenomena.

To study the large scale structure of the Universe, we embarked on a rudimentary study of geometry and in particular, the concept of curved spacetime. We then came to the remarkable conclusion that gravity is the manifestation of the geometry of spacetime. Two extreme cases of gravity were studied, that is the geometry of black holes and the cosmological structure of the Universe. The Universe was further found to be largely invisible, made up of mostly dark matter and dark energy fields.

The Universe is composed of galaxies, stars and planets and it is the interplay of gravity and matter that results in the astronomical objects that inhabit the sky. The matter content, in tandem with gravity, drives the formation and evolution of large aggregates of matter, with the nucleus of atoms providing the source of energy that powers stellar burning. It was shown that deep inside the core of stars is the kitchen where many of the elements of the Universe are 'cooked' via nuclear processes.

We then analyzed the constituents of the nucleus, and found that the electrons, protons and neutrons that form the atom are a small piece of a larger jigsaw puzzle, that finds its completion in the Standard Model. The quantitative description of the constituents and interactions of matter led us to the formalism of quantum field theory, one of the leading notions of theoretical physics.

The synthesis of the macroscopic with the microscopic is the impelling force behind the effort to unify all forms of the physical Universe, and led us to the study of superstring theory. Superstring theory is a mathematical construct that is at the frontier of Physics and its empirical validity is still not established. It was, nevertheless, included in the book to provide a prime exemplar of the manner in which one grapples with the unknown. The material in this book prepares the reader to follow the future developments of superstrings.

So where does the invisible Universe come into all of this? An underlying template in all our discussions has been guidance from the leading ideas of Physics. On a closer examination of these leading ideas, one realizes that the theoretical constructs are only indirectly connected to perceptible and tangible entities that we can perceive through our five senses. Moreover, we have discussed many natural phenomena that, in principle, are impossible to directly observe.

The interior of black holes can never be observed from outside the black hole and the curvature of spacetime can never be directly perceived by our five senses. The process of tunneling that is driving stellar nucleosynthesis is invisible; the quantum fields of the Standard Model are indeterminate and beyond direct observations. All these invisible entities and processes culminate in the construction of the superstring, an object that is thought to be inconceivably small and which exists in 10 or higher spacetime dimensions — all beyond direct human perception.

The visible Universe is vast and the invisible Universe is even more immense. In making a voyage to the invisible and unseen realms of Nature, a reliable and powerful guide is the principles of Physics. The reader can reflect on the ideas that were employed to understand Nature, and enjoy the unfolding of the fascinating, inexhaustible and invisible domains of the Universe.

Appendix

Laws

Name	Equation	Page
Ideal gas law	$PV = nRT = Nk_BT$	8
Coulomb's law	$F = k_e\frac{q_1q_2}{r^2}$	11
Hubble's law	$v = H_0d$	132
Newton's laws		
First law	Inertia	
Second law	$F = ma$	68
Third law	Action = −Reaction	
Law of gravitation	$F = -G\frac{m_1m_2}{r^2}$	66
Gauss' law (in vacuum)	$\nabla \cdot E = 0$	17
Gauss' law (for enclosed charges)	$\nabla \cdot E = \rho/\epsilon_0$	460
Faraday's law	$\nabla \times \mathbf{E} = -\frac{\partial \mathbf{B}}{\partial t}$	17
Ampere's law	$\nabla \times \mathbf{B} = \frac{1}{c^2}\frac{\partial \mathbf{E}}{\partial t}$	17
Wien's law	$\lambda_m T = \text{constant}$	143

Equations

Name	Equation	Page
Doppler effect	$f' = \left(\frac{v}{v+v_r}\right) f$	164, 244
Friedmann equation	$\left(\frac{dR}{dt}\right)^2 = \frac{8\pi\rho G}{3c^2} R^2 - k$	134
Momentum	$p = mv$	
Kinetic energy	$T = \frac{1}{2} mv^2$	
Gravitational potential energy	$U = mgh$	
Mass–energy equivalence	$E = mc^2$	
Planck relation	$E = h\nu$	
Einstein curvature tensor	$\mathcal{G} = \frac{8\pi G}{c^4} \mathcal{T}$	81
Schwarzschild radius	$R_S = \frac{2GM}{c^2}$	91
Entropy	$S = k_B \ln \Gamma$	117

Maxwell's Equations

A summary of Maxwell's field equations and the symbols used is given below.

$\nabla \cdot E = \dfrac{\rho}{\epsilon_0}$	Gauss's law
$\nabla \cdot B = 0$	Gauss's law for magnetism
$\nabla \times E = -\dfrac{\partial B}{\partial t}$	Faraday'a law of induction
$\nabla \times B = \mu_0 \left(J + \epsilon_0 \dfrac{\partial E}{\partial t} \right)$	Ampere's law extended by Maxwell

Symbol	Meaning	Units
E	Electric field	Volt per meter (V/m)
B	Magnetic flux density	Tesla, weber per square meter (T)
ρ	Free electric charge density	Coulomb per cubic meter (C/m^3)
ϵ_0	Permittivity of free space	Coulomb per newton per square meter (C/Nm^2)
μ_0	Permeability of free space	Newton per ampere squared (N/A^2)
J	Free electric current density	Ampere per square meter (A/m^2)
$\nabla\cdot$	Divergence operator	Per meter (1/m)
$\nabla\times$	Curl operator	Per meter (1/m)

The divergence operator $\nabla\cdot$ gives the tendency of a vector field to go to, or come from, a source while the curl operator $\nabla\times$ gives the vector field's rate of rotation.

Gauss's law: Sources of electric fields are electric charges.
Gauss's law for magnetism: There are no magnetic monopoles.
Faraday's law of induction: A changing magnetic field creates an electric field.
Ampere's law extended by Maxwell: An electric current and/or a changing electric field creates a magnetic field.

Spacetime Metrics

Name	Spacetime metric	Page
Minkowski	$ds^2 = -c^2dt^2 + (d\mathbf{x})^2 = -c^2dt^2 + dx^2 + dy^2 + dz^2$	55
Minkowski (polar)	$ds^2 = -c^2dt^2 + dr^2 + r^2\big[d\theta^2 + \sin^2\theta d\phi^2\big]$	55
Friedmann–Robertson–Walker	$ds^2 = -c^2dt^2 + R^2(t)\big[dx^2 + dy^2 + dz^2\big]$	56, 78
Schwarzschild	$ds^2 = -c^2\left(1 - \frac{2GM}{c^2r}\right)dt^2 + \frac{dr^2}{(1 - \frac{2GM}{c^2r})} + r^2\big[d\theta^2 + \sin^2\theta d\phi^2\big]$	92
Kerr	$ds^2 = -c^2dt^2 + \frac{\rho^2}{\Delta}dr^2 + \rho^2d\theta^2 + (r^2 + a^2)\sin^2\theta d\phi^2$	111
	$+ \frac{2GMr}{c^2\rho^2}(a\sin^2\theta d\phi - cdt)^2$	
Reissner–Nordstrom	$ds^2 = -\left(1 - \frac{r_S}{r} + \frac{r_Q^2}{r^2}\right)c^2dt^2 + \frac{1}{1 - \frac{r_S}{r} + \frac{r_Q^2}{r^2}}dr^2 + r^2\big(d\theta^2 + \sin^2\theta d\phi^2\big)$	115

Units

Système International (SI) base units

Symbol	Name	Quantity
m	meter	length
kg	kilogram	mass
s	second	time
A	ampere	electric current
K	kelvin	temperature
mol	mole	amount of substance
cd	candela	luminous intensity

Selected Système International (SI) derived units

Symbol	Name	Base units	Quantity
J	joule	$\text{kg m}^2\ \text{s}^{-2} = \text{N m}$	energy, work
N	newton	kg m s^{-2}	force
Pa	pascal	$\text{kg m}^{-1}\ \text{s}^{-2} = \text{N m}^{-2}$	pressure
T	tesla	$\text{kg s}^{-2}\ \text{A}^{-1}$	magnetic field
C	coulomb	s A	electric charge
V	volt	$\text{m}^2\ \text{kg s}^{-3}\ \text{A}^{-1} = \text{J C}^{-1}$	electrical potential difference
Ω	ohm	$\text{m}^2\ \text{kg s}^{-3}\ \text{A}^{-2} = \text{V/A}$	electrical resistance
F	farad	$\text{s}^4\ \text{A}^2\ \text{m}^{-2}\ \text{kg}^{-1} = \text{C V}^{-1}$	capacitance
W	watt	$\text{m}^2\ \text{kg s}^{-3} = \text{J s}^{-1} = \text{V A}$	power

Constants

Symbol	Name	Value	Units
c	speed of light in vacuum	2.998×10^8	m s^{-1}
G	gravitational constant	6.6742×10^{-11}	N m^2 kg^{-2}
g	standard gravity	9.80665	m s^{-2}
h	Planck constant	6.626×10^{-34}	J s
		4.136×10^{-15}	eV s
k	Boltzmann constant	1.381×10^{-23}	J K^{-1}
N_A	Avogadro's number	6.022×10^{23}	mol^{-1}
R	molar gas constant	8.314	J mol^{-1} K^{-1}
e	electric charge	1.602×10^{-19}	C
m_e	electron rest mass	9.109×10^{-31}	kg
		0.5110	MeV/c^2
m_p	proton rest mass	1.673×10^{-27}	kg
		938.3	MeV/c^2
m_n	neutron rest mass	1.675×10^{-27}	kg
		939.6	MeV/c^2
μ_0	magnetic constant (vacuum permeability)	$\pi \times 10^{-7}$	N A^{-2}
ϵ_0	electric constant (vacuum permittivity)	8.854×10^{-12}	F m^{-1}
		$= \dfrac{1}{\mu_0 c^2}$	
k_e	electrostatic constant	8.988×10^9	m^2 N C^{-2}
		$= \dfrac{1}{4\pi\epsilon_0}$	
T_0	absolute zero	-273.15	°C
		0	K

Periodic Table

Atomic Number: 1
Element Symbol: H
Element Name: Rutherfordium
Atomic Mass (u): 1.00794
Shells: K 2, L 8, M 18, N 32, O 32, P 18, Q 8

- s-Block
- p-Block
- d-Block
- f-Block

State at room temperature (298 K)

- X Liquid
- X Solid
- X Gaseous

Group / Period	1	2	3	4	5	6	7	8	9	10	11	12	13	14	15	16	17	18
1	1 H Hydrogen 1.00794 (1)																	2 He Helium 4.00260 (2)
2	3 Li Lithium 6.941 (2, 1)	4 Be Beryllium 9.01218 (2, 2)											5 B Boron 10.811 (2, 3)	6 C Carbon 12.0107 (2, 4)	7 N Nitrogen 14.0067 (2, 5)	8 O Oxygen 15.9994 (2, 6)	9 F Fluorine 18.9984 (2, 7)	10 Ne Neon 20.1797 (2, 8)
3	11 Na Sodium 22.98976 (2, 8, 1)	12 Mg Magnesium 24.3050 (2, 8, 2)											13 Al Aluminium 26.9815 (2, 8, 3)	14 Si Silicon 28.0855 (2, 8, 4)	15 P Phosphorus 30.9738 (2, 8, 5)	16 S Sulfur 32.065 (2, 8, 6)	17 Cl Chlorine 35.453 (2, 8, 7)	18 Ar Argon 39.948 (2, 8, 8)
4	19 K Potassium 39.0983 (2, 8, 8, 1)	20 Ca Calcium 40.078 (2, 8, 8, 2)	21 Sc Scandium 44.9559 (2, 8, 9, 2)	22 Ti Titanium 47.867 (2, 8, 10, 2)	23 V Vanadium 50.9415 (2, 8, 11, 2)	24 Cr Chromium 51.9961 (2, 8, 12, 2)	25 Mn Manganese 54.9380 (2, 8, 13, 2)	26 Fe Iron 55.845 (2, 8, 14, 2)	27 Co Cobalt 58.9332 (2, 8, 15, 2)	28 Ni Nickel 58.6934 (2, 8, 16, 2)	29 Cu Copper 63.546 (2, 8, 17, 2)	30 Zn Zinc 65.38 (2, 8, 18, 2)	31 Ga Gallium 69.723 (2, 8, 18, 3)	32 Ge Germanium 72.64 (2, 8, 18, 4)	33 As Arsenic 74.9216 (2, 8, 18, 5)	34 Se Selenium 78.96 (2, 8, 18, 6)	35 Br Bromine 79.904 (2, 8, 18, 7)	36 Kr Krypton 83.798 (2, 8, 18, 8)
5	37 Rb Rubidium 85.486 (2, 8, 18, 8, 1)	38 Sr Strontium 87.62 (2, 8, 18, 8, 2)	39 Y Yttrium 88.9059 (2, 8, 18, 9, 2)	40 Zr Zirconium 91.224 (2, 8, 18, 10, 2)	41 Nb Niobium 92.9063 (2, 8, 18, 12, 1)	42 Mo Molybdenum 95.96 (2, 8, 18, 13, 1)	43 Tc Technetium 97.9072 (2, 8, 18, 14, 1)	44 Ru Ruthenium 101.07 (2, 8, 18, 15, 1)	45 Rh Rhodium 102.906 (2, 8, 18, 16, 1)	46 Pd Palladium 106.42 (2, 8, 18, 18, 0)	47 Ag Silver 107.868 (2, 8, 18, 18, 1)	48 Cd Cadmium 112.411 (2, 8, 18, 18, 2)	49 In Indium 114.818 (2, 8, 18, 18, 3)	50 Sn Tin 118.710 (2, 8, 18, 18, 4)	51 Sb Antimony 121.760 (2, 8, 18, 18, 5)	52 Te Tellurium 127.60 (2, 8, 18, 18, 6)	53 I Iodine 126.904 (2, 8, 18, 18, 7)	54 Xe Xenon 131.293 (2, 8, 18, 18, 8)
6	55 Cs Caesium 132.905 (2, 8, 18, 18, 8, 1)	56 Ba Barium 137.327 (2, 8, 18, 18, 8, 2)	57 La Lanthanum 138.905 (2, 8, 18, 18, 9, 2)	72 Hf Hafnium 178.49 (2, 8, 18, 32, 10, 2)	73 Ta Tantalum 180.948 (2, 8, 18, 32, 11, 2)	74 W Tungsten 1.00794 (2, 8, 18, 32, 12, 2)	75 Re Rhenium 186.207 (2, 8, 18, 32, 13, 2)	76 Os Osmium 190.23 (2, 8, 18, 32, 14, 2)	77 Ir Iridium 192.217 (2, 8, 18, 32, 15, 2)	78 Pt Platinum 195.084 (2, 8, 18, 32, 17, 1)	79 Au Gold 196.967 (2, 8, 18, 32, 18, 1)	80 Hg Mercury 200.59 (2, 8, 18, 32, 18, 2)	81 Tl Thallium 204.3833 (2, 8, 18, 32, 18, 3)	82 Pb Lead 207.2 (2, 8, 18, 32, 18, 4)	83 Bi Bismuth 208.980 (2, 8, 18, 32, 18, 5)	84 Po Polonium 208.982 (2, 8, 18, 32, 18, 6)	85 At Astatine 209.987 (2, 8, 18, 32, 18, 7)	86 Rn Radon 222.018 (2, 8, 18, 32, 18, 8)
7	87 Fr Francium 223 (2, 8, 18, 32, 18, 8, 1)	88 Ra Radium 226 (2, 8, 18, 32, 18, 8, 2)	89 Ac Actinium 227 (2, 8, 18, 32, 18, 9, 2)	104 Rf Rutherfordium 261 (2, 8, 18, 32, 32, 10, 2)	105 Db Dubnium, 262 (2, 8, 18, 32, 32, 11, 2)	106 Sg Seaborgium 266 (2, 8, 18, 32, 32, 12, 2)	107 Bh Bohrium 264 (2, 8, 18, 32, 32, 13, 2)	108 Hs Hassium 277 (2, 8, 18, 32, 32, 14, 2)	109 Mt Meinerium 268 (2, 8, 18, 32, 32, 15, 2)	110 Ds Darmstadtium 271 (2, 8, 18, 32, 32, 17, 1)	111 Rg Roentgenium 272 (2, 8, 18, 32, 32, 18, 1)	112 Uub Ununbium 285 (2, 8, 18, 32, 32, 18, 2)	113 Uut Ununtrium 284 (2, 8, 18, 32, 32, 18, 3)	114 Uuq Ununquadium 289 (2, 8, 18, 32, 32, 18, 4)	115 Uup Ununpentium 288 (2, 8, 18, 32, 32, 18, 5)	116 Uuh Ununhexium 292 (2, 8, 18, 32, 32, 18, 1)	117 Uus Ununseptium 290? (2, 8, 18, 32, 32, 18, 7)	118 Uuo Ununoctium 294 (2, 8, 18, 32, 32, 18, 8)

58 Ce Cerium 140.116 (2, 8, 18, 19, 9, 2)	59 Pr Praseodymium 140.908 (2, 8, 18, 21, 8, 2)	60 Nd Neodymium 144.242 (2, 8, 18, 22, 8, 2)	61 Pm Promethium 145 (2, 8, 18, 23, 8, 2)	62 Sm Samarium 150.36 (2, 8, 18, 24, 8, 2)	63 Eu Europium 151.964 (2, 8, 18, 25, 8, 2)	64 Gd Gadolinium 157.25 (2, 8, 18, 25, 9, 2)	65 Tb Terbium 158.925 (2, 8, 18, 27, 8, 2)	66 Dy Dysprosium 162.500 (2, 8, 18, 28, 8, 2)	67 Ho Holmium 164.930 (2, 8, 18, 29, 8, 2)	68 Er Erbium 167.259 (2, 8, 18, 30, 8, 2)	69 Tm Thulium 168.934 (2, 8, 18, 31, 8, 2)	70 Yb Ytterbium 173.04 (2, 8, 18, 32, 8, 2)	71 Lu Lutetium 174.967 (2, 8, 18, 32, 9, 2)
90 Th Thorium 232.038 (2, 8, 18, 32, 18, 10, 2)	91 Pa Protactinium 231.036 (2, 8, 18, 32, 20, 9, 2)	92 U Uranium 238.029 (2, 8, 18, 32, 21, 9, 2)	93 Np Neptunium 237 (2, 8, 18, 32, 22, 9, 2)	94 Pu Plutonium 244 (2, 8, 18, 32, 24, 8, 2)	95 Am Americum 243 (2, 8, 18, 32, 25, 8, 2)	96 Cm Curium 247 (2, 8, 18, 32, 25, 9, 2)	97 Bk Berkelium 247 (2, 8, 18, 32, 27, 8, 2)	98 Cf Californium 251 (2, 8, 18, 32, 28, 8, 2)	99 Es Einsteinium 252 (2, 8, 18, 32, 29, 8, 2)	100 Fm Fermium 257 (2, 8, 18, 32, 30, 8, 2)	101 Md Mendelevium 258 (2, 8, 18, 32, 31, 8, 2)	102 No Nobelium 259 (2, 8, 18, 32, 32, 8, 2)	103 Lr Lawrencium 262 (2, 8, 18, 32, 32, 9, 2)

Index